VLSI
ARRAY
PROCESSORS

**PRENTICE HALL INFORMATION
AND SYSTEM SCIENCES SERIES**

Thomas Kailath, Editor

VLSI
ARRAY
PROCESSORS

S. Y. Kung

Department of Electrical Engineering
Princeton University

Prentice Hall
Englewood Cliffs, New Jersey 07632

Library of Congress Cataloging-in-Publication Data

Kung, S. Y. (Sun Yuan)
  VLSI array processors.

  Bibliography
  Includes index.
    1. Integrated circuits—Very large scale integration
—Design and construction. 2. Array processors—
Design and construction. 3. Signal processing.
I. Title.
TK7874.K86 1987   621.395   87–17508
ISBN 0-13-942749-X

Editorial/production supervision: Richard Woods
Cover design: Photo Plus Art
Manufacturing buyer: Richard Washburn
Page layout: Martin J. Behan

 © 1988 by Prentice Hall
A Division of Simon & Schuster
Englewood Cliffs, New Jersey 07632

Printed in the United States of America

10 9 8 7 6 5 4 3 2 1

ISBN 0-13-942749-X 025

Prentice-Hall International (UK) Limited, *London*
Prentice-Hall of Australia Pty. Limited, *Sydney*
Prentice-Hall Canada Inc., *Toronto*
Prentice-Hall Hispanoamericana, S.A., *Mexico*
Prentice-Hall of India Private Limited, *New Delhi*
Prentice-Hall of Japan, Inc., *Tokyo*
Simon & Schuster Asia Pte. Ltd., *Singapore*
Editora Prentice-Hall do Brasil, Ltda., *Rio de Janeiro*

To my wife
Se-Wei
and my sons
Li and Charles
for their love, support, and inspiration.

To my parents
Dr. and Mrs. Ku-Sheng Kung
for a joyful dawning in the quest
for truth and knowledge.

CONTENTS

7 IMPLEMENTATION OF ARRAY PROCESSORS   452

8 APPLICATIONS TO SIGNAL AND IMAGE PROCESSING   537

PREFACE

A more complete title of this book might have been *VLSI Array Processors for Signal/Image Processing and Scientific Computing,* since it addresses how to design VLSI arrays for handling the extremely stringent real-time processing requirements in those applications. The book may be considered unique in its cohesive and cross-disciplinary study of *applicational, algorithmic, architectural,* and *technological* aspects of VLSI array processors. It advocates a vertically integrated VLSI system design methodology, covering technology constraints, algorithm analyses, parallelism extractions, architecture design, system development, and application understanding.

In VLSI, memory and processing power are relatively cheap and the main emphasis of the design is shifted to reducing the overall interconnection complexity and keeping the overall architecture highly regular, parallel, and pipelined. Algorithm based architectures have become a very promising trend for future supercomputing technology. The important roles of *parallel algorithm analysis and mapping* in array processor design are demonstrated throughout the book. It is specifically noted that most signal processing algorithms are well structured and they share the common attributes of regularity, recursiveness, and local communication. These properties can be effectively exploited in the innovative systolic and wavefront array processors. Such arrays maximize the strength of VLSI in terms of intensive and pipelined computing and yet circumvent its main limitation on communication. Moreover, they offer massive concurrent computing which is essential to real-time signal/image processing.

This book begins with *algorithm analysis,* followed by *architecture design,* and concludes with *applicational systems,* which is the ultimate objective of the VLSI design. To make the study of the book somewhat more systematic, a *road map* is provided in the next page which displays the relationship between the eight chapters and their major sections.

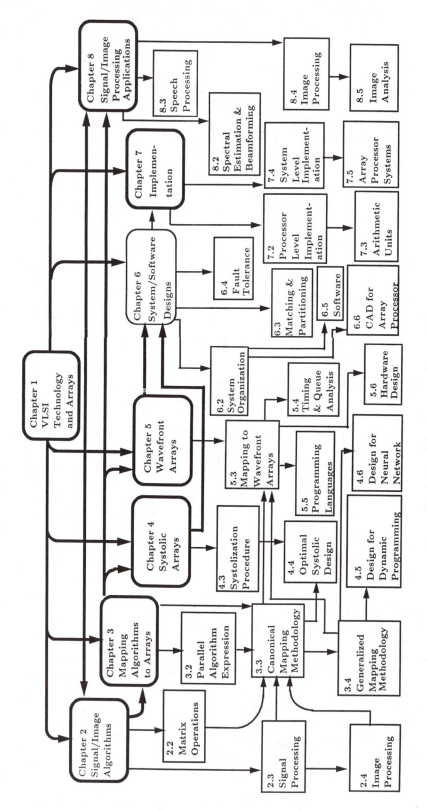

Road map of the chapters and key sections.

THE USE OF THE BOOK

The book introduces *VLSI array processors* from the aspects of *VLSI design, computer engineering,* and *signal/image processing.* This book, however, is not intended to be a foundation text in any one of these specific areas and indeed many excellent texts currently exist for them. The intention of this book is to cohesively study all of these disciplines and the interplay between them, and so act as a complimentary text to existing literature. Therefore, it is hoped that this book could be incorporated into a course for *VLSI system design* or *parallel processing architecture* or *signal processing hardware.* The book is intended for use both as a textbook at the graduate level and as a reference work for electronic engineers and computer scientists. This book may be used as a textbook for one semester (approximately 40-50 lecture hours) or two semesters of graduate courses.

- Suitable for one semester in *computer architecture* are:
  Sections *1.2, 1.3, 3.2, 3.3, 3.4, 4.3, 4.4, 5.3, 5.4, 6.2, 6.3, 6.4, 8.4.*
  (Sections *2.2, 2.3, 2.4* are optional.)

- Sections for one semester in *VLSI signal processing* are:
  Sections *1.2, 1.3, 3.2, 3.3, 4.3, 5.3, 6.2, 7.2, 7.3, 8.2, 8.4.*
  (Sections *2.2, 2.3, 2.4* are optional.)

- For a combined two semester course the suggested sections are:
  Semester 1: Sections *1.2, 1.3, 2.2, 2.3, 2.4, 3.2, 3.3, 3.4, 4.3, 4.4, 4.5, 8.2, 8.3.*
  Semester 2: Sections *5.3, 5.4, 6.2, 6.3, 6.4, 7.2, 7.3, 7.4, 7.5, 8.4, 8.5.*

The exact course outline will depend on the students' background as well as their specific interests. The *road map* below should assist instructors in deciding the most suitable contents for their needs and any necessary prerequisites from different sections of the book. An important part of this text is a collection of approximately 140 homework problems. These problems have been designed to test the reader's understanding of the text, to illustrate real applications, and to develop awareness of the potential usefulness of the theorems and results given in the book.

Finally, many of the topics presented here are still very active and expanding areas of research. In conjunction with the comprehensive bibliography provided, it is hoped that some future research in the areas of VLSI array processors and special purpose supercomputers may be inspired from studying this text.

ACKNOWLEDGMENTS

The experience of writing this book could never have been more rewarding and fruitful to me. Not only that the subject *VLSI Array Processors* has been immensely challenging and inspiring, but also that it has been a great pleasure to collaborate with a group of very spirited young researchers. In particular, I am grateful to my Ph.D. students at the University of Southern California; C. W. Chang, E. Chow, W. C. Fang, J. N. Hwang, S. N. Jean, P. S. Lewis, S. C. Lo, J. C. Lien, E. Manolakos, S. L. Peng, R. W. Stewart, S. W. Sun, J. Vlontzos, and many others. These individuals have provided boundless energy and selfless assistance in debating and researching the earlier drafts which have gradually converged to this final version. They have made major research contributions that have provided foundation for many sections of the book, and in this sense they have effectively coauthored most of the book. I am also greatly indebted to many of my colleagues and friends whose timely encouragement and critical advice have proven to be inspiring and invaluable in the course of this seemingly endless writing process. In particular, I would like to thank Professors K. S. Arun of the University of Illinois, P. Cappello of the University of California, Santa Barbara, P. Dewilde of Delft University, Y. H. Hu of the Southern Methodist University, K. Hwang of USC, C. W. Jen of the National Chiao-Tung University, T. Kailath of Stanford University, D.V.B. Rao of the University of California, San Diego, K. Yao of the University of California, L.A., and Drs. Y. C. Jenq of Tektronix Inc., J. T. Johl of Hughes Aircraft Corp., J. McWhirter of RSRE, R. Raghavan of Lockheed Missiles & Space Company, and S. Rao of AT&T Bell Laboratories. I am also grateful to the Signal and Image Processing Institute (SIPI) of the University of Southern California for providing facilities for the preparation of the manuscript and Ms. Linda Varilla of SIPI for her outstanding administrative assistance.

The research work presented in this book was supported in part by the National Science Foundation under Grant ECS-82-13358, by the Office of Naval Research under Selected Research Opportunity Program N00014-81-K-0191, by the Innovative Science and Technology Office of the Strategic Defense Initiative Organization and was administered through the Office of Naval Research under Contract No. N00014-85-K-0469 and N00014-85-K-0599, and by Semiconductor Research Corporation under SRC-USC Program 86-01-075.

S. Y. Kung

Chapter 1

AN OVERVIEW

1.1 Introduction

The increasing demands of speed and performance in modern signal and image processing applications necessitate a revolutionary super-computing technology. The availability of low-cost, high-density, high-speed very large scale integration (VLSI) devices and emerging computer-aided design (CAD) facilities presages a major breakthrough in the design and application of massively parallel processors. In particular, VLSI microelectronics technology has inspired many innovative designs in array processor architectures. This trend has now become a major focus of attention for government, industry, and the university community. In the last decade, there has been a dramatic worldwide growth in research and development efforts on mapping various signal/image processing applications onto such VLSI architectures.

In this book, we stress the need of high-speed and massive computing capabilities for signal and image processing and scientific computation applications. Modern signal/image processing technology depends critically on the device and architectural innovations of the computing hardware. Sequential systems will be inadequate for future real-time processing systems, and the additional computational capability available through VLSI concurrent array processors will become a necessity. In most real-time digital signal processing (DSP) applications, general-purpose parallel computers cannot offer satisfactory processing speed due to severe system overheads. Therefore, special-purpose array processors will become the only appealing alternative.

Vertically Integrated VLSI Array Design   In this book, we address the issue of algorithm-oriented array processor design. A cross-disciplinary design approach will involve three main areas: *DSP application and algorithms* and *VLSI system design* [Oppen75], [Hwan84a], [Mead80]. Corresponding to these areas, there are three different types of representations: *functional, structural, and geometrical,* as summarized by the **Y**-chart in Figure 1.1.

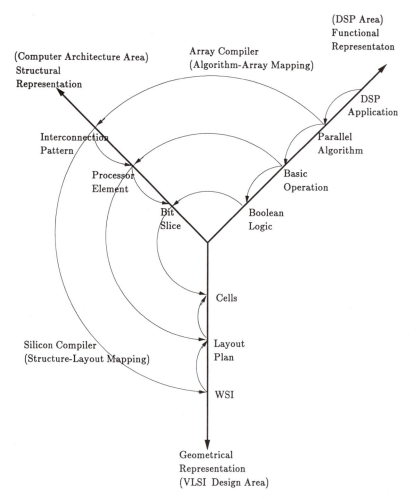

Figure 1.1: **Y**-chart for array processor design.

The chart also helps to illustrate the complementary roles of an array compiler and a silicon compiler. An array compiler maps an algorithm to a dependence graph and to an array structure representation. A silicon compiler maps structural description to logic/circuit, stick diagram, and layout, such as a Caltech intermediate form (CIF) file.

More precisely, given an algorithm, *how can a dedicated array processor be systematically derived?* A fundamental issue on mapping algorithms onto array processors is *how to express parallel algorithms in a notation that is easy to understand by the human designer and possible to compile into efficient VLSI array processors.* The ultimate design should begin with a powerful algorithm notation to express the recurrence and parallelism associated with the description of the space-time activities. Next, this description is converted into a VLSI hardware description or into executable array processor machine code.

1.2   Array Processors for Signal and Image Processing

Until the middle of the 1960s, most signal processing tasks were performed with specialized analog (especially optical) processors because of the hardware complexity, power consumption, and lower speed of digital systems. However, digital processors can provide better (and sometimes indispensable) precision, dynamic range, long-term memory and other flexibilities, such as programmability and expandability, to accommodate changing requirements. Ultimately, a system designer must choose the best available device technology, pipeline technique, and parallel processing to achieve satisfactory performance. The feasibility of VLSI array processors enables the processing speed achievable via digital processing to increase by several orders of magnitude.

There are four main points of attack in designing VLSI array processors: *applications, algorithms, architectures,* and *technology.* In striving for a cohesive exploration of the design of VLSI array processors, cross-disciplinary discussions on applications, algorithms, architectures and technology is necessary.

1.2.1   Applications

The applicational domain of VLSI array processors covers image processing, computer vision, nuclear physics, structure analysis, speech, sonar, radar, seismic, weather, astronomical, medical signal processing applications, and so on. A successful array processor design requires an understanding of the signal and image formation process, the algorithm class involved, and the specifications of the intended applicational system. For example, let us examine some of the application requirements in a real-time vision processing system. The job is to recognize an object and check its geometric and physical properties against some given specifications to determine if the object is a target or not. This is a real-time application with a rate of 512 × 512 pixel image frames per millisecond. Therefore, the processing speed (million operations per second, or MOPs) required will be

10 ops/pixel × (512 × 512 pixels/frame) × 1000 frames/s ≈ 2500 MOPs.

Even for a commercial application such as digital video processing, the processing speed required will be approximately

10 ops/pixel × (512 × 512 pixels/frame) × 24 frames/s ≈ 60 MOPs.

It is quickly apparent that special-purpose parallel processing architectures are indispensable.

1.2.2 Algorithms

Digital signal and image processing encompasses a wide variety of mathematical and algorithmic techniques. However, most signal and image processing algorithms are dominated by transform techniques, convolution/correlation filtering and some key linear algebraic methods. Some sample image processing algorithms are as follows [Pratt78]:

- *Point type*: gray scale transformation, histogram equalization, requantization, intensity mapping, and so on.

- *Filtering type*: template matching (Prewitt, Kirsch or Hueckel operators), window techniques (Bartlett, Hamming windows), convolution/correlation, linear-phase filtering (low-pass high-pass, band-pass, band-stop), median filtering, inverse filtering, Wiener filtering, Kalman filtering, adaptive filtering, and so on.

- *Matrix algebra type*: singular value decomposition (SVD), geometric rotation/display, maximum entropy estimation, maximum likelihood estimation, pseudo-inverse calculation and image restoration, stochastic parameter estimation, and so on.

- *Transform type*: Fourier transform, number theoretic transform, Haar transform, cosine transform, geometric distortion correction, Hough transform, Hadamard transform, K-L transform, and so on.

- *Sorter type*: merge sort, bitonic sort, and so on.

The dominating aspects in signal and image processing requirements are essentially *enormous throughput rates* and *huge amounts of data and memory*. A computation rate in excess of 1000 million operations per second may frequently become necessary for real-time performance. Fortunately, most of these algorithms possess common properties such as *regularity, recursiveness and locality*, which are very useful for array processor design. Some typical algorithm examples encountered in the image processing area and their throughput requirements are illustrated in Table 1.1.

Here we assume an image quality equivalent to that of television, with a spatial resolution equivalent to 512×512 pixels and a frame rate of 30 frames a second, resulting in a data rate of 10^7 samples per second. This implies that during the feature extraction stage for a subwindow of size 3×3, a memory access rate of the order of 10^8 per second is required. For larger windows, say with 64×64 pixels, access rates may be of the order of 10^{11} per second. Hence, even for linear operations such as spatial filtering,

Processing function	Necessary throughput
Linear operations, $O(N)$ – spatial filtering – convolution – edge detection	$10^2 - 10^5$ MOPs
Second-order operations, $O(N^2)$ – sorting operations – median filtering – nearest-neighbor classification	$10^3 - 10^7$ MOPs
High order operations – matrix based – spectral processing – adaptive operations	$10^4 - 10^8$ MOPs

Table 1.1: Throughput requirements for DSP algorithms.

convolution, and edge detection, the throughput requirement will range from 10^2 to 10^5 MOPs, as shown in Table 1.1.

1.2.3   Architectures

Current parallel computers can be divided into three structural classes: vector processors, multiprocessor systems, and array processors [Hwan84a]. The first two classes belong to the general-purpose computer domain. The development of these systems requires a complicated design of control units and optimized schemes for the allocation of machine resources. The third class, however, belongs to the domain of special-purpose computers, and the design of such systems requires a broad knowledge of the relationship between parallel computing algorithms and optimal computing hardware and software structures.

It is the last class that we shall focus upon, since it offers a promising solution to meet real-time processing requirements. In particular, locally interconnected computing networks, such as systolic and wavefront arrays, are well suited to efficiently implement a major class of signal processing algorithms, due to their massive parallelism and regular data flow [HTKun82], [Kung82a]. Such architectures promise real-time solutions to a large variety of advanced computational tasks.

Pipelining, array processing, and multiprocessing represent standard methods in computer organization which are commonly used for high-speed processing, to reduce the inherent complexity in the design of large-scale multiprocessor arrays. Various solutions have been proposed, which all hinge upon imposing a certain degree of special-purpose restriction upon the application. For example, the use of only localized communication significantly simplifies the design of interconnection architecture at the expense of a somewhat restricted class of applications. Single instruction multiple data stream (SIMD) computers, multiprocessors, and VLSI arrays are representatives of various attempts at solving the parallel processing problems. Typical examples are the ILLIAC IV for the SIMD array processing, dataflow machines for multiprocessor systems, and systolic/wavefront processors for VLSI arrays.

1.2.3.1   SIMD Arrays

SIMD computers (see Figure 1.2) are implemented as an array of arithmetic processors, with the local connectivity between them and local memory associated with each [Flynn66]. Instructions are broadcast from a host, and

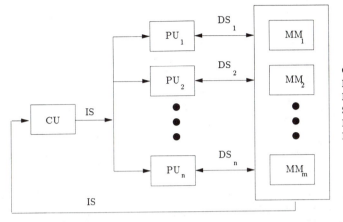

CU: control unit
PU: processor unit
MM: memory module
SM: shared memory
IS: instruction stream
DS: data stream

Figure 1.2: SIMD architecture, adapted from [Hwan84a].

all processors execute each instruction simultaneously. SIMD array processors usually allow explicit expression of parallelism in user programs. The compiler (in the host), upon detection of parallel processing operations, will generate object code to be loaded into the processing element (PE) array and the control unit for execution. The concepts of SIMD systems were studied by Unger (1958) and later by Von Neumann (1966), but the first SIMD machine constructed for practical use was the ILLIAC IV in the 1970s [Hwan84a].

The ILLIAC IV system has been a typical example for SIMD array computers. Examples of other large-scale SIMD array computers implemented in LSI technology are NASA's massively parallel processor (MPP) and ICL's distributed array processor (DAP). The instructions of these systems are stored in a global main memory together with data. The central control unit directs the operations of all PEs, communicates with them via a global broadcasting network, and synchronizes all the processor executions.

1.2.3.2   MIMD Arrays

Multiple instruction multiple data stream (MIMD) computers consist of a number of processing elements, each with its own control unit, program, and data (see Figure 1.3) [Flynn66]. The main feature of an MIMD machine is that the overall processing task may be distributed among the processing elements for the purpose of increasing processing parallelism. Understandably, mapping of algorithms onto an MIMD array is usually performed at the task or processor level. In general, MIMD machines may encounter com-

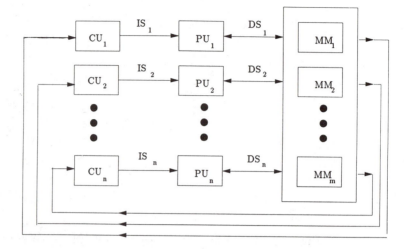

Figure 1.3: MIMD architecture, adapted from [Hwan84a].

munication bottlenecks when multiple PEs attempt to access shared system resources simultaneously. This phenomenon usually results in decreased processing throughput. There are also potential problems of synchronizing tasks among distributed processing elements. These factors have somewhat unfavorably affected the popularity of MIMD architectures as compared to SIMD architectures [Hwan84a]. Nevertheless, the flexibilities in MIMD architectures are often essential in order to deal with irregularly structured algorithms, such as those appearing in intelligent image processing and vision analysis applications.

1.2.3.3   VLSI Array Processors

The practicality of any DSP algorithm will ultimately be determined by its computational feasibility. Real-time signal and image processing depends critically on the speed and volume capabilities offered by the state-of-the-art parallel processing computers. Due to severe system overheads, general-purpose supercomputers are often not very suitable for real-time signal/image processing. Consequently, a new approach based on VLSI array processors is becoming increasingly competitive. Note that the key attributes processed by the above-mentioned algorithms can be exploited in a special type of VLSI array processor architecture. These arrays maximize the strength of VLSI in terms of intensive and pipelined computing and yet circumvent its main limitation on communication. This provides the theoretical footing for the design of locally interconnected VLSI arrays such as systolic arrays [HTKun78] and wavefront arrays [Kung82a]. *The massive concurrency in systolic/wavefront arrays is derived from pipeline processing, parallel processing, or both.* This is illustrated in Figure 1.4. *Parallel processing* means that all processes defined in terms of the data **D** and the instructions **i** can directly access the m processors in parallel and keep all processors busy. *Pipeline processing* means that a process is decomposed into many subprocesses, which are pipelined through m processors aligned in a chain, and each subprocess is processed one after another. For each subprocess coming out of the array, there will be a processor vacant and ready to receive and handle its subprocess immediately. Therefore, all m processors can be kept busy all the time by utilizing the pipeline technique.

1.2.4   Technology

There are currently two popular semiconductor device technologies: bipolar and metal-oxide semiconductor (MOS). Although new technologies combining features of both are emerging, some important differences between bipolar and MOS devices exist. While bipolar technology is faster, MOS technology, offers higher density integration and consumes less power. In

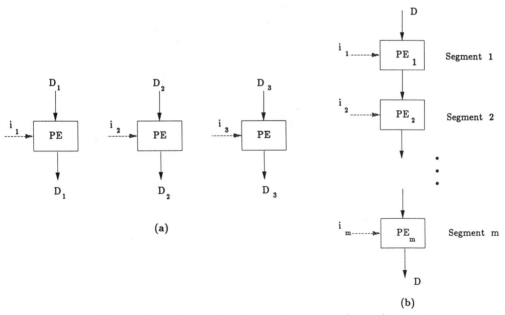

Figure 1.4: Parallel and pipeline processing. Array processors derive a massive concurrency from both (a) *parallel processing* and (b) *pipeline processing* [Hayes78].

Requirement	VHSIC Phase 1	VHSIC Phase 2
Lithographic feature size	1.25 μm	0.5 μm
Functional throughput rate	5×10^{11} (gate-Hz/cm^2)	1×10^{13} (gate-Hz/cm^2)
On-chip clock speed	25 MHz	100 MHz

Table 1.2: Major specifications for the VHSIC program.

the mid 1980s, the DoD (U.S. Department of Defense) very high-speed integrated circuits (VHSIC) program successfully reached its Phase 1 goal (see Table 1.2), and some chips based on this technology are now commercially available. The Phase 2 program is currently in progress; its specifications are also listed in Table 1.2.

VLSI device technology offers promising potential but creates new design constraints. More details on VLSI architectural principles are given in Section 1.2. The key design criteria for VLSI architectures are summarized as follows [Rande82]:

- Critical design complexity and essential CAD tools

- Modularity and effective utilization of building blocks

- Simple and regular data and control paths

- Localized or reduced interconnections

- Balance between input/output (I/O) and computation

- Extensive concurrency (i.e., pipeline and/or parallel processing)

- Synchronous versus asynchronous implementation consideration

- Programmability for DSP applications

- Adequate reconfigurability and fault tolerance

- Balanced array and chip partitioning.

As long as communication in VLSI remains restrictive, locally interconnected arrays will be of great importance. An increase in efficiency can be expected if the algorithm arranges for a balanced distribution of work load while observing the requirement of locality, i.e., short communication paths. These properties of load distribution and information flow serve as a guideline to the designer of VLSI algorithms and eventually lead to new VLSI architecture designs.

When the array architecture designs are ready to be realized in hardware, some additional issues are technology, cost, speed, flexibility, transportability, and (high-level) programmability. The system obtained will have to go through final verification by being tested in the intended applications and having its performance evaluated. Additional modification at this stage may be required to achieve a satisfactory performance.

1.3   VLSI Architecture Design Principles

VLSI architectures should exploit the potential of the VLSI technology and also take into account the cost of silicon area and I/O pins. A major layout constraint is the interconnection cost in terms of area and time. Communication is costly because wires occupy the most space on a circuit and communication degrades clock time. When the delay time of the circuit depends largely on the interconnection delay (instead of the logic gate delay), minimal and local interconnections become an essential factor for an effective realization in VLSI. A second constraint on VLSI implementation is the circuit complexity, which results in high design cost. The burden can be alleviated by the use of regular, repetitive architectural structures. Consequently, VLSI architecture design principles should include modularity, regularity, local communication, massive parallelism, and minimized I/O. The building block concept and the scaling effects are very instrumental and important in VLSI design. Also, in order to handle design complexity, a well-defined systematic CAD tool for design, simulation, and verification is required. More detailed guidelines pertaining to the array processor design are discussed next.

1.3.1   VLSI Technology

Recently, the trend of MOS technology has shifted from n-channel MOS (NMOS) to complementary MOS (CMOS). This is because CMOS may be scaled easily to small feature sizes and offer high performance at lower power. Basically, CMOS circuits are constructed from n-channel and p-channel transistors. An n-channel transistor is made in a p-type substrate, whereas a

p-channel transistor is made in an n-type substrate. In an n-channel transistor, the drain and source regions are created by n-type diffusion. The
gate is made of a conductor (polysilicon) over a thin oxide layer covering
the region between the drain, source, and substrate voltages, electrons are
attracted to the surface of the substrate. Above a certain threshold, the
number of electrons is so large that the electrons form a conducting channel
between the source and the drain. In principle, a CMOS inverter draws no
power in steady state due to the fact that normally only one of the transistors is on. The primitive CMOS cells are the inverter and transmission
gate. Their circuits and physical layout diagrams are depicted in Figure
1.5. The inverter and its combinational logic (e.g., NAND, NOR, or PLA)

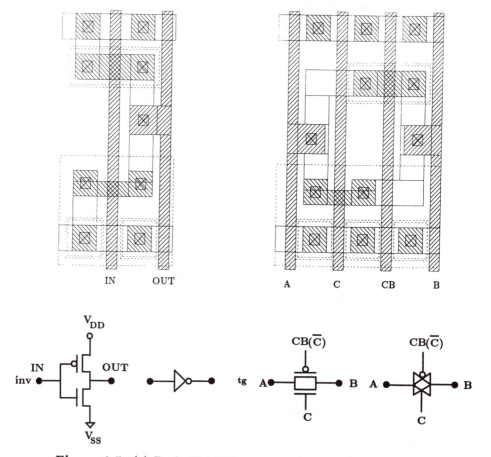

Figure 1.5: (a) Basic CMOS inverter and transmission gate and
(b) their circuit layout diagrams.

constitute all the conventional CMOS logic. However, the transmission gate is becoming increasingly popular in modern CMOS circuit design.

1.3.1.1 Scaling Effects

The exponential growth of IC complexity and capability since the birth of VLSI technology has been caused by a combination of a down scaling of the minimum feature size achievable and an up scaling of the maximum chip size, both subject to the constraint of reasonable yield. VLSI architecture enjoys a major advantage of being very scalable technologically [Mead80]. This means that the efforts of architectural redesign will be minor when the device technology is scaled down to the submicron level.

Although the scaling process has many beneficial effects, in contrast, interconnection problems become very severe due to increased chip size. Eventually, chip cost, performance, and speed are determined primarily by interconnect delay and area. Therefore, VLSI device technology does not simply offer a promising future but also raises some new design constraints. To cope with the constraints, the use of modular building blocks and the alleviation of the burden of global interconnection are often essential in VLSI design.

In the scaling of geometry, we often assume that all the dimensions as well as the voltages and currents on the chip are scaled down by a factor α (an α greater than 1 implies that the sizes or levels are shrinking). When scaling down the linear dimensions of a transistor by α, the number of transistors that can be placed on a chip of given size is scaled up by α^2. Figure 1.6 depicts the effect of scaling down a conductor and a MOS field-effect transistor (MOSFET) by a factor α.

The switching delay of a transistor is scaled down at least by α due to the fact that the channel length is decreased by a factor α. Scaling also affects the interconnections between devices. Since the cross-sectional area of the conductor is decreased by a factor α^2, the resistance per unit length is increased by a similar factor. If the length of the conductor is scaled by α (as simple scaling implies), then the net increase of resistance is in proportion to α. At the same time, scaling implies changes of the capacitance of the interconnection. Regarding the conductor as one plate of a parallel-plate capacitor, scaling down of both linear dimensions of the plate by α implies a decrease of the capacitance by α^2. However, scaling down also implies a decrease by α of the thickness of the oxide insulating layer separating the plates of the capacitor. Hence the capacitance of a fixed interconnection scales down by α. We see that the scaling up of resistance and down of

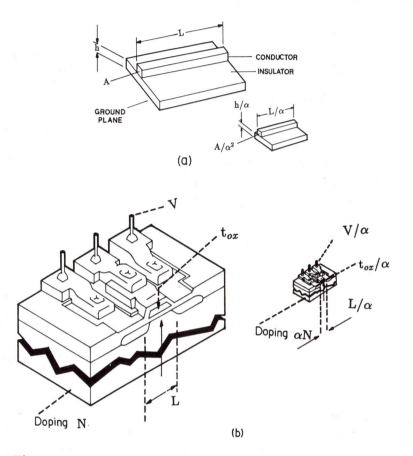

Figure 1.6: Scaling of a MOSFET transistor and a conductor; (a) scaling of a conductor, α is the scaling factor, (b) scaling of a MOSFET transistor.

capacitance exactly cancel, leaving the RC (resistor-capacitor) time constant and the interconnect delay unchanged.

1.3.2 Interconnection and I/O Constraints

It is clear that since gate delays decrease with scaling, whereas interconnection delays remain constant with scaling, eventually the speed at which a circuit can operate will be dominated by interconnect delays rather than device delays. However, the situation is actually somewhat worse than the above considerations imply, due to the factor of *stuffing*. Stuffing means that the lengths of the interconnections do not scale down with the inverse of α, as was assumed in scaling. In practice, as the complexity of the circuit increases, the distances over which interconnections must be maintained on a chip of fixed area may stay roughly constant. It has been argued from statistical considerations [Keyes79] that a good approximation to the maximum length L_{max} of interconnection required is given by

$$L_{max} = \frac{A^{1/2}}{2}$$

where A represents the area of the chip. Therefore, when stuffing occurs the average interconnection delay may actually increase.

Note that if scaling occurs and chip size is also increased, the interconnect problem becomes further exacerbated. When the delay time of the circuit depends largely on the interconnection delay (instead of the logic gate delay), minimal and local interconnections will become an essential factor for an effective realization of the VLSI circuits.

Architectures which balance communication and computation and circumvent communication bottlenecks with minimum hardware cost will eventually play a dominating role in VLSI systems. A very critical factor here is that of the limited number of I/O pins. Very often parallel signal communication in many directions is required. (For example, $\log_2 N$ communication wires for each node in a hypercube computer with N PEs.) A good design should take into account the constraints on I/O pins and resultant costs in terms of area and time.

In accordance with the hierarchical nature of design, a complicated VLSI chip can be regarded as consisting of a multitude of subunits of circuitry, called "blocks". As scaling proceeds, the complexity of the blocks can be made greater, and the number of blocks that can be realized in a system also grows. As the number of elements in a block increases, the number of interconnections required from that block to other blocks also increases. There is a well-known empirical relation, known as Rent's rule, which specifies that the number of interconnections M required for a block consisting of N devices grows as approximately

$$M = N^{2/3}$$

For example, a circuit block consisting of 100,000 gates requires about 2000 interconnections. For a 10 mm $\times$ 10 mm chip this works out to a connection pad every 20 μm. It should be noted that Rent's rule applies only to circuits consisting of random logic elements and VLSI array processors may be exempted from such stringent requirements if the principle of locality is complied with at every level of VLSI design [Seitz84]. From array architecture perspective, typical localities are *local data communication* and *distributed control*. Indeed, most recursive signal processing algorithms permit both locality features, which can be fully exploited in systolic and wavefront array architectures. This important fact will also have a major impact on wafer-scale integration and future ultra-submicron technology.[1]

1.3.3   Regularity and Modularity

In VLSI design, the overall architecture should be as *regular* and *modular* as possible, thus reducing design error, time, and cost. Memory and some special-function chips are becoming very inexpensive due to their high regularity and modularity. Even with global communication, any form of regularity, as derived by a careful algorithmic design, may prove useful for mapping algorithms onto architectures.

The building-block concept (e.g., macrocell) is often instrumental and important in VLSI design. The building block approach may be combined with high-level tools, such as silicon compilers, to give the VLSI designer a tremendous amount of flexibility to cope with the ever-increasing complexity of VLSI design.

1.3.4   Pipeline and Parallel Processing

Throughput rate is the overriding factor dictating the system performance. In order to optimize throughput, real-time signal processing requires extensive concurrency by pipeline and parallel processing. Furthermore, a design choice different from that of minimizing the total processing time (latency) is often made. Suitable pipelining techniques are now well established and are quite popular in many DSP algorithms.

For signal processing arrays, pipelining at all levels should be pursued. Pipelining may bring about an extra order of magnitude in performance with very little additional hardware. Although most of the current array processors stress only word-level pipelining, the new trend is to exploit the potential of multiple-level pipelining (i.e., combined pipelining in all the bit-level, word-level, and array-level granularities).

---

[1]As the number of devices realizable on a single chip grows, the assumption underlying Rent's rule may become invalid, and the exponent associated with that rule may fall to less than 0.5.

1.3.5 Globally Synchronous versus Asynchronous Systems

For VLSI systems, the design for clocking is a very demanding task. Therefore, the choice between synchronous or asynchronous array processing becomes critical [Kung82b], [Frank82]. The concept of asynchronous processing has been adopted in dataflow computers [Denni80], which use the flow of data to initiate the execution of an instruction (unlike the Von Neumann, or conventional stored-program computer, which is based on the availability of control flow). Thus in data flow, an instruction is activated for execution only after all the operands of the instruction have arrived. This approach *eliminates the need for global control and global synchronization*, since all interactions among the modules in a dataflow network are asynchronous. It also neatly handles data dependencies: data flow processing is in general dictated by a dataflow graph which displays the exact dependency of data [Allen85].

In the globally synchronous scheme, there is a global clock network that distributes the clock signal over the entire array [Fishe83]. For very large systems, the clock skew incurred in global clock distribution is a nontrivial factor, causing unnecessary slowdown in the clock rate. In fact, a detailed analysis [Kung82b] indicates that clock skew may grow with the array size at a much higher than linear rate. An immediate conclusion from this analysis is that while for a small array size, a globally synchronized array may be easier to implement, for large array size, an asynchronous system will become more favorable. Moreover, complete synchrony of all the PEs in a large array also implies high instantaneous peak power. As the progress on wafer-scale and submicron-device technology continues, the clock skew and the peak power problems incurred in global clock distribution will be more acute. Therefore, locally synchronous systems will become more appealing in the future, and global synchronization should be avoided via architectural design whenever possible.

1.3.6 Programmability

Two types of array processors are of interest: one is characterized by inflexible and highly dedicated structures and the other allows some flexibility, such as programmability and reconfigurability. Hard-wired dedicated processors offer high processing speed but suffer from long design time and high design cost. With the advent of modern algorithm/architecture analysis, the programmable array processors will become not only more economical but also more appealing in coping with constant changes of system specifications.

According to [Hoare78], the main objective of parallel language research should be to find the simplest possible mathematical theory with the following desirable properties:

1. It should describe a wide range of interesting algorithms covering signal/image processing and most scientific computing applications.

2. It should be capable of efficient implementation on a variety of networks of communicating PEs.

3. It should provide clear assistance to the programmer in his or her tasks of specification, design, implementation, verification, and validation of complex computer systems.

Even though the original idea of the systolic array suggests a tendency for dedicated design, software and programming will be essential, especially for wavefront arrays. Therefore, it is important to develop a complete set of software packages for the proposed array architecture or even a formal algorithmic notation and programming language. Two critical aspects in expressing array processing are *concurrency* and *communication*. To simplify the language design, a high-level programming language, such as Occam, usually ignores exact timing of occurrences of events.

1.3.7   Reconfigurability and Fault Tolerance

By reconfigurability we mean the ability to alter the interconnection patterns between the PEs for certain intended applications such as multi-function or fault tolerance. Two types of reconfiguration strategies are applicable: *Static reconfiguration* is used to establish a preprocessing step wherein the network is configured prior to the initiation of the tasks, and *dynamic reconfiguration* is used to reconfigure the execution paths during run time. This latter capability is particularly desirable in applications where the communication patterns are nondeterministic. The choice of either static or dynamic reconfiguration is constrained by application-specific goals, such as real-time response, reliability, or both.

To enhance the yield and reliability of computing systems, array processor architectures demand a special attention to fault tolerance. Both the fabrication-time and real-time fault-tolerant designs should be investigated. In a large two-dimensional array processor constructed on a large chip or wafer, the probability that a fatal flaw will occur increases exponentially with the area. This necessitates some kind of fault tolerance within the network, usually on the processor level, since each processor is identical. The classical approach to this problem is to employ redundancy and route data around faulty modules. Some means must be devised to employ redundant PEs and have them connected into the array to replace bad ones. Because of the communication constraints, fault-tolerance design features such as reconfiguration and rollback may become very involved. Recently, for real-time signal and image processing applications, several algorithm-oriented fault tolerant approaches have been explored. In short, for fault-tolerance consideration, the array design strategy should (1) exploit the regularity, locality, and self-timing features of systolic and wavefront array processors, and (2) explore the structural properties of array algorithms.

1.3.8   Array and Chip Partitioning

For any given array, there often arise situations where the size of the computing array is not matched to the size of the problem. In the case of systolic/wavefront arrays, it is very common that the problem size is much larger than the array size. Therefore, the data given in the problem has to be somehow partitioned into blocks of data to fit into the array. This is often referred to as "partitioning".

Another important issue is partitioning of a function into more than one VLSI chip if the entire function cannot be accommodated on a single chip. The chip boundaries have the following major effects: (1) they impose hard limits on data bandwidth on and off the chip, and (2) they create a substantial disparity between on-chip and off-chip communication delays. Consequently, the chip-partitioning strategy is impacted by two issues – number of pinouts and on-chip versus off-chip memory trade-off. As an example, off-chip memory for a reasonably sized PE array would need hundreds of I/O pins and require correspondingly high driver power.

1.3.9   Hierarchical Design and CAD Techniques

The growth in VLSI complexity has made hierachical CAD techniques necessary. Such approaches are imperative due to the extremely large number of MOS gates per chip in current technology. The complexity associated with designing VLSI circuits amounts often to 50,000 devices (as in today's technology) or even up to 1,000,000 devices in the future. The Mead-Conway

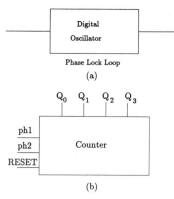

Phase Lock Loop

(a)

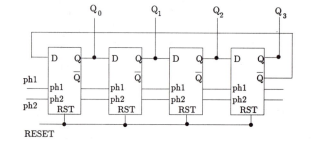

(b)

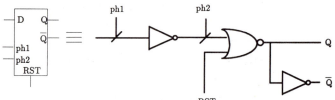

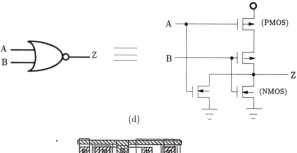

(c)

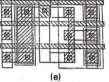

(d)

(e)

Figure 1.7: Five description levels of design: (a) algorithmic level; (b) architectural level; (c) register/logic level; (d) electrical/circuit level; (e) geometric/layout level.

design automation concept provides a basic methodology, which may be extended to a hierarchical design system involving circuit, logic, timing simulation and design rule checking.

1.3.9.1   Description Levels

In order to keep the design description comprehensible it is necessary to introduce different levels of system description. Five such description levels are as follows (see Figure 1.7):

1. *Algorithmic level*, specifying algorithms in a high-level and formal language/notation.

2. *Architectural level*, giving an accurate description of the system behavior and structure.

3. *Register/logic level*, starting from the architectural specification to give a more detailed description of the function/logic behavior in terms of data transfers and operations on data between registers.

4. *Electrical/circuit level*, specifying the detailed electrical structure of the circuit in terms of transistors, capacitors, and resistors.

5. *Geometric/layout level*, specifying the photolithography mask features of the circuit.

The major complication with this approach in VLSI design is that all the levels will be interwoven. This is especially clear in leaf-cell design, which combines electric transistor circuit design and layout design. These two levels are difficult to separate because the parasitics that affect the electrical behavior depend strongly on the layout design.

At each of the description levels the cells are hierarchically specified to decrease the complexity of the description. To reduce the design costs, a modular design approach is used; it is less expensive to implement a general module that can be used in a number of different places than to implement a specific module that can be used only once.

The main three components in a digital system are: *storage, control, and processing* [Bensc83]. Corresponding examples are memory, control unit, and arithmetic unit. There are several possible basic building-blocks for each component. For example, memory is built from the basic cell, which is a flip-flop; control logic can be implemented by a programmable logic array

(PLA); and the arithmetic unit can be implemented by using full adders (FAs). This building-block approach is illustrated in Figure 1.8. A special logic module PLA can be used to implement any set of Boolean equations and, if combined with a state register, can even implement a complete finite state machine (FSM). A PLA can be generated directly from a register transfer level specification. In general, with the increasing use of high-level design aids, we see more and more programs that are able to synthesize large portions of a VLSI circuit from a high-level (register transfer) specification.

1.3.9.2   Design Styles

The VLSI design methods can be classified as *semicustom, full custom and silicon compilation*.

Semicustom Design   Semicustom design comprises two distinct forms of circuit implementation: *gate arrays and standard cells.*   A gate array is a matrix of transistors. Customization is accomplished by laying down different interconnect paths between the transistors. Semiconductor vendors can stockpile basic arrays and individualize them at the customer's request. Advantages of gate arrays include rapid turnaround, since only partial processing is required, and reduced processing cost, since the same base layers

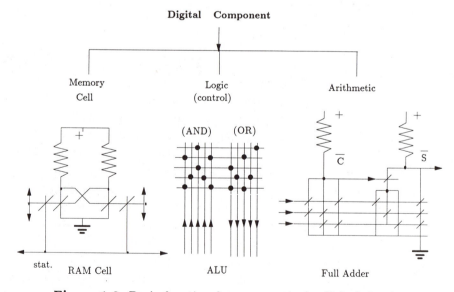

Figure 1.8: Basic functional components of a digital circuit.

can be used in many gate arrays. Their major disadvantage is that some transistors are not used, causing chip area to be wasted and resulting in increased costs.

Chips designed with standard cells overcome this disadvantage. A standard cell chip consists of predesigned blocks of common height but variable width, which realize specific logic functions. Standard cell design places the desired blocks in rows and then connects the inputs and outputs to perform the required function. The resulting chip area is not as optimized as a full custom chip, for which each circuit element can be individually tailored, but design effort is greatly reduced. Standard cell designs thus fall between gate array and full custom designs in terms of the design effort and production cost that are necessary.

Full Custom Design   The objectives of custom design are to minimize the silicon area and to optimize the performance. The key aspects of custom design includes cell design; cell library development and use; automated routing and layout of PLA, ROM, and RAM; simulation at functional, logical, and timing levels; verification; testing for design verification, electrical characterization, and chip release to production.

Silicon Compilation   Silicon compilation has captured the imagination of designers with its promise of automatic layout generation from the high level system/circuit description [Gajsk85]. Silicon compilation includes layout, circuit models, and compilation issues such as layout cells, compilation with "one-sided" cells, reorganizing to improve the layout, and electrical modeling. The maturing of silicon compiler techniques promises to revolutionize VLSI design.

While silicon compilations are still largely at the development stage, cell compilers have already been adopted in industry. A cell compiler defines a set of logic functions in a database library. For example, if it is required to design a 4-input NAND gate with output loading at 1pf and 5ns output delay, all we must do is to select a NAND gate and supply the specified parameters. The cell compiler automatically generates the silicon layout. Although the area used will be somewhat greater than the optimal (often by a factor 10 to 15% more than full custom design), this is quite acceptable.

In summary, for array processor designs, the system specifications required by the applications may change significantly if the development cycle is too long. Therefore, fast turn around implementation is critical, and CAD tools for all levels of array processor design are essential. The development of hierarchical and structured design methodology and simplified VLSI design rules have already allowed VLSI chips to be designed quickly. Although silicon compiler technology is becoming more and more mature, it is important that a high-level (array processor) structured design and description tools be developed. It is essential that they be compatible with the existing low-level CAD tools and silicon compilers. This leads to the concept of the *array compiler*, which is discussed shortly (see Figure 1.1).

1.4   Overview of the Chapters

This book is designed to provide a fundamental background regarding VLSI array processor design, particularly for real-time signal/image processing applications. An important and unique feature of the book lies in its synergistic and cohesive exploration of the applicational, algorithmic, and architectural aspects of VLSI array processors.

1.4.1   Integrated Study on VLSI Array Processors

The basic discipline in a vertically integrated VLSI system design methodology, as depicted in Figure 1.9, depends on a fundamental understanding of algorithm, architecture, and application. Therefore, array processor design involves a very broad spectrum of disciplines, including algorithm analyses, parallelism extractions, array architectures, programming techniques, functional primitives, structural primitives, and numerical performance of DSP algorithms. To present a complete and consistent study on the subject of VLSI array processors, the following selected chapters are covered:

1. Introduction – An Overview

2. Signal and Image Processing Algorithms

3. Mapping Algorithms onto Array Structures

4. Systolic Array Processors

5. Wavefront Array Processors

6. System and Software Design

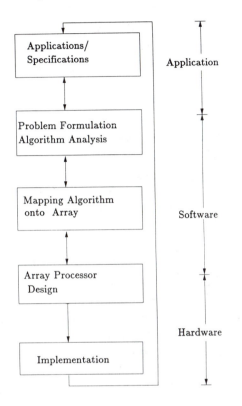

Figure 1.9: Vertically integrated VLSI system design.

7. Implementation of Array Processors

8. Applications to Signal and Image Processing

1.4.2   Chapter 2: Signal and Image Processing Algorithms

An array algorithm is "a set of rules for solving a problem in a finite number of steps by a multiple number of interconnected machines." Therefore, an array algorithm depends on the machine characteristics as well as the interconnection strategies. After most of the image processing routines are examined, some commonalities become apparent. These include intensive computation, matrix appearance, and localized, or perfect-shuffle communications. They all point to a promising systematic design of array architectures. In other words, the solution to real-time signal/image processing hinges upon novel array processors for common signal/image processing functions such as convolution, FFT, and matrix operations.

There are two types of algorithms: the *local communication* type and the *global communication* type. A majority of signal/image processing methods fall into the classes of convolution, filtering, or matrix and transform type algorithms. These algorithms possess some useful common properties, such as regularity, recursiveness, and local data communication, and they span the class of local communication algorithms. Another main class of recursive signal processing algorithms are of *global communication* type, that is, the spatial separations between nodes are beyond a certain limit (a typical example is the fast Fourier transform (FFT) algorithm). An architectural study would indicate that the hardware cost paid for global interconnectivity will become dominant and certainly influence the system performance.

Chapter 2 introduces the commonly used algorithms for signal and image processing applications. These algorithms include matrix algorithms, discrete time systems, one and two dimensional digital filtering, convolution, correlation, and transform methods. Also addressed in the chapter are some advanced algorithmic techniques, including dynamic programming, relaxation techniques and simulated annealing. Finally, VLSI algorithms with special attention on locally/globally recursive algorithm formulation will be discussed.

1.4.3 Chapter 3: Mapping Algorithms onto Array Structures

VLSI array processors may derive a maximal concurrency by using both pipelining and parallel processing. A key question is: *How is the inherent concurrency (i.e., parallel and pipeline processing) in signal/image processing algorithms fully expressed?*

In general, concurrency is often achieved by decomposing a problem into independent subtasks (executable in parallel) or into dependent subtasks executable in a pipelined fashion. The degree of concurrency varies significantly among different techniques. When mapping these algorithms onto parallel processors, the following closely related questions are posed: *How is the array processor design dependent on the algorithm? How is the algorithm best implemented in the array processor?*

A dependence graph (DG) provides a useful first step toward a stationary answer. In deriving an array processor, a series of intermediate design levels are involved. They are creating a (1) DG design, (2) mapping the DG onto a signal flow graph (SFG) array and (3) deriving a systolic array from the SFG.

Stage 1: DG Design   For a given problem, the designer has to identify a suitable algorithm described in terms of a certain convenient expression. A recursive algorithm may be easily transformed to a DG by tracing the associated space-time index space and using proper arcs to display the dependencies in the index space.

Stage 2: SFG Design   The SFG expression primarily consists of *processing nodes, communicating edges*, and *delays*. A simple (although not the only) way of mapping a DG onto an SFG array is by means of projection, which assigns the operations of all nodes along a line to a single PE. For example, the three-dimensional index space of a DG may be projected onto a two-dimensional SFG array.

Stage 3: Array Processor Design   The SFG obtained in Stage 2 can then be mapped to an SIMD, systolic array, wavefront array, or even an MIMD machine. For example, to convert an SFG array into a systolic array, a cut-set based systolization (retiming) procedure may be adopted.

Chapter 3 introduces (1) a *canonical mapping* methodology for mapping homogeneous DGs onto processor arrays; and (2) a *generalized mapping* methodology for mapping heterogeneous DGs onto processor arrays.

For array processor design, there are an enormous number of algorithms that have the useful properties of being totally regular and localizable. For example, matrix multiplication, convolution, autoregressive filtering, discrete Fourier transform (DFT), discrete Hadamard transform, Hough transform, least squares solution, sorting, perspective transform, median filtering, LU decomposition, and QR decomposition all belong to this important class. By exploiting this regularity, the array processor design for such algorithms can be greatly simplified. This canonical mapping methodology is suitable to treat such a class of algorithms, which can be expressed by shift-invariant dependence graphs. This mapping method consists of three design stages, each utilizing an appropriate canonical form.

The generalized mapping methodology allows us to treat a broader class of algorithms and the corresponding dependence graphs. There are many other important algorithms that are not completely regular (i.e., not totally shift-invariant) but exhibit a certain degree of regularity. This semi-regularity very often proves to be useful for an efficient mapping methodology. The generalized mapping methodology allows us to deal with an extended DG classification and to have options by linear or nonlinear assignment/schedule. More flexibilities are also created by using multiple projections, allowing global communication, and treating totally irregular DG structures. The generalized mapping methodology can provide effective designs to many algorithms including Gauss-Jordan elimination, shortest-path problems, transitive closure, simulated annealing, partial differential equation (PDE) problems, singular value decomposition (SVD), FFT, and Viterbi decoding.

1.4.4   Chapter 4: Systolic Array Processors

In a VLSI context, memory and processing power are relatively cheap. The emphasis is on keeping the overall architecture regular and reducing the relative complexity. Good VLSI architectures are highly pipelined and hierarchical. Hence they require well-structured algorithms with predictable results. As long as interconnection in VLSI remains restrictive, the locality of a recursive algorithm will be of great concern. An increase in efficiency can be expected if the algorithm arranges for a balanced distribution of work load while observing the requirement of locality, that is, short communication paths.

The first result based on such a consideration is the design of the systolic array. *A systolic system is a network of processors that rhythmically compute and pass data through the system.* Every processor regularly pumps data in and out, each time performing some short computation, in order that a regular flow of data is kept up in the network [HTKun78]. Systolic arrays are amenable to VLSI implementation, since they feature the important properties of modularity, regularity, local interconnection, and highly pipelined and synchronized multiprocessing.

Chapter 3 addresses systematic methodologies mapping algorithms onto SFG structures, whereas Chapter 4 proposes a cut-set based retiming procedure for converting SFG arrays into synchronous systolic arrays. This chapter also addresses several issues and guidelines for designing optimal systolic arrays, such as maximization of throughput rate, minimizing delay elements required, optimal processor utilization, latency, and block pipeline rate. Systolic designs for many algorithms, such as filtering, convolution, matrix operations, and sorting, are presented. Systolic design with two-level pipelining is also discussed. Optimal designs for some dynamic programming (transitive closure, shortest path) problems will also be proposed.

1.4.5   Chapter 5: Wavefront Array Processors

The fact that the activities in a systolic array must be controlled by global timing-reference "beats" may put systolic designs in a very disadvantageous situation: From a hardware perspective, global synchronization incurs problems of *clock skew, fault tolerance, and peak power.* The burden of having to synchronize the entire computing network will become increasing intolerable for ultra-large-scale or ultra-high-speed VLSI arrays. From a software perspective, severe requirements are imposed on compiler design, in order to synchronize the activities in a systolic array, a simple solution to these problems is to exploit the principle of data flow computing for array processors. This leads to the design of wavefront array processors.

By virtue of the data-driven approach in wavefront array processing, instructions cannot be executed until their operands have become available. In this approach *the arrival of data from neighboring processors will be interpreted as a change of state and will initiate some action.* The wavefront arrays are reminiscent of the action of wavefront propagation, and thus envisage a distributed and globally asynchronous array processing system. This approach substitutes the requirement of correct *timing* by correct *sequencing* and handles the data dependency locally. Thus, it eliminates the need for global control and global synchronization.

In general, there are two approaches to deriving wavefront arrays: a traditional approach is to trace the computational wavefronts and pipeline the fronts on the processor array. Here we propose a different approach, which is based on converting an SFG array into a data flow graph (DFG) array and then into a wavefront array by properly imposing several key elements in dataflow computing.

Chapter 5 stresses the notions of wavefront processing and dataflow computing. A DFG is formally proposed as an abstract model for wavefront array processors. We address first the issue of transforming an SFG into a DFG by an equivalence transformation from SFG to DFG. Then a complete treatment of the timing analysis for generalized (cyclic or acyclic) DFG networks, largely based on a notion of timed petri net [Ramam80], is developed. The analysis also allows the prediction of the minimal number of buffers required on all the edges of the DFG in order to achieve the best possible throughput rate.

Based on the notion of wavefront processing, the design of array processor languages can be somewhat simplified. In Chapter 5, two languages for programming array processors, the wavefront language (MDFL) and Occam, are discussed. MDFL is based on the notion of computational wavefronts. Occam was developed by INMOS. The two schemes are very compatible, since both of them employ a dataflow principle to facilitate the description of parallel data movements and executions in VLSI array processors.

1.4.6   Chapter 6: System and Software Design

Chapter 6 addresses a number of important issues encountered in the array processor design. These include overall system design considerations; algorithmic partitioning if resources are limited; the fault tolerance algorithms for array processors and finally software and CAD tools for array processors are addressed.

As to the overall system design, the major components of an array processor system consists of: (1) host computer, (2) interface system, including buffer memory and control unit, (3) interconnection networks, and (4) processor array, comprising of a number of processor elements with local memory. The host computer supports data storage and formatting and schedule program management. The interface system, connected to the host via the host bus, has the functions of down-loading and up-loading data. Based on the schedule program, the controller monitors the interface system and array processor. The interface system should also furnish adequate hardware support for many common data management operations. Interconnection networks provide a set of mappings between processors and memory modules to accommodate certain common global communication needs. Incorporating certain structured interconnections may significantly enhance the speed performance of the processor arrays.

This chapter discusses the *partitioning* issues in array processor designs. In general, there are two schemes to partition the problem: the LSGP (locally sequential globally parallel) and LPGS (locally parallel globally sequential) methods. In the LSGP scheme, a cluster of neighboring virtual processors are mapped onto a real processor, whereas in the LPGS scheme, a window of the real array size moves through the problem and processes it. The discussion also includes techniques for matching algorithms to some a priori selected array architectures such as mesh or hypercube arrays.

This chapter also treats the very important issue of array fault tolerance, including the fabrication-time and real-time fault tolerant designs. Fabrication-time fault tolerance means that some form of fault tolerance must be included to allow cost-effective yield in the possible presence of fabrication defects. On the other hand, some form of fault tolerance must be included to overcome faults that will occur during the processor's lifetime. This is called real-time (operational) fault tolerance. Two approaches will be proposed in Chapter 6: architectural and algorithmic approaches. The architectural approach is based on reconfiguring the array structure such that the problem can be executed on a reduced size array [Fort85a]. The algorithmic fault-tolerance approach exploits the powerful DG expression and certain algebraic properties inherent in many real-time DSP algorithms to recover the correct data of a faulty processor.

The actual implementation of VLSI array processors can be as either dedicated or programmable. Programmable arrays are often preferred, due to the need of coping with constantly changing application specifications. However, array programming is significantly more complicated than sequential programming due to the difficulty of keeping track of several simultaneously occurring events. It is desirable to have new notations or languages which better fit array processors, instead of using conventional languages. It is also important to develop coherent software techniques for *programming* or *design* of array processors. (For users' convenience, it is useful to make available a set of software packages for the programmable array processors.) Chapter 6 will discuss a variety of programming languages as well as propose an approach to develop a high-level and integrated CAD tool for facilitating mapping or matching algorithms to array architectures.

1.4.7 Chapter 7: Implementation of Array Processors

Array processors implemented with today's VLSI technology can offer a much greater hardware capacity, higher speed, and lower power compared with other existing technologies. Chapter 7 will address two levels of implementational considerations:

- Processor level design: *How can the systolic and wavefront arrays be best implemented in off-the-shelf or custom-designed hardware components?*

- System level design: *How can the enormous array processing capability be incorporated into an overall host computer systems?*

In the processor level design, we shall discuss the basic building blocks of a DSP chip, the specification of these building blocks, and the way in which they are interconnected. We shall also address the key architectural features of DSP processors, such as *speed, precision, pipelining, Harvard architecture, RISC-style architecture, addressing modes, and bit serial versus bit parallel designs.*

A typical set of primitive operations for DSP processors should include primitives for arithmetic and logic operations, data accessing and storage, control, I/O, and communications. Examples of commercially available VLSI chips, worthy of consideration for array processor implementations, are IN-MOS's transputer, NCR's GAPP, NEC's data flow chip μPD7281, TI's programmable DSP chip TMS320, and recent 32-bit processors such as AND 29325, Motorola 68020, and Weitek's 32-bit (or the new 64-bit) floating-point chips [Ware84]. Many DSP applications requires special features such as fast multiply/accumulate, high-speed RAM, fast coefficient table addressing, and so on. They should be incorporated into the development of a series of *application-specific integrated circuits* (ASICs) oriented toward array processors.

In this chapter, some existing array processor systems are presented and compared. Examples include SIMD arrays, systolic arrays, wavefront arrays, hypercube array computers, and some other enhanced array architectures.

1.4.8   Chapter 8: Applications to Signal and Image Processing

Chapter 8 will demonstrate how to apply the array processor design methodology to construct high-speed parallel signal processors using VLSI systolic/wavefront arrays. The applications considered include spectral estimation, speech processing, image processing and image analysis.

The generic algorithm similarities between *spectral estimation, beamforming,* and *Kalman filtering* are exploited in the array processor. The algorithms for these applications all share a common least squares formulation. It can be shown that a triangular array configuration (termed *triarray*) will be very suitable for these types of algorithms. Application of the triarray to beamforming and Kalman filtering will be presented in detail in this chapter.

The main areas in speech processing are *speech analysis/synthesis, speech coding,* and *speech recognition.* Array processor designs for these three areas are presented in this chapter. The design examples cover linear prediction techniques for speech analysis/synthesis, vector quantization for speech coding, and dynamic time warping for speech recognition.

For image processing applications, two classes of problems are considered: *low-level image processing* and *high-level image analysis*. Image processing involves algorithms for transformation, enhancement, restoration, and reconstruction. Image analysis involves classifying segments or features of the image into known classes. As examples for image processing, this chapter discusses array processor examples for median/rank-order filtering for image enhancement, relaxation technique for image restoration, and interpolation techniques for image reconstruction. For image analysis, edge detection for feature extraction, Hough transform for line/curve detection, and template matching for pattern classification.

1.5   Other Closely Related Research Disciplines

1.5.1   VLSI and Wafer Scale Integration

Actual systolic/wavefront array processing hardware is bulky and power consumptive. Even with the state-of-the-art feature size and packaging techniques, there is a severe limitation on the amount of hardware that may be included in a VLSI chip. Order-of-magnitude increases in packing density are now possible using wafer-scale integration (WSI) technology. It is thus essential to investigate the feasibility of dramatic improvements in size, weight, and power factors by implementing massively parallel architectures in WSI.

WSI technology can offer significant performance enhancements over the conventional method, where individually packaged chips are surface mounted on a printed circuit board. The key important advantages include (1) shorter interconnect distances between chips; (2) the ability to mix semiconductor technologies; (3) faster system clock rates and speeds; and (4) feasibility of dynamic interconnection on WSI. The benefits introduced by the development of WSI systolic technology are manyfold: the local interconnection nature of systolic/wavefront architectures makes WSI particularly useful. However, the very critical problem of power dissipation remains largely unresolved.

Two types of WSI technologies are of interest: (1) monolithic wafer-scale processors, and (2) hybrid approaches involving "flip-chips" on monolithic substrates. They are explained below:

Monolithic WSI   It has been argued that the various chips required for a particular system should be processed contiguously and interconnected on a wafer. This scheme is termed *monolithic WSI*. The expected benefits include reduced cost due to savings in dicing and individual packaging expense, improved density due to elimination of interpackage spacing, higher speed due to reduced interconnect length, and improved reliability due to reduction in the number of parts. Although significant strides have been made in laser restructuring of interconnects for monolithic WSI, there is still the very demanding reconfigurable multilevel metal interconnect system required to implement redundancy and ensure productivity and reliability. So far, monolithic WSI technology has not been able to guarantee a satisfactory yield. But a rapid advance in WSI technology is being made, and reasonable yield improvements are expected in the future.

Hybrid WSI   Any scheme that utilizes pretested but unpackaged individual chips, assembles and interconnects them on a substrate material can be termed *hybrid WSI*. In hybrid WSI, since the individual chips are pretested to be fully functional, no elaborate redundancy scheme is needed. This greatly simplifies the required metal interconnect system, resulting in enhanced productivity and yield. One hybrid WSI approach is the so-called flip-chip technology. The substrate material carries double level metal interconnect.[2] Since the chips are not packaged individually, high system density can be achieved. High density also leads to high speed due to reduced interconnect delays. The elimination of cost for individual chip packages may also result in lower system cost, provided the substrate material and processing are optimized.

Development of Array Processors in WSI   With today's VLSI and WSI technologies, the development of fault-tolerant systolic/wavefront array processors using WSI integration becomes a realistic goal. These architectures with respect to fault-tolerant techniques should be investigated. Issues to be considered include techniques for fault detection, techniques for array reconfiguration to detour around faulty elements, topologies for interconnection paths between elements, and methods for insuring reliabil-

---

[2]The I/O pads of the chips to be interconnected are plated with solder bumps. These chips are flipped upside down onto the substrate with the solder bumps facing corresponding pads on the substrate. The substrate is heated to melt the solder, which wets the substrate metal pads, making contact. Other components such as capacitors and resistors can also be mounted alongside the chips.

ity of the central controller [Raffe85], [Jessh85]. The design of the PEs in the array and their associated hardware should take into account an optimum partitioning into modules for fault-tolerant WSI. The design should consist of detailed block-level designs with simulations of critical circuits to verify speed, performance, and power. Furthermore, a system level architecture must be developed that embeds the proposed WSI systolic/wavefront processors into a processing subsystem that includes memory, control structures, host interfacing, and data input and output. In summary, parallel processing architectures and systolic architectures are important for obtaining real-time processing performance. WSI can provide major benefits in this area. Chapter 6 will present a fuller discussion on WSI and fault-tolerance.

1.5.2   VLSI and Optical Processing

Optical processing offers another complementary approach [Goodm84]. Optical computers process information encoded in light beams. Thus optical computing presents, for certain operations, the potential of computation at the speed of light for signal and image processing, with throughput far beyond perceived limits for VLSI GaAs or other electronic processing technologies.

Advantages of Optical Processing   Optical computing has a good number of additional advantages. Optical lenses can perform certain special mathematical calculations much more effectively than their electronic counterparts. Algebraic optical array processors have proven to be technically sound. The properties of a lense may be utilized to perform Fourier transforms, convolutions, and advanced mathematics such as matrix-matrix multiplication in linear algebra. High-density, three-dimensional information storage can be achieved through optical holography. Interconnect space limitation and clock skew due to wire communications are eliminated [Huang84]. (Research programs on fixed interconnects and reconfigurable interconnects are currently in progress [Goodm84].) Finally, optical processing also offers a unique feature on radiation hardness.

Weaknesses of Optical Processing   Compared with digital technology, optical processing provides rather limited flexibility, limited programmability and low precision. Furthermore optical technology is less mature. This is the reason why optical computers are not yet at the commercial stage, with the exception of some devices used in the military for applications such as synthetic aperture radar (SAR). Currently, there exists no optical technology (e.g., bistable optical material) that can perform high speed optical logic and nonlinear operations. Finally, the total system's speed performance is often dampened by the overhead time incurred in the D/A (digital/analog) and A/D conversions [Psalt84].

Hybrid VLSI and Optical Array Processors   Hybrid VLSI and optical array processors represent a new dimension for future high-speed supercomputing technology. A hybrid system can incorporate the advantages of both the complimentary capabilities of optics (efficient linear operations and communications) and electronics (efficient nonlinear operations). Some hybrid designs use systolic arrays to allow parallel processing of digital and analog data. Optical/VLSI pattern-recognition systems also utilize the inherent ability of optical systems to perform transforms for computing high-speed correlations. For a very broad application domain, hybrid systems will offer a significant improvement in cost, size, weight, power consumption, and reliability. In some important areas, hybrid VLSI and optical hybrid array processors have the potential of becoming the next-generation technology, surpassing electronic computers. Applications will include the areas of pattern recognition, Fourier transform, feature extraction, correlation systems, scene classification, image analysis, machine vision, alphanumeric data analysis, and SAR processing.

Optical computing has great potential for solving the computationally demanding problems of future military systems and industrial automation, but the missing technology for practical realization is the ability to integrate the optical and electronic components on a single material system or on a compact hybrid system. Therefore, there is a strong incentive for developments of opto-electronic components and devices. Although not directed at optical array processing, the information and mapping methodologies in Chapter 3, 4, and 5 for VLSI array processors should be useful to the development of optical array processors as this technology becomes more mature.

Table 1.3 summarizes the comparison of digital optical and VLSI processing systems and technologies [Jenki87].

1.5.3   AI-Oriented VLSI Supercomputing

Number crunching has been the prevailing processing requirement in DSP and scientific computations. Because of the regularity of most numerical

	Optical Processing	VLSI Processing
System Characteristics:		
• Physical interconnections	Global and local (e.g., cellular pyramid)	Local only (e.g., systolic/wavefront arrays)
• Algorithm communication	Global and local (e.g., FFT)	Local only (e.g., vector/matrix multiplication, DFT)
• Input/output	Parallel	Pin-out constraints
Physical Characteristics:		
• Number of gates	Large	Large
• Switching energy	pj-fj	pj-fj
• Gate speed	≤ 100 ps	ns

Table 1.3: Comparison of digital optical and VLSI processing systems and technologies (**Courtesy from [Jenki87]**). Note that the parameters on the switching energy and speed are for an individual gate only, and for large arrays, the parameters might change to nj-pj, and ms, respectively.

processing algorithms, VLSI array processors with very local and/or regular communication can be successfully applied to these algorithms.

On the other hand, the symbolic processing used in artificial intelligence (AI) exhibits highly irregular communication patterns. For example, the nodes in a semantic network can have a branching factor (number of links) that varies from 0 to 10,000 or more. In a fixed-connection VLSI array processor with a limited number of links between PEs, it is impossible to allocate all of the related informations to near-neighbor nodes.

Considerable effort has been devoted to finding architectures for AI processing with VLSI [Hwang86], [Moldo86]. An AI-oriented computing machine requires very different architectural features from the VLSI array processors used for number crunching. Their major features are summarized below:

• *Large problem size:* Due to the huge problem size, the number of processor elements in an AI machine has to be very large. For example, the prototype of the connection machine CM-1 [Hilli85] contains 64K processor elements.

- *Frequent memory access:* Memory access is very frequent, causing bottlenecks. This problem may be alleviated by including local memory into each PE. For example, each PE of the CM-1 has dedicated 4K-bit memory cells.

- *Intensive and irregular communication:* Communication between processor elements should be very fast and flexible. For example, in the CM-1, the basic building blocks are custom CMOS chips, each containing 16 processors and 1 router. The router is responsible for routing messages between chips, which are physically connected through a 12-dimension hypercube network.

- *Associative processing:* There are two types of memory structures: one is location addressable and the other is content addressable. It has been recognized that the content-addressable type is more closely related to the way the human brain functions. In fact, a popular approach to its implementation is based on neuron networks. This leads to the notion of an associative processor/memory network, with the main features including recognition and error correction based on partial or cluttered input [Kohoe72].

- *Pattern matching and set operations:* Symbolic processing requires pattern matching and logical set operations (union, intersection, negation) on a large data base. From an architectural perspective, this requires a broadcasting capability to allow matching or unification to be done in parallel.

Recently, connectionist networks are proposed for AI applications [Fahlm87]. The information storage/retrieval process is no longer based on conventional memory cells. Instead, it is accomplished by *altering the pattern of connecting among a large number of primitive cells,* and/or by *modifying certain weighting parameters associated with each connection.* Some examples on Hopfield and Boltzmann networks for some associative retrieval and optimization applications are discussed in Chapter 2. As an architectural example, CM-1 machine is discussed in Chapter 7.

1.5.4 Complementary Roles of General-purpose Supercomputers and Array Processors

VLSI technology has already contributed to the realization of supercomputer systems capable of processing in thousands of MOPS or MFLOPS (mega-floating point operations per second). Commercially available supercomputers now include the Cray 1/2, Cyber 205, and HEP from the United States, and Fujitsu VP200, Hitachi S-810/20, and NEC SX-2 from Japan. General-purpose supercomputers enjoy a great flexibility, with many software packages available for certain scientific computing applications. However, from a signal processing point of view, the excessive supervisory overhead incurred in general purpose supercomputers often severely hampers the processing rates. In order to achieve a throughput rate adequate for real-time image processing, the only effective alternative appears to be massively concurrent processing. Therefore, for special applications (stand alone or peripheral), dedicated array processors will be more cost-effective and perform much faster. The advent of VLSI technology and modern CAD techniques have further facilitated a speedy prototyping and implementation of application-oriented (or algorithm-oriented) array processors. There are two important reasons for a future trend of special-purpose array processors. First, the implementation cost and power consumption for a large-size array processor will become feasible with VLSI technology. Second, the design/implementation turnaround time can be drastically reduced using CAD tools, resulting in a customer delivery schedule that is more acceptable.

For general-purpose supercomputers, the software and programming techniques are generally developed based on a given hardware. On the other hand, for special-purpose (either dedicated or programmable) array processors, the hardware design is heavily based on an extensive algorithmic analysis and specific programming languages.

A reasonable compromise in the choice between selecting supercomputers or (special-purpose) array processors is to let them play complementary roles in real-time signal and image processing systems. In this case, the supercomputer will be the host computer supervising central control, resource scheduling, and database management and, most importantly, providing super-speed I/O interfaces with the (peripheral) array processors. The concurrent array processors will, in turn, speedily execute the (computation-bound) functional primitives such as FFT, digital filtering, correlation, and matrix multiplication/inversion and handle other possible computational bottleneck problems.

1.6   Concluding Remarks

We have witnessed the rapid growth of signal processing and computing technology that followed the invention of the transistor in the 1940s and integrated circuits in the late 1950s. The emergence of new VLSI technology, along with modern engineering workstations, CAD tools, and other hardware and software advances in computer technology, virtually assures a revolutionary information processing era in the near future. Indeed, the pace of research and development in supercomputer technology continues to accelerate. In particular, the systolic and wavefront array processors have shown great promise for a broad range of applications in signal and image processing. The research and development on these array processors will play a key role in defining the trend of the future supercomputer technology.

A fundamental question is, how should the research area on VLSI array processors take shape at this stage and in the near and far future? In a strict sense, VLSI is a device research area, whereas the array processor belongs to the research area of computer engineering. In a broad sense, however, there have been significant input derived through a close interaction between the *VLSI* and *array processor* areas. This perspective prompts a new research area in *VLSI array processors*. Indeed, array processors as implemented in VLSI offer a new and promising opportunity to the development of the state-of-the-art supercomputer. In our opinion, this research area bears even deeper significance than merely a collection of voluminous systolic designs for specific algorithms or applications. The heart of the matter lies in a systematic methodology of mapping applications and algorithms onto array architectures. In other words, the research essence of VLSI array processors hinges upon a new meaning of the word "integration". The term integration should be one that vertically integrates the apparently diversified areas of *applications, algorithms, architectures*, and *technology*.

1.7   Problems

1. *Scaling effects on VLSI*: The first-order MOS scaling theory, which is based upon the constant-field model, indicates that the characteristics of a MOS device can be maintained and the basic operational characteristics preserved if the critical parameters of a device are scaled by a factor α as listed in the table.

   (a) Completely fill in all the blanks in this table, giving a brief explanation for each.

Device/Circuit Parameters	Scaling Factor
Device dimensions (length, width, oxide thickness, junction depth)	$1/\alpha$
Doping concentration	α
Supply voltage	$1/\alpha$
Field across gate oxide	1
Gate oxide capacitance	
Current	
Gate delay	
Power dissipation	
Power-speed product	
Gate area	
Power density	
Line resistance	
Line capacitance	
Line response time	
Line voltage drop	
Line current density	

(b) As a rule, for constant chip size the lengths of some of the signal paths that traverse across the chip do not scale down. This is called the *stuffing effect*. The wire delay can be modeled as an RC product, in which R is the line resistance and C is either the gate input capacitance or the line capacitance (depending on which is larger). Discuss the wire delay scaling effect due to stuffing.

(c) What is the impact on VLSI technology for stuffing wire delay, line current density, and power driving capacity, respectively?

2. *Ware model of parallel processor speedup*: Let

p = number of processors

f = fraction of parallel processable work

and also assume that at any instant that either all p processors are operating or only one processor is operating. Then according to the Ware model, the speedup is equal to

$$S_P = \frac{T_1}{T_P} = \frac{T_1}{T_1[(1-f)+f/p]}$$

Show that unless f is close to 1, the difference on speedup factor S_P will be somewhat limited. Therefore, a good parallel algorithm design (so that f will be close to 1) is more important.

3. *FFT Computation Load*: It takes about $N\log_2(N)$ multiplication time units to perform an N-points FFT. For performing FFT operations on input raster scanned image with sampling rate 25K/s, and 1024-point FFT program are used (some segmentation of the input sequence is required), what is the required computation power (MFLOP/s) in order to get real-time output?

4. *Fabrication and cost analysis*: [Murog82] Given a wafer of diameter d, how many chips can we obtain from this wafer? Assume that each chip is a square with side length a.

   (a) Draw approximate curves of cost versus a, for $a = 10$ to 300 mil, using d as a parameter, with $d = 3$, 4, and 6 in.

   (b) Find the cost of a chip with $a = 300$ mil, assuming that a processed wafer costs \$150. (Simply divide \$150 by the number of chips, ignoring other factors.)

5. *Function-level comparison*: Pick a computer (mainframe, mini or microcomputer) that you are most familiar with. Draw the block diagram of this computer on a function-level. Describe the information flow process from input to output. Discuss and explain which factors will determine the processing speed in this computer.

6. *Terminology comparisons*: Clarify the terminologies used in computer architecture in each of the following groups:

   (a) Parallel and pipeline.

   (b) Synchronous communication and asynchronous communication.

   (c) Control flow and data flow.

7. *Clarification of the terminology array processor*: The *array processor* in this book and that in the Floating Point System's (FPS) "array processor" have different meanings [Hwan84a].

   (a) Sketch the FPS architecture and show that it is a vector processor.

(b) Verify that the FPS array processor in fact means *array of data* rather than *array of processors* as used in our definition.

8. *Power consumption in optical computing*: In the last few years rapid progress has been made in optical bistability for logic gates. Switching energies between 1 and 10 pJ have been experimentally demonstrated for devices that have response time of 1 to 10 ns. Based on this data, what is the expected power consumption of a single device?

9. *Vision understanding system*: An ideal vision understanding system is one that has the visual acuity of a human being but also has the ability to understand its user's requests. There are at least four seemingly distinct operations in what is being suggested: image analysis, pattern recognition, scene analysis, and language understanding. These call for the cooperative effort of at least three styles of architectures: a parallel image array processor (IP), a semiparallel pattern recognizer (PR), and a data base structure (closest to a conventional architecture) that acts as a knowledge store.

   (a) What is the configuration structure's relationship between these three style of architectures?

   (b) From regularity and parallelism points of view, which one (IP or PR) can take advantage of regularity and parallelism? Briefly explain your reasons.

   (c) Two search techniques (depth first and breadth first) can be applied in the data base structure to help the operations of IP and RP. Between these two which one can take more advantage of parallelism for this understanding system. Why?

10. *VLSI and WSI for array processors*: List at least five VLSI or WSI chip designs or implementation for the array processors published in available literature.

11. *Learning and AI*: [Rich83] One criticism of AI is that machines cannot be called intelligent until they are able to *learn* or, in a more advanced term, to *think*. What would be a good definition of learning? What is the definition of thinking? *Think* about it.

Chapter 2

SIGNAL AND IMAGE PROCESSING ALGORITHMS

2.1   Introduction

An *algorithm* is a set of rules for solving a problem in a finite number of steps. There is extensive literature exploring various computational aspects of signal/image processing and scientific computation algorithms, and a considerable number of software packages are now available. The application domain covers adaptive array processing, vision systems, nuclear physics, structure analysis, and digital signal processing for speech, image, seismic, weather, astronomical, and medical applications. For example, LINPACK [Stewa73] and EISPACK [Smith76] are popular packages for many scientific computations (especially those using various types of matrix operations). SPIDER, a portable image processing software package developed by the Electrotechnical Laboratory, MITI, Japan, has an elaborate collection of about 350 kinds of image processing algorithms.

Lists of Algorithms   Some examples of the various operations and algorithms used in signal/image processing are listed next.

1. *Matrix Operations*

   (a) Matrix-vector multiplication, matrix-matrix multiplication.

   (b) Solution of linear systems: matrix triangularization (QR decomposition), solution of triangular linear systems (back-substitution), matrix inversion, pseudoinverse.

   (c) Singular value decomposition (SVD), eigenvalue computation.

   (d) Solution of Toeplitz linear systems.

2. *Basic Digital Signal Processing Operations*

   (a) Filtering type — FIR, IIR filtering, 1-D, 2-D convolution and correlation, 1-D, 2-D interpolation and resampling, 1-D, 2-D median filtering, template matching (Prewitt, Kirsch or Hueckel operators), linear phase filtering (low-pass, high-pass, band-pass, band-stop), inverse filtering, Wiener filtering, Kalman filtering, adaptive filtering, windowing operations (rectangular, Gaussian, Hamming, Hanning, Bartlett), differential filter (gradient, Laplacian), and so on.

   (b) Transform type — discrete Fourier transform, fast Fourier transform (radix 2, mixed), Walsh-Hadamard Transform, fast Hadamard transform, slant transform, Fermat number transforms, Haar transform, discrete cosine transform, Hough transform, Karhunen-Loeve transform, and so on.

3. *Image Processing Algorithms*

   (a) Restoration: inverse filter, Wiener filter, constrained least squares filter, pseudo inverse filter by SVD, and so forth.

   (b) Reconstruction: Radon transform, back-projection, interpolation, regularized optimization, pseudoinverse filter, constrained optimization, algebraic reconstruction techniques, etc.,

   (c) Enhancement and smoothing: histogram transform, iterative enhancement and noise elimination, hysteresis smoothing, median filter, edge- and line-preserving smoothing, and so on.

(d) Edge and line detection: differential edge detection (Laplacian, Roberts, Sobel, Prewitt), template matching type (Prewitt, Kirsch, Robinson), Hough transform, heuristic search method, Heuckel operator, iterative type, and so forth.

(e) Texture analysis: concurrence matrix, difference statistics, local extrema, run length, autoregressive model, autocorrelation, Fourier features, texture edge detection, texture edge preserving smoothing, and so on.

(f) Region segmentation: heuristic method, iterative merging method, iterative thresholding method, split and merge method, and so forth.

(g) Geometrical operation: operations on connected components (labeling, consecutive numbering of labels, adoption or rejection), boundary detection, boundary description (chain code, slope, curvature, Fourier descriptions), shape features (starting point, centroid, circumscribed quadrilateral, area, perimeter, size , elongatedness, moments, Fourier description), expansion and contraction, line thining, shrinking, distance transform and skeleton, cross-section, projection, geometrical warping, and so on.

(h) Other operations: fundamental statistics, histogram computation, thresholding, linear filters, gray-scale translation (normalization, shift, etc), affine transform (with linear interpolations, without interpolation), relaxation labeling requantization, transposition, mode change, replacement and copy, and so on.

4. *Others*

(a) Searching and sorting (merge sort, bitonic sort, etc.).

(b) Graph and geometrical algorithms: transitive closure, minimum spanning trees, and finding of connected components.

(c) Polynomial operations: polynomial multiplication and division, greatest common divisor, and encoding/decoding for error correction.

Two other important aspects concerning algorithmic study are application domains and computation counts. Several examples on application domains are provided in the following table [Casas84]:

Application	Attractive Problem Formulation	Candidate Solutions
high-resolution direction finding	Symmetric eigensystem	SVD
State estimation	Kalman filter	Recursive least squares (Square-root formulation)
Adaptive noise cancellation	Constrained least squares	Triangular or orthogonal decomposition

A second table is given for computation counts [Broml84]:

Order	Name	Examples
N	Scalar	Inner product, IIR filter.
N^2	Vector	Linear transforms, Fourier transform, convolution, correlation, matrix-vector products.
N^3	Matrix	Matrix-matrix products, matrix decomposition, solutions of eigensystems or LAEs or least squares problems.

Modern digital signal processing (DSP) is characterized by a growing interplay among several fields: signal and image processing; matrix operations; and some nonnumerical analysis. There is a broad class of compute-bound processing where multiple operations are performed on each data item in a recursive and regular manner. This is especially true for most of the time-consuming front-end processing that deals with large amounts of data obtained from sensors in various DSP operations. The ever-increasing demands of speed and performance in modern DSP clearly points to the need for a tremendous computation requirement, which may be satisfied only by a revolutionary supercomputing technology.

In this chapter, we discuss some of the important operations and algorithms used in DSP. This background will be essential for the development of special-purpose supercomputers—in particular, VLSI array processors. The discussion of these operations is divided into three main categories: matrix operations, signal processing algorithms, and image processing algorithms.

2.2   Matrix Algorithms

Matrix operations are very prevalent in many signal and image processing applications. An m by n matrix consists of $m \times n$ elements (i.e., scalar data) arranged in m rows and n columns. A vector may be considered as a special type of matrix. An n-dimensional vector consists of n elements, which can be represented as a column vector or a row vector.

2.2.1   Basic Matrix Operations

There are a number of operations that can be performed with matrices. Some of the common matrix related operations are discussed below.

2.2.1.1   Inner Product

The inner product of two n-dimensional vectors $\mathbf{u}$ and $\mathbf{v}$ is obtained as the product of the row vector $\mathbf{u}^T$ and the column vector $\mathbf{v}$, where

$$\mathbf{u}^T = [u_1, u_2, ..., u_n] \text{ and } \mathbf{v} = \begin{bmatrix} v_1 \\ v_2 \\ \cdot \\ \cdot \\ \cdot \\ v_n \end{bmatrix}$$

Mathematically, the inner product $< \mathbf{u}, \mathbf{v} >$ is given by

$$< \mathbf{u}, \mathbf{v} > = u_1 v_1 + u_2 v_2 + \cdots + u_n v_n = \sum_{j=1}^{n} u_j v_j$$

2.2.1.2   Outer Product

The outer product of an n-element column vector $\mathbf{u}$ and an m-element row vector $\mathbf{v}^T$ is a matrix of dimension $n \times m$

$$
\begin{bmatrix} u_1 \\ u_2 \\ \cdot \\ \cdot \\ \cdot \\ u_n \end{bmatrix} \cdot [v_1 \; v_2 \; \ldots v_m] = \begin{bmatrix} u_1 v_1 & u_1 v_2 & \cdots & u_1 v_m \\ u_2 v_1 & u_2 v_2 & \cdots & u_2 v_m \\ u_3 v_1 & u_3 v_2 & \cdots & u_3 v_m \\ \vdots & \vdots & & \vdots \\ u_n v_1 & u_n v_2 & \cdots & u_n v_m \end{bmatrix}
$$

2.2.1.3   Matrix-Vector Multiplication

The multiplication of an $n \times m$ matrix $\mathbf{A}$ and an m-element column vector $\mathbf{u}$ results in

$$
\mathbf{v} = \mathbf{Au} \tag{2.1}
$$

where $\mathbf{v}$ is an n-element column vector. The ith element of $\mathbf{v}$ is

$$
v_i = \sum_{j=1}^{m} a_{ij} u_j \tag{2.2}
$$

where a_{ij} is the element which is in the ith row and the jth column of $\mathbf{A}$.

2.2.1.4   Matrix Multiplication

If $\mathbf{A}$ is an $m \times n$ matrix and $\mathbf{B}$ is an $n \times p$ matrix, then the two matrices can be multiplied and the product, denoted by $\mathbf{C}$, is an $m \times p$ matrix given by $\mathbf{C} = \mathbf{AB}$. Its elements are:

$$
c_{ij} = \sum_{k=1}^{n} a_{ik} b_{kj}
$$

Note that the inner product, the outer product, and the matrix-vector multiplication are all special cases of matrix multiplication.

2.2.2   Solving Linear Systems

An important matrix problem that arises in signal/image processing is solving a set of simultaneous linear equations. The problem consists of solving for n unknowns given n linear equations. In matrix notation, the problem is to find an $n \times 1$ vector $\mathbf{x}$ such that

$$\mathbf{A} \cdot \mathbf{x} = \mathbf{y}$$

where $\mathbf{A}$ is a given $n \times n$ nonsingular matrix and $\mathbf{y}$ is a given $n \times 1$ vector.[1]

The solution to this equation can be obtained by using matrix inversion, that is, $\mathbf{x} = \mathbf{A}^{-1} \cdot \mathbf{y}$. However, the number of computations needed for direct matrix inversion is high and the procedure is numerically unstable. An often used procedure is to first triangularize the matrix $\mathbf{A}$ to get

$$\mathbf{A}' \cdot \mathbf{x} = \mathbf{y}'$$

where $\mathbf{A}'$ is an upper triangular matrix, and then use back-substitution to obtain the solution $\mathbf{x}$.

2.2.2.1   Matrix Triangularization

There are many methods to triangularize a matrix: Gaussian elimination, LU decomposition, and QR decomposition are but a few. Here we discuss a procedure based on QR decomposition. The QR decomposition of a matrix can be obtained by using either the Gram-Schmidt orthogonalization procedure or an orthogonal transformation (e.g., Givens' rotation (GR) and Householder transformation). The GR is introduced here, since it will be frequently used in later chapters. Interested readers may refer to [Stran80] for the other methods.

QR Decomposition: GR   A matrix, $\mathbf{A}$, can be written as the product of a matrix with *orthonormal* [2] columns and an invertible upper triangular matrix, that is, $\mathbf{A} = \mathbf{QR}$, where $\mathbf{Q}$ is a matrix with orthonormal columns and $\mathbf{R}$ is an upper triangular matrix. This decomposition, known as the QR decomposition, can be obtained by a sequence of GRs. The GR is a numerically stable orthogonal operator that performs a plane rotation of the

---

[1] In general, $\mathbf{A}$ is an $n \times m$ matrix ($n \geq m$). If we denote the number of linearly independent column vectors of $\mathbf{A}$ as k, then general simultaneous linear equations solving can be grouped into three cases: (1) $k < m$: underdetermined case; a solution exists, but it is not unique; (2) $k = m$ and $n > m$: solution may not exist, but a generalized least squared error solution can be obtained; (3) $k = m = n$: a unique solution exists as will be shown in this section.

[2] A set of vectors $\{v_1, v_2, \ldots, v_n\}$ is called *orthonormal* if $v_i^T v_j = 1$ whenever $i = j$ and $v_i^T v_j = 0$ otherwise.

matrix **A**. The purpose of these rotations is to annihilate the subdiagonal elements of matrix **A** and reduce it to upper triangular form. Thus, in the Givens' algorithm, the subdiagonal elements of the first column are nullified first, then the elements of the second column, and so forth until an upper triangular form is eventually reached. We now elaborate on this procedure for clarity.

For an invertible matrix **A**, the upper triangular matrix **R** is obtained as follows:

$$\mathbf{Q}^T\mathbf{A} = \mathbf{R},$$
$$\mathbf{Q}^T = \mathbf{Q}_{(N-1)}\,\mathbf{Q}_{(N-2)}\,\cdots\,\mathbf{Q}_{(1)} \tag{2.3}$$

and

$$\mathbf{Q}_{(p)} = \mathbf{Q}^{(p,p)}\mathbf{Q}^{(p+1,p)}\ldots\mathbf{Q}^{(N-1,p)} \tag{2.4}$$

where $\mathbf{Q}^{(q,p)}$ is the Givens' rotation operator used to annihilate the matrix element located at the $(q+1)$st row and pth column and has the following form:

$$qth \qquad (q+1)st$$

$$column \quad column$$

$$\mathbf{Q}^{(q,p)} = \begin{bmatrix} 1 & 0 & & \cdots & & 0 & 0 \\ 0 & 1 & & \cdots & & 0 & 0 \\ 0 & 0 & \ddots & \cdots & & 0 & 0 \\ \vdots & \vdots & & \cos\theta & \sin\theta & & \vdots \\ & & & -\sin\theta & \cos\theta & & \\ 0 & 0 & & & \ddots & & \vdots \\ 0 & 0 & & \cdots & & 0 & 1 & 0 \\ 0 & 0 & & \cdots & & 0 & 0 & 1 \end{bmatrix} \begin{matrix} \\ \\ \\ qth\ \ row \\ (q+1)st\ row \\ \\ \\ \end{matrix}$$

$$\tag{2.5}$$

where $\theta = \tan^{-1}[a_{q+1,p}/a_{q,p}]$ is an abbreviation of the function $\theta(q,p)$. The above operation of creating $\cos\theta$ and $\sin\theta$ is named Givens' generation (GG).

The matrix product $\mathbf{A}' = \mathbf{Q}^{(q,p)}\mathbf{A}$ is then:

$$a'_{q,k} = a_{q,k}\cos\theta + a_{q+1,k}\sin\theta \tag{2.6}$$

$$a'_{q+1,k} = -a_{q,k}\sin\theta + a_{q+1,k}\cos\theta \tag{2.7}$$

$$a'_{jk} = a_{jk} \quad if \ j \neq q, q+1$$

for all $k = 1, \ldots, N$.

The operations in Eq. 2.6 and Eq. 2.7 form the GR. The effects of the GR operations on the qth and $(q+1)$st rows of **A** are as follows:

$$\begin{bmatrix} \cos\theta & \sin\theta \\ -\sin\theta & \cos\theta \end{bmatrix} \cdot \begin{bmatrix} a_{q,1} & a_{q,2} & \cdots & a_{q,N} \\ a'_{q+1,1} & a'_{q+1,2} & \cdots & a'_{q+1,N} \end{bmatrix} = \begin{bmatrix} a'_{q,1} & a'_{q,2} & \cdots & a'_{q,N} \\ 0 & a''_{q+1,2} & \cdots & a''_{q+1,N} \end{bmatrix} \tag{2.8}$$

The full procedure is illustrated by the example in Figure 2.1.

2.2.2.2　Back-substitution

After triangularizing the linear equations, the remaining problem can be stated as follows: Find an $n \times 1$ vector **x** such that

$$\mathbf{A}' \cdot \mathbf{x} = \mathbf{y}'$$

where $\mathbf{A}'$ is an upper triangular matrix.

$$\begin{bmatrix} x & x & x & x \\ x & x & x & x \\ x & x & x & x \\ x & x & x & x \end{bmatrix} \quad \begin{bmatrix} x & x & x & x \\ x & x & x & x \\ x & x & x & x \\ 0 & x & x & x \end{bmatrix} \quad \begin{bmatrix} x & x & x & x \\ x & x & x & x \\ 0 & x & x & x \\ 0 & x & x & x \end{bmatrix} \quad \begin{bmatrix} x & x & x & x \\ 0 & x & x & x \\ 0 & x & x & x \\ 0 & x & x & x \end{bmatrix}$$
$$(a) \qquad\qquad (b) \qquad\qquad (c) \qquad\qquad (d)$$

$$\begin{bmatrix} x & x & x & x \\ 0 & x & x & x \\ 0 & x & x & x \\ 0 & 0 & x & x \end{bmatrix} \quad \begin{bmatrix} x & x & x & x \\ 0 & x & x & x \\ 0 & 0 & x & x \\ 0 & 0 & x & x \end{bmatrix} \quad \begin{bmatrix} x & x & x & x \\ 0 & x & x & x \\ 0 & 0 & x & x \\ 0 & 0 & 0 & x \end{bmatrix}$$
$$(e) \qquad\qquad (f) \qquad\qquad (g)$$

Figure 2.1: Procedure for QR triangularization: (a) the original 4×4 matrix, (b) result after computing Eq. 2.8, with $q = 3$, (c) $q = 2$, (d) $q = 1$, (e),(f) elimination of lower part of the second column, and (g) elimination of lower part of the third column.

This problem can be easily solved by using back-substitution. The following example explains this procedure.

Example 1: Back-Substitution

$$\begin{bmatrix} 1 & 1 & -1 \\ 0 & 3 & -2 \\ 0 & 0 & 1 \end{bmatrix} \cdot \begin{bmatrix} x_1 \\ x_2 \\ x_3 \end{bmatrix} = \begin{bmatrix} -2 \\ -9 \\ 3 \end{bmatrix}$$

Thus

$$x_1 + x_2 - x_3 = -2 \tag{2.9}$$

$$3x_2 - 2x_3 = -9 \tag{2.10}$$

$$x_3 = 3 \tag{2.11}$$

By substituting x_3 into Eq. 2.10, we obtain $x_2 = -1$. Then substituting x_2 and x_3 into Eq. 2.9, we obtain $x_1 = 2$. Thus the solution obtained is $x_1 = 2$, $x_2 = -1$, and $x_3 = 3$.

This procedure can be extended in an obvious manner to solve a general system of equations involving an upper triangular matrix.

2.2.3   Iterative Method

The method of reduction to triangular form and then back-substitution is just one approach among a myriad of techniques. We now discuss another approach that is useful when the matrix involved in the linear system solver has either a large dimension or a sparse structure. To solve this problem, an iterative method appears to be more efficient.

The iterative methods can be applied to X-ray image reconstruction problems, which generally have large and sparse matrices. It can be formulated as a large set of linear equation $\mathbf{g} = \mathbf{Hf}$, where $\mathbf{g}$ represents the physical measurements of the attenuation of the X-ray. The matrix $\mathbf{H}$ is large (typically of the order $10^5 \times 10^5$) and sparse (sometimes binary). To directly find the inverse of the matrix is almost impossible, however the iterative method appears to offer an effective approach. Another very useful application of the iterative linear equation solving is in the partial differential equation (PDE) problem [Arpin86]. PDEs arise in connection with various physical and geometrical problems. One very important equation of physics is Poisson's equation, which can also be written in a linear equation form with a large and sparse matrix involved if we restrict the boundary condition to be the so-called Dirichlet type [Kuo85] [Kung86c]. For this class of linear system solvers, the matrix involved has a larger dimension and is very sparse, hence the iterative method will be most suitable. Due to its importance, we now elaborate farther on this method.

The matrix $\mathbf{A}$ is split into two matrices $\mathbf{S}$ and $\mathbf{T}$ such that

$$\mathbf{A} = \mathbf{S} + \mathbf{T} \tag{2.12}$$

Then, starting with an initial guess $\mathbf{x}_0$, the iteration

$$\mathbf{S}\mathbf{x}_{k+1} = -\mathbf{T}\mathbf{x}_k + \mathbf{y} \tag{2.13}$$

is used to generate a sequence of vectors $\mathbf{x}_{k+1}$, which eventually are expected to converge to $\mathbf{x}$.

For efficiency, the splitting has several requirements:

1. The new vector $\mathbf{x}_{k+1}$ should be easy to compute. This implies $\mathbf{S}$ should be easily invertible. In particular, $\mathbf{S}$ should be a diagonal or a sparse triangular matrix.

2. In practice, the matrix $\mathbf{T}$ is usually very sparse. Therefore, the computation required for each iteration (i.e., $\mathbf{T}$ multiplied with $\mathbf{x}_k$) will be very little. This implys a significant saving in computation. Moreover, the matrix $\mathbf{T}$ will remain constant throughout all of the iterations. This means that no additional data loading is necessary.

3. The sequence should converge. The iterative method is convergent if and only if the absolute value of every eigenvalue of $\mathbf{S}^{-1} \cdot \mathbf{T}$ is less than 1. For more details, interested readers are referred to [Stran80].

Some important iterative methods based on Eq. 2.12 are given below. In these methods, $\mathbf{S}$ and $\mathbf{T}$ are expressed in terms of three types of matrices $\mathbf{L}$ (a lower triangular matrix), $\mathbf{D}$ (a diagonal matrix) and $\mathbf{U}$ (an upper triangular matrix), i.e.,

$$\mathbf{L} = \begin{bmatrix} 0 & 0 & \cdots & 0 & 0 \\ a_{21} & 0 & \cdots & 0 & 0 \\ a_{31} & a_{32} & \cdots & 0 & 0 \\ \vdots & \vdots & \ddots & \vdots & \vdots \\ a_{n1} & a_{n2} & \cdots & a_{n,n-1} & 0 \end{bmatrix}$$

$$
\mathbf{D} = \begin{bmatrix}
a_{11} & 0 & \cdots & 0 & 0 \\
0 & a_{22} & \cdots & 0 & 0 \\
0 & 0 & \cdots & 0 & 0 \\
\vdots & \vdots & \ddots & \vdots & \vdots \\
0 & 0 & \cdots & 0 & a_{n,n}
\end{bmatrix}
$$

$$
\mathbf{U} = \begin{bmatrix}
0 & a_{12} & \cdots & a_{1,n-1} & a_{1n} \\
0 & 0 & \cdots & a_{2,n-1} & a_{2n} \\
\vdots & \vdots & \ddots & \vdots & \vdots \\
0 & 0 & \cdots & 0 & a_{n-1,n} \\
0 & 0 & \cdots & 0 & 0
\end{bmatrix}
$$

2.2.3.1   Jacobi Iteration

In the Jacobi iteration, we let

$$
\mathbf{S_J} = \mathbf{D}, \quad \mathbf{T_J} = \mathbf{L} + \mathbf{U}
$$

then Eq. 2.13 leads directly to the following equation:

$$
x_i^{(k+1)} = \frac{1}{a_{ii}}\left(-\sum_{j=1}^{i-1} a_{ij}x_j^{(k)} - \sum_{j=i+1}^{n} a_{ij}x_j^{(k)} + y_i\right), \quad i = 1, 2, ..., n. \quad (2.14)
$$

2.2.3.2   Gauss-Seidel Iteration

Note that in the Jacobi iteration, the most recently available information is not used when computing $x_i^{(k+1)}$. If we revise the Jacobi iteration by setting

$$
\mathbf{S_G} = \mathbf{D} + \mathbf{L}, \quad \mathbf{T_G} = \mathbf{U}
$$

then, by Eq. 2.13, we can get

$$
(\mathbf{D} + \mathbf{L})\mathbf{x}_{k+1} = -\mathbf{U}\mathbf{x}_k + \mathbf{y}
$$

This leads to the *Gauss-Seidel iteration* [Golub83]

$$
x_i^{(k+1)} = \frac{1}{a_{ii}}\left(-\sum_{j=1}^{i-1} a_{ij}x_j^{(k+1)} - \sum_{j=i+1}^{n} a_{ij}x_j^{(k)} + y_i\right) \quad , i = 1, 2, ..., n. \quad (2.15)
$$

2.2.3.3   Successive Over-Relaxation Iteration

The Gauss-Seidel iteration is very attractive because of its simplicity. Unfortunately, if the largest absolute eigenvalue of matrix $\mathbf{S}_G^{-1}\mathbf{T}_G$ is close to unity, then the convergence rate may be prohibitively slow. In an effort to rectify this, let $w \in R$ (R is the set of real numbers) and consider the following modification of the Gauss-Seidel step: $\mathbf{S_s} = \mathbf{D} + w\mathbf{L}$ and $\mathbf{T_s} = -(1-w)\mathbf{D} + w\mathbf{U}$. This defines the method of *successive over-relaxation (SOR)*.

$$x_i^{(k+1)} = \frac{w}{a_{ii}}\left(-\sum_{j=1}^{i-1} a_{ij}x_j^{(k+1)} - \sum_{j=i+1}^{n} a_{ij}x_j^{(k)} + y_i\right) + (1-w)x_i^{(k)}, \quad i = 1, 2, ..., n.$$

$$(2.16)$$

The relaxation parameter, w, should be chosen to maximize the rate of convergence.

For $w = 1$, the method obviously reduces to the Gauss-Seidel method [Dahlq74]. It can be shown that for symmetric and positive-definite real matrix $\mathbf{A}$, the SOR method converges for all w, $0 < w < 2$. For some classes of matrices of practical importance (for example, symmetric, positive-definite, and block-tridiagonal), the optimal value of w is known and the rate of convergence is evidently much better than with the Gauss-Seidel method.

2.2.4   Eigen Value and Singular Value Decomposition

Another matrix related operation that arises in many signal processing applications is the finding of eigenvalues and eigenvectors and, more recently, singular values and singular vectors.

2.2.4.1   Eigenvalue-Decomposition

Let $\mathbf{A}$ be an $n \times n$ matrix. If there exists a vector $\mathbf{e}$ such that

$$\mathbf{Ae} = \lambda\mathbf{e}$$

then λ is called the eigenvalue, and $\mathbf{e}$ is the corresponding eigenvector. The eigenvalues can be obtained by solving the characteristic equation of $\mathbf{A}$, i.e., $\det|\mathbf{A} - \lambda\mathbf{I}| = 0$. If the eigenvalues are distinct, then

$$\mathbf{AE} = \mathbf{E}\Lambda$$

$$= [\mathbf{e}_1\ \mathbf{e}_2\ \cdots\ \mathbf{e}_n] \cdot \begin{bmatrix} \lambda_1 & 0 & \cdots & 0 \\ 0 & \lambda_2 & \cdots & 0 \\ \vdots & \vdots & \ddots & \vdots \\ 0 & 0 & \cdots & \lambda_n \end{bmatrix}$$

where λ_i's are the eigenvalues of $\mathbf{A}$ and $\mathbf{e}_i$'s are the corresponding eigenvectors. The matrix $\mathbf{E}$ is also invertible, and hence $\mathbf{A} = \mathbf{E}\Lambda\mathbf{E}^{-1}$. For matrices with repeated eigenvalues, interested readers are referred to [Stewa73].

If the matrix $\mathbf{A}$ is an $n \times n$ normal matrix, i.e., $\mathbf{A}^H\mathbf{A} = \mathbf{A}\mathbf{A}^H$, it can be factored into $\mathbf{A} = \mathbf{U}\Lambda\mathbf{U}^T$, where $\mathbf{U}$ is an $n \times n$ unitary matrix and Λ is a diagonal matrix that has all the eigenvalues of $\mathbf{A}$ as the diagonal elements.[3]

This decomposition, often called the spectral decomposition, is useful in control and signal processing. The Karhunen-Loeve (KL) transform, an optimal transform, is based on such a decomposition of the covariance matrix.

2.2.4.2 Singular Value Decomposition (SVD)

Another important decomposition of matrices is the singular value decomposition (SVD). The use of SVD techniques in digital image processing and computer vision is of considerable interest. For example, SVD methods are useful for image coding [Andr76a] and image enhancement [Andr76b]; SVD can also be applied to image restoration and reconstruction processing methods based on the pseudoinverse techniques [Shim81].

It can be shown that any $m \times n$ matrix $\mathbf{A}$ can be factored into $\mathbf{A} = \mathbf{Q}_1\Sigma\mathbf{Q}_2^T$, where $\mathbf{Q}_1$ is an $m \times m$ unitary matrix, $\mathbf{Q}_2$ is an $n \times n$ unitary matrix, and Σ has the special diagonal form[4]

$$\Sigma = \begin{bmatrix} \mathbf{D} & \mathbf{0} \\ \mathbf{0} & \mathbf{0} \end{bmatrix}$$

---

[3] Hermitian, skew-Hermitian, and unitary matrices are all special cases of a normal matrix.

[4] A unitary matrix U is a matrix that has orthonormal columns. That is, $\mathbf{U}^H\mathbf{U} = \mathbf{U}\mathbf{U}^H = \mathbf{I}$ ($\mathbf{U}^H = (\mathbf{U}^*)^T$). A real unitary matrix is called an orthogonal matrix. An important property of unitary matrices is: Multiplication by a unitary matrix has no effect on inner products or angles or lengths. That is, $(\mathbf{U}\mathbf{x})^H(\mathbf{U}\mathbf{y}) = \mathbf{x}^H\mathbf{y}$ and $\|\mathbf{U}\mathbf{x}\| = \|\mathbf{x}\|$, where $\mathbf{x}$ and $\mathbf{y}$ are two column vectors.

where $\mathbf{D} = \mathrm{diag}(\ \sigma_1,\ \sigma_2,\ \cdots,\ \sigma_r\ )$, $\sigma_1 \geq \sigma_2 \geq \cdots \geq \sigma_r > 0$ and r is the rank of $\mathbf{A}$. Such that the SVD formulation can also be written as

$$\mathbf{A} = \mathbf{Q}_1 \Sigma \mathbf{Q}_2^T = \Sigma_{i=1}^r \sigma_i u_i v_i^T \qquad (2.17)$$

where u_i is the column vector of matrix $\mathbf{Q}_1$, and v_i is the column vector of matrix $\mathbf{Q}_2$.

This is called the SVD of $\mathbf{A}$. The singular values of $\mathbf{A}$, i.e., σ_1, σ_2, ... , σ_r, are the square roots of the nonzero eigenvalues of $\mathbf{A}^T \mathbf{A}$ (or $\mathbf{A}\mathbf{A}^T$). The column vectors of $\mathbf{Q}_1$ and the column vectors of $\mathbf{Q}_2$, i.e., the singular vectors of $\mathbf{A}$, are the eigenvectors of $\mathbf{A}\mathbf{A}^T$ and $\mathbf{A}^T \mathbf{A}$, respectively. These facts can be easily verified by noting that

$$[\mathbf{A}\mathbf{A}^T]\mathbf{Q}_1 = \mathbf{Q}_1 \Sigma \Sigma^T$$

$$[\mathbf{A}^T \mathbf{A}]\mathbf{Q}_2 = \mathbf{Q}_2 \Sigma^T \Sigma$$

The SVD has recently been used in a number of signal and image processing applications. For instance, it can be used for solving the least squares problem, described in the next section, under less restrictive conditions [Stran80], [Stewa73]. The singular values can also be used to determine the rank of a matrix in a numerically reliable manner. Furthermore, they can be used to find a good low-rank approximation to the original matrix. Such low-rank approximations have been very beneficial in a number of signal and image processing problems. The decomposition is illustrated in the following example.

Example 2: SVD

Consider the 3 × 2 matrix

$$\mathbf{A} = \begin{bmatrix} 1 & 1 \\ 2 & 2 \\ 2 & 2 \end{bmatrix}$$

$$= \begin{bmatrix} \frac{1}{3} & \frac{-2}{\sqrt{5}} & \frac{2\sqrt{5}}{15} \\ \frac{2}{3} & \frac{1}{\sqrt{5}} & \frac{4\sqrt{5}}{15} \\ \frac{2}{3} & 0 & \frac{-5\sqrt{5}}{15} \end{bmatrix} \begin{bmatrix} 3\sqrt{2} & 0 \\ 0 & 0 \\ 0 & 0 \end{bmatrix} \begin{bmatrix} \frac{1}{\sqrt{2}} & \frac{1}{\sqrt{2}} \\ \frac{1}{\sqrt{2}} & \frac{-1}{\sqrt{2}} \end{bmatrix}$$

Note that $\mathbf{A}^T\mathbf{A}$ has eigenvalues 18 and 0 and $\mathbf{A}\mathbf{A}^T$ has eigenvalues 18, 0, and 0. The singular value of $\mathbf{A}$ is the square root of 18. (Note $\mathbf{A}$ has rank 1.)

2.2.5   Solving Least Squares Problems

The well known linear least squares technique is useful in control, communication and signal processing. This technique is applied in many DSP areas, such as equalization, spectral analysis, digital whitening, adaptive arrays, interchannel interference mitigation, and digital speech processing [Giord85]. Many image restoration and reconstruction algorithms can be naturally formulated as a least squares problem. The problem differs from the linear system solving in that there are more equations than unknowns. In matrix notation, the problem can be formulated in the following manner:

Given an $n \times p$ $(n > p)$ full rank (rank=p) observation data matrix $\mathbf{A}$ and an n-element desired data vector $\mathbf{y}$, find $\mathbf{w}$, a p-element vector of weights, which minimizes the Euclidean norm of the residual vector[5] $\mathbf{e}$. The vector $\mathbf{e}$ is defined by

$$\mathbf{e} = \mathbf{y} - \mathbf{A} \cdot \mathbf{w}$$

2.2.5.1   Unconstrained Least Squares Algorithm

The Euclidean norm of a vector remains unchanged if it is premultiplied by an orthonormal matrix. Hence, analogous to the linear system solving, orthonormal matrices can be used to reduce $\mathbf{A}$ to an upper triangular form. A sequence of GR rotation matrices are used to premultiply the residual vector $\mathbf{e}$ resulting in the transformed vector:[6]

$$\mathbf{Q} \cdot \mathbf{e} = \mathbf{Q} \cdot \mathbf{y} - [\mathbf{Q} \cdot \mathbf{A}] \cdot \mathbf{w} = \mathbf{y}' - \mathbf{A}' \cdot \mathbf{w}$$

---

[5]Some weight factor can be imposed on the residual to generalize the problem, i.e., the Euclidean norm of $\mathbf{b}^T \cdot \mathbf{e}$ will be minimized instead of the norm of $\mathbf{e}$. This object function is introduced to weight the observed data and to emphasize the present observed data against the preceding data. This generalized problem is termed the weighted least squares problem and can be solved by slightly modifying the mathematical formulation [Stran80].

[6]The orthogonal triangularization process may be carried out by using GR, Householder transformation, or Gram-Schmidt orthogonalization. The GR method is discussed in Sec. 2.2.2.1.

where $\mathbf{Q}$ is an $n \times n$ orthonormal matrix and $\mathbf{A}'$ is an $n \times p$ matrix with structure:

$$
\begin{bmatrix}
x & x & x & \cdots & x & x \\
0 & x & x & \cdots & x & x \\
0 & 0 & x & \cdots & x & x \\
\vdots & \vdots & \vdots & \ddots & \vdots & \vdots \\
0 & 0 & 0 & \cdots & x & x \\
0 & 0 & 0 & \cdots & 0 & x \\
0 & 0 & 0 & \cdots & 0 & 0 \\
\vdots & \vdots & \vdots & & \vdots & \vdots \\
0 & 0 & 0 & \cdots & 0 & 0
\end{bmatrix}
$$

In other words, $\mathbf{A}'$ can be represented as the following,

$$
\mathbf{A}' = \begin{bmatrix} \mathbf{R} \\ \mathbf{0} \end{bmatrix}
$$

where $\mathbf{R}$ is an upper triangular square matrix of dimension p.

To minimize the Euclidean norm of $\mathbf{y}' - \mathbf{A}' \cdot \mathbf{w}$, the optimal weight vector $\mathbf{w}_{opt}$ can be easily obtained by noting that $\mathbf{w}$ has no influence upon the lower parts of the vector difference. Therefore,

$$
\mathbf{R} \cdot \mathbf{w}_{\text{opt}} = \mathbf{y}'
$$

where $\mathbf{y}'$ is a vector consisting of the p leading elements of $\mathbf{y}'$.

Since $\mathbf{R}$ is an upper triangular matrix, the optimal weight vector $\mathbf{w}_{\text{opt}}$ can be obtained by using back-substitution, as shown in Sec. 2.2.2.2.

2.2.5.2 Constrained Least Squares Algorithm

In addition to the least squares problem, linearly constrained least squares estimation is an important problem. For instance, this problem arises naturally in adaptive antenna arrays. Details of this problem and its applications are discussed in Chapter 8.

2.3 Digital Signal Processing Algorithms

In this section, basic DSP operations, such as convolution, correlation, filtering, and digital transforms, are introduced.

2.3.1   Discrete Time Systems and the Z-transform

In DSP, discrete time signals and systems are extensively dealt with. A discrete time signal is represented by a discrete sequence. The discrete time signal may arise naturally, or as is often the case, may be obtained by sampling an analog waveform (continuous time signal). A common operation is to pass the sequence through a discrete time system, resulting in another discrete time sequence. Usually these discrete time systems are designed with a specific purpose in mind and belong to a special class of systems termed linear time invariant (LTI) systems. An interesting fact is that LTI systems are completely characterized by their response, $h(n)$, to the unit sample sequence $\delta(n)$ where

$$\delta(n) = \begin{cases} 1 & \text{if } n = 0 \\ 0 & \text{otherwise} \end{cases}$$

Here, $h(n)$ is called the unit sample response or the impulse response of the system. It can be shown that the output, $y(n)$, of an LTI system (refer to Problem 7) is given by

$$y(n) = \sum_{k=-\infty}^{\infty} x(k)h(n-k) \tag{2.18}$$

Eq. 2.18 is referred to as the convolution operation and is denoted by

$$y(n) = x(n) * h(n)$$

Further details regarding the convolution operation will be given in Sec. 2.3.2. LTI systems are usually classified into two groups: finite impulse response (FIR) and infinite impulse response (IIR) systems. If $h(n)$ is finite in duration, the system is said to have an FIR. If $h(n)$ is not finite in duration, the system is said to have an IIR.

So far our discussion has been on characterizing signals and systems in the time domain. In addition to the time domain representation, it is common to represent these signals and systems in a transform domain. Analogous to the Laplace transform for continuous time signals, the Z-transform can used in connection with discrete time signals.

Z-transform   The *definition of Z-transform* is as follows:

$$X(z) = Z[x(n)] = \sum_{n=-\infty}^{\infty} x(n)z^{-n}$$

where z is a complex number in a region of the z-plane.

Example 3: *Z*-**Transform**

Let

$$x(n) = \left(\frac{1}{2}\right)^n s(n),$$

where $s(n)$ is the unit sample function defined as

$$s(n) = \begin{cases} 1 & \text{if } n \geq 0 \\ 0 & \text{otherwise} \end{cases}$$

Then

$$\begin{aligned} X(z) &= Z[x(n)] \\ &= \sum_{n=0}^{\infty} \left(\frac{1}{2}\right)^n z^{-n} \\ &= \sum_{n=0}^{\infty} (0.5z^{-1})^n \\ &= \frac{1}{1 - 0.5z^{-1}} = \frac{z}{z - 0.5}, \text{ for } |z| > 0.5 \end{aligned}$$

Two useful properties of the *Z*-transform that will be used later are:

$$\text{(i)} \quad x(n) * h(n) \longleftrightarrow X(z)H(z) \tag{2.19}$$

$$\text{(ii)} \quad x(n + n_0) \longleftrightarrow z^{n_0} \cdot X(z) \tag{2.20}$$

Note that the Fourier transform $X(e^{j\omega})$ of a discrete sequence $x(n)$ defined as

$$X(e^{j\omega}) = \sum_{n=-\infty}^{\infty} x(n)\, e^{-j\omega n}$$

can be obtained by evaluating the *Z*-transform on the unit circle, i.e.,

$$X(e^{j\omega}) = X(z)|_{z=e^{j\omega}}$$

In the Z-domain, the convolution operation reduces to a simple multiplication. Taking the Z-transform of $y(n)$ in Eq. 2.18, it can be shown that

$$Y(z) \;=\; H(z)\, X(z)$$

where

$$H(z) \;=\; \sum_{n=-\infty}^{\infty} h(n)\, z^{-n}$$

and

$$X(z) \;=\; \sum_{n=-\infty}^{\infty} x(n)\, z^{-n}$$

This fact is exploited in performing convolution efficiently. $H(z)$, a characterization of an LTI system in the Z-domain, is called the *transfer function* of the system and is a ratio of the transforms of the output and input sequences.

2.3.2 Convolution

The convolution of two sequences $u(n)$ and $w(n)$ is defined as

$$y(n) \;=\; \sum_{k=-\infty}^{\infty} u(k)w(n-k),$$

and is represented by

$$y(n) \;=\; u(n) * w(n)$$

If $u(n)$ and $w(n)$ are causal sequences and each is of finite length N, i.e., $n \;=\; 0,1,2,...,N-1$, the (linear) convolution of these two sequences is a causal sequence, computed as

$$y(n) \;=\; \sum_{k=0}^{N-1} u(k)w(n-k) \qquad (2.21)$$

where $n \;=\; 0,1,\cdots,2N-2$.

When convolution is used for digital filtering, $u(n)$ denotes the input sequence to be processed and $w(n)$ represents the impulse response function of the processing digital filter. In general, these two sequences can be of different length. The sequence $y(n)$ represents the processed signal, which may be an enhanced or smoothed signal, or bear some other kinds of effects, depending on the type of filter adopted.

From a computational point of view, an efficient way of computing convolution is first to transform it to another domain where the operations are simpler and, then apply an inverse transform. The fast Fourier transform is the most commonly used transformation for this application. By using a transform method, the order of computation needed will be reduced from $O(N^2)$ to $O(N \cdot \log N)$.

In communication, the matched filter is a popular implementation of an optimal receiver. In a digital matched filter, the received signal is convolved with a filter whose impulse response is a "reflection" of the transmitted signal waveform.

Convolution is also a basic operation in the problem of interpolation. While processing signals, we often face a situation where the sampling period is larger than desired. Interpolation is useful in this case. The operation of interpolation consists of taking a weighted sum of the neighboring data to obtain a value for the data in between sampling instants. Note that convolution may be also regarded as a special case of matrix-vector multiplication.

Recursive Equations Recursive equations are convenient mathematical forms widely used to describe many signal processing operations. This notation is useful for programming purposes and, for developing sequential architectures; in our case, they also form an important step in developing parallel architectures. The recursive equation for convolution of two causal sequences is

$$y_j^k \; = \; y_j^{k-1} \; + \; u_k \cdot w_{j-k} \qquad (2.22)$$

$$k = 0, 1, ..., j \quad \text{when } j = 0, 1, ..., N-1$$

and $\quad k = j - N + 1, j - N + 2, ..., N - 1 \quad \text{when } j = N, N+1, ..., 2N-2$

2.3.3 Correlation

Correlation is another useful operation that often arises in connection with random processes and has a formula quite similar to convolution. The mathematical operation of correlation is as follows:

$$y(n) = \sum_{k=-\infty}^{\infty} u(k)w(n+k)$$

If $u(n)$ and $w(n)$ are causal sequences and each is of finite length N, i.e., $n = 0, 1, 2, \cdots, N-1$, the correlation of these two sequences is a causal sequence, which is computed as

$$y(n) = \sum_{k=0}^{N-1} u(k)w(n+k) \qquad (2.23)$$

where $n = -N+1, -N+2, \cdots, -1, 0, 1, \cdots, N-2, N-1$.

Comparing Eq. 2.23 with Eq. 2.21, it is clear why fast transform algorithms can also be applied to compute correlation.

2.3.4   Digital FIR and IIR Filters

Digital filters are LTI systems designed to meet specific design requirements. Often they are specified by the transfer function defined on the unit circle, i.e., $H(e^{j\omega})$. As $H(e^{j\omega})$ is a complex function, it can be written as

$$H(e^{j\omega}) = |H(e^{j\omega})|e^{j\theta(\omega)}.$$

$|H(e^{j\omega})|$, the magnitude, and $\theta(\omega)$, the phase response, are often used to characterize a digital filter.

There are two kinds of digital filters: FIR filters and IIR filters. The choice between an FIR filter and an IIR filter depends upon the application domain and requirements.

Representation of Digital Filters   The difference equations play an important role in describing discrete time systems. A linear time-invariant discrete time system, described by the pth order difference equation

$$y(n) = \sum_{k=1}^{p} a_k y(n-k) + \sum_{k=0}^{q} b_k x(n-k) \qquad (2.24)$$

is a common representation of a digital filter. Here $x(n)$ is the input signal and $y(n)$ is the output signal.

Taking the Z-transform of Eq. 2.24 and applying Eq. 2.20 gives:

$$Y(z) = \sum_{k=1}^{p} a_k z^{-k} Y(z) + \sum_{k=0}^{q} b_k z^{-k} X(z)$$

where $Y(z)$ and $X(z)$ are the Z-transforms of $\{y(n)\}$ and $\{x(n)\}$, respectively. This equation can be put into a simpler form, i.e.,

$$Y(z) = X(z)H(z), \tag{2.25}$$

where the transfer function is given by

$$H(z) = \frac{B(z)}{A(z)} \tag{2.26}$$

$$\text{with} \quad B(z) = \sum_{k=0}^{q} b_k z^{-k}$$

$$\text{and} \quad A(z) = 1 - \sum_{k=1}^{p} a_k z^{-k}$$

A digital filter can be classified as either a recursive or a nonrecursive type. If the coefficients a_k are not all zero, the calculation of $y(n)$ requires the values of the past output samples. In this case the filter is said to be of the recursive type. On the other hand, if all the a_k values are zero, i.e., $a_1 = a_2 = \cdots = a_p = 0$, then the calculation of $y(n)$ does not require the past values of the output, and the filter is said to be of the nonrecursive type.

In fact, there are three classes of basic digital filters:
(1) **Moving Average Filter**

A special case of the general filter as defined by Eq. 2.26 is when $A(z) = 1$. Then, the transfer function of the filter is

$$H(z) = B(z) = \sum_{k=0}^{q} b_k z^{-k}$$

and the input-output can be described as follows:

$$y(n) = b_0 x(n) + b_1 x(n-1) + b_2 x(n-2) + \cdots + b_q x(n-q)$$

As is apparent, each output point is a weighted average of the input data, and the filter is called a moving average (MA) filter. Note that the filter is an FIR filter and the impulse response is $\{b_0, b_1, b_2, \ldots, b_q\}$. Thus the output sequence can be obtained as a convolution of the input sequence and the impulse response of the filter.

(2) Autoregressive Filter

Another common filter used for signal/image processing occurs when $B(z) = 1$. The transfer function of the filter is

$$H(z) = \frac{1}{A(z)}$$

where

$$A(z) = 1 - \sum_{k=1}^{p} a_k z^{-k}$$

and the input-output relationship can be described as:

$$y(n) = \sum_{k=1}^{p} a_k y(n - k) + x(n).$$

Here, output $y(n)$ is generated as a linear regression of its past values, and hence such a filter is known as the autoregressive (AR) filter. The AR filter is an IIR filter.

(3) Autoregressive Moving Average Filter

A more general filter as defined by Eq. 2.24 and Eq. 2.26 is a composite of AR and MA filters, so it is appropriately called the autoregressive moving average (ARMA) filter. ARMA filters are IIR filters.

2.3.5 Linear Phase Filter

In many applications it is desirable to design filters that have linear phase, i.e., $\theta(\omega) = \alpha\omega$. In this way, signals in the passband of the filter are reproduced exactly at the filter output except for a time delay corresponding to the slope of the phase. For a linear phase (FIR) filter, the impulse response function $h(n)$ has the following property:[7]

$$h(n) = h(N - 1 - n), \quad n = 0, 1, \cdots, N - 1. \qquad (2.27)$$

---

[7] Refer to Problem 13 which shows that this is a sufficient condition for linear phase.

To take advantage of the symmetry, a different realization which requires only half number of multiplications can be used. Assume that N is odd; then

$$H(z) = \sum_{n=0}^{N-1} h(n)z^{-n}$$

$$= \sum_{n=0}^{(N-1)/2} h(n)z^{-n} + \sum_{n=(N+1)/2}^{N-1} h(n)z^{-n}$$

By changing the indices,

$$n' = (N-1) - n$$

the equation becomes

$$H(z) = \sum_{n=0}^{(N-1)/2} h(n)z^{-n} + \sum_{n'=0}^{(N-1)/2} h(n')z^{-(N-1-n')}$$

$$= \sum_{n=0}^{(N-1)/2} h(n)\left[z^{-n} + z^{-(N-1-n)}\right]$$

A realization of this filter is shown in Chapter 3.

2.3.6   Discrete Fourier Transform (DFT)

Let $\{x(n), n = 0, 1, ..., N-1\}$ be a finite-length sequence. The DFT of $x(n)$ is defined as

$$X(k) = \sum_{n=0}^{N-1} x(n) \cdot W_N^{nk} \qquad (2.28)$$

where $k = 0, 1, 2, ... , N-1$ and $W_N = e^{-j2\pi/N}$.

The DFT is very useful in signal/image processing and can be efficiently computed using an algorithm called the fast Fourier transform (FFT) (see Sec. 2.3.7). The DFT has the following two important properties.

1. The DFT of a sequence can be obtained by uniformly sampling the Fourier transform of the sequence. More precisely, $X(k)$ is obtained

by sampling the Fourier transform at the following N points,

$$\omega \;=\; 0 \,,\; \frac{2\pi}{N} \,,\; 2\cdot\frac{2\pi}{N} \,,\; \dots \,,\; (N-1)\cdot\frac{2\pi}{N}$$

The inverse DFT is defined as follows:

$$x(n) \;=\; \frac{1}{N}\sum_{k=0}^{N-1} X(k)W_N^{-nk} \,,\; n \;=\; 0,1,\dots,N-1$$

For the correctness of this equation, refer to Problem 14.

If the range of n is not restricted to $0 \le n \le N-1$ in this equation, the resulting sequence $\tilde{x}(n)$, where

$$\tilde{x}(n) \;=\; \frac{1}{N}\sum_{k=0}^{N-1} X(k)W_N^{-nk} \,,$$

for all n, is periodic, i.e., $\tilde{x}(n) = \tilde{x}(n+N)$ and $\tilde{x}(n) \;=\; x(n), 0 \le n \le N-1$. $\tilde{x}(n)$ is often referred to as a periodic extension of $x(n)$.

2. The effect of multiplying the DFT of two N-point sequences is equivalent to the circular convolution of the two sequences in the time domain (or space domain for image processing). Mathematically, if

$$X_1(k) \;=\; \text{DFT of}\,[x_1(n)]$$

$$X_2(k) \;=\; \text{DFT of}\,[x_2(n)]$$

Then $X_3(k) \;=\; X_1(k)X_2(k)$, is the DFT of $[\,x_3(n)\,]$, where

$$x_3(n) \;=\; \sum_{m=0}^{N-1} \tilde{x}_1(m)\tilde{x}_2(n-m)$$

and $n \;=\; 0, 1,\dots, N-1$.

Note that in this equation, $x_3(n)$ is of length N and is a linear convolution of two periodic sequences, $\tilde{x}_1(m)$ and $\tilde{x}_2(m)$ (or *circular convolution*

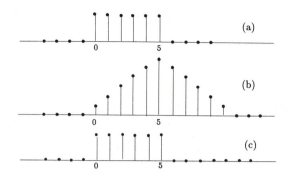

Figure 2.2: (a) Two sequences are given, $x_1(n) = x_2(n)$, (b) The linear convolution of these two sequences, (c) The circular convolution of these two sequences.

of two nonperiodic sequences, $x_1(m)$ and $x_2(m)$). The effect of wraparound due to circular convolution is shown in Figure 2.2. The wraparound effect can be avoided by appending zeros to the original nonperiodic sequences. This technique is called zero-padding and can be used to compute the linear convolution of two sequences. The procedure to compute the linear convolution of two sequences by a transformation method is discussed after the FFT algorithm is introduced.

In addition to the previous equations, the DFT can be represented in a matrix-vector multiplication format. Here we let $W = W_N$.

$$
\begin{bmatrix}
X(0) \\
X(1) \\
X(2) \\
\vdots \\
X(N-1)
\end{bmatrix}
=
\begin{bmatrix}
1 & 1 & 1 & \cdots & 1 \\
1 & W & W^2 & \cdots & W^{N-1} \\
1 & W^2 & W^4 & \cdots & W^{2N-2} \\
\vdots & \vdots & \vdots & \ddots & \vdots \\
1 & W^{N-1} & W^{2N-2} & \cdots & W^{(N-1)(N-1)}
\end{bmatrix}
\begin{bmatrix}
x(0) \\
x(1) \\
x(2) \\
\vdots \\
x(N-1)
\end{bmatrix}
$$

2.3.7   Fast Fourier Transform (FFT)

We first examine the computational complexity in computing the DFT of a sequence using Eq. 2.28 directly. The computation of $x(n)W^{nk}$ requires one complex multiplication (four real multiplications + two real additions). To compute $X(k)$ $\{k = 0, 1, \ldots, N-1\}$, we need N^2 complex multiplications and $N(N-1)$ complex additions (i.e., roughly, N^2 operations). We now examine the FFT algorithm to compute the DFT of a sequence.

Let $N = 2^m$. By utilizing the symmetry and periodicity properties of W_N^{nk}, the number of operations will be reduced from N^2 to $N \log_2 N$.

Based on the assumption that one complex multiplication takes 0.5 μsec., a comparison of the computation times for DFT and FFT is given as follows:

N	T_{DFT}	T_{FFT}
2^{12}	8 sec.	0.013 sec.
2^{16}	0.6 hour	0.26 sec.
2^{20}	6 days	5 sec.

There are two kinds of FFT algorithms, namely, *decimation in time* (DIT) FFT and *decimation in frequency* (DIF) FFT. They have the same complexity of computation. Only the DIT FFT is discussed here. The derivation of the DIF FFT is left as a problem.

DIT FFT

$$X(k) = \sum_{n=0}^{N-1} x(n)W_N^{nk}$$

Separating $x(n)$ into its even and odd numbered points gives

$$X(k) = \sum_{n=even} x(n) \cdot W_N^{nk} + \sum_{n=odd} x(n) \cdot W_N^{nK}$$

Substituting $n = 2r$ for even n, and $n = 2r + 1$ for odd n,

$$X(k) = \sum_{r=0}^{(N/2)-1} x(2r)W_N^{2rk} + \sum_{r=0}^{(N/2)-1} x(2r+1)W_N^{(2r+1)k}$$

$$= \sum_{r=0}^{(N/2)} \left(W_N^2\right)^{rk} + W_N^k \sum_{r=0}^{(N/2)-1} x(2r+1)\left(W_N^2\right)^{rk}$$

since

$$W_N^2 = e^{-2j(2\pi/N)} = e^{-2j\pi/(N/2)} = W_{N/2}$$

$$X(k) = \sum_{r=0}^{(N/2)-1} x(2r)W_{N/2}^{rk} + W_N^k \sum_{r=0}^{(N/2)-1} x(2r+1)W_{N/2}^{rk}$$

$$= G(k) + W_N^k H(k) \qquad\qquad (2.29)$$

In the last equation, both $G(k)$ and $H(k)$ can be obtained via an $(N/2)$-point FFT. Thus an N-point FFT can be achieved by using two $(N/2)$-point FFTs and then combining the results via Eq. 2.29. Applying this kind of decomposition recursively, we may use a number of 2-point FFTs, which involve only addition and subtraction, and combine their results via Eq. 2.29. An 8-point FFT computed in this way is shown in Fig. 2.3. Note that the input data are in bit reversed order. Hence this algorithm is called "decimation in time" FFT.

It can be seen from Figure 2.3 that the computation for the FFT consists of a sequence of special computations, called "butterflies", consisting of one addition, one subtraction, and one multiplication. The reason for the subtraction is

$$W_N^{(r+N/2)} \;=\; W_N^r \cdot W_N^{N/2}$$

and

$$W_N^{N/2} \;=\; \exp(-j2\pi N/2N) \;=\; exp(-j\pi) \;=\; -1$$

A DIT FFT and inverse FFT subroutine, written in FORTRAN, is listed in Figure 2.4 to demonstrate how to program the algorithm. For the decimation in frequency FFT, readers may refer to [Oppen75]. Note

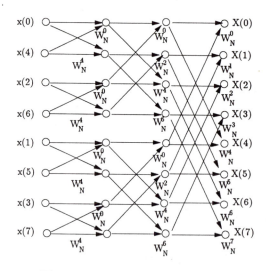

Figure 2.3: DIT FFT ($N = 8$).

```
C                                          J=J-K
C   FFT SUBROUTINE: ( DECIMATION IN TIME )  K=K/2
C                                          GOTO 6
C   MATRIX A IS A COMPLEX MATRIX          7 J=J+K
C   2**M POINT FFT                        C
C   INV > 0 THEN FFT; OTHERWISE, INVERSE FFT  C   BUTTERFLY OPERATION
C                                          C
C                                          PI=3.141592653589793
    SUBROUTINE FFT(A,M,INV)                DO 20 L=1,M
    COMPLEX A(1024),U,W,T                  LE=2**L
    N=2**M                                 LE1=LE/2
C                                          U=(1.0,0.)
C   FFT OR INVERSE FFT ?                   FLE1=FLOAT(LE1)
C                                          W=CMPLX(COS(PI/FLE1),-SIN(PI/FLE1))
    IF (INV.GT.0) GOTO 4                   DO 20 J=1,LE1
    DO 3 I=1,N                             DO 10 I=J,N,LE
  3 A(I)=CONJG(A(I))                       IP=I+LE1
C                                          T=A(IP)*U
C   BIT REVERSAL OPERATION                 A(IP)=A(I)-T
C                                       10 A(I)=A(I)+T
  4 NV2=N/2                              20 U=U*W
    NM1=N-1                                C
    J=1                                    C   FFT OR INVERSE FFT ?
    DO 7 I=1,NM1                           C
    IF (I.GE.J) GOTO 5                     IF (INV.GT.0) GOTO 30
    T=A(J)                                 DO 25 I=1,N
    A(J)=A(I)                           25 A(I)=CONJG(A(I))/FLOAT(N)
    A(I)=T                              30 RETURN
  5 K=NV2                                  END
  6 IF (K.GE.J) GOTO 7
```

Figure 2.4: DIT FFT and inverse FFT subroutine.

that in the program the inverse FFT can be achieved simply by appending some more instructions to the FFT program. This can be explained by the following simple derivation.

$$X(k) = \sum_{n=0}^{N-1} x(n) \cdot W_N^{nk} \quad \text{(DFT formula)}$$

$$x(n) = \frac{1}{N} \sum_{k=0}^{N-1} X(k) W_N^{-nk} \quad \text{(IDFT formula)}$$

$$= \frac{1}{N} \left[\sum_{k=0}^{N-1} X^*(k) W_N^{nk} \right]^*$$

$$= \frac{1}{N} [\text{DFT}\{X^*(K)\}]^*$$

So far we have concentrated on "radix-2" algorithms, where $N = 2^m$. More general algorithms for the computation of the DFT can be utilized when N is decomposed as a product of prime factors [Oppen75]. Because of the drastic speedup of the FFT over the DFT, the FFT algorithm is fre-

quently used to compute convolution, correlation and deconvolution.[8] Thus
the FFT algorithm has very versatile applications. Here the procedure of
applying the transformation method to compute the linear convolution of
two sequences is discussed:

1. If the lengths of the two sequences to be convolved are N and M
   respectively, then the length of the convolution result will be $N+M-1$.
   Let L be the smallest integer that is a power of two and larger than or
   equal to $N + M - 1$.

2. Append zeros to these two sequences to make them both of length L.

3. Apply an L-point FFT to these two zero-appended sequences to get
   two new sequences, both of length L, in the transformed domain.

4. Multiply these two transformed domain sequences point by point to
   get a new sequence, also of length L.

5. Apply an L-point inverse FFT to this new sequence to get a sequence
   of length L. The first $N + M - 1$ points of this sequence is the desired
   output sequence of the linear convolution.

2.3.8 Discrete Hadamard Transform

The preceding discussion indicates the usefulness of transforms. Transform
methods are useful in applications, particularly if the feature of interest can
be characterized in the transform domain. For example, Fourier transforms
play an important role in signal processing. There are a number of transforms, each having its own advantages and disadvantages.

Another useful transform is the Walsh-Hadamard transform (WHT).
The basic functions are Walsh functions that are binary valued with $\{1,
-1\}$. Thus the generation and implementation of Walsh functions is sim-
ple. Furthermore, the only operations involved are addition and subtraction,
which means the cost of computation is low.

The WHT can be formulated as a matrix-vector multiplication similar
to the DFT. The Hadamard matrix is a square array of $+1$s and -1s, whose
rows and columns are orthogonal. This matrix can be defined iteratively as:

---

[8]Given the output sequence and the impulse response of the filter, find the input sequence. This is the deconvolution problem. Two FFT operations followed by division and
one inverse FFT can be used to solve this problem.

$$H_2 = \frac{1}{\sqrt{2}} \cdot \begin{bmatrix} 1 & 1 \\ 1 & -1 \end{bmatrix}$$

and

$$H_{2N} = \frac{1}{\sqrt{2}} \cdot \begin{bmatrix} H_N & H_N \\ H_N & -H_N \end{bmatrix}$$

A Hadamard matrix of size eight is shown next.

$$H_8 = \frac{1}{2\sqrt{2}} \cdot \begin{bmatrix} 1 & 1 & 1 & 1 & 1 & 1 & 1 & 1 \\ 1 & -1 & 1 & -1 & 1 & -1 & 1 & -1 \\ 1 & 1 & -1 & -1 & 1 & 1 & -1 & -1 \\ 1 & -1 & -1 & 1 & 1 & -1 & -1 & 1 \\ 1 & 1 & 1 & 1 & -1 & -1 & -1 & -1 \\ 1 & -1 & 1 & -1 & -1 & 1 & -1 & 1 \\ 1 & 1 & -1 & -1 & -1 & -1 & 1 & 1 \\ 1 & -1 & -1 & 1 & -1 & 1 & 1 & -1 \end{bmatrix}$$

Suppose we have N $(N = 2^n)$ input data represented in vector form, say $\mathbf{x}$. The transformed data is vector $\mathbf{y}$. Then, $\mathbf{y} = H_N \cdot \mathbf{x}$.

The versatility of the Hadamard transform is shown by the fact that it has been applied to signal and image processing, speech processing, word recognition, signature recognition, character recognition, pattern recognition, spectral analysis of linear system, correlation and convolution, filtering, data compression, coding, communication, detection, statistical analysis, spectrometric imaging, and spectroscopy [Pratt78]. Similar to most of the other transforms, the Hadamard transform also has a fast algorithm, called the fast Hadamard transform.

2.3.9   Least Mean-Squares Estimation

So far we have examined operations involving deterministic signals. Discrete random signals and their operations are also important in many signal processing applications. A problem that arises in many signal processing applications is of predicting one random process $\{y_n\}$ from observations of another random process $\{x_n\}$. An estimate $\hat{y}_n$ is sought by taking a finite linear combination of the present and past samples of x_n, i.e., $\hat{y}_n = a_0 x_n + a_1 x_{n-1} + a_2 x_{n-2} + \ldots + a_p x_{n-p}$, such that $E\{[y_n - \hat{y}_n]^2\}$ is minimized.[9]

---

[9] $E\{\cdot\}$ denotes the expectation operator.

2.3.9.1 Wiener-Hopf Equation

The best estimate of $\hat{y}_n$ can be found in a relatively straightforward manner by expanding $E\{[y_n - \hat{y}_n]^2\}$ and setting the partial derivatives with respect to the a_i's equal to zero. This results in the Wiener Hopf equation to be solved for $\mathbf{a}$,

$$\mathbf{R}_x \cdot \mathbf{a} = \mathbf{r}_{xy} \tag{2.30}$$

where

$$\mathbf{a} = [a_0, a_1, a_2, ..., a_p]^T$$
$$\mathbf{R}_x = E\{X \cdot X^T\}$$
$$\mathbf{r}_{xy} = E\{X \cdot y_n\}$$

and

$$X = [x_n, x_{n-1}, ..., x_{n-p}]^T$$

IF $\{x_n\}$ and $\{y_n\}$ are jointly wide-sense stationary then

$$\mathbf{R}_x = \begin{bmatrix} r_{xx}(0) & r_{xx}(-1) & \cdots & r_{xx}(-p) \\ r_{xx}(1) & r_{xx}(0) & \cdots & r_{xx}(-p+1) \\ \vdots & \vdots & \ddots & \vdots \\ r_{xx}(p) & r_{xx}(p-1) & \cdots & r_{xx}(0) \end{bmatrix}$$

and

$$\mathbf{r}_{xy} = [r_{xy}(0), r_{xy}(1), ..., r_{xy}(p)]^T$$

where $r_{xy}(i) = E\{y_n x_{n-i}\}$.

The matrix $\mathbf{R}_x$ has a special shift-invariant structure and is called a *Toeplitz matrix*.

2.3.9.2 Wiener Filtering

In the wide-sense stationary case the matrix multiplication $\mathbf{R}_x \cdot \mathbf{a}$ is just the finite convolution of the autocorrelation sequence $\mathbf{R}_x(n)$ and $\mathbf{a}(n)$. If $\mathbf{a}(n)$ is allowed to be infinite in length and noncausality is allowed, i.e.,

$$\hat{y}_n = \sum_{n=-\infty}^{\infty} a_k x_{n-k} \, ,$$

then Eq. 2.30 can be rewritten as

$$\mathbf{R}_x(n) * \mathbf{a}(n) \ = \ \mathbf{r}_{xy}(n)$$

If we now take the Fourier transform of both sides of this equation we get

$$S_{xx}(e^{j\omega})H(e^{j\omega}) \ = \ S_{yx}(e^{j\omega})$$

Thus we see that the theoretical optimum linear filter for obtaining the minimum mean squares error estimate of y(n) from $x(n)$ has the ideal frequency response

$$H(e^{j\omega}) \ = \ \frac{S_{yx}(e^{j\omega})}{S_{xx}(e^{j\omega})}$$

This filter is commonly known as the Wiener filter. Note that this is a noncausal filter. This filter has found application in image restoration problems.

An infinite-length $\mathbf{a}(n)$ with the additional restriction of causality is yet another problem. The interested reader is referred to [Giord85] for details. The finite-length Wiener filter, solved through the Wiener-Hopf equation, is still popular. For stationary random processes, this solution reduces to a Toeplitz system of equations. Such solutions of Toeplitz systems are also important to many spectrum estimation methods, such as the maximum entropy method and the maximum likelihood method. A solution for a Toeplitz system will be discussed in the next section.

2.3.10   Solving Toeplitz System (Schur Algorithm)

Here we attempt to determine $\mathbf{x}$ when

$$\mathbf{Rx} \ = \ \mathbf{y}. \qquad (2.31)$$

and $\mathbf{R}$ has a Toeplitz structure, i.e., $r(i,j) \ = \ r(|i-j|)$.

This system of equations can be solved using the procedures outlined in the previous section on solving linear systems. However, they have the drawback that they do not exploit the Toeplitz structure. The most popular sequential algorithms for solving a symmetric Toeplitz system is the well-known Levinson algorithm [Levin47] [Kaila74]. However, this algorithm is not well suited for parallel implementation. The Levinson algorithm requires inner product operations which hampers its execution on a linear array of processors [Kung83b]. Since our interest is in parallel processing and parallel architectures, we describe the Schur algorithm. The Schur algorithm avoids the inner product operations and is more suitable for parallel processing.

The major function in solving the Toeplitz system is to perform a triangular decomposition of the matrix $\mathbf{R}$, i.e.,[10],

$$\mathbf{R} = \bar{\mathbf{U}}^T \mathbf{D} \bar{\mathbf{U}} = \bar{\mathbf{U}}^T \mathbf{U} \quad (\mathbf{U} = \mathbf{D}\bar{\mathbf{U}})$$

where $\mathbf{D}$ is a diagonal matrix and $\mathbf{U}$ is an upper triangular matrix. Then the solution $\mathbf{x}$ of Eq. 2.31 can be solved explicitly:

$$\mathbf{x} = \mathbf{R}^{-1}\mathbf{y} = \mathbf{U}^{-1}(\bar{\mathbf{U}}^T)^{-1}\mathbf{y}$$

which can be separated into a forward-substitution step,

$$\mathbf{g} = (\bar{\mathbf{U}}^T)^{-1}\mathbf{y}$$

combined with a back-substitution step:

$$\mathbf{x} = \mathbf{U}^{-1}\mathbf{g}.$$

For simplicity, we use a 4 × 4 matrix to demonstrate the triangularization procedure. Note that the inverse of an upper (lower) triangular matrix is also an upper (lower) triangular matrix. The problem is to find the elements $\{\ell_{ij}\}$ and $\{u_{ij}\}$ such that

$$\begin{bmatrix} 1 & 0 & 0 & 0 \\ \ell_{21} & 1 & 0 & 0 \\ \ell_{31} & \ell_{32} & 1 & 0 \\ \ell_{41} & \ell_{42} & \ell_{43} & 1 \end{bmatrix} \cdot \mathbf{R} = \begin{bmatrix} u_{11} & u_{12} & u_{13} & u_{14} \\ 0 & u_{22} & u_{23} & u_{24} \\ 0 & 0 & u_{33} & u_{34} \\ 0 & 0 & 0 & u_{44} \end{bmatrix} \tag{2.32}$$

denoted as

$$\tilde{\mathbf{L}}\mathbf{R} = \mathbf{U} \quad \text{where} \quad \tilde{\mathbf{L}} = (\bar{\mathbf{U}}^T)^{-1}$$

[10]The overbar "-" of the triangular matrix $\bar{\mathbf{U}}$ indicates that $\bar{\mathbf{U}}$ has unities along its diagonal.

The top rows of $\tilde{\mathbf{L}}$ and $\mathbf{U}$ are determined by the structure. To find the second row, we start with the following equation

$$\begin{bmatrix} 1 & 0 & 0 & 0 \\ 0 & 1 & 0 & 0 \end{bmatrix} \cdot \mathbf{R} = \begin{bmatrix} t_0 & t_1 & t_2 & t_3 \\ t_1 & t_0 & t_1 & t_2 \end{bmatrix}$$

where t_i's are the elements of $\mathbf{R}$ (cf., Toeplitz matrix structure).

Now perform row operations on both sides of this equation:

$$\begin{bmatrix} 1 & K^{(2)} \\ K^{(2)} & 1 \end{bmatrix} \begin{bmatrix} 1 & 0 & 0 & 0 \\ 0 & 1 & 0 & 0 \end{bmatrix} \cdot \mathbf{R} = \begin{bmatrix} v_0^{(2)} & 0 & v_2^{(2)} & v_3^{(2)} \\ 0 & u_1^{(2)} & u_2^{(2)} & u_3^{(2)} \end{bmatrix}$$

where the so-called "reflection coefficients" $K^{(2)}$ is computed as:

$$K^{(2)} = \frac{-t_1}{t_0}$$

This equation can be rewritten as

$$\begin{bmatrix} 1 & K^{(2)} & 0 & 0 \\ K^{(2)} & 1 & 0 & 0 \end{bmatrix} \cdot \mathbf{R} = \begin{bmatrix} v_0^{(2)} & 0 & v_2^{(2)} & v_3^{(2)} \\ 0 & u_1^{(2)} & u_2^{(2)} & u_3^{(2)} \end{bmatrix} \qquad (2.33)$$

This equation implies:

$$\begin{bmatrix} K^{(2)} & 1 & 0 & 0 \end{bmatrix} \cdot \mathbf{R} = \begin{bmatrix} 0 & u_1^{(2)} & u_2^{(2)} & u_3^{(2)} \end{bmatrix}$$

By comparing with the second row of the right hand side (RHS) of Eq. 2.32, it is clear that a zero is created by the row operation and the desired second rows of $\tilde{\mathbf{L}}$ and $\mathbf{U}$ are obtained:

$$\tilde{\mathbf{L}}_2 = [\ell_{21}\ 1\ 0\ 0] = \begin{bmatrix} K^{(2)} & 1 & 0 & 0 \end{bmatrix}$$
$$\mathbf{U}_2 = [0\ u_{22}\ u_{23}\ u_{24}] = \begin{bmatrix} 0 & u_1^{(2)} & u_2^{(2)} & u_3^{(2)} \end{bmatrix}$$

To compute the third rows of the matrices $\tilde{\mathbf{L}}$ and $\mathbf{U}$, the same strategy can be repeated. To be ready for the next recursion, we first *right-shift* the second row on both sides of Eq. 2.33, i.e.,

$$\begin{bmatrix} K^{(2)} & 1 & 0 & 0 \end{bmatrix} \longrightarrow \begin{bmatrix} 0 & K^{(2)} & 1 & 0 \end{bmatrix},$$

$$\begin{bmatrix} 1 & K^{(2)} & 0 & 0 \\ 0 & K^{(2)} & 1 & 0 \end{bmatrix} \cdot \mathbf{R} = \begin{bmatrix} v_0^{(2)} & 0 & v_2^{(2)} & v_3^{(2)} \\ u_{-1}^{(2)} & 0 & u_1^{(2)} & u_2^{(2)} \end{bmatrix} \qquad (2.34)$$

Note that by using the Toeplitz structure of the matrix $\mathbf{R}$, we have $u^{(2)}$ in Eq. 2.33 right-shifted accordingly and the *only new term* is $u_{-1}^{(2)}$, which is equal to $v_2^{(2)}$, since $u_{-1}^{(2)} = k^{(2)} \cdot t_1 + t_2 = v_2^{(2)}$.

Note that through this shift operation, the two 0s created in the previous recursion on the RHS are realigned into the same column. They will remain unaffected by the linear combination of the two rows in the next recursion. With this arrangement, a similar procedure as in the previous recursion can now be repeated:

$$\begin{bmatrix} 1 & K^{(3)} \\ K^{(3)} & 1 \end{bmatrix} \begin{bmatrix} 1 & K^{(2)} & 0 & 0 \\ 0 & K^{(2)} & 1 & 0 \end{bmatrix} \cdot \mathbf{R} = \begin{bmatrix} \mathbf{v}^{(3)} \\ \underline{\mathbf{u}}^{(3)} \end{bmatrix} = \begin{bmatrix} v_0^{(3)} & 0 & 0 & v_3^{(3)} \\ 0 & 0 & u_2^{(3)} & u_3^{(3)} \end{bmatrix}$$

where

$$K(3) = \frac{-v_2^{(2)}}{u_1^{(2)}}$$

By comparing $\underline{\mathbf{u}}^{(3)}$ with the third row on the RHS of Eq. 2.32, clearly, the third rows of the matrices $\tilde{\mathbf{L}}$ and $\mathbf{U}$ are obtained:

$$\begin{aligned} \tilde{\mathbf{L}}_3 &= [\ell_{31} \; \ell_{32} \; 1 \; 0] \\ &= \left[K^{(3)}, \; \left(K^{(3)} K^{(2)} + K^{(2)} \right), \; 1 \; 0 \right] \end{aligned}$$

$$\mathbf{U}_3 = [0 \; 0 \; u_{33} \; u_{34}] = \left[0 \; 0 \; u_2^{(3)} \; u_3^{(3)} \right]$$

This completes the second recursion. By induction, the future recursions can be carried out in the same manner until all the rows of the matrices $\tilde{\mathbf{L}}$ and $\mathbf{U}$ are computed.

Note that the shift operation in each recursion is natural due to the Toeplitz structure of matrix $\mathbf{R}$, since it retains the zeros produced in the previous recursion. The purpose of the shift is to realign these zeros with those of the auxiliary vector $\mathbf{v}$ (cf., Eq. 2.34), such that these zeros will remain unaffected by the upcoming row operations. This also explains the purpose and the necessity of computing the auxiliary vectors $\mathbf{v}$ in each recursion.

2.4   Image Processing Algorithms

Most of the preceding one-dimensional signal processing operations can be naturally extended to two-dimensional or multidimensional processing applications. Some image processing algorithms, which are just extended forms of their one-dimensional counterparts, are discussed in this section.

2.4.1   Two-dimensional Convolution and Correlation

In image processing, since the input data is two-dimensional (2-D) (with two space indices), the operations of convolution and correlation will also be 2-D. The main difference between 1-D and 2-D operations is that the number of indices of the formulas is doubled. The 2-D convolution formula is as follows:

$$y(n_1, n_2) = \sum_{k_1=0}^{n_1} \sum_{k_2=0}^{n_2} u(k_1, k_2) w(n_1 - k_1, n_2 - k_2)$$

where $n_1, n_2 \in \{0, 1, \ldots, 2N - 2\}$

The 2-D correlation formula is:

$$y(n_1, n_2) = \sum_{k_1=0}^{n_1} \sum_{k_2=0}^{n_2} u(k_1, k_2) w(n_1 + k_1, n_2 + k_2)$$

where $n_1, n_2 \in \{-N + 1, -N + 2, \ldots, -1, 0, 1, \ldots, N - 2, N - 1\}$

The number of computations needed for 2-D convolution or 2-D correlation is usually very large. Hence, transform methods are usually used for efficient computation. The method is similar to the 1-D case.

2.4.2   Two-dimensional Filtering

The input-output relation of a 2-D filter can be represented either by a 2-D difference equation in the space domain or by a transfer function in the 2-D frequency domain (or 2-D Z-domain). Occasionally, the computation of a 2-D filter can be achieved by successively using 1-D filtering. However, in general, the computation will involve fast 2-D convolution via the 2-D FFT or directly using the 2-D difference equation. Interested readers may refer to [Rabin75]. A 2-D filter described by a 2-D difference equation is given in the following example.

Example 4: 2-D Filtering

With suitable initial conditions, the following difference equation can be used to describe a 2-D filter.

$$y(n_1, n_2) = x(n_1, n_2) + 0.3 \cdot y(n_1-1, n_2) + 0.3 \cdot y(n_1, n_2-1) - 0.2 \cdot y(n_1-1, n_2-1)$$

2.4.3   2-D DFT, FFT and Hadamard Transform

Two-dimensional transforms can be defined in a manner similar to the 1-D case. The 2-D DFT is defined as follows:

$$X(k_1, k_2) = \sum_{n_1=0}^{N-1} \sum_{n_2=0}^{N-1} x(n_1, n_2) \cdot W_N^{n_1 k_1 + n_2 k_2}$$

where $k_1, k_2 \in \{0, 1, 2, \ldots, N-1\}$ and $W_N = e^{-j2\pi/N}$, as defined before

The 2-D DFT has several important properties:

1. By using 1-D N-point FFTs $2N$ times, the 2-D DFT can be calculated in $O(2N^2 \log_2 N)$ time. This can be derived as follows:

$$
\begin{aligned}
X(k_1, k_2) &= \sum_{n_1=0}^{N-1} \sum_{n_2=0}^{N-1} x(n_1, n_2) \cdot W_N^{n_1 k_1 + n_2 k_2} \\
&= \sum_{n_1=0}^{N-1} \left(\sum_{n_2=0}^{N-1} x(n_1, n_2) \cdot W_N^{n_2 k_2} \right) \cdot W_N^{n_1 k_1} \\
&= \text{DFT}_{n_1} \left(\text{DFT}_{n_2} \left[x(n_1, n_2) \right] \right)
\end{aligned}
$$

Thus, a 2-D DFT can be calculated by first applying 1-D FFTs row-wise (columnwise) N times and then applying, on the transformed sequence, 1-D FFTs columnwise (rowwise) N times. This is the 2-D FFT algorithm.

2. Similar to its 1-D counterpart, 2-D convolution can be quickly computed by using transformation methods, e.g., by using a 2-D FFT, multiplying the transforms, and then using a 2-D inverse FFT.

In a similar fashion, one can define the 2-D Hadamard transform. The interested reader is referred to [Pratt78] for details.

2.5 Advanced Algorithms and Applications for Further Explorations

In the following we shall discuss several important algorithm classes worthy of further advanced exploration, since they have two desirable features (1) each of them represents a certain algorithmic class and important applicational domain and (2) they are potentially very suitable for VLSI array processors.

2.5.1 Divide-and-Conquer Technique

When solving a large size problem, a common technique is to decompose the problem into smaller parts, find solutions for the parts, and then combine the solutions for the parts into the solution for the whole. For general divide-and-conquer techniques, the subproblems can be formulated just like smaller versions of the original problem, thus the same routine may be repeatedly used in different partitions of the problem. Then the divide-and-conquer approach can be used recursively and yields efficient solutions [Aho74]. For example, consider the problem of finding both the maximum element of a set S containing n elements, where n is a power of 2. The divide-and-conquer approach would divide the set S into two subsets S_1 and S_2, each with $n/2$ elements. The algorithm would then find the maximum element of each of the two halves and, obviously, the maximum element of S could be calculated as the larger of the maximum elements of S_1 and S_2. An efficient solution can be realized by recursive application of the algorithm.

One important research subject on array architectures is on the design of interconnection networks for certain classes of divide-and-conquer algorithms. Prominent examples include sorting (bitonic sorting) and the FFT algorithm. By a careful structural analysis, it can be shown that a perfect-shuffle network can effectively facilitate data routing for these algorithms and thus significantly speed up the computation (see Figure 2.6).

2.5.2   Dynamic Programming Method

Dynamic programming techniques are widely used in optimization problems, in which an objective function must be either minimized or maximized subject to a set of constraints. Dynamic programming can be generally applied to problems of this class if they adhere to the *principle of optimality.* Dynamic programming is in essence a bottom up procedure in which the solution to all subproblems are first calculated and the results used to solve the whole problem. The problem is again divided into stages, proceeding from small subproblems to larger ones. The intermediate results are stored and utilized in later stages. Contrasting to top down divide and conquer techniques in which one stage consists of many subproblems, the dynamic programming method enjoys the advantage that in one stage there is only one subproblem. The basic principle of dynamic programming is that the solutions of these subproblems are linked by a recurrence relation [Gondr84].

Such a recursive formulation is instrumental in a successful mapping from algorithm to arrays with local interconnection. There are many important examples of the dynamic programming problems; one example is the shortest path problem, which is to determine the lengths of the shortest paths between all pairs of nodes in the graph. More details will be discussed in Section 4.4 of this book [Kung86a]. Another very similar example is minimum cost path finding problem, which may be applied to speech recognition application for compensating the variation of speaking rates. For this application, it may be solved by the so-called dynamic time warping (DTW) algorithm. A detailed discussion is deferred to Section 8.5 of this book [Kung86c].

2.5.3   Relaxation Technique

A very promising algorithm class from the array processing perspective are those based on the so-called relaxation techniques. The relaxation technique is an iterative approach to many problems, which makes updating in parallel at each point and in each iteration based on the data (all data elements or in most cases neighboring data elements) available from the most recent updating or in the immediate preceding iteration. It is comparatively more powerful than one-shot computational methods, because its initial choices are successively refined based on the the newly available information. It makes tentative, rather than firm, guesses which may be adaptively updated toward a desired or an optimal solution. *The relaxation approach is very suitable for array processors because it is order-independent and can be greatly speeded up by parallel processing.* This is because the updating at every data point in each iteration can be executed in parallel. The most popular relaxation method is the linear type relaxation method, which can be formulated into an iterative linear equation solving problem (see Section 2.2). For more sophisticated problems, some nonlinear type relaxation or stochastic type relaxation will be required to improve the performance.

The relaxation method based on the iterative linear equation solver formulation has been successfully applied in image reconstruction and partial differential equation applications [Kuo85]. Using the stochastic relaxation technique, image restoration from various blurring mechanisms can be effectively performed.

2.5.4   Simulated Annealing via Stochastic Relaxation

For certain optimization problems, the computation complexity may be too great to be solved exactly. Therefore, the iterative improvement method may also be applied to these optimization problems. One severe drawback with this approach is that usually the solution obtained may only be locally optimal and not globally optimal. Such a deficiency is common for most optimization methods, since once a state is in a local optimum, it is trapped there forever due to the "greedy" nature of the iterative improvement.

Simulated annealing is a search technique that allows the possibility of getting out of the trap of the local optimum by introducing a mechanism of flattening the trap and a possibility based on a stochastic decision of accepting an updating which (temporally) corresponds to a *worse* solution. The general formulation of the simulated annealing technique for a finite state, discrete systems is discussed next.

Suppose that an "energy" function E defined on a finite set of states S. The problem is to find a state s^* which would minimize $E(s)$. We assume that for each state $s \in S$, a transition probability $R(s, s')$ is defined as the probability of making a transition from s to s', a stochastic scheme for constructing a sequence of states $s_0, s_1, s_2, \ldots$, (the initial state is set to be s_0) can now be described. Given a present state, say s_k, a potential next state s_k' is chosen from the allowable transition states with probability distribution

$$P[s_k'|s_k] = R(s_k, s_k') \qquad (2.35)$$

Then we set

$$s_{k+1} = \begin{cases} s_k', & \text{with probability } P_k \\ s_k, & \text{otherwise} \end{cases}$$

where

$$P_k = \exp\left\{\frac{-\mathrm{Max}[\triangle E, 0]}{T_k}\right\}. \qquad (2.36)$$

where $\triangle E = E(s_k') - E(s_k)$, and $\{T_k\}$ is the control parameter (called "temperature") for flattening the trap.

This demonstrates how the sequence $s_1, s_2, \ldots$, is chosen. This process of slow cooling is analogous to slow annealing of a metal in order to crystallize it in its lowest energy state. The convergence can be guaranteed if the system has certain properties and a suitable temperature sequence is defined [Hajek85].

To understand the mechanism of how to avoid local minima and guarantee to arrive at the global minimum, let us consider the probability P_k in two cases of the temperature, i.e., high and very low (near zero) temperature. In the beginning of the simulated annealing process, T is very high, such that for a fixed difference $\triangle E$ of the E function, the probability (P_k) of accepting the new state s_k' is large. That means if the state is a local minimum, there is a possibility that it can get out of the local minimum region. On the other hand, during the iteration process the temperature parameter T will be gradually lowered, and it is assumed that by a finite number of iterations (to be estimated) that the state will arrive at the global minimum region. Suppose that by the time the temperature has become sufficiently low (e.g., near zero), then the probability of moving away from the global minimum P_k is very close to zero. This would help the convergence.

Simulated annealing has been applied to the application of image restoration and reconstruction based on the stochastic relaxation technique, and successfully restores images from various degradation mechanisms, e.g., blurring, nonlinear deformation, multiplicative or additive noise [Geman84]. The applications of simulated annealing include optimization, code design for communication systems, certain aspects of artificial intelligence and image restoration [Hajek85]. The detailed design for a simulated annealing array processor [Kung86c] is treated in Chapter 8.

2.5.5   Associative Retrieval

Filing and retrieving data by association appears to be a powerful solution to many high-volume information processing problems. An associative processing system is very adept at recognition and recall from partial information and has remarkable error correction capabilities. A popular associative processing model is based on *content addressable memory* (CAM) which represents one of the simplest collective properties of a simple neuron system. In a CAM, data stored in an associative memory are addressed by their contents and, in this sense, associative memory and CAM can be considered to be equivalent. The major advantages of associative memory over RAM is its capability of performing parallel search and parallel comparison operations. These operations are frequently needed in many important applications, such as the storage and retrieval of a rapidly changing database, radar signal tracking, image processing, computer vision, and artificial intelligence.

Correlation Matrix Model   It is also well known that the neural networks in the eye-brain systems process information in parallel with the aid of large number of simple interconnected processing elements called *neurons*. Since the 1970s, the correlation matrix model of the distributed associative memory network has been gaining popularity. Note that the conception of associative memory can be traced all the way back to Aristotle's work [370 B.C.] on memory and reminiscence. The cybernetic research work in the 1950s on learning digital networks, perception, and conditioned connection crossbars [Farle54], [Rosbl58], [Stein61], have paved way for the modern era of studies on the subject. In the 1970s, the matrix model (as distinguished from the connection model) of memory networks seems to have gained the attention of many researchers [Kohoe72], [Nakan72], [Ander77]. More recently, Hopfield further extended the previous work to structure a computational model by a notion of energy functions with an iterative computation model [Hopfi82]. In the research mentioned above, it has been recognized that the content-addressable type memory is more closely related to how the human brain functions. In fact, a popular approach to its implementation is based on one which resembles that of a neuro-network. Neural signals are trains of pulses with variable frequencies. The task of memory is to reproduce the neural signal at the places where they occurred earlier. This leads to the notion of associative memory networks and associative retrieval, whose main features include recognition and error correction based on partial or cluttered input.

Hopfield Networks   The original Hopfield [Hopfi82] model uses two-state threshold neurons that followed a stochastic algorithm. Each model neuron i has two states, characterized by the output V_i of the neuron having the value V_i^0 or V_i^1. The input of each neuron comes from two sources, external input I_i and states of other neurons. Hopfield identified an energy function for symmetrically connected networks. The energy function of any state is given by

$$E = -\sum_{i<j} T_{ij} V_i V_j - \sum_i I_i V_i \qquad (2.37)$$

where T_{ij} can be biologically viewed as a description of the synaptic interconnection strength from neuron j to neuron i.

Given Hopfield's quadratic definition of energy, the difference of energy function between two different levels of neuron i ($V_i = 1$ when on, and $V_i = 0$ when off), given the current states of the other neurons is

$$E_{i\text{-}on} - E_{i\text{-}off} = \Delta E_i = -\sum_j T_{ij} V_j - I_i \qquad (2.38)$$

If the energy difference ΔE_i is negative, the unit should turn on (or stay on) to minimize the energy. Otherwise it should turn off (or stay off).

The Hopfield model behaves as an associative memory or CAM. With reference to Figure 2.5, each local minimum corresponds to a stored target pattern. The memory is content-addressable because if it is started anywhere close to a particular stable state (local minimum from energy point of view) and far from all others, then it would converge to that stable state (see Figure 2.5).

In the given state space, if the position of a particular stable point is considered to be the information of a specific memory of the overall sys-

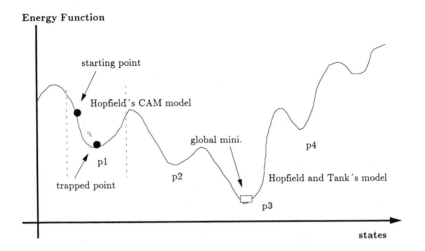

Figure 2.5: The original Hopfield model behaves as an associative memory, the local minimum ($p1$, $p2$, $p3$, and $p4$) correspond to stored target patterns. The Hopfield and Tank's formulation can be used to reach the global minimum state $p3$.

tem, then states in close proximity to that particular stable point include partial information about that memory. A final stable state containing all of the information of the memory can be found from a consideration of an initial state of partial information concerning a memory. By supplying some subpart of the memory in the initial state, the memory can be reached; as opposed to using an address. This constitutes the idea of CAM.

Boltzmann Machine   In fact, there are two ways of viewing the Hopfield formulation. The first, is to simulate the CAM as discussed above. The other formulation which proposed by Hopfield and Tank [Hopfi84], [Hopfi85] adopted an analog computational network, and tries to solve discrete problems in a continuous decision space in which the neural computation operates. Using this network, we can escape from local optima, and reach the global optimum by introducing into the analog circuit some "gain" parameters, which is equivalent to the "temperature" parameter in the Boltzmann machine introduced next.

Using simulated annealing techniques to escape local minima on the Hopfield model, the Boltzmann machine architecture can be realized. Specifically, for each memory unit i, the energy difference $\triangle E_i$ can be calculated from Eq. 2.38, and it can subsequently be proved that the on-off decision mechanism is governed by the following probability function P_i;

$$P_i = \frac{1}{1 + e^{-\Delta E_i/T}}$$

The T term is analogous to a temperature parameter in simulated annealing, and is essentially a scaling factor that controls the "volume" of noise. As T increases, P_i tends towards 0.5, the system therefore assumes states at random, irrespective of the constraints within the network; for $T = 0$ and ΔE_i, the system becomes definite (i.e. $P_i = 1$, and moves downhill to the nearest local minimum as in the pure Hopfield model. When the system has reached thermal equilibrium, the relative probability of residing in state a as opposed to state b at any given T, obeys a Boltzmann distribution:

$$\frac{P_a}{P_b} = e^{-(E_a - E_b)/T}$$

Although the system is in thermal equilibrium, this does not ensure that a particular stable state has been reached. Essentially the probability distribution over states has settled down, but the states are still changing. To establish thermal equilibrium at any given temperature, T, the best route is to start at a higher temperature (making it easy for the system to traverse the energy barriers), and gradually reduce T, increasing the preference for low energy states. By reducing T at a slow enough rate, there is a high possibility of residing in the best (or global optimum) state, or at least a state very close to the desired optimum.

Relationship between Hopfield model and Simulated Annealing
The circuit proposed by Hopfield and Tank [Hopfi85], is a real-valued, analog, nonlinear model used to avoid being trapped in local minimum. And this model approximates simulated annealing, but does not actually simulate a stochastic system.

2.6   VLSI Array Algorithms

An array algorithm is a set of rules for solving a problem in a finite number of steps by a multiple number of interconnected processors. Therefore, an array algorithm depends on the machine characteristics as well as the interconnection strategies. In the subsequent discussion, we shall focus our attention on the interconnection issue.

Concurrency is important for achieving high throughput using VLSI array processors. In general, concurrency is often achieved by decomposing a problem into independent subtasks (executable in parallel) or into dependent subtasks executable in a pipelined fashion. The degree of concurrency varies significantly among different techniques. While mapping algorithms onto parallel processors, the following closely related questions are posed: How does an array processor design depend on the algorithms?, How are these algorithms best implemented in array processors? and more importantly, How do we fully utilize the inherent concurrency (parallel and pipeline processing) in signal/image processing algorithms?

The most crucial issue that affects the array processing efficiency is *communication*, i.e., the scheme of moving data among processor elements (PEs) in a large-scale interconnection network. Correspondingly, a communication-oriented analysis of concurrent algorithms will be most useful for mapping algorithms onto the arrays. To conform with the constraints imposed by VLSI technology, we emphasize a special class of algorithms, namely, *recursive* and *locally dependent* algorithms.[11]

An effective algorithm design must start with a full understanding of the problem specification, mathematical analysis, and (concurrent and optimal) algorithmic analysis. In order to maximize parallel and pipeline processing, a dependence graph provides an effective tool, since it exhibits the full (data) dependencies incurred in the execution of a specific algorithm. These dependencies constitute the main constraints to the sequence of processing. To effectively exploit the potential concurrency available in an array processor environment, a new algorithmic design methodology tailored to array processors must be developed. While algorithm design is still an art that might never be fully automated, we offer some strategies which are useful in devising array algorithms.

2.6.1   Algorithm Design Criteria for VLSI Array Processors

The effectiveness of mapping activities of an algorithm onto a processor array is directly related to the way in which the algorithm is decomposed. *Two dissimilar algorithms with equivalent performance in a sequential computer,*

---

[11] In a recursive algorithm, all processors perform nearly identical tasks, and each processor repeatedly executes a fixed set of tasks on sequentially available data.

(i.e., requiring same number of arithmetic operations) may perform very differently in an array processing environment.

In conventional sequential algorithm analysis, the complexity of an algorithm depends on the computation and storage required, with the most important measure being the computation count. For example, the FFT algorithm is known to be superior to the DFT algorithm in terms of operation counts: $N \cdot \log_2 N$ operations for FFT versus N^2 operations for DFT. This is a very valid comparison in a traditional Von Neumann machine, where each operation is accompanied by a uniform overhead in memory write/fetch. In an array processing environment, however, the overhead is not uniform. Because a large number of PEs are used, the overhead depends critically on the availability and accessibility of resources. Fewer operations does not necessarily imply shorter computation time. In other words, operation count alone can no longer be regarded as an effective measure of processor performance.

2.6.1.1   Area-Time Complexity Theory

Complexity theory attempts to provide systematic information about algorithmic complexity to allow an informed choice to be made between two algorithms. A model for VLSI computation is often based on a "grid model" [Prepa84] [Thomp79]. Area-time complexity measures have received special attention. They depend on two factors, computation time (T) and chip area (A). In particular, the complexity measure AT^2 is very popular in lower-bound analysis of VLSI algorithms. It was originally motivated by the amount of information exchanged between two bisected regions in a VLSI circuit. However, we do *not* emphasize such a complexity measure because of the following two reasons. One is that there are already many good discussions on the topic in the VLSI algorithm literature; see for example, [Ullma84]. The other reason is more critical: an AT^2 measure seems to offer little practical implication in VLSI system design. Practically, any cost-effectiveness measure $f(A, T)$ as a function of the speed performance (T) and area cost (A) depends strongly on individual applications. For example, in a (military) system where very stringent real-time speed is stressed, the function $f(A, T)$ must place heavy weighting on the speed parameter T. On the other hand, those working with consumer products, where speed is less important, might prefer to emphasize the cost parameter (chip area), A, in order to enhance marketability. In any event, little relationship has been established between the special measure AT^2 (as used in the academic community) and a practical measure $f(A, T)$ (as might be derived from a cost-effectiveness consideration). It is probably fair to say that the AT^2 based complexity theories have *not* been recognized as directly applicable to the design of array processors.

2.6.1.2   Design Criteria for VLSI Array Algorithms

It is clear that a new criterion is required to determine the efficiency of an algorithm, since the simplistic measure of operation counts originally adopted for sequential machines is not adequate. In particular, because of the stringent *communication* problems associated with VLSI technology, this new criterion has to take into account the potentially significant *interconnection* costs. The modern array computation criterion should include some additional influential factors, such as parallelism and **pipelining rate**.[12] Generally speaking, in designing array algorithms for array processing, the design criteria should comprise computation, communication, memory and I/O. Their key aspects are as follows:

1. **Maximum parallelism** Two algorithms with equivalent performance in a sequential computer may perform quite differently in parallel processing environments. An algorithm will be favored if it expresses a higher parallelism, which is exploitable by the computing arrays. A very good example is the problem of solving Toeplitz systems, for which the major algorithms proposed in the literature are the Schur algorithm and the Levinson algorithm [Kung83b]. The latter is very popular in many spectrum estimation applications. In terms of sequential processing, both algorithms require the same number of operations. However, in terms of the achievable concurrency when executed in a linear array processor, the Schur algorithm displays a clear-cut advantage over the Levinson algorithm. More precisely, using a linear array of N PEs the Schur algorithm will need only $O(N)$ computation time, compared with $O(N\log_2 N)$ required for the Levinson algorithm. Here N represents the dimension of the Toeplitz matrix involved. For more information see [Kung83b].

2. **Maximum pipelinability** Most signal processing algorithms demand very high throughput rate and are computationally intensive (as compared with their I/O requirements). The exploitation of pipelining is

---

[12] Definition of *pipelining period:* the time interval between two successive input data of a problem instance, denoted as α. Pipelining rate is the reciprocal of pipelining period.

often very natural in regular and locally connected networks; therefore, a major part of concurrency in array processing will be derived
from pipelining. To maximize the throughput rate, we must select the
best among all possible algorithms and arrays. Unpredictable data dependency may severely jeopardize the processing efficiency of a highly
regular and structured array algorithm. Effective VLSI arrays are inherently highly pipelined and hence require well structured algorithms
with predictable data movements. Iterative methods with dynamic
branching, dependent on data produced during the process, are less
well suited for pipelined architectures.

3. **Balance among computations, communications and memory**
   A good array algorithm should offer a sound balance between different
   bandwidths incurred in different communication hierarchies to avoid
   data draining or unnecessary bottlenecks. Balancing the computations
   and various communication bandwidths is critical to the effectiveness
   of array computing. In today's technology, it is not hard to improve
   the computation bandwidth; however, it is much harder to increase
   the I/O bandwidth. In this case, the pipeline techniques are especially suitable for balancing computation and I/O because the data
   tend to engage as many processors as possible before leaving the array.
   This will reduce I/O bandwidth for outside communication. But, for
   certain computation-bound problems, notably matrix multiplication,
   FFT and sorting, if the computation bandwidth is increased while the
   I/O bandwidth is kept constant, the size of local memory has to increase in order to balance the computation with I/O [HTKun86].

4. **Trade-off between computation and communication** To make
   the interconnection network practical, efficient, and affordable, regular
   communication should be encouraged. Key issues affecting the communication regularity include local versus global, static versus dynamic,
   and data-independent versus data-dependent interconnection modules.
   The criterion should maximize the trade-off between interconnection
   cost and throughput. To conform with the communication constraints
   imposed by VLSI, a lot of emphasis has recently been placed on local and recursive algorithms. Let us use the example of DFT versus FFT computing; for which the computation costs are $O(N^2)$ and
   $O(N\log_2 N)$, respectively. The FFT is favored by almost one order of
   magnitude computation-wise (depends on N). On the other hand, the

DFT enjoys simple communication needs because it belongs to a "locally recursive" class, whereas the FFT computation requires a global interconnection [Thomp83]. This leads to a contrasting trade-off, especially in an array processing environment. As another example, an algorithm requiring only a static network is preferable to one requiring a dynamic network, since a static interconnection network is physically easier to construct.

5. **Numerical performance and quantization effects** Numerical behavior depends on many factors, such as the word length of the computer and the algorithms used. As an example, a QR decomposition (based on a GR) is often preferred over an LU decomposition for solving linear systems, since the former has a more stable numerical behavior. The price, however, is that QR takes more computations than LU decomposition. Quite often, additional computations may be wisely utilized to improve the overall numerical performance. However, the tradeoff between computation and numerical behavior is very problem dependent and there is no general rule to apply[13]. A prominent example is that of the FFT computation, which is computationally cost effective and at the same time numerically well behaved.

2.6.2   Locally Recursive Algorithms and Globally Recursive Algorithms

The previous discussion in this chapter provides a basic mathematical analysis of signal and image processing algorithms. The issues of interest now are, How do signal/image processing applications dictate the design of array processors? and Practically, how can these algorithms be implemented in VLSI hardware? On examination of many signal processing routines, some features become apparent. These include intensive computation, matrix operations, and localized, or perfect-shuffle, communications. These all point to promising systematic design methods for array architectures. The solution to real-time digital signal/image/vision processing hinges upon novel array processors for the common functions such as convolution, FFT, and matrix operations.

For proper communication in an interconnected computing network, each PE in the array should know **where**, **when**, and **how** to send or

---

[13]Except that the extra computation can always be used to increase the word-length (double or quadruple precision) and thus assures an improved performance.

fetch data. When mapping a locally recursive algorithm onto a computing network, there is a simple solution to the question of **where** to send the data, because the data movements can be confined to nearest neighbor PEs. Therefore, locally interconnected computing networks will suffice to execute the algorithm with high performance.

The conventional approach to the second question, **when** to send the data, is to use a globally synchronous scheme, where the timing is controlled by a sequence of "beats" [Ullma84]. A prominent example is the systolic array [HTKun78] [HTKun82]. However, the notion of locality has two meanings in array processor designs: localized data transactions and/or localized timing scheme (i.e., using self-timed, data-driven control). In fact, the class of locally recursive algorithms permits both locality features, which may be exploited in architectural designs [Kung82a]. As to **how** to communicate, it depends very much on the interconnection network adopted in the system.

2.6.2.1 Inherent Local or Global Communications in Algorithms

The performance of concurrent processing depends quite critically on the communication cost for data transactions and it is therefore necessary to have a formal way of characterizing the communication requirement and evaluating the cost. Each PE will be assigned a location index such that the distance between two PEs may be defined as the difference of their location indices. The communication cost can be effectively characterized by the distance between the PEs involved in the data transactions.

To describe fully the recursive activities in a concurrent processing environment, we need a **time index** as well as a **spatial index** to indicate **when** and **where** each computation of the algorithm takes place. There are two major classes of recursive algorithms, those with *local interconnections* and those with *global interconnections*:

1. A recursive algorithm is said to be of *local type* if the space index separations incurred in the same recursion are within a given limit. A great majority of signal processing algorithms possess this recursive and locally data-dependent property. One typical example is a class of matrix algorithms, which are most useful for signal processing and applied mathematical problems. This class of algorithms is the focus of our discussion.

2. If the recursion involves globally separated space indices, the algorithm is said to be of *global type*; and it always calls for globally intercon-

nected computing structures. For example, FFT and sorting are two important signal processing operations that entail global interconnections between the elements of the computing array.

Some prominent examples for locally recursive algorithms and globally recursive algorithms are listed in the accompanying table.

Locally Recursive Algorithms	Globally Recursive Algorithms
Matrix multiplication Convolution Back-substitution IIR filtering Selection sorting Toeplitz system solution Transitive closure	FFT Bitonic sorting Viterbi decoding

2.6.2.2   Locally Recursive Algorithms

A majority of the algorithms used in signal and image processing share the common traits of *localized operations, intensive computation (as compared with I/O), and matrix operations.* The key feature of these algorithms is that when mapped onto an array structure only local communications are required. This can be exploited to simplify the architectural and software requirements in the design of special-purpose signal/image processing array processors. The next chapter is devoted mainly to the discussion of the class of locally recursive algorithms. In the meantime, a short digression will be taken to explore the class of globally recursive algorithms.

2.6.2.3   Globally Recursive Algorithms: An FFT Example

A typical example for the class of globally recursive algorithms is the FFT algorithm. The FFT inherently requires global communication to support the perfect shuffling of data between stages of computation [Singl67], [Stone71]. A detailed derivation of array processors for FFT computation, based on a perfect-shuffle interconnection network [Stone71], is presented here. The configuration is shown in Figure 2.6. Note that in the FFT, the total number of butterfly operations is $(N/2) \cdot \log_2 N$ and each butterfly operation consists of four real multiplications and four real additions. In the single-stage

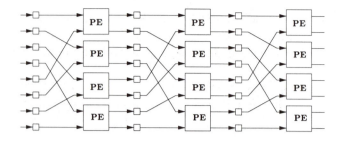

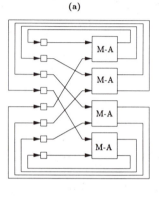

(b)

Figure 2.6: Array configurations for FFT computation, with the perfect-shuffle interconnection network: (a) multistage array; (b) single-stage array.

configuration, we use a linear array of $N/2$ PEs, which needs $\log_2 N$ time units to complete an FFT of N points.

To show that the FFT can be implemented by shuffle-exchange networks, decimation in time FFT is used as an example.

Perfect-Shuffle Permutation   The name *perfect-shuffle* is derived from an analogy to card shuffling; when a deck of playing cards is shuffled it is divided into equal halves, and recombined into a single deck by interleaving the cards (see Figure 2.7(a)). This permutation corresponds to a single-bit left shift of the binary representation of index x.

That is, if the binary representation of x is

$$x = \{b_n, b_{n-1}, ..., b_1\}$$

then the perfect shuffle operation is as follows:

$$\sigma(x) \; = \; \{b_{n-1}, b_{n-2}, ..., b_1, b_n\}.$$

Exchange Permutation   The exchange permutation can be defined as follows:

$$\varepsilon_{(k)}(x) \; = \; \{b_n, \ldots, b'_k, \ldots, b_1\}$$

The b'_k denotes the complement of the kth bit. The exchange permutations are shown in Figure 2.7(b).

FFT via Shuffle-Exchange Network   Note that the exchange network shown in Figure 2.7(b) is exactly the same as that used in a 8-point decimation in time FFT (See Figure 2.3). In fact, the interconnection network of a 8-point decimation in time FFT can be represented as follows:

$$\rho\left[\varepsilon_{(1)}\left[\varepsilon_{(2)}\left[\varepsilon_{(3)}\right]\right]\right] \tag{2.39}$$

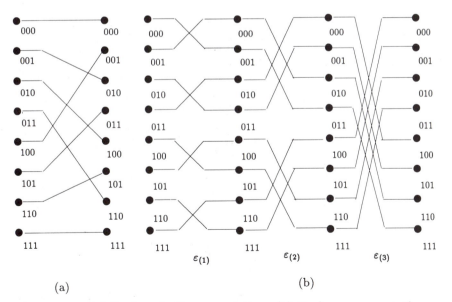

Figure 2.7: (a) Perfect shuffle permutation. (b) Exchange permutations.

The interconnection network for an in-place computation has to provide an exchange permutation $(\varepsilon_{(k)})$ and a bit-reversal permutation (ρ). Note that the sequence of operations is to apply $\varepsilon_{(3)}$ first and $\varepsilon_{(2)}$ second, and so on. This means that the sequence of operations is from the right-hand side of Eq. 2.39 to the left-hand side. These may be illustrated by considering a special case of N, for example, $N = 2^3 = 8$.

We can consider computing $X(k)$ by separating $x(n)$ into two $N/2$-point sequences consisting of the even-numbered points in $x(n)$ and the odd-numbered points in $x(n)$.

$$X(k) = \sum_{n=0}^{7} x(n) W_N^{nk} \qquad k = 0, 1, \ldots, 7$$

Both k and n are represented by 3-bit binary numbers,

$$n = (n_3 n_2 n_1) = 4n_3 + 2n_2 + n_1,$$
$$k = (k_3 k_2 k_1) = 4k_3 + 2k_2 + k_1.$$

Using the binary representations of n and k, we obtain

$$X(k_3 k_2 k_1) = \sum_{n_3=0}^{1} \sum_{n_2=0}^{1} \sum_{n_1=0}^{1} x(n_3 n_2 n_1) W_N^{(4k_3+2k_2+k_1)(4n_3+2n_2+n_1)}$$

$$= \left(\sum_{n_3=0}^{1} \sum_{n_2=0}^{1} x(n_3 n_2 0) W_{N/2}^{(2k_2+k_1)(2n_3+n_2)} \right) W_{N/2}^{4k_3(2n_3+n_2)}$$

$$+ \left(\sum_{n_3=0}^{1} \sum_{n_2=0}^{1} x(n_3 n_2 1) W_{N/2}^{(2k_2+k_1)(2n_3+n_2)} \right) \cdot W_N^{4k_3+2k_2+k_1}$$

But since $W_{N/2}^{4k_3(2n_3+n_2)} = W_4^{4k_3(2n_3+n_2)} = 1$, the above equation can be simplified as follows:

$$X(k_3 k_2 k_1) = \sum_{n_3=0}^{1} \sum_{n_2=0}^{1} x(n_3 n_2 0) W_{N/2}^{(2k_2+k_1)(2n_3+n_2)}$$

$$+ \left(\sum_{n_3=0}^{1} \sum_{n_2=0}^{1} x(n_3 n_2 1) W_{N/2}^{(2k_2+k_1)(2n_3+n_2)} \right) \cdot W_{N/2}^{4k_3+2k_2+k_1}$$

Due to the in-place index replacement, i.e., the input data and output data share the same storage, it can be seen (from the above equation) that n_1 is replaced by k_3. That is, by separating input data into two groups, $x(n_1 = 0)$ and $x(n_1 = 1)$, and output data into two groups, $X(k_3 = 0)$ and $X(k_3 = 1)$, the above equation shows how to relate these groups of data via the butterfly structure.(see Figure 2.3) Furthermore,

$$\sum_{n_3=0}^{1} \sum_{n_2=0}^{1} x\,(n_3 n_2 0)\, W_{N/2}^{(2k_2+k_1)(2n_3+n_2)} \;=\; \sum_{n_3=0}^{1} x\,(n_3 00)\, W_{N/4}^{k_1 n_3}$$

$$+ \left(\sum_{n_3=0}^{1} x\,(n_3 10)\, W_{N/4}^{k_1 n_3} \right) \cdot W_{N/2}^{2k_2+k_1}$$

$$\sum_{n_3=0}^{1} \sum_{n_2=0}^{1} x\,(n_3 n_2 1)\, W_{N/2}^{(2k_2+k_1)(2n_3+n_2)} \;=\; \sum_{n_3=0}^{1} x\,(n_3 01)\, W_{N/4}^{k_1 n_3}$$

$$+ \left(\sum_{n_3=0}^{1} x\,(n_3 11)\, W_{N/4}^{k_1 n_3} \right) \cdot W_{N/2}^{2k_2+k_1}$$

Thus n_2 is replaced by k_2. Similarly, it can be shown n_3 is replaced by k_1. If $(n_3 n_2 n_1)$ is the binary representation of the index of the sequence $x(n)$, then the sequence $x(n_3 n_2 n_1)$ is stored in the array position $X(k_1 k_2 k_3)$. That is, in determining the position of $x(n_3 n_2 n_1)$ in the input, we must reverse the order of the bits of the index n (cf., Figure 2.3 and the following table).

Original Index		Bit-reversed Index	
$x(0)$	000	$x(0)$	000
$x(1)$	001	$x(4)$	100
$x(2)$	010	$x(2)$	010
$x(3)$	011	$x(6)$	110
$x(4)$	100	$x(1)$	001
$x(5)$	101	$x(5)$	101
$x(6)$	110	$x(3)$	011
$x(7)$	111	$x(7)$	111

Theorem 2.1: The DIT FFT computation may be implemented by using a static single-stage perfect-shuffle network, as shown in Figure 2.6(b).

Proof: First we want to show that the following relationship is true.

$$\varepsilon_{(k)} = \sigma^{-(n-k+1)} \varepsilon_{(1)} \sigma^{n-k+1}$$

The kth exchange permutation is defined by complementing the kth bit of the binary representation. We define σ^{n-k+1} as the operation that left shifts the binary representation $n - k + 1$ times and $\sigma^{-(n-k+1)}$ is the reverse operation of σ^{n-k+1}. Then, following the procedure $\sigma^{-(n-k+1)} \varepsilon_{(1)} \sigma^{n-k+1}$, we shift the kth bit to the least significant bit (LSB) position, complement it and shift it back to the original position. This procedure is actually the same procedure as $\varepsilon_{(k)}$.

Then Eq. 2.39 can be rewritten as follows:

$$\rho\left[\sigma^{-3}\varepsilon_{(1)}\sigma^3 \left[\sigma^{-2}\varepsilon_{(1)}\sigma^2 \left[\sigma^{-1}\varepsilon_{(1)}\sigma^1\right]\right]\right] = \rho\left[\sigma^{-3}\varepsilon_{(1)}\sigma \left[\varepsilon_{(1)}\sigma \left[\varepsilon_{(1)}\sigma\right]\right]\right]$$

$$= \rho\left[\varepsilon_{(1)}\sigma \left[\varepsilon_{(1)}\sigma \left[\varepsilon_{(1)}\sigma\right]\right]\right]$$

This is actually the connection network of Figure 2.6(a). If we fold the stages of Figure 2.6(a) into only one stage, we obtain the structure of Figure 2.6(b).

Note that the perfect shuffle network is equivalent to a "circulant shift to left" in terms of binary spatial indexing. As a result, $X(b_3b_2b_1)$ and $X(b_3'b_2b_1)$ are now brought next to each other. (In the binary address representation, two objects are next to each other if their binary addresses differ *only in the rightmost bit.* In other words, the shuffle network helps *localize* the physical distance of every pair of interacting data.

Fast Fourier, fast cosine, and fast Hadamard transforms are very similar and important operations in digital signal processing. From a functional perspective, the computation involved in the Fourier transform is complex multiplication and addition. In the discrete cosine transform it is real multiplication and addition, and in the Hadamard transform it is real addition. However, from a structural perspective, they share the same communication property: They require global interconnections between the elements of the computing array.

2.7   Concluding Remarks

Among a diversity of topics of the book, i.e., *applications, algorithms, architectures*, and *technology*, this section covers a broad spectrum of algorithms. In our opinion, for special purpose array processors, it is important to first understand the properties and classifications of algorithms before engaging the issues on architecture and implementation. Therefore, this section stresses the basic background for algorithmic understanding. Although the algorithms studied in this section are primarily for signal and image processing, they are naturally suitable for many scientific computing applications. In addition, this section has also introduced some advanced algorithmic techniques, which will become increasingly attractive and popular with the advent of VLSI array processor technology.

Several critical tasks on algorithmic understanding remain to be explored:

1. To ensure that the algorithms behave numerically the same (at least approximately) as they are designed to do.

2. To provide a systematic methodology of mapping/matching algorithms onto array architectures.

3. To devise algorithms suitable for an array processor environment. This task is more critical and harder than architectural design and implementation, at least from a theoretic understanding perspective.

The study on numerical properies of algorithms is very important but, unfortunately, will not be covered in this book. The methodology of mapping and/or matching algorithms to locally-interconnected arrays will be the subject of Chapter 3. The utimate goal for algorithm design methodology is to develop an automatic design (or a definitive methodology). However by the very heuristic nature of algorithm design, this is a difficult task.

2.8   Problems

1. *Computation count for matrix multiplication*: Compute the number of multiplications and additions involved when a 3 × 4 matrix is post-multiplied by a 4 × 12 matrix.

2. *Convergence of iterative method*: Prove that the iterative method for linear system solver is convergent if and only if the absolute value of every eigenvalue of $S^{-1}T$ is less than one.

3. *Convergence criterion of SOR method*: Use the description of the SOR iterative linear equation solver:

    (a) Based on the value given to S_S and T_S, find the result of $S_S + T_S$ and the relationship between this result and matrix A. Give an intuitve solution why this result can help the convergence of the iterative algorithm.

    (b) Start from the Eq. 2.13 and verify the correctness of Eq. 2.16.

    (c) Show that for the iteration matrix $B_S = S_S^{-1}T_S$, we have the following criterion, and the method can converge only for $0 < w < 2$:

$$\rho(B_s) \geq |w - 1|$$

    where $\rho(B_S)$ defines the largest absolute eigenvalue of matrix B_S.

4. *Gaussian elimination algorithm for LU decomposition*: Use Gaussian elimination to solve the following linear system [Stran80]:

$$\begin{aligned} 2u + v + w &= 1 \\ 4u + v &= -2 \\ -2u + 2v + w &= 7 \end{aligned}$$

    Hint: First triangularize the system.

5. *Back-substitution*: Use back-substitution to solve the following equation:

$$\begin{bmatrix} 4 & 2 & 1 & 0 \\ 0 & 4 & 2 & 1 \\ 0 & 0 & 4 & 2 \\ 0 & 0 & 0 & 4 \end{bmatrix} \cdot \begin{bmatrix} x_1 \\ x_2 \\ x_3 \\ x_4 \end{bmatrix} = \begin{bmatrix} 6 \\ 3 \\ 6 \\ -4 \end{bmatrix}$$

6. *Schur algorithm for Toeplitz system*: Use the Schur algorithm to solve the following Toeplitz system.

$$\begin{bmatrix} 4 & 2 & 1 & 0 \\ 2 & 4 & 2 & 1 \\ 1 & 2 & 4 & 2 \\ 0 & 1 & 2 & 4 \end{bmatrix} \cdot \begin{bmatrix} 1 \\ a \\ b \\ c \end{bmatrix} = \begin{bmatrix} 1 \\ -5 \\ -1 \\ -4 \end{bmatrix}$$

7. *Operation of a linear time-invariant system as a convolution*: Show that for a linear time-invariant (LTI) system, the output $\{y(n)\}$ is related to the input $\{x(n)\}$ via the following operation.

$$y(n) = \sum_{k=-\infty}^{\infty} x(k)h(n-k)$$

where $\{h(n)\}$ is the impulse response of the system.

8. *Linear convolution computation*: What is the resultant sequence when,

   (a) $\{ 1, 2, 3, 4 \}$ convolves with $\{ 1, -1, 1, -1, 1 \}$?
   (b) $\{ 1, 0, 3 \}$ convolves with $\{ 1, -1, 1, 0, 1, 2 \}$?

9. *Circular Convolution computation*: What is the resultant sequence when,

   (a) $\{ 1, 2, 3, 4 \}$ convolves with $\{ 1, -1, 1, -1 \}$?
   (b) $\{ 1, 0, 3, 2, 4, 1 \}$ convolves with $\{ 1, -1, 1, 0, 1, 2 \}$?

10. *Region of convergence for the Z-transform*: In general, the Z-transform is defined only on some region of the Z plane called *the region of convergence* such that the geometrical series of the Z-transform converges.

    Find the region of convergence for the following sequences.

    (a) $x(n) = 2^n u(n)$
        where $u(n)$ is the unit sample function, i.e., $u(n) = 1$ for $n = 0$, $1, 2, \ldots$ and $u(n) = 0$ for $n < 0$
    (b) $x(n) = -\left(\frac{1}{2}\right)^n u(-n-1)$

11. *Properties of the Z-transform*: Show that the following properties of the Z-transform are correct.

    (a) $x(-n) \longleftrightarrow X(z^{-1})$

(b) $a^n x(n) \longleftrightarrow X(a^{-1}z)$

(c) $nx(n) \longleftrightarrow -z\left(\frac{dX(z)}{dz}\right)$

12. *Eigenfunction of an LTI system*: An eigenfunction of a system is a function which when applied to a system results in an output that is identical to the input except for some scaling. Show that $\{e^{j\omega n}\}$ is the eigenfunction of a discrete LTI system.

13. *Condition for linear phase filter*: In a linear phase (FIR) system, the impulse response sequence $\{h(0), \ldots, h(n-1)\}$ has the property (assuming that, without loss of generality, N is odd)

$$h(n) = h(N-1-n), \quad n = 0, 1, \ldots, N-1.$$

(a) Show that this is a sufficient condition for linear phase.
    Hint: Denote

$$h_1(n) = h\left(n + \frac{N-1}{2}\right).$$

Note that $h_1(n)$ is symmetrical, implying that its Fourier transform $H_1\left(e^{j\omega}\right)$, is real.

(b) Show that (for the case, N is odd) the following condition is also a sufficient condition for linear phase.

$$h(n) = h(N-1-n), \quad n = 0, 1, \ldots, N-1.$$

14. *Correctness of the inverse DFT formula*: Show that by applying the inverse DFT formulation to the $X(k)$ sequence (the computed DFT sequence of $x(n)$), the $\tilde{x}(n)$ obtained is the same as $x(n)$ for $n = 0, 1, 2, \ldots, N-1$.

15. *DFT and convolution*: Consider two finite-duration sequences $x(n)$ and $y(n)$ where both are zero for $n < 0$ and with

$$x(n) = 0, \quad n \geq 10$$
$$y(n) = 0, \quad n \geq 18$$

The 20-point DFTs of each of the sequences are multiplied and the inverse DFT is computed. Let $r(n)$ denote the inverse DFT. Specify which points in $r(n)$ correspond to points that would be obtained in a linear convolution of $x(n)$ and $y(n)$.

16. *DIF FFT*: An 8-point DIF FFT is shown in Figure 2.8. Show that the computation is equivalent to that of the DFT.

    Hint:

    $$X(k) = \sum_{n=0}^{(N/2)-1} x(n)W_N^{nK} + \sum_{n=N/2}^{N-1} x(n)W_N^{nK}$$

    $$= \sum_{n=0}^{(N/2)-1} x(n)W_N^{nK} + W_N^{(N/2)k} \sum_{n=0}^{(N/2)-1} x\left(n + \frac{N}{2}\right)W_N^{nK}$$

17. *Comparisons of DFT and DHT*: Compare DFT (FFT) and DHT (discrete Hadamard transform) (FHT) from computation complexity, orthogonality of transform, etc., point of views.

18. *Discrete Hadamard transform*: What is the discrete Hadamard transformed sequence of the following vector?

    $$\begin{bmatrix} 2 \\ 3 \\ 6 \\ -4 \end{bmatrix}$$

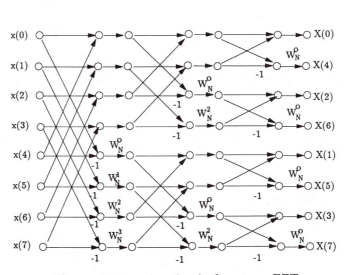

Figure 2.8: Decimation in frequency FFT.

19. *Fast Hadamard transform*: Show that a Hadamard matrix of size four can be factored into the product of two sparse matrices as shown below. How many addition (subtraction) operations are required to compute the Hadamard transform by using these two sparse matrices? Compare with the requirement for the original Hadamard transform. Draw a computation flow graph (see Figure 2.3) for this computation.

$$
\begin{bmatrix}
1 & 1 & 0 & 0 \\
0 & 0 & 1 & 1 \\
1 & -1 & 0 & 0 \\
0 & 0 & 1 & -1
\end{bmatrix}
\cdot
\begin{bmatrix}
1 & 1 & 0 & 0 \\
0 & 0 & 1 & 1 \\
1 & -1 & 0 & 0 \\
0 & 0 & 1 & -1
\end{bmatrix}
$$

20. *Fast Hadamard transform (DIF)*: Draw a computation flow graph (cf., Figure 2.8) by noting that an N-point Hadamard transform can be decomposed in the following way.

$$
\mathbf{H}_N = \frac{1}{\sqrt{2}} \cdot
\begin{bmatrix}
\mathbf{H}_{\frac{N}{2}}\mathbf{x}_1 + \mathbf{H}_{\frac{N}{2}}\mathbf{x}_2 \\
\mathbf{H}_{\frac{N}{2}}\mathbf{x}_1 - \mathbf{H}_{\frac{N}{2}}\mathbf{x}_2
\end{bmatrix}
= \frac{1}{\sqrt{2}} \cdot
\begin{bmatrix}
\mathbf{H}_{\frac{N}{2}}(\mathbf{x}_1 + \mathbf{x}_2) \\
\mathbf{H}_{\frac{N}{2}}(\mathbf{x}_1 - \mathbf{x}_2)
\end{bmatrix}
$$

21. *Two-dimensional convolution*: Show that, given a 5 × 5 matrix $\mathbf{A}$, it is always possible to find solutions to the following equation:

$$
\mathbf{A} = \mathbf{c}_1 * \mathbf{d}_1 + \mathbf{c}_2 * \mathbf{d}_2 + \mathbf{c}_3 * \mathbf{d}_3 + \mathbf{c}_4 * \mathbf{d}_4
$$

where the * stands for convolution and all the matrices $\mathbf{c}_i$ and $\mathbf{d}_i$, $i = 1, 2, 3, 4$, are 3 × 3 matrices.

Hint: All the matrices $\mathbf{d}_i$, $i = 1, 2, 3, 4$, consist of 1 and 0 elements only.

22. *Mathematical definition of various terminologies*: Give a mathematical definition for each of the following:

   (a) Solution of linear systems
   (b) Singular value decomposition
   (c) 2-D correlation
   (d) 2-D DFT
   (e) Linear phase filter

23. *Locally and globally recursive algorithm*: Differentiate between a locally recursive algorithm and a globally recursive algorithm.

24. *Perfect-shuffle permutation*: What is the sequence that results from doing perfect shuffle three times on the sequence

$$\{1, 2, 3, 4, 5, 6, 7, 8\}.$$

25. *Exchange permutation*: What is the sequence that results from doing exchange permutation $\varepsilon_{(1)}$ followed by $\varepsilon_{(2)}$ and then $\varepsilon_{(3)}$ on the sequence $\{1, 2, 3, 4, 5, 6, 7, 8\}$.

26. *Planar graph of an FFT array*: Following the array configuration in Figure 2.6, plot the linear array for a 16 point FFT. If we regard each processing element as a node, is this SFG a planar graph? (A graph is planar if and only if it can be mapped to another topologically equivalent graph that has no crossover of wires.)

Chapter 3

MAPPING ALGORITHMS ONTO ARRAY STRUCTURES

3.1 Introduction

Localized operations, intensive computation, and matrix operations are features of many algorithms used in signal and image processing. The common features of these algorithms should be exploited to facilitate the design of special-purpose signal/image processing array processors. Therefore, we concentrate on the *expression* and *transformation* of this special class of algorithms.

Algorithm expression is a basic tool for a proper description of an algorithm for parallel and pipeline processing. There are already quite a number of research efforts devoted to the formal description of the space-time activities in array processors [HTKun78], [Chen83]. One natural approach is to describe the actual space-time activities in terms of snapshots, which display data activities at a particular time instant, or as a recursive space-time program. Although these descriptions are often rather complex, they may be used as a starting step leading to a more systematic and formal description, such as a dependence graph. So the following are natural questions to ask now:

110

1. What are the main considerations in providing a formal and powerful description (expression) of the algorithm?

2. What are the proper guidelines for VLSI algorithm design and revision?

3. What is a systematic method to transform an algorithm description to an array processor?

4. How is optimization of parallel algorithms achieved?

In this chapter, we introduce a variety of algorithm expressions and different transformations for converting one expression into another. The design criteria and design guidelines for achieving an optimal algorithm/array design are treated in Chapter 4.

3.2   Parallel Algorithm Expressions

Parallel algorithm expressions may be derived by two approaches:

1. Vectorization of sequential algorithm expressions.

2. Direct parallel algorithm expressions, such as snapshots, recursive equations, parallel codes, single assignment code, dependence graphs, and so on.

3.2.1   Vectorization of Sequential Algorithm Expressions

Conventionally, an algorithm is written as sequential code. High-level languages provide concise algorithm expression and have been used as machine independent programming tools. Two of the most common languages, FORTRAN and ALGOL, were originally developed in the 1950s and thus reflect the earlier computer structures, which perform sequences of operations on scalar data. Programming in these languages therefore requires the decomposition of an algorithm into a sequence of steps, each of which performs an operation on a scalar object. The resulting ordering of the calculation is often arbitrary.

For example, consider a mathematical expression of the matrix addition $\mathbf{C} = \mathbf{A} + \mathbf{B}$:

$$\mathbf{C}(i, j) = \mathbf{A}(i, j) + \mathbf{B}(i, j) \quad \text{for all } i \text{ and } j \qquad (3.1)$$

The corresponding FORTRAN code can be written as (where **A**, **B** and **C** are 4 × 4 matrices):

```
        DO  10  J=1,4
        DO  10  I=1,4
        C(I,J)=A(I,J)+B(I,J)
   10   CONTINUE
```

Here the elements of **A** and **B** are accessed in column major order, which is, by definition, the order in which they are stored. On many computers, if this order is reversed, the program will not execute as efficiently. In this example, because no ordering is required by the algorithm, it is unwise to encode an ordering in the program. Yet an ordering must be implied in a sequential programming language. *This ordering may prevent the algorithm from being executed efficiently in parallel.*

If no ordering is encoded, the compiler may choose the most efficient ordering for the target computer. Moreover, should the target computer contain parallelism, then some or all of the operations may be performed concurrently, without analysis or ambiguity. *Since ordering is unavoidable when using sequential code, parallel expression of an algorithm is very desirable.*

There exist abundant sequential codes for signal/image processing and scientific computing. Due to high software cost, users may not wish to rewrite them manually in parallel constructs. Therefore, a conventional approach to extracting concurrency is by using a vectorizing compiler, although its performance is rarely ideal. *A vectorizing compiler processes a source code written in a sequential language and, where possible, generates parallel machine instructions.* Sometimes, a more compact intermediate form, such as the directed graph representation of the source code for the Texas Instruments Advanced Scientific Computer (ASC NX), may also be compilable. The most likely place in which to find suitable sequences of operations for vectorization is within repetitive – i.e., recursive – calculations, such as DO loops in FORTRAN. Thus a vectorizing FORTRAN compiler will focus on DO loops. For example, the ASC NX compiler can analyze nests of three DO loops and can produce one machine instruction to execute the triple loop when the dependencies permit. The Burroughs Scientific Computer (BSP) vectorizer can analyze nested DO loops and reorder the loops if one of the innermost loops contains a dependency [Hockn83]. In fact, detecting and analyzing the dependencies between the statements within loops is the major task in vectorization [Hockn83].

For example, consider the FORTRAN code for a matrix-vector multiplication problem (cf., Eq. 2.1),

$$\mathbf{c} = \mathbf{A}\mathbf{b}$$

The ith element of $\mathbf{c}$ is

$$\mathbf{c}_i = \sum_{j=1}^{m} \mathbf{A}_{ij}\mathbf{b}_j$$

A FORTRAN code for matrix-vector multiplication can be written as

```
      DO  10  I=1,4
      C(I)=0.
      DO  10  J=1,4                                    (3.2)
      C(I)=C(I)+A(I,J)*B(J)
   10 CONTINUE
```

In this program, C(2) is calculated after C(1). Furthermore, it is not necessary to calculate A(I,1)*B(1) before A(I,2)*B(2) and vice versa. This kind of ordering is definitely unnecessary and is to be avoided for the purpose of concurrency. Exploiting these facts, a vectorizing compiler may translate this FORTRAN code into the following parallel code.

```
      IN  PARALLEL  FOR 1 < I < 4, 1 < J < 4 DO  BEGIN
         TEMP(I,J)=A(I,J)*B(J)
      END  IN  PARALLEL  DO

      IN  PARALLEL  FOR 1 < I < 4 DO  BEGIN
         C(I)=0
         DO  10  J=1,4
         C(I)=C(I)+TEMP(I,J)
      10 CONTINUE
      END  IN  PARALLEL  DO
```

3.2.2 Direct Expressions of Parallel Algorithms

Since a vectorizing compiler may not be sufficiently effective in extracting the inherent concurrent (parallel and pipeline) processing, it is advantageous that a user/designer use parallel expressions to describe an algorithm in the first place. *This is a key step leading to an algorithm-oriented array processor design.* Many different expressions may be used to represent a parallel algorithm, including snapshots, recursive algorithms with space-time indices, parallel codes, dependence graphs (DGs), or signal flow graphs (SFGs). A major factor in selecting an algorithm expression is that it should express algorithms clearly and concisely.

Single Assignment Code   Let us now introduce an instrumental notion, that of single assignment code. A **single assignment code** is *a form where every variable is assigned one value only during the execution of the algorithm.*

A single assignment code is in a sharp contrast to a conventional Fortran code, which is, in general, not written in a single assignment form. For example, consider the matrix-vector multiplication case (see Eq. 3.2). Note that in this program, C(I) is overwritten many times to save storage space. Thus the value of C(I) is assigned more than once. To transform the above program to a **single assignment** code, the number of indices of the vector C is increased; the FORTRAN program thus obtained is as follows (where A and C are 4 × 4 matrices and B is a 4 × 1 vector):

```
      DO  10  I=1,4
      C(I,1)=0.
      DO  10  J=1,4
      C(I,J+1)=C(I,J)+A(I,J)*B(J)
  10  CONTINUE
```

Each element of the vector C will be assigned one value only. Thus this program is indeed a single assignment code. If we use the indices I and J to span a 2-D index space, the index space is rectangular with 16 (i.e., 4 × 4) index points. At each index point, three variables A, B, and C are defined with no ambiguity, since the program is a single assignment code. (Assume the value of B at each index point is independent of the I-index.) Later, we show that single assignment code is critical to the derivation of DGs.

3.2.2.1   Recursive Algorithms

A convenient and concise expression for the representation of many algorithms is to use recursive equations. The derivation of recursive equations is often straightforward.[1] The recursive equation for the matrix-vector mul-

---

[1] Some more detailed derivations are discussed in Problem 1.

tiplication $\mathbf{c} = \mathbf{A}\mathbf{b}$ is:

$$\mathbf{c}_i^{(j+1)} = \mathbf{c}_i^{(j)} + \mathbf{a}_i^{(j)}\mathbf{b}_i^{(j)}$$

where j is the recursion index, $j = 1, 2, \ldots, N$, and

$$\mathbf{c}_i^{(1)} = 0$$

$$\mathbf{a}_i^{(j)} = \mathbf{A}(i, j)$$

$$\mathbf{b}_i^{(j)} = \mathbf{B}(j)$$

A recursive equation with space-time indices uses one index for time and the other indices for space. By doing so, the activities of a parallel algorithm can be adequately expressed. The preceding equation can be viewed as a recursive equation with the j-index as the time index and i-index as the space index. These roles could just as well be reversed. From a mathematical point of view, the superscripts and subscripts should not make any difference. In other words, i and j may be treated equally. Therefore, in the following discussion, we shall adopt an *index space*, which nondiscriminatingly embraces both the time and the space indices.

We note that *a recursive algorithm is inherently given in a single assignment formulation.*

Snapshots   A *snapshot* is a description of the activities at a particular time instant. Snapshots are perhaps the most natural tool an algorithm designer can adopt to check or verify a new array algorithm. Sample snapshots for a systolic matrix-vector multiplication algorithm are depicted in Figure 3.1.

3.2.2.2   Dependence Graph (DG)

To achieve the maximal parallelism in an algorithm, we must carefully study the data dependencies in the computations. In the special case when the operations of a sequential algorithm have no data dependencies between each other, they can be executed at the same time in a parallel computer. For example, it is apparent that all the additions shown in Eq. 3.1 have no mutual dependency, therefore, they can be executed simultaneously. However, in general, there is always a certain degree of dependency which dictates the sequence of computation.

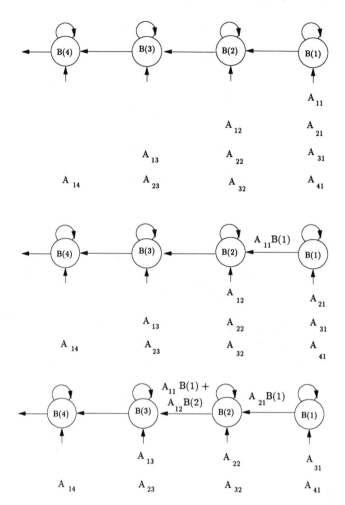

Figure 3.1: Snapshots for a systolic matrix-vector multiplication algorithm.

Graph Terminologies A *graph* $G = [N, A]$ is a set N whose elements are called *nodes* and a set A whose elements are called *arcs, or edges.* Each arc, $a \in A$, connects a pair of nodes $i, j \in N$ and is written $a = i \xrightarrow{a} j$. Here i is the *initial endpoint* of a and j is the *terminal endpoint* of a. An arc whose endpoints are the same node is called a *loop.* A *chain* is a sequence of arcs, $L = \{a_1, \ldots, a_q\}$, such that arc a_r $(2 \le r \le q - 1)$ has one endpoint common with arc a_{r-1} $(a_r \ne a_{r-1})$ and its second endpoint common with arc a_{r+1} $(a_r \ne a_{r+1})$, without regard to the direction of the arcs. The "free" endpoints of the first and last arcs of a chain are the endpoints of the chain. A *path* is a chain, all of whose arcs are directed the same way. Note that a path has well-defined initial and terminal endpoints. If the endpoints of a path are the same node, then it is a *cycle*. If the endpoints of a chain are the same node then it is an *undirected cycle*. An *elementary* chain, path, undirected cycle, or cycle is one in which the same node is not encountered twice (with the exception of the endpoints). A graph is *connected* if there exists a chain between every pair of nodes. It is *strongly connected* if there exists a path from each node to all other nodes.

Dependence Graph   A *dependence graph is a graph that shows the dependence of the computations that occur in an algorithm. A DG can be considered as the graphical representation of a single assignment algorithm.*

In the previously-mentioned single assignment algorithm, $C(I, J + 1)$ is said to be *directly dependent upon* $C(I, J)$, $A(I, J)$, and $B(J)$. By viewing each dependence relation as an arc between the corresponding variables located in the index space, a DG as shown in Figure 3.2(a), will be obtained.

Note that only the dependencies between nodes are shown in Figure 3.2. *The operations inside each node are deliberately ignored* in the DG, since they will be assigned to the same processing element when the DG is used to map an algorithm to an array processor. However, it is straightforward to extend the DG concept to include the operations inside each node. This DG, which we call a *complete DG*, specifies all the dependencies between all variables in the index space. An algorithm is *computable* if and only if its complete

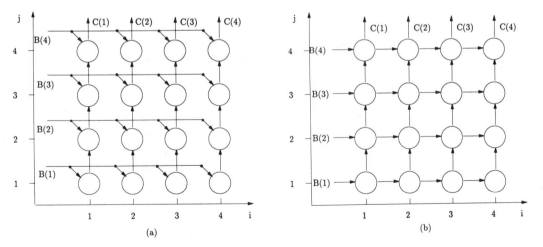

Figure 3.2: DG for matrix-vector multiplication (a) with global communication and (b) with only local communication.

DG contains no loops or cycles. Therefore, the complete DG is very useful for studying issues related to computability. However, in this book we rarely need to display the complete DG explicitly.

Localized Dependence Graph   In Figure 3.2(a), the value B(J) of each element of vector **b** should be "broadcast" to all the index points having the same J-index. This kind of data is termed *broadcast data*. In general, this means that global communication is involved in array processor design. In many cases, such broadcasting can be avoided and replaced by local communication.

An algorithm is localized if all variables are (directly) dependent upon the variables of neighboring nodes only. As an example, a localized DG is shown in Figure 3.2(b), where the B(J) is "propagated" step by step, without being modified, to all the nodes with the same J-index. This kind of data, which is propagated without being modified, is called *transmittent data*. Otherwise, it is called *nontransmittent data*. The corresponding (localized) FORTRAN is shown next (B(1, J) = b(J), J = 1, 2, 3, 4):

```
      DO  10  I=1,4
      C(I,1)=0.
      DO  10  J=1,4
      B(I+1,J)=B(I,J)
      C(I,J+1)=C(I,J)+A(I,J)*B(I,J)
  10  CONTINUE
```

In this single assignment algorithm, B(I+1,J) is directly dependent upon B(I,J), and C(I,J+1) is directly dependent upon C(I,J), A(I,J), and B(I,J).

The idea of local and regular DGs was explored by Karp, Miller, and Winograd [Karp67] in their "systems of uniform recurrence equations". They first use an *index space* display to show the complete dependency of locally recursive algorithms. Along this line, some instrumental notions such as reduced DG and separating "equitemporal" hyperplanes, and degree of parallelism are introduced. Furthermore, a graph theoretical approach and integer programming techniques for analyzing locally recursive algorithms were also proposed [Karp67], [Rao85]. Another intuitive and popular approach to the construction of a DG is based on *snapshots*: If successive snapshots are put together along one direction, we may view each data position as an index point for the DG. By tracing the data movement in the successive snapshots, we can determine the direction of data flow and thus the data dependencies, which will eventually lead to the derivation of the DG.

Having introduced the idea of a DG, we now give a definition of a locally recursive algorithm. A *locally recursive algorithm* is an algorithm whose corresponding DG has only *local dependencies*, i.e., the length of each dependency arc is independent of the problem size, and *most nodes of the DG consists of the same kind of operations* [2] [Kung87].

Since the data dependencies are explicitly expressed in the dependence graph, a systematic approach to derive an array processor implementation by using such regular DGs is possible [Moldo83], [Miran84], [Quint84], [Cappe84], [Rao85], [Chen85b], [Chen85c], [Moldo86], [Kung86a], [Kung87]. This subject will be discussed in a moment.

3.3   Canonical Mapping Methodology

We have so far introduced several algorithm expressions. We shall now consider the issue of transforming such expressions to an array processor design. Figure 3.3 sketches, from top to bottom, the three key design levels involved in the design process. This is basis for a CAD tool which provides a direct graphic interface between different levels.

Graph-Based Design Methodology

- Stage 1: DG Design

  For a given problem, the user first identifies a suitable algorithm. Then the designer identifies a suitable algorithm expression and generates a DG for it. Since the structure of a DG greatly affects the final array design, further modifications on the DG are often desirable in order to achieve a better design.

- Stage 2: SFG Design

  Based on different mappings of the DG onto array structure, a number of SFGs can be derived from the DG.

---

[2]Sometimes, the operations performed in some boundary nodes of the DG may be different from those performed in the interior nodes, such as in LU decomposition (see Section 3.3.4.3.)

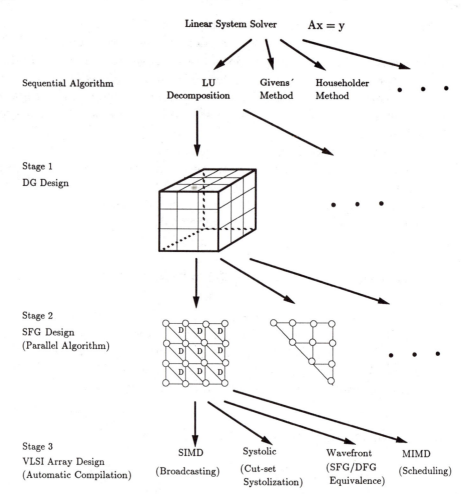

Figure 3.3: Different levels in designing systolic algorithms and arrays.

- **Stage 3: Array Processor Design**

  The SFG obtained in Stage 2 can physically realized in terms of an SIMD, systolic, wavefront, or MIMD array. The methodology for this stage is treated in greater detail in Chapters 4 and 5.

  In general, the choice of a particular DG for an algorithm, the direction of the projection as well as the schedule can greatly affect the performance of the resulting array. This motivates the investigation into the classifications of DGs and the possible assignment and schedule schemes. This chapter introduces (1) a *canonical mapping* methodology for mapping homogeneous

DGs onto processor arrays; and (2) a *generalized mapping* methodology for mapping heterogeneous DGs onto processor arrays. They are outlined in Figure 3.4.

In this section, we concentrate only on the canonical mapping methodology. There are a large number of algorithms that may be expressed in terms of a very regular and localized DG. For example, matrix multiplication, convolution, autoregressive filtering, DFT, discrete Hadamard transform, Hough transform, least squares solution, sorting, perspective transform, median filtering, LU decomposition, and QR decomposition all belong to this important class. By exploiting this regularity, the array processor design for such algorithms may be greatly simplified. Briefly speaking, in the canonical mapping methodology, the regularity of the DGs is exploited and linear projection and scheduling schemes are adopted, so that simple and regular array structures may be achieved.

3.3.1   Stage 1 Design: Mapping Algorithm to DG

Note that although many methods have been proposed to construct a DG from sequential code, a formal and automatic methodology remains a major open research problem. Briefly speaking, an extraction of the partial ordering of operations in an algorithm will yield a DG expression. For example, the derivation of a DG from a certain class of simple algorithms, such as a sorting algorithm, is rather straightforward. However, to derive a DG from a general recursive algorithm, some guidelines are proved useful. They will

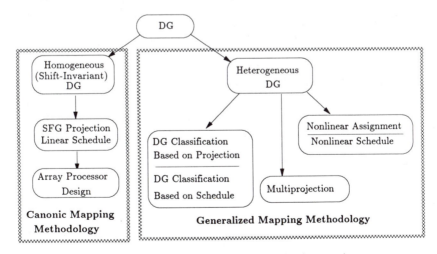

Figure 3.4: Canonical and generalized mapping.

be illustrated momentarily by means of the convolution and AR filtering examples.

Our approach to the construction of a DG will be based on the *space-time indices in the recursive algorithm*: Corresponding to the space-time index space in the recursive algorithm, there is a natural lattice space (with the same indices) for the DG, with one node residing on each grid point. Then the data dependencies in the recursive algorithm may be explicitly expressed by the arcs connecting the interacting nodes in the DG, while its functional description will be embedded in the nodes. One earlier example of such lattice model is the DG for the matrix-vector multiplication, as shown in Figure 3.2.

3.3.1.1   Shift-Invariance (Homogeneity) of DG

A DG is *shift-invariant* if the dependence arcs corresponding to *all* nodes in the index space are independent of their positions. Formally, this means that for index vectors $\mathbf{i}_1$ $\mathbf{i}_2$, and $\mathbf{j}$, if a variable at $\mathbf{i}_1$ depends on a variable at $\mathbf{i}_1 - \mathbf{j}$, then a variable at $\mathbf{i}_2$ will depend on a variable at $\mathbf{i}_2 - \mathbf{j}$. Note that the node functions can be different and I/O to the border nodes are exempted from the shift-invariant requirement. Shift-invariance of a DG is a very basic assumption for the canonical mapping methodology.

Example 1: Shift-Invariant DG for Sorting Algorithm

Sorting is one of the most frequently used procedures in data processing and scientific computation. The problem of sorting can be stated as follows: Given a sequence $\{x(i)\}$, the problem is to obtain a new sequence $\{m(i)\}$, consisting of the elements of $\{x(i)\}$, rearranged into a sorted order. For example, the sequence $m(i)$ can be considered to be sorted in a decreasing order if $m(i) \geq m(k)$ when $i < k$.

Sorting algorithms are easily formulated into a locally recursive form. The single assignment program of a selection sorting algorithm can be written as:

For i **from** 1 **to** N

For j **from** 1 **to** i

$$m(i+1, j) \longleftarrow max\,[x(i, j), m(i, j)]$$

$$x(i, j+1) \longleftarrow min\,[x(i, j), m(i, j)]$$

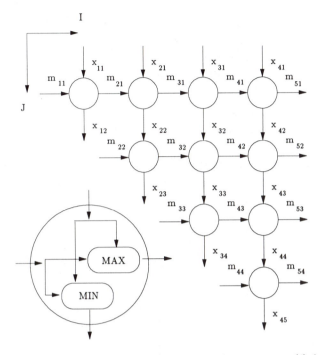

Figure 3.5: DG for sorting; The original sequence $x(i)$ is input from the upper row as $x(i, 1)$. The data $m(i, i)$ along the diagonal nodes is set to be $-\infty$. The sorted sequence $m(j)$ is output from the rightmost column as $\{m(N,j), j = 1, 2,\ldots, N\}$.

Here the $x(i, 1)$ is initialized to the original unsorted sequence $x(i)$, and $m(i, i)$, is set to $-\infty$, and the sorted output sequence $m(j)$ is equal to $\{m(N, j), j = 1, 2,\ldots, N\}$ which is in decreasing order.

The DG corresponding to the single assignment program is shown in Figure 3.5 [Rao85]. By inspection, the DG is clearly shift-invariant since the dependency structures do not vary from node to node.

3.3.1.2 Localization of DG: Broadcast vs. Transmittent Data

In canonical mapping, the DGs treated have the feature of being locally dependent. The recursive algorithms, as popularly adopted, are usually not directly given in a localized form. A nonlocalized recursive algorithm, when mapped onto an array processor, is likely to result in an array with global interconnections. Although, in certain instances, such global arcs can be avoided by using a proper projection direction in the mapping schemes. To guarantee a locally interconnected array, a localized recursive algorithm and, equivalently, a localized DG would be derived.

Currently, some methods to localize the recursive algorithm do exist. Examples may be found in [Quint84], [Chen85b], [Chen85c], and [Li85]. Although the basic concepts are mostly understood, the problem of deriving a locally recursive algorithm from a nonlocally recursive algorithm is not yet completely solved. In fact, most existing approaches to deriving locally recursive algorithms are rather heuristic. In the following, a more formal (but admittedly still somewhat heuristic) strategy is proposed.

Broadcast Data: If a variable is to be broadcast, then there exists, a corresponding set of index points on which the same data value repeatedly appears. This set is termed a *broadcast contour*. The data are called *broadcast data*. A DG may contain a number of such broadcast contours.

Transmittent Data: The key point is that instead of broadcasting the (public) data along a global arc, the same data may be *propagated* via local arcs and thus become *transmittent data*. By replacing the broadcast contours by local arcs, a global DG, can be converted to a localized version. This is demonstrated by the following example.

Example 2: DG for Convolution Algorithm

The problem of convolution is defined as follows: Given two sequences $u(j)$ and $w(j)$, $j = 0, 1,\ldots, N - 1$, the convolution of the two sequences is

$$y(j) \; = \; \sum_{k=0}^{j} u(k)w(j - k)$$

or

$$y_j \; = \; \sum_{k=0}^{j} u_k w_{j-k}$$

where $j = 0,1,\ldots,2N-2$

The derivation of a DG for convolution is similar to the matrix vector multiplication case. The first step of deriving the recursive equation is to introduce a recursive variable y_j^k. Then the convolution equation can be rewritten in terms of a recursive form:

$$y_j^k \; = \; y_j^{k-1} \; + \; u_k \cdot w_{j-k} \tag{3.3}$$

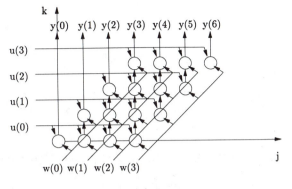

(a)

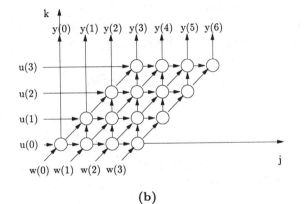

(b)

Figure 3.6: DG for convolution: (a) global; (b) localized.

for $j = 0, 1, \ldots, N-1$, and $k = 0, 1, \ldots, j$ and for $j = N, N+1, \ldots, 2N-2$, and $k = j - N + 1, j\ N + 2, \ldots, N - 1$

Note that Eq. 3.3 is already in a single assignment form; therefore, the DG can be readily sketched, as shown in Figure 3.6(a). Equation 3.3 is an expression with global data dependencies and it is therefore not a locally recursive algorithm.

By replacing the broadcast contours by local arcs, the (global) DG, as shown in Figure 3.6(a), can be easily converted to a localized version as shown in Figure 3.6(b), which obviously performs the same algorithm. The localized DG has a corresponding locally recursive algorithm, as follows:

$$ y_j^k = y_j^{k-1} + u_j^k \cdot w_j^k \, , \ \ y_j^0 = 0 $$

$$u_j^k \;=\; u_{j-1}^k \;, \;\; u_0^k \;=\; u_k$$

$$w_j^k \;=\; w_{j-1}^{k-1} \;, \;\; w_{j-k}^0 \;=\; w_{j-k}$$

for $j = 0, 1,\ldots, N-1$, and $k = 0, 1,\ldots, j$ and for $j = N, N+1, \ldots, 2N-2$, and $k = j - N + 1, j \; N + 2,\ldots, N-1$

3.3.1.3   Reversible Arcs for Associative Operations

Note that if the operation used in the recursion is associative, then the directions of the arcs may be reversible. This can be illustrated by a simple example. Since the operation used in the preceding recursion is addition (which is associative), the following two equations will give the same result if suitable initial conditions are given:

$$y_j^k \;=\; y_j^{k-1} \;+\; u_k \cdot w_{j-k}$$

$$y_j^k \;=\; y_j^{k+1} \;+\; u_k \cdot w_{j-k}$$

This means that the arcs in the DG are reversible. The rule will be also found useful in localizing an AR filtering DG as discussed below.

3.3.1.4   Localization with Intermediate Variables Involved

There exists a subtle difference among the different broadcast-data variables encountered in a recursive equation. A variable in a set of recursive equations is said to be an *intermediate variable* if it appears on both the right-hand side and the left-hand side of the recursive equations (naturally with different subscripts or superscripts). Otherwise, the variable is termed an *external variable*. The data of an intermediate variables is *produced* by a node in the DG and will be *utilized* by some other node in the DG. Obviously, the distance between the producing node and the utilizing node has some effect on the communication required. Therefore, it affects the localization scheme used. This is illustrated in the following AR filtering example.

Example 3: DG for AR Filtering Algorithm

To illustrate the general procedure involved in the localization of a recursive algorithm, we present a more sophisticated example of AR filtering.

An AR filtering algorithm is described by the following difference equation:
For $j = 1, 2, \ldots$,

$$y(j) = \sum_{k=1}^{N} a_k y(j - k) + u(j)$$

Note that the output of an AR filter is an indefinite-length sequence even if the input sequence $u(j)$ has a finite duration. To derive a DG for the AR algorithm, one might follow the formal procedure given earlier. To simplify the illustration, here we take advantage of the derivation used for the convolution algorithm and propose in a similar fashion a preliminary DG for the AR algorithm as shown in Figure 3.7(a). Here we assume that $N = 4$. Note that the key differences are: (1) The data $y(c)$, where $c = 0, 1, 2, \ldots$, are assumed available without showing where do they come from; (2) Note that it does not indicate the direction of data flow along the vertical lines because both directions are viable (cf., Figure 3.7(b)), according to the associative operation principle. Consequently, the input data $u(i)$ may be input either from the top or the bottom of the vertical lines (not shown in the figure).

Recall that Figure 3.7(a) does not indicate exactly how $y(c)$ is produced and utilized. To see how this may be produced, let us verify for example the operations in the fifth column ($j = 4$),

$$y(4) = u(4) + a_4 y(0) + a_3 y(1) + a_2 y(2) + a_1 y(3) \qquad (3.4)$$

Intermediate and External Variables We observe that in the DG, there are two types of broadcast contours. One is comprised of index points of a constant k-index, transmitting the *broadcast-data* a_k. The other comprised of the index points defined by $j - k =$ constant, transmitting the *broadcast-data* $y(j-k)$. Recall that a variable in a set of recursive equations is said to be an *intermediate variable* if it appears on both the right-hand side and the left-hand side of these recursive equations. Otherwise, the variable is termed an *external variable*. Thus, in the AR filtering algorithm, $y(j-k)$ is considered an intermediate variable while a_k is an external variable.

Just like the procedure for localization of the convolution DG discussed previously, the broadcast-data a_k, which is an external variable, can be localized simply by replacing a global arc by many local arcs. However, the broadcast-data $y(j - k)$, which is an intermediate variable, requires more thought to achieve a best localized DG.

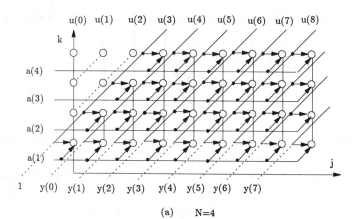

(a)   N=4

Associative
Summation

Broadcast
Contours

(b)

(c)

Figure 3.7: (a) A preliminary (global) DG for AR filtering; (b) a detailed node;
(c) a spiral DG; (d) a localized DG.

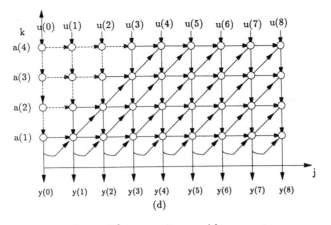

Figure 3.7: cont'd

Recall that in the preliminary DG we have not decided where the data $y(c)$ should be produced. It in fact depends on where the produced data will be utilized. Note that the data $y(c)$ may be injected at any of the nodes on the broadcast contour $j - k = c$. However, it can be shown that the bottom node$(c+1, 1)$ is the most suitable place to inject the data. [3] For example, if $c = 4$, then the data $y(4)$ will be injected at node$(5,1)$. Referring to Eq. 3.4, the data y(4) can be produced either at the top node$(4,4)$ or the bottom node$(4,1)$ of the vertical line ($j = 4$). The latter is of course closer to the node$(5,1)$ where the produced data will be utilized.

In general, there are two possible positions where the intermediate variable $y(c)$ may be produced, which correspond to two different designs proposed below:

- *Spiral Communication Approach:*

  If the data y(c) is produced at the top nodes of the vertical lines, then the recursive algorithm for AR filtering is

$$y_j^k = y_j^{k-1} + a_k \cdot y_{j-k}^N \qquad (3.5)$$

$$\text{Input}: \quad y_j^0 = u(j); \quad \text{output}: \quad y(j) = y_j^N$$

  To verify Eq. 3.5, let us again display the node activities (from bottom up) along the vertical line for $j = 4$,

---

[3]In order to obtain a DG which has some permissible linear schedules (to be discussed in a moment), it turns out that node$(c + 1, 1)$ is the only choice.

$$
\begin{aligned}
y_4^1 &= y_4^0 + a_1 y_3^4 \\
y_4^2 &= y_4^1 + a_2 y_2^4 = y_4^0 + a_1 y_3^4 + a_2 y_2^4 \\
y_4^3 &= y_4^2 + a_3 y_1^4 = y_4^0 + a_1 y_3^4 + a_2 y_2^4 + a_3 y_1^4 \\
y_4^4 &= y_4^3 + a_4 y_0^4 = y_4^0 + a_1 y_3^4 + a_2 y_2^4 + a_3 y_1^4 + a_4 y_0^4
\end{aligned}
$$

The produced result at the top node is

$$
y(4) = y_4^4 = u(4) + a_1 y(3) + a_2 y(2) + a_3 y(1) + a_4 y(0)
$$

This shows that in general $y(c)$ is produced at node(c, N) $(y_c^N = y_c^{N-1} + a_N y_{c-N}^N)$. (since $y(c)$ depends on y_c^N). A spiral communication arc will be required to feed the results of node(c, N) to node$(c+1, 1)$. This leads to a DG with *spiral communication* (see Figure 3.7(c)).

- *Local Communication Approach:*

Note that the operation used in the recursion Eq. 3.5 is addition and is thus *associative* (cf., Figure 3.7(b)). Therefore, an alternative recursive equation (via a reversed direction) is also possible. This is corresponding to the case when the data y(c) is produced at the bottom nodes of the vertical lines. Correspondingly, the recursive algorithm now becomes

$$
y_j^k = y_j^{k+1} + a_k \cdot y_{j-k}^1 \tag{3.6}
$$

$$
\text{Input}: \quad y_j^{N+1} = u(j); \quad \text{output}: \quad y(j) = y_j^1
$$

Thus $y(c)$ is produced at node$(c, 1)$ and only a *local communication* arc is required to send the data to node$(c+1, 1)$. The corresponding DG is shown in Figure 3.7(d), which is completely localized. Obviously, this localized DG is more preferable than the spiral version. The locally recursive algorithm for this DG can now be obtained:

$$
y_j^k = y_j^{k+1} + a_j^k \cdot B_j^k
$$

$$
a_j^k = a_{j-1}^k, \quad a_{-1}^k = a(k)
$$

$$B_j^k \;=\; B_{j-1}^{k-1} \,, \quad B_j^0 \;=\; y_j^1$$

$$y_j^{N+1} \;=\; u(j) \,, \text{ and } y(j) \;=\; y_j^1$$

where $j = 0, 1, 2, \ldots, \infty$, $k = N, N-1, N-2, \ldots, 1$.

Note that, because the variable y_j^k has already been used in the $j = c$ contour to denote the data involved in the iterative summation process, a new variable B_j^k is introduced to propagate the data $y(c)$ along the broadcast contour $j - k = c$.

3.3.1.5   Index transformation and reindexing of DG

For the sake of reshaping the DG it is always possible to apply a coordinate transformation to the index space. One such example is illustrated in Problem 6. However, such a transformation does not lead to any new SFG in Stage 2. In other words, Stage 2 design basically covers index transformation as a special situation. Therefore, index transformation is not emphasized in this mapping methodology.

A more useful technique for modifying the DG is by the so-called *reindexing* scheme. Examples of reindexing are plane-by-plane shifting or circular shifting in the index space. For instance, when there is no permissible linear schedule (see next section) or systolic schedule for the original DG, it is often desirable to modify the DG so that such a desired schedule may be obtained. Very often, such methods will need to make use of the notion of transmittent data so that the DG may be localized. That is also why it is useful to properly label the transmittent data to allow more flexibility in the future design process. An example of a reindexing scheme will be provided in Chapter 4 when we discuss systolic designs for the shortest-path problem.

3.3.2   Stage 2 Design: Mapping DG to SFG

To determine a valid array structure for a locally recursive algorithm, one straightforward design method is to designate one processing element (PE) for each node in a DG. This however in general leads to very inefficient utilization of the PEs, since each PE can be active only for a small fraction of the computation time. In order to improve PE utilization, it is often desirable to map the nodes of the DG onto a fewer number of PEs. To achieve this it is useful to map the DG first to an intermediate expression, i.e., in a SFG form.

3.3.2.1 Signal Flow Graph (SFG)

The SFG offers a powerful abstraction and graphical representation for problems in scientific and signal processing computations. The SFG expression, which consists of processing *nodes*, communicating *edges*, and *delays*, is shown in Figure 3.8. In general, a *node* is often denoted by a circle representing an arithmetic or logic function performed with *zero delay*, such as multiply or, add. An *edge*, on the other hand, denotes either a dependence relation or a delay. Unless otherwise specified, the following conventions are adopted for convenience. When an edge is labeled with a capital letter **D** (or **D'**, **2D**,···), it represents a time delay operator with delay time **D** (or **D'**, **2D**,···). The SFG representation derives its power from the fact that the computations are assumed to be delay-free, i.e., they take no time at all. Consequently, the burden of tracing the detailed time-space activities associated with pipelining are eliminated. Moreover, any delay in the system has to be explicitly introduced in the form of so-called delay branches. These delay branches allow history-sensitive systems to be described in a clear and unambiguous way.

A complete SFG description should include both functional and structural description parts. The functional description defines the behavior within a node, whereas the structural description specifies the interconnection (edges and delays) between the nodes. Theoretically, the structural part of an SFG can be represented by a finite directed graph, $G = <V, E, D(E)>$. The vertices V model the nodes. The directed edges E model the interconnections between the nodes. Each edge e of E connects an output port of

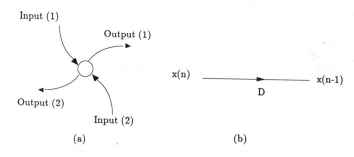

Figure 3.8: SFG notations: (a) an operation node; (b) an edge as a delay operator.

a node to an input port of some node and is weighted with a delay count $D(e)$. The delay count is the number of delays along the connection. Often, input and output ports are referred to as *sources* and *sinks*, respectively.

As compared with the DG, the SFG has the following properties:

1. The SFG can be viewed as a simplified graph. That means the SFG is a more concise representation than the DG.

2. The SFG is more specific, i.e., it is closer to hardware level design. Therefore, the SFG also dictates the type of arrays that will be obtained.

3. While there are no loops in any DG, the SFG can have loops, as long as there is at least one delay **D** on each loop.

Example 4: SFG for Convolution

An SFG for convolution is shown in Figure 3.9. Note that the existence of the delay-free edges means that the data will be propagated in "zero time". The computation in each node is also assumed to be done in "zero time".

3.3.2.2 Processor Assignment and Scheduling

There are two basic considerations for mapping from a DG to an SFG :

1. To which processors should operations be assigned? (A criterion for example might be to minimize communication/exchange of data between processors.)

2. In what ordering should the operations be assigned to a processor? (A criterion might be to minimize total computing time.)

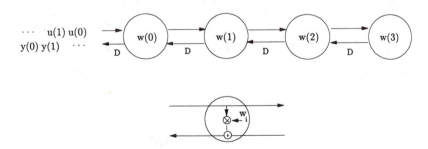

Figure 3.9: An SFG for convolution.

Therefore, two steps are involved in mapping a DG to an SFG array. The first step is the *processor assignment*. Once the processor assignment is fixed, the second step is the *scheduling*. The allowable processor and schedule assignments can be very general; however, in order to derive a regular systolic array, the regularity of the DG can be used to advantage. Therefore, regular assignments and scheduling attract more attention. It is common to use a *linear projection* for processor assignment, in which nodes of the DG in a certain straight line are projected (assigned) to a PE in the processor array, (see Figure 3.10(a)), and a *linear scheduling* for schedule assignment, in which nodes in a parallel hyperplane in the DG are scheduled to be processed at the same time step (see Figure 3.10(b)).

Processor Assignment: As a simple example, a projection method may be applied, in which nodes of the DG along a straight line are assigned to a common PE. Since the DG of a locally recursive algorithm is very regular, the projection maps the DG onto a lower dimensional lattice of points, known as the *processor space*. Mathematically, a linear projection is often represented by a *projection vector* $\vec{d}$. The results of this projection are is represented by an SFG. As an example, with reference to Figure 3.10, the 2-D index space of matrix-vector multiplication may be decomposed into a direct sum of a 1-D *processor space* and 1-D *delay space*. The delay space is related to the scheduling as explained below.

Scheduling: The projection should be accompanied by a *scheduling* scheme, which specifies the sequence of the operations in *all* the PEs. A schedule function represents a mapping from the N-dimensional index space of the DG onto a 1-D schedule (time) space. A linear schedule is based on a set of parallel and uniformly spaced hyperplanes in the DG. These

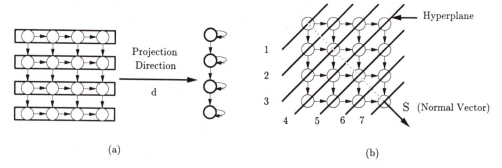

(a)   (b)

Figure 3.10: Illustration of (a) a linear projection with projection vector $\vec{d}$; (b) a linear schedule $\vec{s}$ and its hyperplanes.

hyperplanes are called *equitemporal hyperplanes*, all the nodes on the same hyperplane must be processed at the same time. Mathematically, *the schedule can be represented by a (column) schedule vector $\vec{s}$, pointing to the normal direction of the hyperplanes.* For any index point **i** in the DG, its time step is $\vec{s}^{\mathrm{T}}\mathbf{i}$. A set of (linear schedule) hyperplanes and the schedule vector are illustrated in Figure 3.10(b).

There is a *partial ordering* among the computations, inherent in the algorithm, as specified by the DG. More specifically, if there is a directed path from node x to node y, then the computation represented by node y must be executed after the computation represented by node x is completed. The feasibility of a schedule is determined by the *partial ordering* and the *processor assignment* scheme. The necessary and sufficient conditions are stated below.

Permissible Linear Schedules   Given a DG and a projection direction $\vec{d}$, we note that not all hyperplanes qualify to define a valid schedule for the DG. Some of them violate the precedence relation of computation specified by the dependence arcs. The allowable directions of the hyperplanes actually define the class of *permissible linear schedules*. In order for the given hyperplanes to represent a permissible linear schedule, it is necessary and sufficient that the normal vector $\vec{s}$ satisfies the following two conditions:

$$(1) \quad \vec{s}^{\mathrm{T}}\vec{e} \geq 0, \quad \text{for any dependence arc } \vec{e}. \tag{3.7}$$

$$(2) \quad \vec{s}^{\mathrm{T}}\vec{d} > 0. \tag{3.8}$$

Both the conditions Eq. 3.7 and Eq. 3.8 can be checked by inspection. In short, *the schedule is permissible if and only if (1) all the dependency arcs flow in the same direction across the hyperplanes; and (2) the hyperplanes are not parallel with projection vector $\vec{d}$.* The first condition means that a causality should be enforced in a permissible schedule. Namely, if node **p** depends on node **q**, then the time step assigned for **p** can *not* be less than the time step assigned for **q**. The second condition implies that nodes on an equitemporal hyperplane should *not* be projected to the same PE. The permissible hyperplane directions defined by the first condition is the same as the notion of the time cone in [Karp67] and [Delos86].

Types of Schedules   Given a DG and a *projection vector*, $\vec{d}$, the most likely used schedules for the SFG projection are the following:

1. *Default schedule* The corresponding hyperplanes are orthogonal to the projection direction $\vec{d}$; or the normal direction of hyperplanes, $\vec{s}$, is parallel to the projection direction, $\vec{d}$.

2. *Recursion schedule* The schedule vector $\vec{s}$ is parallel to one of the axes in the index space of the DG. Usually, the one corresponding to the recursion numbering is used.

3. *Systolic schedule* This is the schedule for a systolic array. It means that there is at least one delay on each edge of the resulting SFG. This will be treated further in Chapter 4.

4. *Optimized schedule* The schedule is determined by the data dependencies, processor availability and the design criteria. The use of computer-aided-design tools may be required.

In this chapter, only the first two schedules are discussed. The systolic schedule and the optimized schedule are addressed in Chapter 4.

3.3.2.3   TI SFG and STI SFG

The most interesting type of SFG is the one that is both structurally and functionally time-invariant. For convenience, this chapter will treat two different classes of SFGs. The first is time-invariant (TI) SFGs, *whose node functions and edge structures remain invariant all the time.* The second class is structurally time-invariant (STI) SFGs, *whose edge structures remain invariant.*[4] The notion of SFG can be easily extended to cover a much broader domain, including linear and nonlinear, time-varying and time-invariant, and multidimensional systems. However, the generalized notion also implies more complicated physical implementation. For example, the introduction of time-varying functions in the STI SFGs will complicate the control hardware of the corresponding array design.

Conditions for STI SFG Projection   The following statement guarantees that if a DG is shift-invariant then there exists a corresponding STI SFG by projecting the DG along any direction.

---

[4]Note that in the notion of TI and STI we focus the attention on functional and structural descriptions of the SFG. The I/O description is purposefully excluded from the invariance conditions for TI or STI.

A DG is shift invariant if and only if, for any vector $\vec{v}$, the DG may be projected onto an STI SFG with a permissible linear schedule, along the projection direction $\vec{d} = \vec{v}$ or $\vec{d} = -\vec{v}$.

Proof *Sufficiency:* If the DG, of dimension n, is not shift invariant, then there must exist at least one "particular" node, which has different dependency arcs from the other nodes. Then we may find a specific projection direction such that the projected SFG is not STI.

Necessity: Since the DG is shift invariant, the projected SFG for any projection direction must be STI. The only thing left to be proved is the existence of a permissible linear schedule. (1) Since the graph is acyclic (i.e., no loops or cycles), the DG must be a biparti graph – i.e., there exists at least one surface to cut the DG into two subgraphs, and the arcs are all from one subgraph to the other subgraph. Moreover, these surfaces may be chosen to be planes using the shift invariance property. (2) Since the graph is shift-invariant, there should be a sequence of such planes, parallel with each other. Thus the nodes between two neighboring planes can be regarded as an equi-temporal hyperplane. Based on (1) and (2), *a linear schedule, which satisfies Eq. 3.7, can thus be defined in this DG.* Since there is only finite number of arcs for each node, a new schedule vector, which also satisfies Eq. 3.7, can be created by perturbing the direction of the schedule vector with an infinitesimal angle. Therefore there exist at least two linear schedules for a shift invariant DG. For any vector $\vec{v}$ at least one of these two linear schedules will satisfy Eq. 3.8.

Remark: Now the question is that, among the many possible projection directions, how to select the optimal one. The answer is rather involved and remains an open problem despite of the attempts made by many researchers. More on this subject will be treated in Chapter 4.

3.3.2.4 SFG Projection Procedure

In general, the projection procedure involves the following steps:

1. For any projection direction, a processor space is orthogonal to the projection direction. A processor array may be obtained by projecting the index points to the processor space. A default schedule is fixed for the projection direction. If the default schedule is not a permissible one, i.e., it violates the conditions in Eq. 3.7 and Eq. 3.8, then a recursion schedule may be used. We can thus choose either a default schedule or a recursion schedule.

2. Replace the arcs in the DG with zero or nonzero delay edges between their corresponding processors. The number of delays on each edge is determined by the timing and is equal to the number of time steps needed for the corresponding arcs.

3. Since each node has been projected to a PE and each input (or output) data is connected to some nodes, it is now possible to attach the input and output data to their corresponding processors.

Example 5: Matrix-Vector Multiplication

Consider the multiplication of a matrix $\mathbf{A}$ of size 4×4 and a vector $\mathbf{b}$ of size 4×1, i.e., $\mathbf{c} = \mathbf{Ab}$. The DG for this problem is shown in Figure 3.2(b). By using the projection along the $[0\ 1]$ direction (and the default schedule), we have a 1-D SFG, as shown in Figure 3.11.

Referring to the DG in Figure 3.2(b), we first find a linear schedule function $\vec{s} = [1\ 1]^{\mathrm{T}}$. This schedule, as shown in Figure 3.12, is a *systolic schedule*.

Next, we find a projection vector $\vec{d} = [1\ 0]^{\mathrm{T}}$. An SFG topology, as shown in Figure 3.13(a), is obtained by this projection. If we map the systolic schedule to the SFG, the timing is obtained, and the corresponding systolic array is displayed in Figure 3.13(b). The pipelining period α for this systolic array is one.

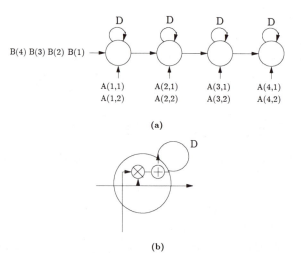

(a)

(b)

Figure 3.11: (a) SFG for matrix-vector multiplication algorithm; (b) detailed configuration of a node.

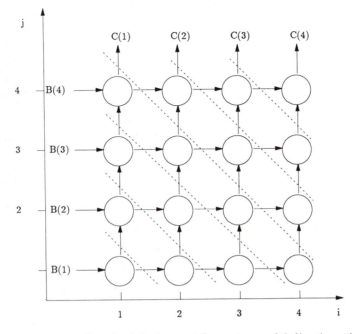

Figure 3.12: A systolic schedule for matrix-vector multiplication, $\vec{s} = [1\ 1]^{\mathrm{T}}$.

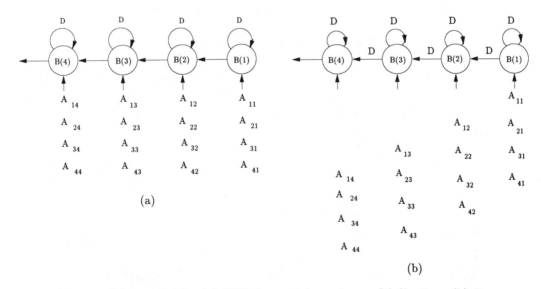

Figure 3.13: (a) SFG for matrix-vector multiplication; (b) its corresponding systolic array ($\alpha = 1$).

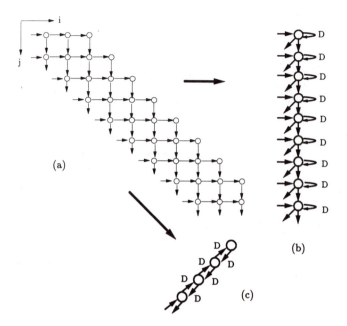

Figure 3.14: Band matrix vector multiplication: (a) DG; (b) SFG obtained along [1 0]; (c) SFG obtained along [1 1].

Example 6: Multiplication of Band Matrix and Vector

Figure 3.14(a) shows a DG for the multiplication of a band matrix **A**, of size 9 × 9 and bandwidth 4, and a vector **b** of size 9 × 1. By using the projection along the [1 0] direction and the default schedule, we can obtain the 1-D SFG as shown in Figure 3.14(b). However, if we use the projection along the [1 1] direction (and the default schedule), an SFG of a smaller size can be derived as shown in Figure 3.14(c). In general, *the array size may be reduced or minimized by choosing a suitable projection direction.*

3.3.2.5 Algebraic Approach for SFG Projection

The above graph-based projection procedure may be formally described in terms of algebraic transformations. Given a DG of dimension n, a projection vector, $\vec{d}$, and a permissible linear schedule, $\vec{s}$, then an SFG may be derived based on the following mappings:

1. *Node mapping* This mapping assigns the node activities in the DG to processors. The index set of nodes of the SFG are represented by the mapping

$$\mathbf{P} : R \longrightarrow I^{n-1};$$

where R is the index set of the nodes of the DG, and I^{n-1} is the Cartesian product of $(n-1)$ integers. The mapping of a computation **c** in the DG onto a node **n** in the SFG is found by:

$$\mathbf{n} = \mathbf{P}^{T}\mathbf{c}$$

where the *processor basis* P, denoted by an $n \times (n-1)$ matrix, is *orthogonal* to $\vec{d}$. Mathematically,

$$\mathbf{P}^{T}\vec{d} = \mathbf{0}$$

2. *Arc mapping* This mapping maps the arcs of the DG to the edges of the SFG. The set of edges $\vec{e}$ into each node of the SFG and the number of delays $D(\vec{e})$ on every edge are derived from the set of dependence edges $\vec{b}$ at each point in the (shift invariant) DG by:

$$\begin{bmatrix} D(\vec{e}) \\ \cdots \\ \vec{e} \end{bmatrix} = \begin{bmatrix} \vec{s}^{T} \\ \cdots \\ \mathbf{P}^{T} \end{bmatrix} \begin{bmatrix} \vec{b} \end{bmatrix}$$

3. *I/O mapping* The SFG node position, **n**, and time, $t(\mathbf{c})$, of an input of the DG computation **c** is derived as:

$$\begin{bmatrix} t(\mathbf{c}) \\ \cdots \\ \mathbf{n} \end{bmatrix} = \begin{bmatrix} \vec{s}^{T} \\ \cdots \\ \mathbf{P}^{T} \end{bmatrix} \begin{bmatrix} \mathbf{c} \end{bmatrix}$$

A similar mapping applies to output nodes.

Remark: The elements of $\vec{s}$, $\vec{d}$ and $\mathbf{P}$ are integers. Clearly, it is desirable to have these vectors (matrices) represented by the smallest integers whenever possible. To avoid confusion, from now on the elements of $\vec{s}$ are restricted to be coprime. That is, *the greatest common divisor of all the elements of $\vec{s}$ is 1*. The elements of $\vec{d}$ are also assumed to be coprime, as are the elements of each column vector of $\mathbf{P}$.

Example 7: Projection to Insertion Sorter SFG[Rao85]

For the sorting example, the projection shown in Figure 3.15(b) can be described by an algebraic transformation as follows:

$$\vec{d}^{\mathrm{T}} = \vec{s}^{\mathrm{T}} = [1\ 0], \quad \mathbf{P}^{\mathrm{T}} = [0\ 1]$$

Node mapping:

$$[0\ 1] \cdot \begin{bmatrix} i \\ j \end{bmatrix} = j$$

Arc mapping:

$$\begin{bmatrix} 1 & 0 \\ 0 & 1 \end{bmatrix} \cdot \begin{bmatrix} 0 \\ 1 \end{bmatrix} = \begin{bmatrix} 0 \\ 1 \end{bmatrix}$$

$$\begin{bmatrix} 1 & 0 \\ 0 & 1 \end{bmatrix} \cdot \begin{bmatrix} 1 \\ 0 \end{bmatrix} = \begin{bmatrix} 1 \\ 0 \end{bmatrix}$$

I/O mapping:

$$\text{Input}: \quad \begin{bmatrix} 1 & 0 \\ 0 & 1 \end{bmatrix} \cdot \begin{bmatrix} i \\ 1 \end{bmatrix} = \begin{bmatrix} i \\ 1 \end{bmatrix}$$

$$\text{Output}: \quad \begin{bmatrix} 1 & 0 \\ 0 & 1 \end{bmatrix} \cdot \begin{bmatrix} N \\ j \end{bmatrix} = \begin{bmatrix} N \\ j \end{bmatrix}$$

The resulting SFG is a (vertical) linear array shown in Figure 3.15(b), which corresponds to the *insertion sorter*. Note that the input data is sequentially read in at the first node; however, the output data is obtained

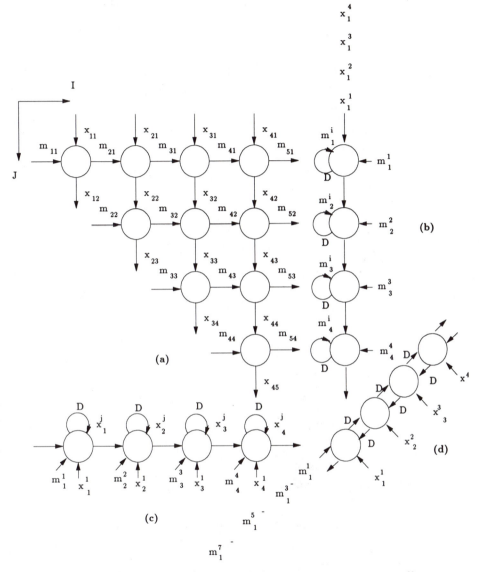

Figure 3.15: Sorting (a) DG; and the SFG arrays corresponding to (b) insertion sorting; (c) selection sorting; (d) bubble sorting.

at different nodes by the output mapping. They may either be output sequentially or fetched in parallel through an output bus. Notice there are N inputs for $m(j, j)$ which can be neglected since $m(j, j)$ is always $-\infty$. The results stay in each individual node. The unsorted sequence is fed in from the top of the array, and the sorted sequence stored in $m(i, j)$ circulates within PE_i.

Example 8: Projection to Selection Sorter SFG

By choosing different directions of projection of the DG, several different linear arrays can be derived.[Rao85] For example, a second choice of the sorting example corresponds to the *selection sorter*. This is shown as a (horizontal) linear array in Figure 3.15(c). This array is derived via a projection along the j-axis ($\vec{d} = [0\ 1]^T$, with the processor basis $\mathbf{P} = [1\ 0]^T$). The schedule is the default schedule, i.e., $\vec{s} = [0\ 1]^T$. The three mappings are derived as:

$$\vec{s}^T = [0\ 1],\ \mathbf{P}^T = [1\ 0]$$

$$\text{Node mapping}: [1\ 0]\begin{bmatrix} i \\ j \end{bmatrix} = [i]$$

$$\text{Arc mapping}: \begin{bmatrix} 0 & 1 \\ 1 & 0 \end{bmatrix}\begin{bmatrix} 0 \\ 1 \end{bmatrix} = \begin{bmatrix} 1 \\ 0 \end{bmatrix};\ \begin{bmatrix} 0 & 1 \\ 1 & 0 \end{bmatrix}\begin{bmatrix} 1 \\ 0 \end{bmatrix} = \begin{bmatrix} 0 \\ 1 \end{bmatrix}$$

$$\text{Input}: \begin{bmatrix} 0 & 1 \\ 1 & 0 \end{bmatrix}\begin{bmatrix} i \\ 1 \end{bmatrix} = \begin{bmatrix} 1 \\ i \end{bmatrix};\ \text{Output}: \begin{bmatrix} 0 & 1 \\ 1 & 0 \end{bmatrix}\begin{bmatrix} N \\ j \end{bmatrix} = \begin{bmatrix} j \\ N \end{bmatrix}$$

Note that in this SFG, the m variables flow from left to right, whereas $x(i, j)$ circulates within PE_j. The initial x's need to be input in parallel, but the result is output sequentially.

Example 9: Projection to Bubble Sorter SFG

A third design can be derived by projecting in the $[1, 1]^T$ direction, see Figure 3.15(d). Again, the schedule used is the default schedule. This corresponds to the *bubble-sorter*. The three mappings are:

$$\vec{s}^T = [1\ 1],\ \mathbf{P}^T = [-1\ 1]$$

$$\text{Node mapping}: [-1\ 1]\begin{bmatrix} i \\ j \end{bmatrix} = [-i+j]$$

Arc mapping : $\begin{bmatrix} 1 & 1 \\ -1 & 1 \end{bmatrix} \begin{bmatrix} 0 \\ 1 \end{bmatrix} = \begin{bmatrix} 1 \\ 1 \end{bmatrix}$; $\begin{bmatrix} 1 & 1 \\ -1 & 1 \end{bmatrix} \begin{bmatrix} 1 \\ 0 \end{bmatrix} = \begin{bmatrix} 1 \\ -1 \end{bmatrix}$

Input : $\begin{bmatrix} 1 & 1 \\ -1 & 1 \end{bmatrix} \begin{bmatrix} i \\ 1 \end{bmatrix} = \begin{bmatrix} i+1 \\ -i+1 \end{bmatrix}$;

Output : $\begin{bmatrix} 1 & 1 \\ -1 & 1 \end{bmatrix} \begin{bmatrix} N \\ j \end{bmatrix} = \begin{bmatrix} N+j \\ -N+j \end{bmatrix}$

Again, the output data is generated in every node. The m and x variables move in opposite directions, and the PE utilization is reduced to 50%.

Remark: Note that the three sorters obtained above have the same number of processors, i.e., N. In certain situations, mapping with a nonlinear assignment scheme may offer more flexible designs and unique advantages. For example, the *rebound sorter*, which uses only about $N/2$ processors, may be obtained by a nonlinear assignment for the sorting problem (see Section 3.4.4.2).

3.3.2.6   Construction of DG from SFG

It is possible to reconstruct the DG from a given SFG. The solution is however not unique. Suppose that the SFG is N-dimensional and there are m iterations involved. Then the reconstructed DG is defined over an $(N + 1)$-dimensional index space, with the added, i.e., the $(N + 1)$st, dimension representing the dimension along the projection direcion. (In this case the projection direction is equivalent to the iteration direction.) This direction should be perpendicular to the N-dimensional SFG space. The dependencies between index points can be determined by the following rules. First, the zero-delay edges in the SFG form an N-dimensional graph. We repeat the same graph m times along the iteration direction. For the n-delay edges, $n = 1,2,...$, the same idea applies, except that the initial endpoints and the terminal endpoints of the reconstructed arcs in the DG should have index increment of n along the iteration direction. By these rules, the reconstructed DG, when projected along the iteration direction, will look like the original SFG.

3.3.3 Stage 3 Design: Mapping an SFG onto an Array Processor

The descriptions of array processing activities, in terms of the SFG representations, are often easy to comprehend. The abstraction provided by the SFG is very powerful to use, and yet the transformation of an SFG description to an SIMD, systolic array, or wavefront array can often be accomplished automatically.

3.3.3.1 Single Instruction Multiple Data Stream (SIMD) Array

A single instruction multiple data stream (SIMD) array is a synchronous array of processing elements under the supervision of one control unit. All PEs receive the same instruction broadcast from the control unit but operate on different data sets from distinct data streams. In general, broadcasting of data is allowed in an SIMD array.

The global DG for matrix-vector multiplication as shown in Figure 3.2(a) can be projected along the j-axis direction to realize a global SFG, as shown in Figure 3.16. Since broadcasting is allowed, such an SFG can be directly implemented on an SIMD array. In general, an SIMD array may serve as a naturally physical realization of the SFG if transmittent data are propagated. In many cases, the data processed by the boundary PEs are transmitted to the other PEs without further modification (that is, the data are transmittent after the processing of the boundary PEs). In this case, a "boundary-enhanced" SIMD (BESIMD) array appears to be more suitable.

3.3.3.2 Systolic Array

Systolic arrays are very amenable to VLSI implementation. They are especially suitable to a special class of computation-bound algorithms with regular, localized data flow.

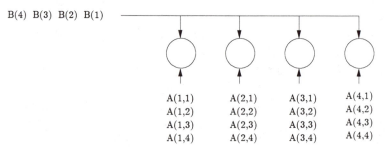

Figure 3.16: A global SFG for matrix-vector multiplication.

A systolic system is a network of processors which rhythmically compute and pass data through the system. Physiologists use the word "systole" to refer to the rhythmically recurrent contraction of the heart and arteries which pulses blood through the body. In a systolic computing system, the function of a processor is analogous to that of the heart. Every processor regularly pumps data in and out, each time performing some short computation, so that a regular flow of data is kept up in the network [HTKun78].

For example, it is shown in [HTKun78] that some basic "inner product" PEs – each performing the operation $Y \leftarrow Y + A \cdot B$ – can be *locally* connected together to perform digital filtering, matrix multiplication, and other related operations. In general, the data movements in a systolic array are prearranged and are described in terms of the "snapshots" of the activities.

A systolic array often represents a direct mapping of computations onto processor arrays. It is usually used as an attached processor of a host computer. The basic principle of systolic design is that all the data, while being "pumped" regularly and rhythmically across the array, can be effectively used in all PEs. The systolic array features the important properties of modularity, regularity, local interconnection, a high degree of pipelining, and highly synchronized multiprocessing. It is also scalable architecturally, i.e., the size of the array may be indefinitely extended, as long as the system synchronization can be maintained. There is extensive literature on the subject of systolic array processing; the reader is referred to [Fish85b] and the references therein. In fact, *a systolic array can be considered as an SFG array in combination with pipelining and retiming.* A systolic array for convolution (derived by an SFG retiming technique that is detailed in the next chapter) is shown in Figure 3.17.

3.3.3.3   Wavefront Array

One problem in a systolic array is that the data movements are controlled by global timing-reference "beats". In order to synchronize the activities in

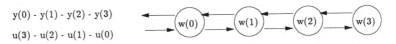

$y(0) - y(1) - y(2) - y(3)$
$u(3) - u(2) - u(1) - u(0)$

Figure 3.17: A systolic array for convolution.

a systolic array, extra delays are often used to insure correct timing. More critically, the burden of having to synchronize the entire computing network eventually becomes intolerable for very-large-scale or ultra-large-scale arrays.

A simple solution to the previously mentioned problems is to take advantage of the control-flow locality in addition to the data flow locality inherently possessed by most algorithms of our interest. This permits a data-driven, self-timed approach to array processing. Conceptually, this approach substitutes the requirement of correct timing for correct sequencing. This concept is extensively used in data flow computers and wavefront arrays.

A data flow multiprocessor [Denni80] is an asynchronous, data-driven multiprocessor that runs programs expressed in data flow graph form. Since the execution of its instructions is data-driven, i.e., – the triggering of instructions depends only upon the availability of operands and resources required – unrelated instructions can be executed concurrently without interference. The principal advantages of data flow multiprocessors are the simple representation they give of concurrent activity, relative independence of individual PEs, greater use of pipelining, and reduced use of centralized control and global memory.

However, for a general-purpose data flow multiprocessor, the interconnection and memory conflict problems remain critical. Such problems can be greatly alleviated if *modularity* and *locality* are incorporated into data flow multiprocessors. This motivates the concept of the wavefront array processor (WAP).

The derivation of wavefront processing consists of three steps: (1) express the algorithm in terms of a sequence of recursions; (2) map each of the recursions to a corresponding computational wavefront; and (3) successively pipeline the wavefronts through the processor array.

Note that the major difference distinguishing a wavefront array from a systolic array is the data-driven property. By relaxing the strict timing requirement, there are many advantages gained, such as speed and programming simplicity.

As a justification for the name wavefront array, we note that the computational wavefronts are similar to electromagnetic wavefronts, since each processor acts as a secondary source and is responsible for the propagation of the wavefront. The pipelining is feasible because the wavefronts of two successive recursions never intersect, thus avoiding any contention problems. It is even possible to have wavefronts propagating in several different fashions; for instance, in the extreme case of nonuniform clocking, the wavefronts are actually crooked. What is necessary and sufficient is that the order of task sequencing must be correctly followed. The correctness of the sequencing of the tasks is ensured by the wavefront principle [Kung82a].

Wavefront processing utilizes both the localities of data flow and control flow inherent in many signal processing algorithms. Since there is no need to synchronize the entire array, a wavefront array is truly architecturally scalable. In fact, it may be stated that *a wavefront array is a systolic array in combination with the data flow principle.*

3.3.3.4   MIMD Array

For sake of completeness, we introduce the MIMD design. It should be noted that, for a regular DG, there is little need to utilize such a complicated design. An MIMD array consists of a number of processing elements, each with its own control unit, program, and data. The main feature of an MIMD machine is that the overall tasks may be distributed among the processing elements for the purpose of increasing parallelism. It is especially promising to exploit the flexibilities in MIMD architectures for handling structurally complicated algorithms, such as those often encountered in intelligent image processing and vision analysis applications. Understandably, mapping of algorithms onto MIMD arrays or programming codes is usually performed at the task level. This subject is not treated here.

3.3.4   Examples: Mapping Algorithms onto SFG Arrays

The assumption of recursive and locally data-dependent algorithms, as used in canonical mapping methodology, incurs little loss of generality, as a great majority of signal processing algorithms possess these properties. In fact, this class of recursive algorithms covers a very broad range of operations, such as 1-D and 2-D convolutions, matrix-vector and matrix-matrix multiplications, LU decomposition (for solving linear systems), QR decomposition, back-substitution, and many others. In the following, some selected mapping examples are discussed to show how to cope with different situations and design criteria.

3.3.4.1   Digital Filters

AR Filter   The locally recursive algorithm for the AR filter was developed previously, and its DG is shown in Figure 3.7(c). Since the index space for the AR filter is not finite, the projection direction to realize an SFG is unique. The projected SFG for the AR filter is shown in Figure 3.18. It is

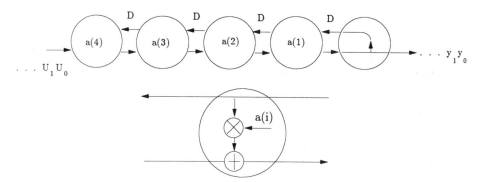

Figure 3.18: SFG for the AR filter.

interesting to note that this corresponds exactly to the standard AR forms found in basic signal processing texts such as [Oppen75].

ARMA Filter Generally, an ARMA filter is defined by a transfer function

$$H(z) = \frac{\sum_{i=1}^{q} b_i z^{-i}}{1 - \sum_{i=1}^{p} a_i z^{-i}}$$

Note that FIR filtering (linear convolution or transversal filtering) is simply a special case when $a_i = 0$, $i = 1, \ldots, p$. The corresponding recursive algorithm is often given in terms of the difference equation:

$$y(k) = \sum_{m=1}^{q} x(k-m)b(m) + \sum_{m=1}^{p} y(k-m)a(m)$$

The ARMA filter can be represented by a cascade of an AR filter followed by an MA filter. Since the operation of an MA filter is simply a convolution and SFGs of both AR and convolution have been derived before, an SFG for an ARMA filter can be easily obtained and is shown in Figure 3.19. Note that this implementation is in direct form. Noticing the fact that the operations of both AR and MA use the same data, we may implement the ARMA filter in canonical form, as shown in Figure 3.20, which uses less storage than the direct form implementation [Oppen75], [Rabin75], [Kaila80].

Linear Phase Filter A direct-form realization of a linear phase filter that takes advantage of the symmetry of the impulse response was introduced in

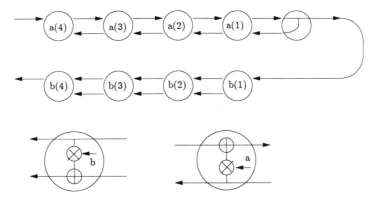

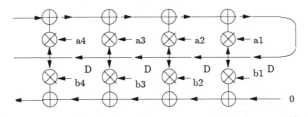

Figure 3.19: Direct form SFG for the ARMA filter.

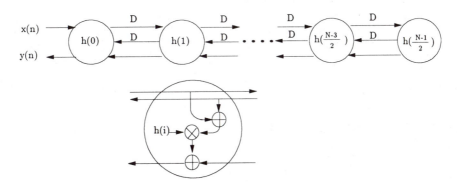

Figure 3.20: Canonical form SFG for the ARMA filter.

Figure 3.21: SFG for the linear phase filter.

Chapter 2. An SFG for a filter of odd order is shown in Figure 3.21. Note that this SFG can be derived by using a DG for convolution and noting the fact that the filter coefficients are symmetric (See Problem 9).

MA and AR Lattice Filters From a filter stability point of view, the ARMA or AR canonical filters are not the most suitable. A more desirable

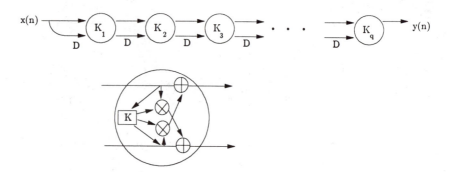

Figure 3.22: SFG for the MA lattice filter.

design is a special structure known as the lattice filter (see Figure 3.22), which has very good stability properties [Marke76]. For the 1-D case, digital lattice filters have many important applications in speech and seismic signal processing. Furthermore, a similar form of the lattice SFG can be adopted for solving a special Toeplitz linear system.

The SFG for the autoregressive lattice filter can be derived similarly, and it is shown in Figure 3.23.

3.3.4.2   Matrix Multiplication

Multiplication of Two Full Matrices   A simple matrix multiplication can be represented as $\mathbf{C} = \mathbf{A} \cdot \mathbf{B}$. Mathematically this expression means

$$c_{ij} = \sum_{k=1}^{N} a_{ik} b_{kj}$$

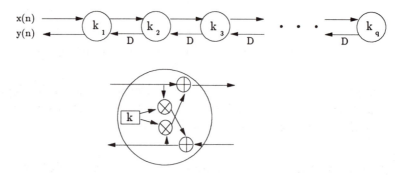

Figure 3.23: SFG for the AR lattice filter.

This equation implies that all the multiplications involved can be calculated at the same time since they do not have any dependencies between each other. To get the maximal parallelism, we need to propagate input data to N^3 multipliers; two input links for each multiplier. That means at least $2N^3$ communication links are necessary.

This motivates the adoption of some new variables, so that the matrix multiplication algorithm may be written in a more useful recursive form:

$$c_{ij}^{(k)} = c_{ij}^{(k-1)} + a_{ik}b_{kj}$$

where k is the recursive index.

Or it can be written as a sequential code, as follows:

For i from 1 to N_1

For j from 1 to N_2

For k from 1 to N_3

$$c(i,j,k) = c(i,j,k-1) + a(i,k)b(k,j)$$

Note that this algorithm is expressed in a global form; i.e., $a(i, k)$ is broadcast to all index points that have the same i-index and k-index. Similarly, $b(k, j)$ is broadcast to all index points that have the same k-index and j-index. Therefore, global communication is implicitly implied by the code.

We note the similarity between a recursive algorithm and its corresponding (sequential) code. Therefore, the same localization strategy used in the recursive algorithm may be adopted to convert the sequential code into a more preferable (i.e., localized) one.

Thus, a locally recursive algorithm, programmed as follows, is derived. (The program shown here is also a single assignment code.)

For i from 1 to N_1

For j from 1 to N_2

For k from 1 to N_3

$$a(i,j,k) = a(i,j-1,k)$$

$$b(i, j, k) = b(i - 1, j, k)$$

$$c(i, j, k) = c(i, j, k - 1) + a(i, j, k)b(i, j, k)$$

The DG corresponding to this algorithm is shown in Figure 3.24. Note that the functional operations executed in each node are also shown in this figure.

Obtained by different DG projections, two SFGs for matrix multiplication are shown in Figure 3.25 and Figure 3.26. In Figure 3.25, the projection direction is $[0\ 0\ 1]$. A delay denotes an advance in the k-index. Note that the insertion of delays is necessary to express the dependency in the $[0\ 0\ 1]$ direction. However, in Figure 3.26, the projection direction is along $[0\ 1\ 0]$, and a delay denotes an advance in the j-index.

Multiplication of a Band Matrix and a Rectangular Matrix Let us look at a slightly different, but commonly encountered, type of matrix multiplication problem. This involves an $N \times N$ band matrix $\mathbf{A}$, with bandwidth P, and an $N \times Q$ rectangular matrix $\mathbf{B}$. In most applications, $N \gg P$ and $N \gg Q$, making the use of $N \times N$ arrays for computing $\mathbf{C} = \mathbf{A} \cdot \mathbf{B}$ very inefficient.

By a slight modification to the boundary of the DG for (rectangular) matrix multiplication, as shown in Figure 3.24, we can derive a DG for the multiplication of a band matrix and a rectangular matrix. This is shown in Figure 3.27(a). Note that one dimension of the DG is not shown in this figure. With the projection along the diagonal direction of the band matrix, the same speed up will be achieved with only a $P \times Q$ rectangular array as opposed to an $N \times Q$ array. The corresponding SFG is shown in Figure 3.27(b). Now, the left memory module stores the matrix $\mathbf{A}$ along

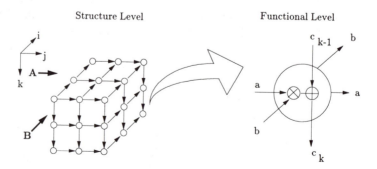

Figure 3.24: DG for matrix multiplication.

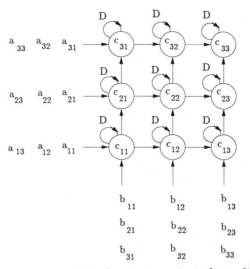

Figure 3.25: Type I SFG (projection along [0 0 1]) for matrix multiplication.

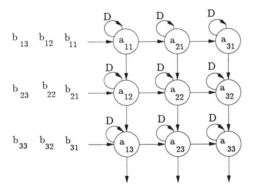

Figure 3.26: Type II SFG (projection along [0 1 0]) for matrix multiplication.

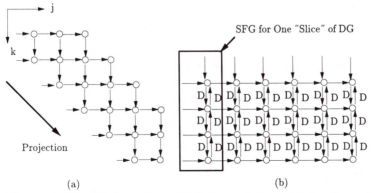

Figure 3.27: (a) One slice of DG for matrix multiplication with one band matrix, note that the i-th dimension is not shown here. (b) SFG obtained from projection.

155

the band direction and the upper module stores B in the same manner as before.

Note that the major modification to the array is that the partial sums should be shifted upwards between the recursions of outer products. This is because the input matrix **A** is loaded in a skewed fashion. The final result, matrix **C**, is also output from the I/O ports of the top row PEs.

Multiplication of Two Band Matrices   Another interesting case is the situation when both **A** and **B** are band matrices, with bandwidths **P** and **Q**, respectively. (See Problem 8.) Let us assume that $N \gg P$ and $N \gg Q$. Then it is possible to achieve full parallelism with only a $P \times Q$ rectangular array (as opposed to an $N \times N$ array).

The left- and upper-memory modules store the matrices **A** and **B** (respectively) along the band direction (see Figure 3.28). The delayed feedback edge (with partial sum of the outer products) is along the diagonal (north west) direction. This is because both **A** and **B** are stored in the skewed version of Figure 3.25.

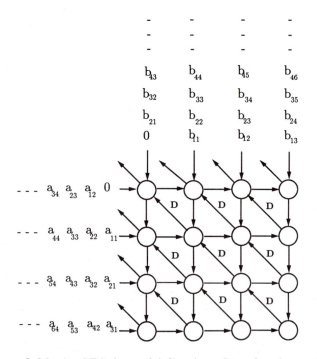

Figure 3.28: An SFG for multiplication of two band matrices.

3.3.4.3   LU Decomposition

The SFGs derived by different projection directions may have substantially different properties. In order to find an optimal projection, we may try several directions and see how the results are. Here, we demonstrate two 2-D arrays for the LU-decomposition example by using different projection directions.

In the LU decomposition, a given matrix $\mathbf{C}$ is decomposed into

$$\mathbf{C} = \mathbf{A} \cdot \mathbf{B} \tag{3.9}$$

where $\mathbf{A}$ is a lower triangular and $\mathbf{B}$ is an upper triangular matrix. The recursions involved are

$$b_j^{(k)} = C_{kj}^{(k-1)}$$

$$a_i^{(k)} = \frac{C_{ik}^{(k-1)}}{C_{(kk)}^{(k)}} \tag{3.10}$$

$$C_{ij}^{(k)} = C_{ij}^{(k-1)} - a_i^{(k)} b_j^{(k)},$$

for $k = 1, 2, \ldots, N;\ k < i < N;\ k < j < N$.

Verifying the procedure by tracing back Eq. 3.10, we note that

$$\mathbf{C}_{ij} = C_{ij}^{(0)} = \sum_{k=1}^{N} a_i^{(k)} b_j^{(k)} \tag{3.11}$$

where $\mathbf{A} = \{a_{mn}\} = \{a_m^{(n)}\}$ and $\mathbf{B} = \{b_{mn}\} = \{b_n^{(m)}\}$ are the outputs of the array processing.

Comparing with Eq. 3.9, Eq. 3.11 is basically a reversal of the matrix multiplication recursions.

DG   A single assignment formulation of the preceding recursive algorithm is shown here.

For k **from 1 to** N

For i **from** k **to** N

For j **from** k **to** N

$$b(i,j,k) \longleftarrow \begin{cases} c(i,j,k-1) & \text{if } i = k, \\ b(i-1,j,k) & \text{otherwise} \end{cases}$$

$$a(i,j,k) \longleftarrow \begin{cases} c(i,j,k-1)/b(i,j,k) & \text{if } j = k, \\ a(i,j-1,k) & \text{otherwise} \end{cases}$$

$$c(i,j,k) \longleftarrow c(i,j,k-1) - a(i,j,k) * b(i,j,k)$$

Initially, $c(i,\ j,\ 0) \longleftarrow \mathbf{C}_{ij}$.

The DG for the LU decomposition algorithm is shown in Figure 3.29. (The additional dependence lines in the k-direction that run from points $(i,\ j,\ k)$ to $(i,\ j,\ k+1)$ are not drawn.) In each plane the points that serve as the source of the row and column values are marked as dark nodes.

Based on the SFG projection of the DG, there are two natural mappings from the locally recursive algorithm onto arrays:

Version I (Projection in $[1\ 1\ 1]^{\mathrm{T}}$ direction)   First, we choose the projection vector $\vec{d} = [1\ 1\ 1]^{\mathrm{T}}$ and choose processor basis $\mathbf{P}$, which is orthogonal to $\vec{d}$. The mappings are:

$$\vec{d}^{\mathrm{T}} = [1\ 1\ 1], \quad \mathbf{P}^{\mathrm{T}} = \begin{bmatrix} 1 & 0 & -1 \\ 0 & 1 & -1 \end{bmatrix}$$

Node mapping:

$$\begin{bmatrix} 1 & 0 & -1 \\ 0 & 1 & -1 \end{bmatrix} \cdot \begin{bmatrix} i \\ j \\ k \end{bmatrix} = \begin{bmatrix} i-k \\ j-k \end{bmatrix}$$

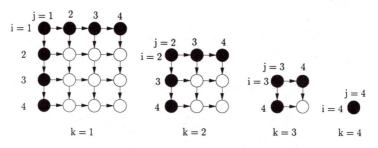

Figure 3.29: DG for LU-decomposition.

Arc mapping:

$$
\begin{bmatrix} 1 & 1 & 1 \\ 1 & 0 & -1 \\ 0 & 1 & -1 \end{bmatrix} \cdot \begin{bmatrix} 1 & 0 & 0 \\ 0 & 1 & 0 \\ 0 & 0 & 1 \end{bmatrix} = \begin{bmatrix} 1 & 1 & 1 \\ 1 & 0 & -1 \\ 0 & 1 & -1 \end{bmatrix}
$$

I/O mapping:

$$
\text{Input}: \quad \begin{bmatrix} 1 & 1 & 1 \\ 1 & 0 & -1 \\ 0 & 1 & -1 \end{bmatrix} \cdot \begin{bmatrix} i \\ j \\ 1 \end{bmatrix} = \begin{bmatrix} i+j+1 \\ i-1 \\ j-1 \end{bmatrix}
$$

$$
\text{Output}: \quad \begin{bmatrix} 1 & 1 & 1 \\ 1 & 0 & -1 \\ 0 & 1 & -1 \end{bmatrix} \cdot \begin{bmatrix} i & k \\ k & j \\ k & k \end{bmatrix} = \begin{bmatrix} i+2k & j+2k \\ i-k & 0 \\ 0 & j-k \end{bmatrix}
$$

The SFG array derived from this projection (Figure 3.30(a)), is a hexagonal array with pipelining period $\alpha = 3$. Note that in this SFG array, the functions of PEs are fixed, i.e., the (0,0) PE performs division and the first column PEs perform multiplication, the first-row PEs only pass data, and the inner PEs perform the multiply-and-add operation. Output matrices $\mathbf{A}$ and $\mathbf{B}$ are obtained from the first-column and first-row PEs.

Version II (Projection in $[1\ 0\ 0]^{\mathrm{T}}$ Direction) First, we choose the projection vector $\vec{d} = [1\ 0\ 0]^{\mathrm{T}}$ and choose the processor basis $\mathbf{P}$, which is orthogonal to $\vec{d}$. The mappings are:

$$
\vec{d}^{\mathrm{T}} = [1\ 0\ 0], \quad \mathbf{P}^{\mathrm{T}} = \begin{bmatrix} 0 & 1 & 0 \\ 0 & 0 & 1 \end{bmatrix}
$$

Node mapping:

$$
\begin{bmatrix} 0 & 1 & 0 \\ 0 & 0 & 1 \end{bmatrix} \cdot \begin{bmatrix} i \\ j \\ k \end{bmatrix} = \begin{bmatrix} j \\ k \end{bmatrix}
$$

Arc mapping:

$$
\begin{bmatrix} 1 & 0 & 0 \\ 0 & 1 & 0 \\ 0 & 0 & 1 \end{bmatrix} \cdot \begin{bmatrix} 1 & 0 & 0 \\ 0 & 1 & 0 \\ 0 & 0 & 1 \end{bmatrix} = \begin{bmatrix} 1 & 0 & 0 \\ 0 & 1 & 0 \\ 0 & 0 & 1 \end{bmatrix}
$$

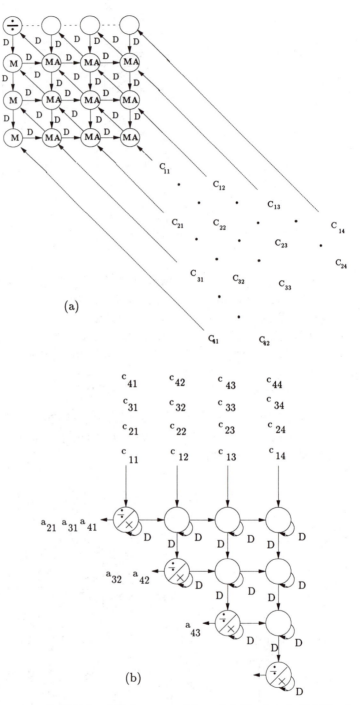

(a)

(b)

Figure 3.30: SFG for LU decomposition: (a) Version I; (b) Version II.

I/O mapping:

$$
\text{Input}: \quad
\begin{bmatrix} 1 & 0 & 0 \\ 0 & 1 & 0 \\ 0 & 0 & 1 \end{bmatrix}
\cdot
\begin{bmatrix} i \\ j \\ 1 \end{bmatrix}
=
\begin{bmatrix} i \\ j \\ 1 \end{bmatrix}
$$

$$
\text{Output}: \quad
\begin{bmatrix} 1 & 0 & 0 \\ 0 & 1 & 0 \\ 0 & 0 & 1 \end{bmatrix}
\cdot
\begin{bmatrix} i & k \\ k & j \\ k & k \end{bmatrix}
=
\begin{bmatrix} i & k \\ k & j \\ k & k \end{bmatrix}
$$

The SFG array derived from this projection (Figure 3.30(b)) is a triangular array. This SFG array uses half the PEs of the previous version and thus saves hardware. Note that the functions of PEs have to change with time. The input is a full matrix C of size $n \times n$. The B matrix remains in the nodes, and the matrix A is output from the diagonal nodes.

The first row of the array processes the first recursion ($k = 1$), the second row processes the second recursion ($k = 2$), and so on. Note that during the process of triangularization, the size of the matrix is shrinking and hence the first recursion has size n, the second has size $(n - 1)$, and so on. This is the reason for the formulation of a triangular array. Note also that although the function changes from division to multiplication for diagonal PEs, we can still easily implement it if a CORDIC arithmetic unit [Ahmed85] is used (cf., Section 7.3). All that is necessary is to change one control bit.

Now we can compare the two versions of LU decomposition array algorithms. Version I has the following advantages: (1) There is no need for processor reprogramming (no change of processor functions), and (2) it is easily adaptable to the band matrix LU decomposition problem. On the other hand, Version II also offers very attractive advantages: (1) There is no need for diagonal connections, and (2) the PE utilization is more efficient since it yields the same throughput rate with only 50% of the PEs. Moreover, as we discuss in the next chapter, the systolized array of Version II enjoys a pipelining period (α) of 1, compared with a value of 3 for Version I (although they both result in the same latency).

3.3.4.4   QR Decomposition

DG for QR Decomposition   The QR decomposition algorithm, described in Section 2.2.2, can be written in a sequential program as follows:

For k **from** 1 **to** N

For i **from** $N-1$ **to** k

$\theta \longleftarrow tan^{-1}(a_{i+1,k}/a_{i,k})$

For j **from** k **to** N

$temp1 \longleftarrow a_{i,j}cos\theta + a_{i+1,j}sin\theta$

$temp2 \longleftarrow -a_{i,j}sin\theta + a_{i+1,j}cos\theta$

$a_{i,j} \longleftarrow temp1$

$a_{i+1,j} \longleftarrow temp2$

where the temporary variables are used to store the new values of $a_{i,j}$ and $a_{i+1,j}$. The input matrix **A** is loaded into the $a_{i,j}$ locations first, with the final value of the $a_{i,j}$ locations yielding the **R** matrix.

This sequential algorithm can be converted to single assignment formulation; however, extra care has to be taken. First, we decide to group the plane rotation of two elements into one node in the DG. In this way, the rotation angle θ is broadcast along the row only once. More importantly, this grouping simplifies the dependencies, since more variables are assigned to a node. Another point to notice is that for each recursion in k, each row (except the Nth row) is updated twice instead of once. To transform the above sequential algorithm into a single assignment form, we use four variables for the **A** matrix at each node (i, j, k). We denote the "old" value of two elements in a rotation as $ox(i, j, k)$ and $oy(i, j, k)$ and the "new" value of the two elements as $nx(i, j, k)$ and $ny(i, j, k)$. The single assignment form is:

For k **from** 1 **to** $N-1$

For i **from** $N-1$ **to** k

For j **from** k **to** N

$ox(i, j, k) \longleftarrow nx(i, j, k-1)$

$$oy(i,j,k) \longleftarrow \begin{cases} ny(i,j,k-1) & \text{if } i = N-1 \\ nx(i+1,j,k) & \text{if } i \neq N-1 \end{cases}$$

$$\theta(i,j,k) \longleftarrow \begin{cases} tan^{-1}(oy(i,j,k)/ox(i,j,k)) & \text{if } j = k \\ \theta(i,j-1,k) & \text{if } j \neq k \end{cases}$$

$$nx(i,j,k) \longleftarrow ox(i,j,k)cos(\theta(i,j,k)) + oy(i,j,k)sin(\theta(i,j,k))$$

$$ny(i,j,k) \longleftarrow -ox(i,j,k)sin(\theta(i,j,k)) + oy(i,j,k)cos(\theta(i,j,k))$$

where $nx(i,j,0) = a_{i,j}$, the input matrix; $ny(N-1,j,0) = a(N,j)$; Row i (except the last row) of the final $\mathbf{R}$ matrix will be in $nx(i,j,i)$, for $i \leq j \leq N$. The Nth (last) row of the $\mathbf{R}$ matrix is in $ny(N-1,j,N-1)$.

The DG of this single assignment program can be easily obtained by comparing the index differences between each assignment statement. The DG is shown in Figure 3.31, which includes the dependence arcs in each k-plane. The dependence arcs between k-planes are harder to draw. There is only one kind of arc between k-planes, from $(i, j, k-1)$ to (i, j, k).

Projection in [0 −1 1] Direction   From the DG, we can project along some direction and realize an SFG array. The projected SFG from a projection along the [0 −1 1] direction is shown in Figure 3.32. This is a regular SFG in which the GG (Givens' Generation) operations are all in the first column PEs, and all the rest of the nodes perform GR operation.

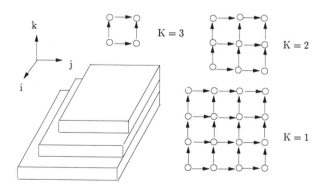

Figure 3.31: DG of QR decomposition for the $N = 4$ case. Note that the dependence arcs between k-planes are not shown.

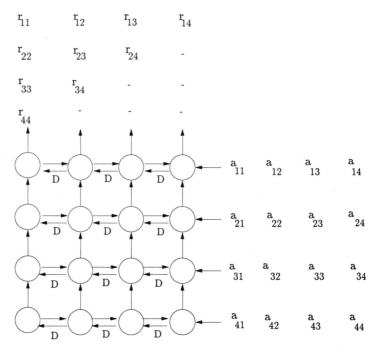

Figure 3.32: SFG of QR decomposition by projection in $[0 \ -1 \ 1]$ direction.

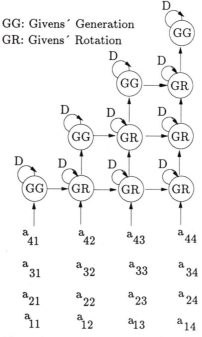

GG: Givens´ Generation
GR: Givens´ Rotation

Figure 3.33: SFG of QR decomposition by projection in $[1 \ 0 \ 0]$ direction.

164

Projection in [1 0 0] Direction If we project in the [1 0 0] direction, the SFG realized is shown in Figure 3.33. This SFG is a triangular array, which saves half of the hardware in comparison to the SFG corresponding to projection in [0 −1 1] direction.

3.4   Generalized Mapping Methodology from DG to SFG

Note that the class of homogeneous DGs (i.e., shift-invariant) already covers a broad range of algorithms. Nevertheless, further generalization is possible and desirable to many other important algorithms that are not totally regular (i.e., shift-invariant) but still have a certain degree of regularity. This semiregularity very often proves to be useful for an efficient mapping methodology. This motivates us to explore a generalized mapping methodology. The new approach will allow an extended DG classification to be dealt with and to have options on linear or nonlinear assignment/schedule. More flexibility is also possible when using multiple projections, allowing global communication as well as treatment of totally irregular DG structures. The generalized mapping methodology can provide effective designs for many algorithms, including Gauss-Jordan elimination, the shortest-path problem, transitive closure, simulated annealing, PDE problems, SVD, FFT, and Viterbi decoding.

Node Assignment and Scheduling To broaden the array design options, we would have to expand the classes of

- The permissible projection direction (node assignments),

- The permissible schedule (schedule assignment).

As to the schemes of node assignment, there are two basic types:

1. *Linear assignment (projection)* A linear assignment of a DG is a linear mapping of the nodes of the DG to PEs, in which nodes along a straight line are mapped to a PE. Mathematically, a linear assignment, denoted by an $(n-1) \times n$ matrix $\mathbf{L}$, maps a node index $\mathbf{I}$ to a vector, $\mathbf{S}$, of $n-1$ elements, where $\mathbf{S(I)} = \mathbf{L} \cdot \mathbf{I}$.

2. *Nonlinear assignment* If the node assignment is not a linear projection, then it is termed a nonlinear assignment.

The scheduling involves specifying the execution time for all the nodes in the DG. The scheduled execution time of a node is represented by a time index (i.e., an integer). Two classes of schedules may be defined:

1. *Linear schedule* A linear schedule maps a set of parallel equitemporal hyperplanes to a set of linearly increased time indices. That is, the time index of the nodes can be mathematically represented by $\vec{s}^T \mathbf{I}$, where $\mathbf{I}$ denotes the index space of the nodes and $\vec{s}$ is the normal vector of the equitemporal hyperplanes.

2. *Nonlinear schedule* If a schedule is not linear, then it is a nonlinear schedule.

The generalized mapping methodology covers all of the options just mentioned. The organization of the section is depicted in Figure 3.34.

3.4.1 Directional Classification of DG

In order to examine the property of the SFGs derived from mapping in a certain projection direction, let us focus on direction specific properties of the DG. This leads to the following extended classification of algorithms based

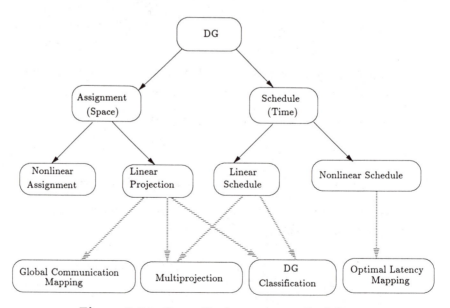

Figure 3.34: Generalized mapping methodology.

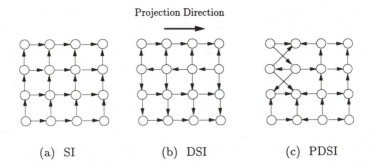

Figure 3.35: Illustration of (a) SI, (b) DSI, and (c) PDSI projections.

on their shift-invariance characteristics. This classification covers algorithms such as Gauss-Jordan elimination, transitive closure, and shortest-path problems.

1. *Shift-invariance* If the dependence arcs corresponding to *all* nodes in the index space do not change with respect to the node positions, then the DG is said to be shift-invariant (SI). An example is given in Figure 3.35(a).

2. *Directional shift-invariance* Given a direction $\vec{d}$, if the dependence arcs corresponding to the nodes along $\vec{d}$ remain invariant with respect to the node position, then the DG is said to be *directionally shift-invariant* (DSI) along $\vec{d}$. Note that SI is equivalent to DSI in all directions. An example of DSI is given in Figure 3.35(b).

3. *Pseudo DSI* Given a projection direction $\vec{d}$, if the projected components of the dependence arcs (which are obtained by applying the projection along $\vec{d}$) are invariant for the nodes along $\vec{d}$, then the DG is said to be *pseudo DSI* (PDSI) along $\vec{d}$. Note that the dependence arcs along $\vec{d}$ may have global or even opposite components. The components will, however, not affect the projected arcs, since they correspond to zero components after projection. An example is given in Figure 3.35(c).

Recall that an STI SFG has static interconnection, i.e., the interconnection arcs are invariant with time. (However, by definition, in an STI SFG, the functions performed within the nodes may be time-varying.) A SI DG is mapped into an STI SFG by projection in any direction. From a broader perspective, an STI SFG can be obtained as long as the corresponding projections of the dependency arcs are directionally shift-invariant. In other words, an STI SFG can be obtained by projecting in any direction that is at least PDSI. These properties may be exemplified by the Gauss-Jordan elimination algorithm discussed below.

Example 1: Gauss-Jordan Elimination

Given an $N \times N$ nonsingular matrix $\mathbf{A}$, the inverse of $\mathbf{A}$ can be found by applying elementary row operations on $\mathbf{A}$ and successively reducing $\mathbf{A}$ to an identity matrix $\mathbf{I}$. Then the product of all basic row operations applied on $\mathbf{A}$ is the inverse of $\mathbf{A}$, i.e., $\mathbf{A}^{-1} \mathbf{A} = \mathbf{I}$. If this same set of row operations is applied to the identity matrix, then the result will be $\mathbf{A}^{-1}$. This is the basic idea of Gauss-Jordan elimination method which is very similar to the Gaussian elimination. The only difference is that: in the Gaussian elimination only the entries below the diagonal are eliminated, thus the resulting matrix is an upper triangular matrix; while in the Gauss-Jordan elimination the entries both below and above the diagonal are eliminated and eventually the matrix is reduced to an identity. (We assume that no pivoting is needed.) Mathematically, the Gauss-Jordan elimination starts initially with an augmented matrix $\mathbf{X}^0 = [\mathbf{A} \mid \mathbf{I}]$ and finally ends with $\mathbf{X}^N = [\mathbf{I} \mid \mathbf{A}^{-1}]$. The recursive equation is described in the following form:

For k from 1 **to** N

For i from 1 **to** N

For j from k **to** $2N$

$$x_{ij}^k \longleftarrow \begin{cases} x_{ij}^{k-1} \cdot (x_{kk}^{k-1})^{-1} & \text{if } i = k, \\ x_{ij}^{k-1} - x_{ik}^{k-1} \cdot (x_{kk}^{k-1})^{-1} \cdot x_{kj}^{k-1} & \text{otherwise.} \end{cases}$$

where $\mathbf{X}^0 = [\mathbf{A} \mid \mathbf{I}]$ and $\mathbf{X}^N = [\mathbf{I} \mid \mathbf{A}^{-1}]$

Suppose that for the purpose of solving a linear system, we want to compute $\mathbf{A}^{-1}\mathbf{b}$. We can do it simply by initializing $\mathbf{X}^0 = [\mathbf{A} \mid \mathbf{b}]$ instead of $[\mathbf{A} \mid \mathbf{I}]$, and the final result $\mathbf{X}^N$ will be $[\mathbf{I} \mid \mathbf{A}^{-1}\mathbf{b}]$.

DG (Stage 1 Design) The preceding dependencies can be localized by adding propagating (transmittent) variables. One is the row transmittent variable r, which propagates data to all the nodes in a row. The other is the column transmittent variable c, which propagates data to all the nodes in a column.

For k from 1 to N

For i from 1 to N

For j from k to $2N$

$$r(i,j,k) \longleftarrow \begin{cases} x(i,j,k-1) & \text{if } j = k \\ r(i,j-1,k) & \text{if } j > k \end{cases}$$

$$c(i,j,k) \longleftarrow \begin{cases} 1 & \text{if } i = k \text{ and } j = k \\ x(i,j,k) & \text{if } i = k \text{ and } j > k \\ c(i+1,j,k) & \text{if } i < k \\ c(i-1,j,k) & \text{if } i > k \end{cases}$$

$$x(i,j,k) \longleftarrow \begin{cases} x(i,j,k-1) \cdot r(i,j,k)^{-1} & \text{if } i = k \\ x(i,j,k-1) - r(i,j,k) \cdot c(i,j,k) & \text{otherwise.} \end{cases}$$

where input $\mathbf{X}^0 = [\mathbf{A} \mid \mathbf{I}]$ and output $\mathbf{X}^N = [\mathbf{I} \mid \mathbf{A}^{-1}]$.

The 3-D DG is shown in Figure 3.36, with each ij-plane drawn separately. (Not drawn are the additional dependence lines in the k-direction that run from points $(i,\ j,\ k)$ to $(i,\ j,\ k+1)$.) In each plane the points that serve as the *source* of the row and column are marked as dark nodes.

Starting from this DG, we present three different projections, each having different properties, and the resulting SFGs.

Case 1: Projection in the [1 1 1] Direction − A DSI projection
The [1 1 1] is a DSI direction in the DG. If we project the DG along this direction and use the recursion schedule in the k-direction, the resulting SFG will be a TI SFG as shown in Figure 3.37(a). The function of each node is fixed and the structure of the SFG does not change with time.

Case 2: Projection in the j-Direction − A DSI Projection The projection direction in the j-axis is a DSI direction, and the resulting SFG array is an STI SFG, which can be scheduled linearly. The SFG obtained

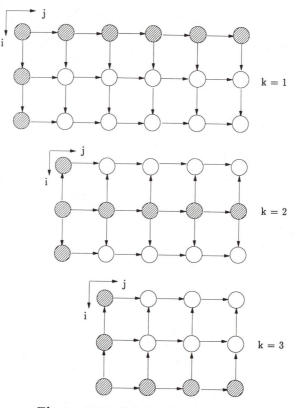

Figure 3.36: DG for the $N = 3$ case.

from this projection is shown in Figure 3.37(b), in which a default schedule is used.

Case 3: Projection in the i-Direction – A PDSI Projection The projection direction in the i-axis is a PDSI direction. The schedule we are using is still the default schedule. However, due to the PDSI projection, the SFG does not have a linear schedule. This SFG is shown in Figure 3.37(c).

Note that if the projection direction chosen is the k-axis, then the projection does not exhibit either DSI or PDSI properties. Therefore, the corresponding SFG is no longer regular.

Remark: The matrix multiplication, LU decomposition, and Gauss-Jordan elimination all share a common recursive formulation:

$$x_{ij}^k \; = \; x_{ij}^{k-1} \; + \; x_{ik}^{k-1} \cdot (x_{kk}^{k-1})^* \cdot x_{kj}^{k-1} \qquad (3.12)$$

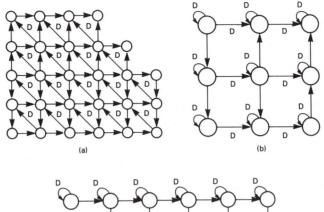

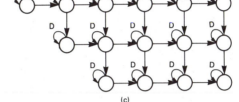

Figure 3.37: (a) A TI SFG derived by projection in the [1 1 1] direction. (b) An STI SFG derived by projection in the j-direction. (c) An STI SFG derived by projection in the i-direction.

where +, ·, and * denote algebraic operators to be specified by the application. This generalized formulation covers a broad and useful application domains. In addition to the matrix operations mentioned above, they also cover transitive closure, shortest-path problems, and many others. The option of adopting different projection directions as well as some reindexing schemes allows many varieties of SFGs to be derived. It is essential for achieving an optimal systolic array. This subject will be treated in Chapter 4.

3.4.2   Mapping to Arrays Without Interior I/O

In many cases projecting a DG onto an SFG will cause the resulting I/O to fall on interior nodes of the SFG array, resulting in a large number of I/O ports. For example, Figure 3.38(a) shows a two dimensional $N \times N$ DG in which inputs occur at the top and left sides, and outputs occur at the bottom and right sides. Although all I/O for this DG occurs on the boundaries, any projection of this DG onto a linear array will result in I/O on the array's

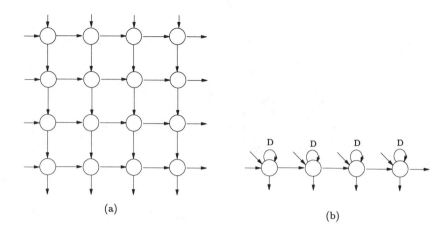

(a)

(b)

Figure 3.38: (a) Dependence graph; (b) SFG from vertical projection.

interior nodes. This is illustrated in Figure 3.38(b) which shows a vertical projection onto a linear SFG array.

To circumvent this problem, it is possible to *extend* the index space of the DG so that all I/O occurs at points that will be mapped to the boundary nodes of the SFG. This extension corresponds to defining the communication schemes (1) *to transport input data from the boundary nodes of the SFG to the interior nodes where it is needed,* and (2) *to transport output data to the boundary nodes of the SFG from the interior nodes where it is produced.* Extending index space to eliminate interior I/O was proposed in [Rao85] for restricted projection directions. Here this concept is generalized to deal with arbitrary projection directions. Given a DG with a projection direction $\vec{d}$, the procedure of index space extension is as follows:

1. *Decide the I/O border of the DG.* Since the structure of the SFG is determined by $\vec{d}$, the boundary nodes of the SFG is well defined and the intended I/O positions in the SFG are known. Some borderlines (or planes), which will be projected to the boundary nodes of the SFG, may be decided in the DG. Figure 3.39(a) shows borderlines for the previous example with $\vec{d} = [1\ 1]^T$ and Figure 3.39(b) shows borderlines with $\vec{d} = [0\ 1]^T$.

2. *Extend the DG with some NOP (no operation) nodes.* The purpose of these NOP nodes is to transmit the input (output) data from (to) the borderlines to (from) the destination nodes. *To simplify the control*

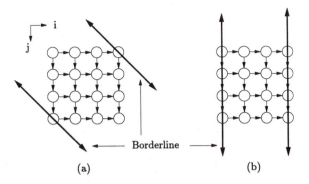

Figure 3.39: (a) Borderlines with $\vec{d}=[1\ 1]^T$; (b) borderlines with $\vec{d}=[0\ 1]^T$.

of the resulting array, no extra dependency edge directions (other than those of the original DG) should be used for these NOP nodes. The insertion of NOP nodes for the previous example with $\vec{d}=[1\ 1]^T$ and its resulting SFG (using the default schedule) is shown in Figure 3.40. Note that only single dependency edge direction is used for each NOP node. For comparison, the insertion of NOP nodes for the case of $\vec{d}=[0$

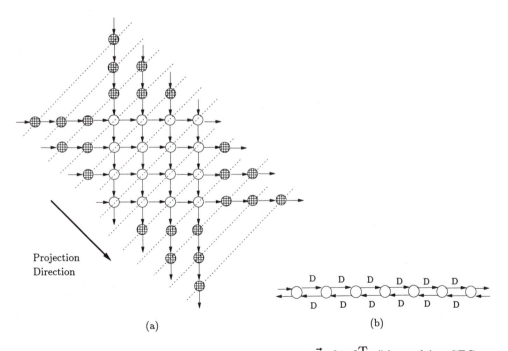

Figure 3.40: (a) Index space extension with $\vec{d}=[1\ 1]^T$; (b) resulting SFG.

1]^{T} and the corresponding SFG (with the schedule $\vec{s} = [1\ 1]^{\mathrm{T}}$) is shown in Figure 3.41. Note that a single dependency edge direction for each NOP node is not enough for this projection direction and thus two directions are used.

Comment: The detailed I/O operation of the array in Figure 3.40 can be stated as follows. The data are first loaded from boundary PEs one by one. Once the destination PE is reached, the computation starts. Similarly, the output data will go toward the boundary PEs once they are produced without waiting until *all* the output data are produced. In this array each boundary PE handles both input and output. The detailed I/O operation of the array in Figure 3.41 is as follows. Each PE takes its required input data, lets the other data pass through, and starts to do computation after all the data with other PEs as the destination pass through. Once the computation is started, the succeeding data are treated as normal incoming data (no more passing through unless the data is transmittent). In this array the input and output is handled by different boundary PEs.

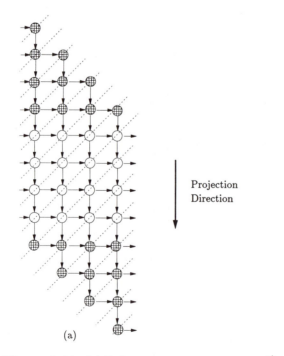

(a)

Projection
Direction

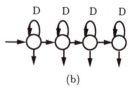

(b)

Figure 3.41: (a) Index space extension with $\vec{d} = [0\ 1]^{\mathrm{T}}$; (b) resulting SFG.

The data I/O may also be accomplished by preloading of data, where the computation is started only after *all* the input data reach their destination PEs, or post-dumping the output data after all computation is done. Similar kinds of control mechanisms to the NOP DG derived arrays will be required. The implementation of preloading (or post-dumping) is easier because the whole array is performing the same operation. However, the overall operation time is in general longer. Hence there is a trade-off between speed and hardware.

3.4.3   Multiprojection

In general, it is possible to map an N-dimensional DG directly onto an $(N - k)$-dimensional SFG ($k = 1, 2,..., N - 1$) [Wong85] [Rao85] [Saal86]. In the previous sections we have proposed a methodology that maps an N-dimensional DG to an $(N - 1)$-dimensional SFG. In principle, the method can be applied k times and thus reduces the dimension of the array to $N - k$. More elaborately, a similar projection method can be used to map an $(N-1)$-dimensional SFG into an $(N - 2)$-dimensional SFG, and so on. This scheme is called the *multiprojection* method.

To facilitate the discussion on the multiprojection technique, we introduce the notion of *DSI subspace*, which extends the previous notion of DSI direction. Given a DG, a k-dimensional subspace is a DSI subspace if and only if the DG is DSI in any direction in the k-dimensional subspace. (Equivalently, a k-dimensional subspace is a *DSI subspace* of a DG if and only if the DG is DSI in k independent vectors (directions) in the subspace.) To simplify the discussion, let us assume that the DG is N-dimensional, the SI subspace is 2-D and is on the ij-plane, and, after one projection (say, in the i direction), the resulting SFG is $(N - 1)$-dimensional. The question is *how to further project the SFG to an (N − 2)-dimensional SFG?*

The potential difficulties of this mapping are (1) the presence of delay edges in the $(N - 1)$-dimensional SFG and (2) the possibilities of loops or cycles in an SFG, although no loops or cycles can exist in a DG. Therefore, mapping this SFG to an $(N - 2)$-dimensional SFG will require additional care.

To handle the cycle problem, an *instance graph* (at certain time $t = t_0$) is defined as *the SFG with all the delay edges removed*. Hence the *instance graph* has no loops or cycles. According to the SFG schedule, all the nodes in the instance graph are executed simultaneously at $t = t_0$. An activity instance (at t_0) can be defined as the nodes represented by the instance graph at $t = t_0$. According to the SFG schedule, there will be no overlap in time between two consecutive activity instances (one at $t = t_0$ and the next instance at $t = t_0 + D$).

Recall that a mapping methodology should consist of a projection part and a schedule part.

- As to the projection part, since the assumption of ij DSI subspace implies that the instance graph is itself DSI in the j-direction, the same projection method proposed in Section 3.3.2.2 may be applied in the j- direction. *The original SFG (with delay edges included) is now projected along the $\vec{d}$ (i.e., j-) direction.*

- The schedule part is somewhat more complicated. First, it is always possible to find a valid SFG schedule vector, $\vec{s}$, for the instance graph, since there is no loop or cycle in an instance graph. To project the $(N-1)$-dimensional instance graph to an $(N-2)$-dimensional graph, it is necessary to create a new type of delay, denoted by τ. Note that the *global* delay D and *local* delay τ are intimately related. The relationship depends upon the constraints imposed by both the processor availability and the data dependency (i.e., data availability). *The original delay edges with βD map to an edge bearing delay weight $\beta D + \vec{s} \cdot \vec{e}\tau$ (see Figure 3.42(b)).*

Now let us more closely examine the relationship between D and τ. *First*, to ensure *processor availability*, an activity instance must have adequate time to be complete before the next activity instance starts. So the following condition is necessary:

(a) $D \geq \tau + (M - 1)(\vec{s} \cdot \vec{d})\tau$

where M is the maximal number of nodes along the $\vec{d}$-direction in the instance graph (see Figure 3.42(a)). Note that this condition also guarantees that there is no time overlap between two activity instances.

Second, to ensure *data availability*, conditions (b) and (c) are necessary for all edges $\vec{e}$:

(b) $\beta D + (\vec{s} \cdot \vec{e})\tau \geq 0$

(c) $\beta D + (\vec{s} \cdot \vec{e})\tau \geq \tau$ for at least one edge in every cycle.

Condition (b) is necessary in order to satisfy the causality condition of the SFG. Condition (c) is necessary to ensure that every cycle in the new

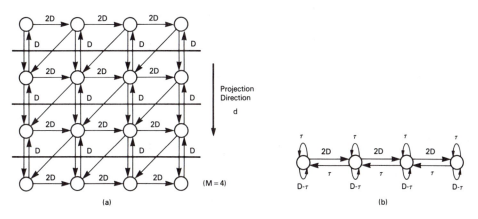

Figure 3.42: (a) The original SFG; (b) The further projected SFG.

SFG has at least one delay element (τ) in the cycle. Note that these two conditions are necessary conditions for a graph to be qualified as an SFG.

Conditions (a), (b), and (c) together guarantee both the processor and data availability for the new SFG. Therefore, the mapping is complete. Note that condition (a) is very often the dominant constraint. In fact, condition (a) would be sufficient to ensure conditions (b) and (c) whenever a locally interconnected SFG is considered.

Obviously, this mapping methodology may be directly applied to map from $(N-2)$-dimensional SFG to $(N-3)$-dimensional SFG. The method can be applied k times (i.e., using k steps of simple projection) to reduce the dimension of the array to $N-k$.

Example 2: Band Matrix Multiplication

To demonstrate the procedure, a 2-D SFG for *band matrix multiplication* and its corresponding projected SFG are shown in Figure 3.43(a) and (b). In Figure 3.43(b), note that the matrix **B** is input in parallel and will lead to I/O bandwidth of order M which is usually not desirable. Since there is only one data in M time steps for each parallel input edge, these data can be interleaved and input from the boundary PE, as shown in Figure 3.43(c). An alternative way to obtain this SFG is to extend the index space (see Section 3.4.2) of the SFG and to overlap the execution of the extended index space (for I/O) with the next instance graph (for computation). By a similar technique, the output directly from interior nodes (cf., Figure 3.43(c)) may also be avoided. The detail procedure is left to the reader.

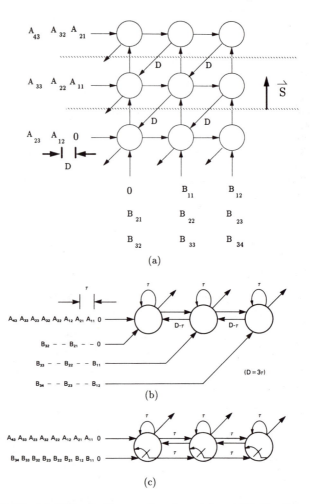

Figure 3.43: Multiprojection on band matrix multiplication: (a) A 2-D SFG and a "good" schedule; (b) the 1-D projected SFG; (c) the final SFG using $\mathbf{D} = 3\tau$.

3.4.4   Nonlinear Schedule and Nonlinear Assignment

3.4.4.1   Piecewise Linear Schedule

If the projection vector $\vec{s}$ is in a DSI direction, there are two kinds of mappings: linear schedule mapping (LS-DSI) and nonlinear schedule mapping (NLS-DSI). An LS-DSI mapping maps a DG to an SFG with a linear schedule $\vec{s}$, which satisfies the following conditions:

(1) $\vec{s}^{\mathrm{T}} \vec{e} \geq 0$,   for any dependence arc $\vec{e}$.

(2) $\vec{s}^{\mathrm{T}} \vec{d} > 0$.

An example is given in Figure 3.44(a). If it is other than an LS-DSI mapping, it is called an NLS-DSI mapping.

Usually, an NLS-DSI mapping can be viewed as a combination of many relaxed LS-DSI mappings, where the condition (2) is replaced by

$$\vec{s}^{\mathrm{T}} \vec{d} \neq 0.$$

In this case, such a mapping is called a *piecewise linear schedule* (PWLS). An example is given in Figure 3.44(b).

Since a PWLS-DSI mapping can be viewed as a combination of many LS-DSI mappings, the corresponding SFG should require only simple control. To prove that the circuit design will be regular and modular, we note that the schedules are linear in both sides of a partitioning "interface" plane (see Figure 3.45). Actually, a linear schedule may be used for each of the LS-DSI DG regions and each region may be projected to an STI SFG with simple control. What is worth noting is the design of interface modules required for handling data crossing the interface plane. There are two possible types of relationships between the two neighboring linear schedules. The first possibility is that both the linear scheduling time indices increase along the $\vec{d}$ direction. The second possibility is that the two linear schedules progress in an opposite direction with respect to $\vec{d}$. Correspondingly, there are two possible interface models. For the first case, for the purpose of adjusting and balancing the speeds of data processing rates, a buffer (FIFO: first-in-first-out) suffices to provide a proper interface between the two regions. On the other hand, for the second case, a stack (LIFO: last-in-first-out) suffices

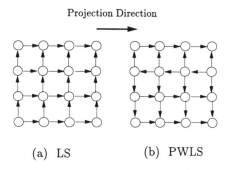

(a) LS   (b) PWLS

Figure 3.44: (a) An LS-DSI DG mapping. (b) A PWLS-DSI DG mapping.

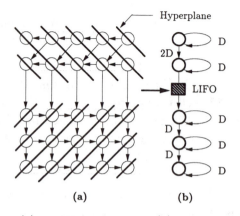

Figure 3.45: (a) A PWLS-DSI DG. (b) Its corresponding SFG with FIFOs and LIFOs.

to handle the interface. Note that all the FIFOs and LIFOs mentioned require only simple control. A PWLS-DSI DG to SFG mapping and its corresponding SFG with FIFOs and LIFOs are shown in Figure 3.45.

Projections in non-DSI directions generally allow only nonlinear schedules. However, the projected SFGs may have other advantages, such as fewer PEs, faster pipelining, or higher utilization of the array. If an advantageous trade-off can be reached, a nonlinear schedule mapping may become preferred.

3.4.4.2   Nonlinear Assignment

By *nonlinear assignment* we mean that multiple nodes *not necessarily along a straight line* are assigned to a PE. Nonlinear assignment to map a DG to an SFG usually incurs the expense of somewhat sophisticated control. In certain special circumstances, a nonlinear assignment mapping may offer some unique flexibility and advantages. In the following, we shall discuss two such examples.

Example 3: Rebound Sorter

As an example, a *rebound sorter* [Chen78] can be derived by using a nonlinear assignment. The DG (derived in Section 3.3.1.1, Figure 3.5) is mapped to SFG by a nonlinear assignment shown in Figure 3.46. In this mapping, node (i, j) is mapped to $f(i, j)$ cell in the SFG, where $f(i, j)$ is a nonlinear function $(f(i, j) = \min\{N + 1 - i, j\})$.

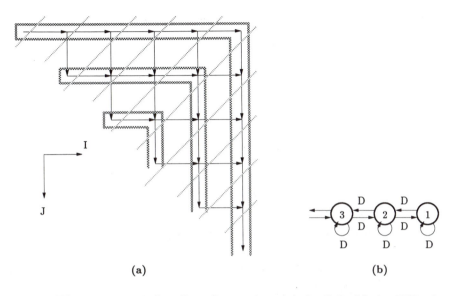

Figure 3.46: A "nonlinear" mapping: (a) the DG; (b) the SFG of rebound sorter.

Consequently, only $N/2$ processors are needed in the SFG array, and the processor utilization is maximal. The name *rebound sorting* comes from the fact that input data comes from the left side of the array, and when it reaches the right side of the array it "rebounds" and travels backward until it is output from the left side of the array. As a result of nonlinear assignment mapping, the control design for this SFG has to be somewhat more complicated.

Example 4: Consecutive Matrix-Vector Multiplication

By using a nonlinear assignment, a more flexible design can be devised and a broader range of algorithms can be covered. One such example is an algorithm represented by *cascaded DG*, in which the algorithm is comprised of a group of DGs connected in cascade. A simple example is shown in Figure 3.47, where three DGs for matrix-vector multiplication are cascaded to compute $e = ABCd$. Here, A, B, and C are 3×3 matrices and both e and d are 3×1 vectors.

For this cascaded DG, all the DGs involved are the same. A nonlinear assignment can be applied to this cascaded DG (see Figure 3.47). In this case, the nonlinear mapping allows handling of the three DGs together in

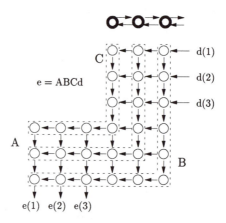

$$e = ABCd$$

Figure 3.47: Cascaded DG for matrix-vector multiplication.

one piece rather than separately. This will ease the data reformating and increase the pipelining rate.

3.4.5   Linear Projection onto SFG with Global Communication

The notion of linear projection and linear schedule on DSI DG may be similarly applied to a global DG. This leads to an SFG with static (but global) interconnection. To illustrate this, the multi-stage algorithm for FFT array computation as discussed in Chapter 2 is reproduced in Figure 3.48(a). Note that this can be viewed as a DSI DG and a projection along the horizontal direction may be applied to yield a linear schedule mapping. The projected SFG array with a perfect-shuffle network is shown in Figure 3.48(b). It is

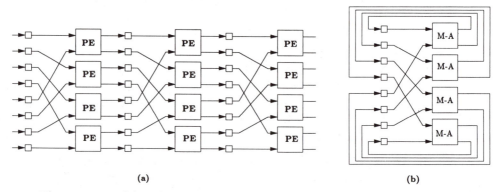

(a)   (b)

Figure 3.48: (a) Multistage array for FFT computation. (b) Projected array.

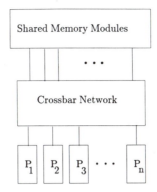

Figure 3.49: A fully connected MIMD machine example.

worth noting that perfect shuffle networks have found applications to many problems such as sorting, coding, and FFT. Another interesting example on the Viterbi decoding is given in Problem 15.

3.4.6   Minimal Latency Mapping for General DGs

In the previous discussion, the DGs treated are highly regular and the nodes of such DGs are often fine grained, such as the MAC (multiplier-and-accumulator) node. In general, a DG can be totally irregular, especially when the nodes are coarse grained, e.g., one task per node. This leads to a totally different kind of computing machines. In general, a very irregular DG may have to be implemented on a fully interconnected, shared-memory MIMD machine (as shown in Figure 3.49). Because of its lack of regularity, neither the linear projection nor the linear schedule may be deemed as effective. A more general mapping strategy will have to be considered. Several points concerning a general DG mapping will now be addressed, including the following:[5]

1. Finding a schedule to minimize the latency subject to *no constraints* on the number of PEs or on local communication. (*Latency* is the time interval between loading the first input and unloading the last output of a problem instance into/from the processor array.) Determining the minimum number of required PEs (and the optimal node assignment) sufficient to achieve the minimal latency schedule.

---
[5]It is assumed that the cost of a global communication and a local communication are the same.

2. Given a fixed number of PEs, how to find the optimal latency schedule and node assignment. Given a fixed number of PEs and a fixed node assignment, how can the optimal latency schedule be determined?

3.4.6.1 Minimal Latency Schedule

Let us first address Point 1. Under the assumption that a sufficient number of PEs are available and all the PEs are fully connected, the minimal latency schedule can be determined rather straightforwardly. The method is based on the partial ordering of the nodes in the DG. In fact, this kind of problem has arised in many operations research applications and several well-known algorithms have been developed and reported [Taha82].

The minimal latency schedule can be determined in two phases. The first phase is a forward-pass phase: the timing calculation begins from the source nodes, i.e., those nodes with no precedents, and travels along all possible paths to the sink nodes. By this scheme, the earliest possible completion time (T_E) for every node can be computed. The minimal latency time (T_m) is thus equal to

$$\max\{T_E\}$$

In the second phase, the paths may be traced back to determine the latest completion time (T_L) for each node so that the minimal latency T_m is still achievable. *A critical path consists of all the nodes whose computations have to be completed by T_E (i.e., $T_L = T_E$ for those nodes) if the minimum latency is to be achieved.*

Now we are ready to address the problem of minimizing the total number of PEs in Point 1. The nodes that are scheduled to be executed at the same time should be assigned to separate PEs so that they may be processed in parallel and the minimal latency schedule may be accomplished. Therefore, the minimum number of PEs required in order to comply with the given schedule will be equal to

$$\aleph \; = \; \max_t\{\text{the number of simultaneously active nodes at } t\} \qquad (3.13)$$

From the perspective of scheduling strategy, note that there are two kinds of active nodes at any time (say, $t = t_0$): (1) those on the critical path and hence with a fixed schedule; and (2) those not on the critical paths that have a flexible schedule. The range of permissible adjustment of timing depends on the difference $T_L - T_E$, and the node execution time is called the total float of that node.

To minimize the number of PEs required, the execution time for those nodes not on the critical paths can be adjusted to minimize the integer $\aleph$ as defined in Eq. 3.13 and yet not to increase the latency. In operations research, this problem is called the *resource leveling problem*. No technique for an optimal solution has yet been developed. However, there exists some heuristic approaches, and the reader is referred to [Taha82].

3.4.6.2   Integer Programming Formulation for Minimal Latency Given a Fixed Number of PEs

Suppose that the problem is to be executed with m PEs. There are two different situations. First, when $m \geq \aleph$, then the approach just given applies directly. The more interesting and practical case is to minimize the latency given a fixed number of PEs and $m < \aleph$. This problem can be regarded as a special case of a *general scheduling problem* in combinatorial optimization, which can be formulated as an integer programming problem [Papad83].

Given a DG of nodes $\{v_1, \ldots, v_n\}$, with node i being assigned an integer τ_i as its processing time.

This problem consists of two parts: How to do the assignment and how to do the schedule.

1. *Node assignment* The node assignment function is denoted by a mapping function p_{ij} (for $i = 1, \ldots, n$, $j = 1, \ldots, m$) with $p_{ij} = 1$ if i is executed on machine j and 0 otherwise.

2. *Scheduling* The problem is to determine a schedule function (denoted by starting time t_i for v_i), for $i = 1, \ldots, n$ (assume that the t_i's are integer, and that $\min\{t_i\} = 0$).

The overall objective is to minimize the latency f which is equal to

$$f = \max_i \{t_i + \tau_i\}$$

Let us investigate the scheduling constraints involved in this problem.

1. *PE availability constraints* Since no two nodes may be executed simultaneously in the same PE, whenever $p_{ij} = p_{kj} = 1$ and node v_i executes before v_k, we have $t_i + \tau_i \leq t_k$.

2. *Data dependency constraints* Due to data dependency, we have $t_i + \tau_i - t_k \leq 0$ if v_k depends on v_i.

If we let $T = \sum_{j=1}^{n} \tau_j$, the minimal latency problem may be reformulated to an integer programming problem [Papad83]:

$$\min f$$

subject to the following conditions (1)–(8) on the nonnegative integers t_i, τ_i, p_{ij}, and δ_{ij}:

1.

$$f \geq t_i + \tau_i \qquad i = 1, \ldots, n$$

2.

$$\sum_{j=1}^{m} p_{ij} = 1 \qquad i = 1, \ldots, n \qquad (3.14)$$

3.

$$p_{ij} \leq 1, \qquad j = 1, \ldots, m, \quad i = 1, \ldots, n \qquad (3.15)$$

4.

$$t_k - t_i \leq \delta_{ik} \cdot T \qquad i \neq k, \quad i, k = 1, \ldots, n$$

5.

$$\delta_{ik} + \delta_{ki} = 1 \qquad i < k, \quad i, k = 1, \ldots, n$$

6.

$$t_i + \tau_i - t_k \leq T \cdot (3 - \delta_{ik} - p_{ij} - p_{kj}) \qquad (3.16)$$
$$j = 1, \ldots, m, \qquad i \neq k, \quad i, k = 1, \ldots, n$$

7.

$$\delta_{ik} = 1 \qquad \text{if } v_k \text{ depends on } v_i$$

8.

$$t_i + \tau_i - t_k \leq 0 \qquad \text{if } v_k \text{ depends on } v_i$$

Minimal Latency Schedule with Fixed Node Assignment Sometimes the assignment is already fixed when the problem is formulated, then the problem to find the optimal latency schedule can also be formulated in a way similar to the preceding integer programming problem. The only modification is that the p_{ij} will be constants instead of variables (i.e., Eq. 3.14 and Eq. 3.15 will be removed and p_{ij} and p_{kj} in Eq. 3.16 are prespecified and are regarded as constants).

3.5 Concluding Remarks

VLSI arrays derive a maximal concurrency by using both pipelining and parallel processing. An important question thus is: *How to fully express the parallel and pipeline processing inherent in algorithms?* Firstly, a fundamental issue is a concise representation of the data structure of the algorithms. For our application, a dependence graph (DG) provides a useful graphic expression and permits certain structured modifications on the graph. Secondly, a critical step in architectural design is mapping data structures to array structures. Starting with an algorithm expressed by its DG, a designer may follow the guidelines provided by the mapping methodologies proposed in this chapter. A basic approach is the canonical mapping methodology that handles regular DGs. To broaden the domain of array mapping, generalized mapping methodology is also proposed to handle certain irregular DGs and more flexible node/schedule assignments. Either the canonical or the generalized mapping methodology can be divided into three stages. In the first stage, the DG is derived and modified to satisfy certain structural constraints. In the second stage, an SFG can be derived from that DG by a simple projection method. In the last stage, the SFG obtained can be mapped to SIMD, systolic, wavefront, or even MIMD array.

The systematic transformation from DG to systolic arrays greatly simplifies the understanding of systolic design and provides a means to verify the design. As to *optimal* systolic design, it depends on the performance criteria. Obviously, the selection of a particular DG for the algorithm and the types of the projection and scheduling can greatly affect the performance of the resulting array processors. This issue will be addressed further in Chapter 4. The DG mapping technique is potentially applicable to problems on partitioning and fault tolerance. This topic is currently being investigated. Another important problem is that, when the array architecture is fixed, how can algorithms be executed effectively by that specific array processor.

This is in fact the problem of matching algorithmic data structures to array architectural structures. This will involve a more general methodology, which still under investigation. In Chapter 6, we shall discuss some preliminary results with several examples on partitioning and structural matching technique.

As clearly indicated by the road map in the preface of the book, Chapter 3 can indeed be considered a pivotal chapter. It paves the ground for the design and analysis of the systolic and wavefront array processor (Chapters 4 and 5), and provides a good understanding of software development on programming/design/simulation (Chapter 6).

3.6 Problems

1. *Derivation of a recursive algorithm*: The following formula is for convolution:

$$y(n) = \sum_{k=0}^{n} u(k)w(n-k)$$

   (a) Write a FORTRAN code to obtain $y(n)$. Assume that both input sequences have the same sequence length N.

   (b) Rewrite the code in a single assignment form, which will lead to the following recursive algorithm.

$$y_j^k = y_j^{k-1} + u_k \cdot w_{j-k}$$

2. *Localization of a recursive algorithm*: Is it possible to localize the following recursive algorithm? If possible, localize it. Give a reason for your answer.

$$y_j^k = y_j^{k-1} + a(k) \cdot y(j-k)$$
$$y(j) = y_j^N, \quad y_j^0 = u(j)$$

   where $j = 0, 1, 2,\ldots, \infty$, $k = N, N-1, N-2,\ldots, 1$.

3. *Localization from a sequential code*: The FORTRAN code given below is for the convolution of two sequences. The sequence length is four

for both sequences. Convert the code to a single assignment code and localize it.

```
        DO  10  J=1,4
        DO  10  K=1,J
        Y(J)=Y(J)+U(K)*W(J − K)
   10   CONTINUE
        DO  20  J=5,7
        DO  20  K=J − 3, 8
        Y(J)=Y(J)+U(K)*W(J − K)
   20   CONTINUE
```

4. *Derivation of DGs*: For the following algorithms, derive suitable DGs.

   (a) Correlation

   (b) Back-substitution

   Show that they belong to the class of locally recursive algorithms.

5. *SFGs for correlation*: Derive three different kinds of SFGs for correlation. (They do not necessarily correspond to the same DG.)

6. *Index transformation and different SFGs for convolution*:

   The DG for convolution is shown in Figure 3.6(b). Taking into account the shape of the DG, we adopt the following index transformation.

   $$\begin{bmatrix} j' \\ k' \end{bmatrix} = \begin{bmatrix} 1 & -1 \\ 0 & 1 \end{bmatrix} \cdot \begin{bmatrix} j \\ k \end{bmatrix}$$

   The result is a rectangular shaped DG as shown in Figure 3.50(a). Note that the $w(i)$ coefficients along the columns remain unchanged. This means that the coefficient $w(i)$ may be stored as a constant in the ith processor, as shown in Figure 3.50(b). Another SFG, obtained by projection along [1 0], is shown in Figure 3.50(c).

   (a) For the DG as shown in Figure 3.6(b), if the projection direction is along [1 1], show the SFG obtained. Compare this SFG with the one obtained from an index-transformed DG, as shown in Figure 3.50(b).

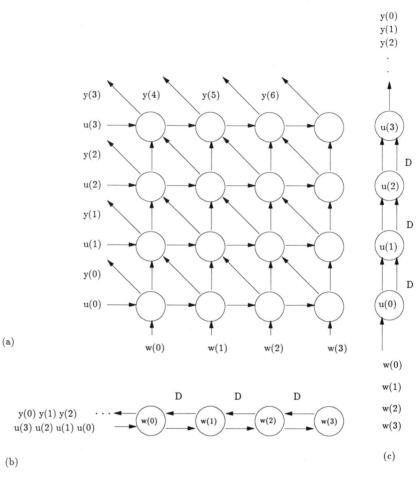

(a)

(b)

(c)

Figure 3.50: (a) Index transformed DG for convolution; (b) SFG obtained by projection along [0 1]; (c) SFG obtained by projection along [1 0].

(b) Note that the operations of addition involved in the convolution are associative, so the direction of dependence arcs can be reversed. Can you get any "new" (or useful) SFGs by using this technique?

7. *Binary number multiplication*: For the problem of binary number long multiplication, assume that the multiplicand is of length 5 bits and the multiplier is of length 4 bits.

   (a) Derive a DG for this problem. The granularity of the node operation of this DG should be in full adder level.

(b) Which projection direction can be applied to obtain an SFG with a serial multiplicand input? How many PEs are needed?

(c) Which projection direction can be applied to obtain an SFG with a serial multiplier input? How many PEs are needed?

(d) Is it possible to obtain a projected SFG that has the property that both multiplicand and multiplier are input serially?

(e) Is it possible to use index transformation (see Problem 6) to reduce the number of PEs needed?

8. *Band matrix multiplication*: Assume that $N \gg P$ and $N \gg Q$, where P and Q are bandwidths for matrices $\mathbf{A}$ and $\mathbf{B}$, respectively. Then it is possible to achieve full parallelism with only a $P \times Q$ rectangular array (as opposed to an $N \times N$ array). This fact can be explained by using the DG and applying suitable projection to get the SFG. Derive the DG for two band matrix multiplication and give a suitable projection to get a $P \times Q$ rectangular array.

9. *SFG derivation of linear phase filter*: Derive an SFG for the linear phase filter of length $N = 5$ by using a DG for convolution.

   Hint: Note that there is a symmetry property in the DG and a folding of the DG is required.

10. *Householder transformation*: The Householder transformation is used as an elementary step in a number of basic transformations, e.g., for decomposing a matrix $\mathbf{A}$ into an orthogonal matrix $\mathbf{Q}$ and an upper triangular matrix $\mathbf{R}$ (the so-called QR decomposition algorithm) [Stran80]. The transformation is implemented by first applying the following decomposition,

$$
\mathbf{Q}(1)\mathbf{A} = \begin{bmatrix} x & x & x & \cdots & x \\ \hline 0 & & & & \\ 0 & & \mathbf{A}_s & & \\ \vdots & & & & \\ 0 & & & & \end{bmatrix}
$$

where $\mathbf{Q}(1) = \mathbf{I} - 2 \dfrac{\mathbf{u}\mathbf{u}^{\mathrm{T}}}{\|\mathbf{u}^{\mathrm{T}}\mathbf{u}\|}$, $\mathbf{u} = \mathbf{A}_1 + \|\mathbf{A}_1\|\mathbf{z}$, and $\mathbf{A}_1$ is the first column vector of $\mathbf{A}$, and $\mathbf{z} = [1 \ 0 \ 0 \ \ldots \ 0]^{\mathrm{T}}$. Then repeatedly apply similar decompositions, namely, $\mathbf{Q}(i)$, $i = 2, 3, \ldots, N - 1$. $\mathbf{Q} = \mathbf{Q}(N - 1) \cdots$

Q(1). (Here **Q**(i) modifies only the lower diagonal submatrix of size $N - i + 1$ and N denotes the number of columns of **A**.)

Show that the above recursive algorithm can be mapped onto a parallel processing SFG network, as shown in Figure 3.51. This figure shows that after one recursion, the submatrix **A**$_s$ will be moved to the left. The delays **D** represent the state of the SFG and contain the columns of the submatrix **A**$_s$ ready to be processed in the next recursion.

Hint: To illustrate the notion of state in the Householder QR algorithm and its SFG representation, note that the algorithm transforms the initial matrix **A** into an upper triangular matrix **R** by means of an orthogonal transformation **Q**, **Q A** = **R**. The orthogonal transformation is implemented by repeatedly applying similar decompositions, namely, **Q**(i), i = 1, 2, 3, ..., $N - 1$. Figure 3.51 shows that after one recursion, the submatrix **A**$_s$ represents the information needed for future processing (i.e., "state"). According to Figure 3.51, **A**$_s$ should be moved to the left in order to be ready for the next recursion. As depicted in Figure 3.51, the delays **D** represent the storage needed for the state of the SFG computing network, which contains the submatrix **A**$_s$ that will be processed in the next recursion.

11. *Guidelines for localization (pictorial approach)*: Following the guideline given below, draw a localized DG for convolution.

    (a) Starting from a single assignment equation (sometimes, a recursive equation), the first step is to specify the boundary of the index space.

    (b) For those variables that have all the superscripts and subscripts

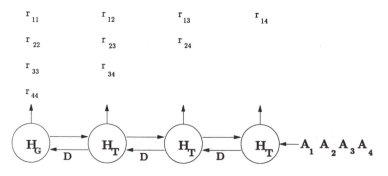

Figure 3.51: SFG for the Householder algorithm.

and are already localized, use arcs to connect their corresponding index points.

(c) For those variables that have any absent (superscript or subscript) indices, two approaches may be applied.

    i. For external variables, use local arcs to propagate the data along the corresponding public-data contours. The directions of propagation can be arbitrarily assigned and should be decided with other design considerations.

    ii. For intermediate variables, use local arcs to propagate the data along the corresponding public-data contours. Now the directions of propagation are no longer arbitrarily assigned. Where the intermediate variables are produced must be taken into consideration.

12. *Guidelines for localization (algebraic approach)*: Using the guidelines (a) and (b) below, derive a locally recursive algorithm for autoregressive (AR) filter, where the following equations are to be localized:

$$y_j^k = y_j^{k+1} + a(k) \cdot y(j - k)$$
$$y(j) = y_j^1, \quad y_j^{N+1} = u(j)$$

(a) For those external variables, use the same method as the pictorial one.

(b) For intermediate variables that should be localized:

    i. Retain the indices as a subscript[6] and append additional superscripts such that all the superscripts and subscripts are explicitly present. The superscripts are appended in such a manner that these variables will be the corresponding output variables.

    ii. The differences of subscript and superscript point to a global dependency. This difference vector is a global dependence vector.

    iii. Express the global dependence vector as linear combination of some localized vectors, which are independent of indices. Use

---

[6]For simplicity, we assume that the mixed indices can be represented as a subscript.

these localized vectors to propagate the data. If necessary, use some dummy variables to propagate the data.

Hint: The equations may be reformulated as follows:

$$y_j^k = y_j^{k+1} + a^k \cdot y_{j-k}^1$$

$$y(j) = y_j^1, \quad y_j^{N+1} = u(j), \quad a^k = a(k)$$

To deal with y_{j-k}^1, note that the dependence vector between y_j^k and y_{j-k}^1 is

$$\begin{bmatrix} k \\ j \end{bmatrix} - \begin{bmatrix} 1 \\ j-k \end{bmatrix} = \begin{bmatrix} k-1 \\ k \end{bmatrix} = k \cdot \begin{bmatrix} 1 \\ 1 \end{bmatrix} + \begin{bmatrix} -1 \\ 0 \end{bmatrix}$$

Thus we may use both $(1, 1)$ and $(0, -1)$ as the propagation vectors. If a dummy propagation variable B_j^k is introduced, the following equations may be obtained:

$$B_j^k = B_{j-1}^{k-1}, \quad B_{j-k}^0 = y_{j-k}^1$$

13. *Canonical mapping*: For the DG as shown in Figure 3.52,

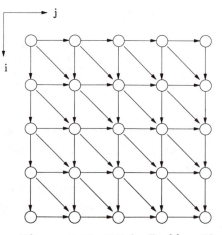

Figure 3.52: DG for Problem 13.

   (a) Which of the following sets of scheduling and projection is permissible?

      i. $\vec{s} = [1\ 0]^{\mathrm{T}}$, $\vec{d} = [1\ 0]^{\mathrm{T}}$
     ii. $\vec{s} = [1\ -1]^{\mathrm{T}}$, $\vec{d} = [1\ 0]^{\mathrm{T}}$
    iii. $\vec{s} = [1\ 1]^{\mathrm{T}}$, $\vec{d} = [1\ 0]^{\mathrm{T}}$
    iv. $\vec{s} = [1\ 1]^{\mathrm{T}}$, $\vec{d} = [1\ -1]^{\mathrm{T}}$

   (b) Give the projected SFG for each permissible set.

14. *Generalized mapping*: For the DG as shown in Figure 3.53, which of the following sets of scheduling is permissible?

   (a) $\vec{s} = [1\ 0]^{\mathrm{T}}$

   (b) $\vec{s} = [1\ -1]^{\mathrm{T}}$

   (c) $\vec{s} = [1\ 1]^{\mathrm{T}}$

15. *Viterbi decoding*: The Viterbi algorithm is well known to be an efficient method for the realization of maximum likelihood decoding of convolutional codes. Here it will be used to be an example of linear projection of global communication.

Consider a convolutional code of code rate $1/n$ and constraint length

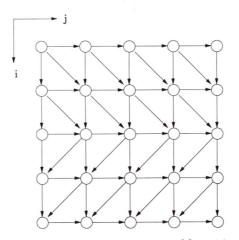

Figure 3.53: DG for Problem 14.

State

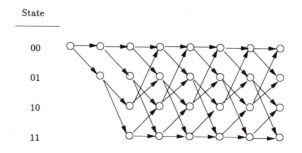

00

01

10

11

Figure 3.54: Trellis diagram (k = 3) (or DG) for Viterbi decoding.

k. Let $N = 2^{k-1}$ be the number of states. A Trellis diagram for the decoding, which can be viewed as a DG, is shown in Figure 3.54.

(a) What is the SFG obtained via a projection along the state iteration direction? What kind of SFG is this? Note that the communication for this SFG is global but static, so it can be implemented by a static interconnection network [Cain84].

(b) What is the SFG obtained via a projection along the vertical direction? What kind of SFG is this?

By using global communication, the PE utilization is very high. Although some research has been done on how to use local communication to do Viterbi decoding [Chang86], due to the communication constraint, the parallelism efficiency is still much worse than that of using global communication.

Chapter 4

SYSTOLIC ARRAY PROCESSORS

4.1 Introduction

In Chapter 3, we proposed methods of mapping algorithms onto SFG structures and suggested that incorporation of pipelining into an SFG array will lead to a systolic design. In this chapter, we first review the mapping methodology and propose the modification necessary for mapping DGs directly to systolic arrays. Then we discuss systematic methods for converting SFG arrays into systolic arrays based on a cut-set systolization (retiming) procedure. We present systolic design examples for many algorithms, such as filtering, convolution, matrix operations, and sorting. We then address the key issues in designing optimal systolic arrays, which include the options available at the different levels of mapping, such as optimization of throughput rate, computation time, block pipelining period and processor utilization. In the last two sections, we present systolic arrays for two important classes of problems that can be solved by (1) the dynamic programming method and (2) artificial neural networks. The first class of problems includes the transitive closure problem and the shortest-path problem in graph theory, Gauss-Jordan elimination in linear algebra and others. The

197

second class includes many image processing and combinatorial optimization problems.

4.2 Systolic Array Processors

Systolic processors [HTKun78], [HTKun82] are a new class of pipelined array architectures. According to Kung and Leiserson [HTKun78], "A systolic system is a network of processors which rhythmically compute and pass data through the system". For example, it can be shown that some basic "inner product" PEs ($y \leftarrow y + a \times b$) can be *locally* connected together to perform digital filtering, matrix multiplication, and other related operations. The systolic array features the important properties of modularity, regularity, local interconnection, a high degree of pipelining, and highly synchronized multiprocessing. The data movements in a systolic array are often described in terms of the snapshots of the activities.

The systolic array design differs from the conventional Von Neumann computer in its highly pipelined computation. More precisely, once a data item is brought out from the memory it can be used effectively at each cell it passes while being "pumped" from cell to cell along the array. This is especially appealing for a wide class of compute-bound computations, where multiple operations are performed on each data item in a repetitive manner. This avoids the classic memory access bottleneck problem commonly incurred in Von Neumann machines.

Computational tasks can be conceptually classified into two families: *compute-bound computations and I/O-bound computations* [HTKun82]. In a computation, if the total number of operations is larger than the total number of input and output operations, then the computation is compute-bound, otherwise it is I/O-bound. For example, ordinary matrix multiplication is compute-bound, whereas adding two matrices is I/O-bound. Speeding up the I/O-bound computations requires an increase in memory bandwidth, which is difficult in current technologies. Speeding up a compute-bound computation, however, may often be accomplished by using systolic arrays. The basic configuration of a systolic array, is illustrated in Figure 4.1. By

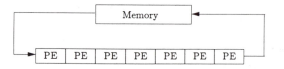

Figure 4.1: Basic configuration of systolic arrays.

replacing a single processor by a 1-D or 2-D array of processors, a higher computation throughput can be achieved without increasing memory bandwidth.

4.2.1   Definition of Systolic Arrays

There are a number of "definitions" of systolic arrays [HTKun78], [Ullma84], [Kung84a]. To give a coherent definition for further discussion, we adopt the following:

Definition: A systolic array is a computing network possessing the following features:

- *Synchrony* The data are rhythmically computed (timed by a global clock) and passed through the network.

- *Modularity and regularity* The array consists of modular processing units with homogeneous interconnections. Moreover, the computing network may be extended indefinitely.

- *Spatial locality and temporal locality* The array manifests a locally-communicative interconnection structure, i.e., spatial locality. There is at least one unit-time delay allotted so that signal transactions from one node to the next can be completed, i.e., temporal locality. Therefore, if an SFG, $G = <V, E, D(E)>$ (cf. Section 3.3.2), represents a systolic design then $D(e) \geq 1$ for all edges, e.

- *Pipelinability* (i.e., O(M) execution-time speedup) The array exhibits a *linear rate pipelinability*, i.e., it should achieve an O(M) speedup, in terms of processing rate, where M is the number of processing elements (PEs). Here the efficiency of the array is measured by the following:

$$\text{speedup factor} = \frac{T_S}{T_P}$$

where T_S is the processing time in a single processor, and T_P is the processing time in the array processor.

Example 1: Systolic array for Convolution   There are many ways to implement convolution using systolic arrays (by different projections of the DG as mentioned in Section 3.3). Here, we show only one particular version, which corresponds to the SFG in Figure 3.50(b). As shown in Figure 4.2,

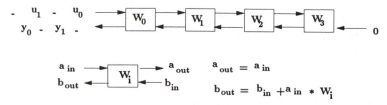

Figure 4.2: Systolic array for convolution.

the sequence $u(n)$ is input from the left, and weights $\{w(n)\}$ are preloaded and stay in the nth PEs in the same order.

The sequence $\{y(n)\}$ is initially pumped in from the right side – with initial values 0 – at the same speed as $u(n)$. The final result is pumped out from the left end of the array.

Example 2: Hexagonal Array for Multiplication of Two Band Matrices   To handle multiplication of two band matrices, $\mathbf{C} = \mathbf{AB}$, a hexagonal array has been proposed [HTKun78]. The general data arrangement is depicted in Figure 4.3.

The question now is how such a scheme actually produces the desired multiplication results. The easiest way to answer this question is to display the space-time activities in the first few consecutive beats, and then derive the remaining activities (and the general rule) by induction and inspection.

In this example, a 2-D hexagonal array is seen to be a natural topology for the band matrix multiplication problem. As shown in Figure 4.3, the input data (from matrices $\mathbf{A}$ and $\mathbf{B}$) are prearranged in an orderly sequence. The output data (of the matrix $\mathbf{C}$) will be pumped from the other side of the array, meeting the right data and collecting all the desired products. The detailed description of data movements and computations is often furnished in terms of the snapshots of the activities [HTKun78].

Interestingly, the LU decomposition (a critical procedure for matrix inversion) can be computed in a very similar fashion. The snapshots for the LU decomposition are just the reversal of those for the matrix multiplication. For more details, the reader is referred to [HTKun78] and [HTKun82].

4.2.2   Properties of Systolic Architectures

4.2.2.1   Why Systolic Architectures?

The major factors of adopting systolic arrays for special-purpose processing architectures are: *simple and regular design, concurrency and communication, and balancing computation with I/O* [HTKun82].

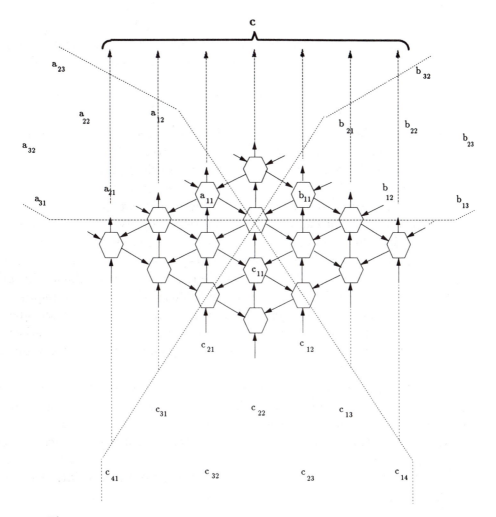

Figure 4.3: A hexagonal systolic array for band matrix multiplication.

Simple and Regular Design   In integrated-circuit technology, the cost of components is dropping dramatically; however, the cost of design grows with the complexity of the system. By using a regular and simple design and exploiting the VLSI technology, great savings in design cost can be achieved. Furthermore, simple and regular systems are likely to be modular and therefore adjustable to various performance goals.

Concurrency and Communication   An important factor in the potential speed of a computing system is the use of concurrency. For special-purpose systems, the concurrency depends on the underlying algorithms employed by the system. When a large number of processors work together, communication becomes significant. In VLSI technology, routing costs dominate the power, time, and area required to implement a computation [Mead80]; therefore, regular and local communication in systolic arrays is advantageous.

Balancing Computation with I/O   A systolic array is typically used as an attached array processor, and it receives data and outputs results through a host computer. Therefore, I/O considerations have to be taken into account in the overall performance. The ultimate performance goal of an array processor system is *a computation rate that balances the available I/O bandwidth with the host.* With the relatively low bandwidth of current I/O devices, to achieve a faster computation rate it is necessary to perform multiple computations per I/O access. However, the repetitive use of a data item requires it to be stored inside the system for a sufficient length of time. In other words, the I/O problem influences not only the required I/O bandwidth but also the required internal memory. The important algorithmic array design issue then is how to arrange a computation, together with an appropriate memory structure and I/O bandwidth, so that computation time is balanced with I/O time.

The I/O problem becomes especially severe when the computation of a problem of large dimension is performed on a small array. Inevitably, it involves a partitioning problem, i.e., the computation must be decomposed. In practice, this is often the case, and therefore, questions such as how a computation can be decomposed and how buffer memory can be arranged to minimize I/O are critical to the practical design of an array processor system.

4.2.2.2   Clock Distribution Schemes

According to [Seitz80], synchronization of the activities in a large synchronous system (such as a systolic array) is controlled by a system wide clock signal. The clock signal serves two purposes; as a sequence reference and also as a time reference. As a sequence reference, the clock transitions serve the purpose of defining successive instants at which system state changes may occur. As a time reference, the period between clock transitions accounts for wiring and element delays in paths from the output to input of clocked elements. "The dual role of clock signal contributes to some ease of design of digital systems; however, combining sequencing and timing so closely results in timing being the source of numerous difficulties in the design, maintenance, modification, and reliability of synchronous systems" [Seitz80].

Since the clock signal dictates the activities of the entire system, clock distribution is very crucial for systolic arrays. There are some difficulties in synchronizing a large array. The problems come mainly from *clock skew*, i.e., each PE in the array may not receive the clock signal at the same time. This may be due to the different path lengths from the clock generator to each PE, or other reasons, such as process variations for different clock paths. To overcome clock path problems, an H-tree scheme can be used to distribute the clock signal to regular arrays such that every PE has the same distance from the clock generator. H-tree layouts for linear arrays, square arrays and hexagonal arrays are shown in Figure 4.4.

Though H-tree schemes can solve the clock path length problem, the issue of the clock skew problem is not completely resolved. Fisher [Fish85a] has shown that an arbitrarily large linear systolic array can be synchronized by a global clock by the use of pipelined clocks. However, an attempt to synchronize a 2-D array usually encounters a clock skew proportional to the size of the array. In fact, the concern over such a clocking problem leads to the development of (asynchronous) wavefront array processors discussed in Chapter 5. In an asynchronous system, the clocking problem can be alleviated, because only correct sequencing (but not timing) is concerned.

4.2.2.3   Systolic versus SIMD and SFG Arrays

A mesh type SIMD array (e.g., ILLIAC IV) is shown in Figure 4.5(a), while a systolic array is shown in Figure 4.5(b). An SIMD array is a synchronous array of processing elements (PEs) under the supervision of one control unit [Hwan84a] and all PEs receive the same instruction broadcast

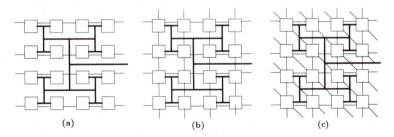

Figure 4.4: H-tree layouts for clocking (a) linear array; (b) square array; and (c) hexagonal array.

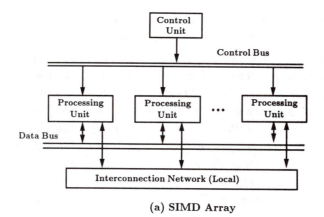

(a) SIMD Array

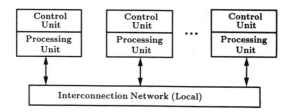

(b) Systolic Array

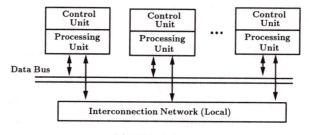

(c) SFG Array

Figure 4.5: (a) A mesh type SIMD array; (b) a systolic array; (c) an SFG array.

from the control unit but operate on different data sets from distinct data streams. Note that an SIMD array usually loads data into its local memories before the computation starts, while systolic arrays usually pipe data from an outside host and also pipe the results back to the host. Dew and Manning [Dew86] show a comparison of SIMD arrays and systolic arrays for a vision preprocessing application. It was reported that local windowing operations can be effectively implemented on both systolic and SIMD arrays.

However, for data dependent operations such as the binary search correlator the utilization of the SIMD array will be inferior to the systolic array. The efficiency of the systolic array is due to the fact that the host is responsible for the image storage and can select the desired data and pipe them into the array.

Let us use this opportunity to also compare the SFG array with the SIMD and systolic arrays. Note that the SFG has for a long time been a popular expression in the DSP literature. Moreover, the SFG array in its own right implies a new array architecture, which is different from either the SIMD or the systolic design. Similar to the SIMD design and in contrast to the systolic design, broadcasting of data in SFG is allowed (see Figure 4.5(c)), which is useful for rapidly communicating *transmittent data*. Just like the systolic array, the SFG array has local instruction code, which is not available to the SIMD arrays. *Potentially, the SFG array represents a very realistic and promising architecture for the optical array computing.*

4.3   Mapping DGs and SFGs to Systolic Arrays

There has been considerable effort in the development of systematic methods for synthesizing systolic arrays based on algorithm-oriented analyses. Let us briefly review the works that are most relevant to the discussion in this section. The work of Karp, Miller and Winograd [Karp67] is discussed in Section 3.2. Cappello and Steiglitz [Cappe83] introduced a geometric interpretation of the linear transformation on index space, which provides an insightful look into how several systolic designs for the same algorithm relate to each other. Along the same line, Moldovan [Moldo83] addresses the issue of mapping cyclic (loop) algorithms onto systolic arrays. The cyclic algorithms are specified in a high-level language, such as FORTRAN, in the form of DO loops. The mapping procedure is based on a linear transformation of index sets and data-dependence vectors. Necessary and sufficient conditions for the existence of valid transformations are given for algorithms with constant data dependencies. To find an *optimal* design (given the cost function), a heuristic procedure is used to search for the best one among many feasible transformations. This mapping can be extended to the partitioning problem [Moldo86]. Independently, Miranker and Winkler [Miran84] developed results similar to that of Moldovan.

Rao [Rao85] defines a class of algorithms, namely the *regular iterative algorithms*, which is similar to the systems of uniform recurrence equations defined by Karp et al. [Karp67]. It is shown that a subclass of the regular iterative algorithms has the characteristics of the systolic algorithms and the corresponding systolic architectures may be systematically derived. The notion of *locally recursive algorithms*, proposed by Kung [Kung83c], stresses the locality of spatial and time indices in a recursive (or iterative) algorithm and therefore is expressible in terms of a spatially local DG or SFG. Recently, there are many insightful research explorations on this subject, e.g. [Delos86] and [Chen86]. A more detailed review can be found in [Fort85a] and [Rao85].

As proposed in the canonical mapping methodology in Section 3.3, the design of systolic arrays can be divided into the following three stages:

- **Stage 1**: Derive a (localized) DG from the algorithm.

- **Stage 2**: Map the DG to an SFG array.

- **Stage 3**: Transform the SFG to a systolic array (i.e., *systolization*).

To obtain an optimal systolic design, it is crucial to know the options available at each of these three stages. We first point out the difference between the SFG schedule and the systolic schedule. Then we discuss a method of mapping a DG directly to a systolic design, i.e., combining Stage 2 and Stage 3. Another situation is: given an SFG, the designer is to transform it to a systolic design, i.e., systolization, for which a cut-set retiming scheme is proposed. This section places more emphasis on Stage 3 – the systolic array design – and makes frequent references to the other two design stages which are already treated extensively in Chapter 3.

4.3.1 Direct Mapping from DGs to Systolic Designs

Many research works on systolic designs are based on direct mapping from DGs onto systolic arrays [Moldo83], [Cappe83], [Rao85]. The approach basically follows the SFG mapping methodology proposed in Section 3.3.2. The only modification being on the schedule constraint discussed below:

Systolic Schedule Mapping a DG to a systolic design in fact means mapping a DG to a special class of SFGs which exhibit a systolic schedule. Recall that, given a projection direction $\vec{d}$, the permissible SFG schedule satisfies the conditions (cf. Section 3.3.2)

$$\vec{s}^T \vec{e} \geq 0 \quad \text{and} \quad \vec{s}^T \vec{d} > 0 \tag{4.1}$$

where $\vec{d}$ is the designated projection vector and $\vec{e}$ is any dependence vector in the DG. Equation 4.1 guarantees $D(e) \geq 0$ in the formal SFG representation (cf. Section 3.3.2). For systolic design, however, the schedule vector $\vec{s}$ in the projection procedure (Section 3.3.2) must satisfy the following stronger conditions:

$$\vec{s}^T \vec{e} > 0 \quad \text{and} \quad \vec{s}^T \vec{d} > 0 \tag{4.2}$$

This means that *every edge of the resulting SFG will have one or more delay elements*, i.e., $D(e) \geq 1$, satisfying the *temporal locality* condition in the definition of the systolic array. In other words, the SFG array qualifies the condition of a systolic array, if the DG is local and (directionally) shift-invariant.

4.3.2   A Cut-set Systolization Procedure

There are several reasons that one might want to first derive an SFG array and then convert it into a systolic array: (1) the SFG offers a concise expression for parallel algorithms, (2) the SFG defines the structure of the array with minimum constraints on timing, and (3) formal transformations from an SFG to a systolic array can be developed.

Obviously, in the mapping from DGs onto SFGs, not all SFG schedules $\vec{s}$ complying with Eq. 4.1 satisfy the conditions of the systolic schedule in Eq. 4.2. Hence retiming is sometimes needed to systolize the SFGs (cf. Section 4.3.2). To better appreciate the difference between the timing requirements of SFGs and systolic arrays, let us use an illustrative example. We note that the convolution SFG array in Figure 3.9 is *spatially localized* but not *temporally localized*. According to the configuration in Figure 3.9, propagating a data u from, say, the leftmost node to the rightmost node uses zero time. More precisely, the SFG imposes the requirement that the data u has to be *broadcast* to all the nodes on the upper path. This is undesirable from systolic design and circuit implementation considerations.

The preceding SFG for convolution is already very close to a systolic array. The major gap is that most SFGs are not given in temporally localized form. In other words,

systolic array = SFG array + pipeline retiming

Here, retiming [Leise81] is the procedure to transform an SFG to an equivalent and temporal localized form. The topic of imposing temporal locality into a computing network has been a focal point of research. There are a series of publications by Fettweis [Fettw76], and Leiserson et al. [Leise81], [Leis83a] , [Leis83b], and, more recently, the works reported in [Kung83c], [Kung84a], [Lev-A83], [Jagad83], [Carai84], and [Barnw83]. In this section, we present a cut-set retiming procedure (largely based on [Kung84a]) to sys-tolize SFGs. The main advantage of this scheme lies in the simplicity of use and the straightforward proof, based on the *colored arc lemma* in graph the-ory (see Theorem 4.1 below). There is a direct relationship between rotating schedule vector in the DGs and the retiming of the SFGs, which is discussed in Section 4.3.3.

4.3.2.1   Cut-set Retiming Procedure

The objective of the cut-set retiming procedure is to convert an SFG into a temporally localized form so that all the edges between modular sections have at least one delay element.

A cut-set in an SFG is a minimal set of edges, which partitions the SFG into two parts. The cut-set retiming procedure is based on two simple rules:

Rule 1: Time-scaling All delays $\mathbf{D}$ may be scaled, i.e., $\mathbf{D} \longrightarrow \alpha \mathbf{D}'$, by a single positive integer α, which is also known as the *pipelining period* of the SFG. Correspondingly, the input and output rates also have to be scaled by a factor of α (with respect to the new time unit $\mathbf{D}'$).

Rule 2: Delay-Transfer Given any cut-set of the SFG, which parti-tions the graph into two components, we can group the edges of the cut-set into *inbound edges* and *outbound edges*, as shown in Figure 4.6, depending upon the directions assigned to the edges. Rule 2 allows advancing k $(\mathbf{D}')$

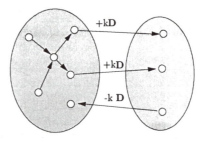

Figure 4.6: Illustration of delay-transfer rule. Notice that the computation performed by a component is not changed if we ad-vance the inputs and delay the outputs by the same number of time units.

time units on all the outbound edges and delaying k time units on the inbound edges, or vice versa. It is clear that, for a (time-invariant) SFG, the general system behavior is not affected, because the effects of lags and advances cancel each other in the overall timing. Note that the input-input and input-output timing relationships also remain exactly the same only if they are located on the same side of the cut-set. Otherwise, they should be adjusted by a lag of $+k$ time units or an advance of $-k$ time units. In other words, if there is more than one cut-set involved and if the input and output are separated by more than one cut-set, then such adjustment factors should be accumulated.

We shall refer to these two basic rules as the (cut-set) *retiming* rules. We prove shortly that any computable SFG (one without zero delay loops or cycles) can be systolized by the cut-set procedure. (*An SFG is meaningful only when it is computable, i.e., there exist no zero-delay loops or cycles in the SFG.*) To show that the basic cut-set retiming rules are sufficient for the systolization procedure, we assert the following theorem.

Theorem 4.1: *All computable SFGs can be made temporally local by following the cut-set retiming rules. Consequently, a spatially local and regular SFG array is always systolizable.*

Proof: We claim that the retiming Rules 1 and 2 can be used to localize any (targeted) zero-delay edge, i.e., to convert it into a nonzero-delay edge. This is done by choosing a *good* cut-set and applying the rules to it. A good cut-set, including the target edge, should not include any *bad edges*, i.e., those zero-delay edges in the opposite direction of the target edge. This means that the cut-set will include only (1) the target edge, (2) nonzero delay edges going in either direction, and (3) zero-delay edges going in the same direction. Then, according to Rule 2, the nonzero delays of the opposite-direction edges can give one or more spare delays to the target edge (in order to localize it). If there are no spare delays to give away, all delays in the SFG are simply scaled according to Rule 1 to create enough delays for the transfer needed.

Therefore, the only thing left to prove is that such a good cut-set always exists. For this, we refer to Figure 4.7, in which only the zero-delay *successor edges* and the zero-delay *predecessor edges* connected to the target edge have been kept, and all the other edges have been removed from the graph. In other words, Figure 4.7 depicts the bad edges, which should not be included in the cut-set. As shown by the dashed lines in Figure 4.7, there must be openings between these two sets of bad edges – otherwise, some set of

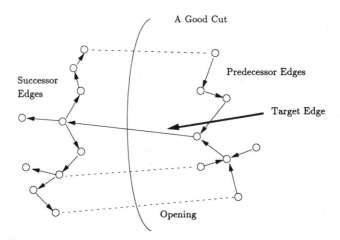

Figure 4.7: Proof of cut-set procedure: There always exist openings between bad edges. Openings between bad edges ensure the existence of a good cut-set. These may be used as a clue for selecting a good cut-set.

zero-delay edges would form a zero-delay cycle, and the SFG would not be computable.

Obviously, any cut-set cutting through the openings is a good cut-set, thus the existence proof is completed. Note that the existence proof discussed earlier is in fact a result known as the *colored arc lemma* in graph theory. It is clear that repeatedly applying Rule 2 (and 1, if necessary) on the cut-sets will eventually lead to a temporally localized SFG.

4.3.2.2 Systolization Procedure

A regular SFG array can be easily systolized to become a systolic array by the following systolization procedure which is based essentially on the cut-set retiming rules:

1. *Selection of basic operation modules* The choice may not be unique. In general, the finer the granularity of the basic modules, the more efficient (in speed) a systolic array is. (A comparison of two possible lattice modules for a systolic array is discussed in this section.)

2. *Applying retiming rules* If the given SFG is regular, i.e., modular and spatially local, then regular cut-sets can be selected and the above

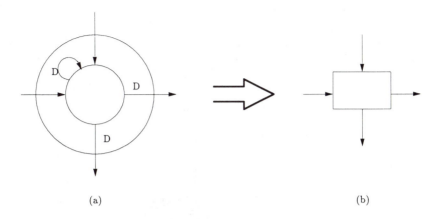

(a)   (b)

Figure 4.8: Illustration of combining delays into module operations. (a) Module operation with delays in the circle. (b) The corresponding processing element.

rules can be used to derive a regular and temporally localized SFG. In order to preserve the modular structure of the SFG (a basic feature of systolic design), the cut-set retiming should be applied uniformly across the network.

3. *Combination of delay and operation modules* To convert such an SFG into a systolic form, it is only necessary to successfully introduce a delay into each of the operation modules, such as $a + b \times c$. The delay can then be combined with the module operation to form a basic systolic element. All the extra delays are modeled as pure delays without operations. Since self-loops are implemented as registers in the PE, they are also combined into the PE. See Figure 4.8 for a detailed illustration.

Example 1: Systolic Lattice Filters

An example of the systolization procedure is the systolic implementation of a digital lattice filter, which has many important applications in speech and seismic signal processing. For this example, let us apply the transformation rules to the SFG for an AR lattice filter, as shown in Figure 4.9(a). There are two possible choices of basic operation modules for the lattice array: (1) a lattice operation module;[1] and (2) a multiply-accumulate

---

[1] That is, the lattice operation is now treated as a single module which is very suitable for CORDIC implementation.

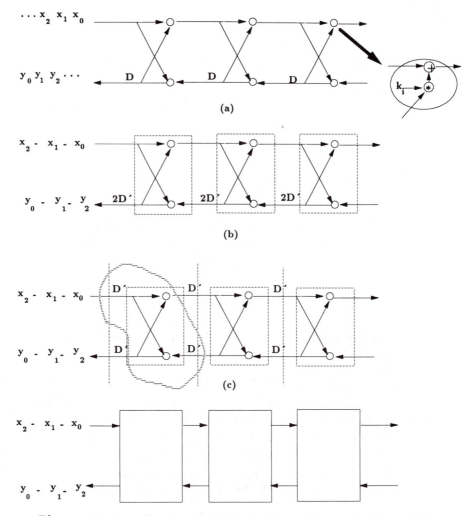

Figure 4.9: Systolization of AR lattice filters (Type A). (a) An SFG
for AR lattice filters. (b) Time-rescaled SFG for AR lattice filters
(Type A). (c) Localized SFG for AR lattice filters (Type A). (d)
Systolic Array for AR lattice filters (Type A).

(MAC) basic module. Note that in each lattice operation there are two MAC
operations – implying that the lattice operation requires twice the time of
the MAC operation.

Lattice Systolic Array (Type A) By retiming Rule 1, we first double
each delay, i.e., $\mathbf{D} \longrightarrow 2\mathbf{D}'$, as shown in Figure 4.9(b). Apply the cuts
uniformly to the SFG and subtract one delay from each of the left-bound

edges and, correspondingly, add one delay to each of the right-bound edges in the cut-sets. This yields Figure 4.9(c). Finally, by combining the delays with the lattice module, we have the final systolic structure, as in Figure 4.9(d). Note that, because of the time-scaling, the input sequence $\{x(i)\}$ will be interleaved with blanks to match the adjusted delays. It is clear that $\alpha = 2$ and the array can yield an $M/2$ execution-time speedup over the single processor case.

Lattice Systolic Array (Type B)   By retiming Rule 1, we first triple each delay; i.e., $\mathbf{D} \longrightarrow 3\mathbf{D}'$. The resultant SFG is shown in Figure 4.10(a). Apply uniform cut-sets to the SFG as shown in Figure 4.10(b). Subtract two delays from each leftbound edge, and, correspondingly, add two delays to every rightbound edge in the cut-sets. This yields Figure 4.10(b). Now use a cut-set partitioning the upper edges from the lower edges, as shown in Figure 4.10(c). Transfer one delay from the down-going edges to the up-going ones. The result is depicted in Figure 4.10(c). Finally, by combining the delays with the MAC module (Figure 4.10(d)), the systolic structure as shown in Figure 4.10(e) is obtained. Note that because of time-scaling, the input sequence $\{x(i)\}$ is interleaved with two blanks to match the adjusted rates. It is clear that the array can achieve an $2M/3$ execution-time speedup over the single MAC case [2]. In terms of speed, this systolic design is superior to the Type A design.

Example 2: Systolic Arrays for Matrix Multiplication

Multiplication of Two Full Matrices   As shown in Figure 4.11, a simple matrix multiplication SFG is one letting columns $\mathbf{A}_i$ and rows $\mathbf{B}_i$ be broadcast instantly along the square array, with the partial sum of outer products fed back via a loop with delay.

Let us apply Rule 2 (Section 4.3.2.1) to the cut-sets shown in Figure 4.11. The systolized SFG has one delay assigned to each edge and thus represents a localized network. According to Rule 2, the inputs from different columns of $\mathbf{B}$ and rows of $\mathbf{A}$ have to be adjusted by a certain number of delays before arriving at the array. By counting the cut-sets involved in Figure 4.11, it is clear that the first column of $\mathbf{B}$ needs no extra delay, the second column needs one delay, the third needs two (because of the two cut-

---

[2]Note that since the upper and lower MAC modules in each PE are never simultaneously active, only one MAC hardware module suffices to serve both functional needs.

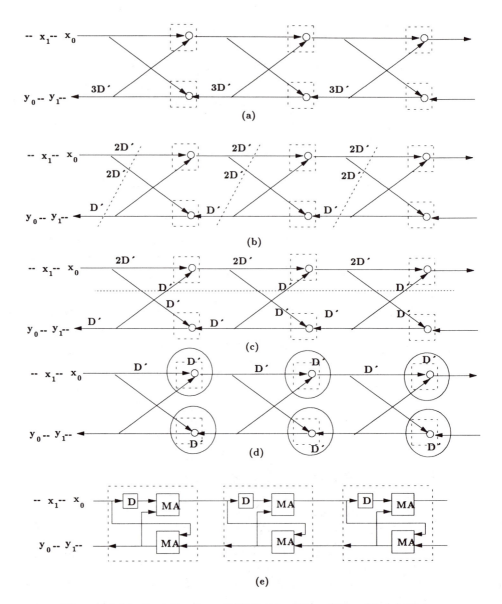

Figure 4.10: Systolization of AR lattice filters (Type B) (a) Time-rescaled SFG (and cut-sets) for AR lattice filters (Type B). (b) Partially localized SFG - first cut-set delay transfer. (c) Localized SFG - second cut-set delay transfer. (d) Localized SFG – all operations are ready to merge with the corresponding unit-time delays D'. (e) Systolic array for AR lattice Filters (Type B) with the small squares denoting pure delays.

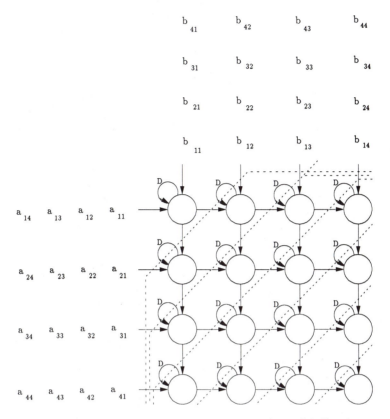

Figure 4.11: Cut-sets of the SFG for matrix multiplication.

sets separating the third column input and the adjacent top-row processor), and so on. Therefore, the **B** matrix will be skewed, as shown in Figure 4.12. A similar arrangement must be applied to **A**.

Multiplication of Two Band Matrices   Refer to the band-matrix multiplication SFG in Figure 3.28. Applying the systolization procedure to the cut-sets as shown in Figure 4.13 calls for a triple scaling of $\mathbf{D} \longrightarrow 3\mathbf{D}'$. (This is because each northwest bound delay edge is *cut* twice.) The procedure leads to the array configuration depicted in Figure 4.14, which is topologically equivalent to the 2-D hexagonal array proposed in [HTKun78] and [HTKun82].

Example 3: Two Systolic Arrays for LU-Decomposition

We can systolize the two SFGs for the LU decomposition corresponding to the projections in the $[1\ 1\ 1]^T$ and the $[1\ 0\ 0]^T$ directions, as shown in

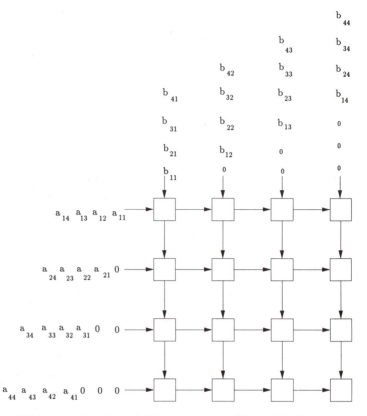

Figure 4.12: A systolic array for matrix multiplication.

Figure 3.30 (a) and (b), by the cut-set retiming procedure. Note that the SFG of the $[1\ 1\ 1]^T$ projection is already systolized. The resulting systolic array (version I) is shown in Figure 4.15. The pipelining period α is 3 for this array, since there are three delays in a cycle formed by PEs (1, 1), (2, 1) and (2, 2). This array configuration is topologically equivalent to the 2-D hexagonal array proposed in [HTKun78], and [HTKun82]. After applying retiming Rule 2 to the SFG of the $[1\ 0\ 0]^T$ projection, version II systolic array is derived, as detailed in Figure 4.16. This systolic array has $\alpha = 1$.

4.3.3   Relationship Between Rotating Schedule Vector in DGs and Cut-set Retiming in SFGs

We assert that Stage 2 and Stage 3 combined together should at least cover the design achieved by the direct mapping methods. This is based on the following proposition.

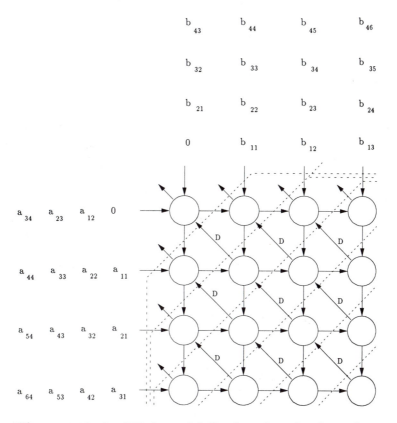

Figure 4.13: An SFG for multiplication of two band matrices.

Proposition : *All systolic arrays obtained from linear projections of the DG can be derived by the following two steps:*

(a) Mapping the DG onto SFGs by the SFG Projection procedure (Stage 2).

(b) Mapping the SFG onto a systolic array by the cut-set retiming (Stage 3).

This proposition establishes the relation between the present method and the direct mapping methods. To see the reason for this, we note that for a given projection vector the only aspect distinguishing a systolic design from a general SFG array is the more strict systolic schedule of Eq. 4.2, as opposed to Eq. 4.1. Let us use $\vec{s}$ and $\vec{s'}$ to denote an SFG and a systolic schedule respectively. Since $\vec{s'}$ is a rotated vector of $\vec{s}$, there is a relationship

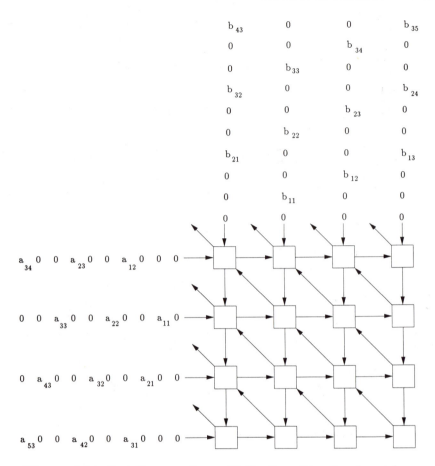

Figure 4.14: Systolic array for multiplication of two band matrices.

between the rotation of schedule vector in DG and cut-set retiming in SFG. This subject is further explained by the following example.[3]

Example 4: Retiming in the Sorting Systolic Arrays

The systolic arrays for the different sorting algorithms can be derived by applying cut-set retiming to the various SFG arrays derived in Section 3.3 (see Figure 3.15). This is shown in Figure 4.17. For insertion and selection sorters, the delay transferred on each cut-set is only one. Note that the SFG projection for the bubble-sorter yields a systolic array directly; so no further retiming is required.

---

[3]In certain cases, we allow the time-scaling factor to be a rational number in order to find the unique cut-set retiming.

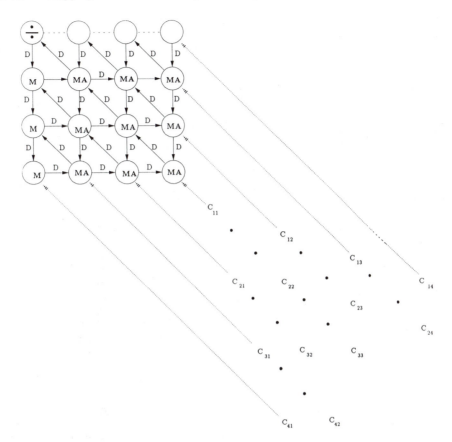

Figure 4.15: (Version I) Hexagonal systolic array for LU decomposition.

The rotation of schedule vectors is shown in Figure 4.18. This figure also shows the default schedule, the desired schedule (hyperplanes), and the cut-set delay transfer. This shows the relation between the rotation of two schedule vectors in the DG and the cut-set retiming in the SFG.

Example 5: Retiming in Convolution Systolic Arrays

We first apply the systolization procedure for the two convolution SFG arrays in Figure 3.50(b) and (c). In Figure 4.19, the two convolution SFG arrays are shown along with the cut-sets. In Figure 4.19(a), we first scale the delay by a factor of two, i.e., $D \longrightarrow 2D'$. Then, one delay can be transferred from the left-going edges to right-going edges in the cut-sets.

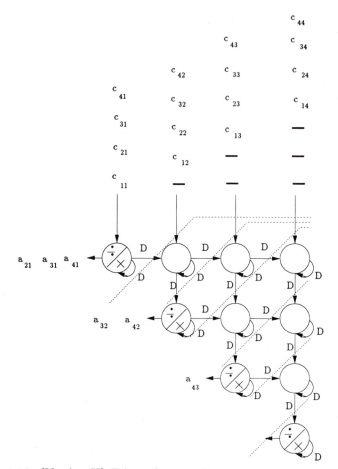

Figure 4.16: (Version II) Triangular systolic array for LU decomposition.

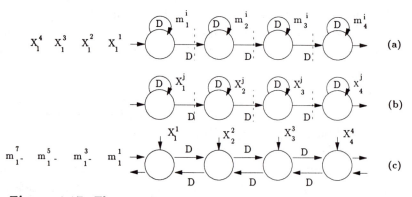

Figure 4.17: The systolic arrays corresponding to: (a) insertion sorter; (b) selection sorter; (c) bubble-sorter.

220

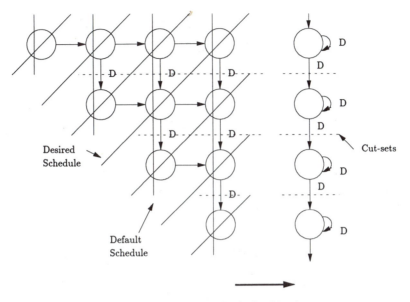

Figure 4.18: The relation of hyperplanes and the cut-set for insertion sorter.

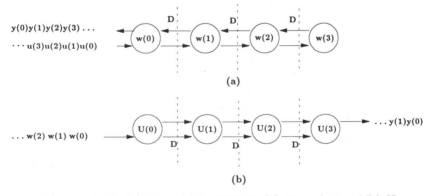

Figure 4.19: (a) Convolution SFG I. (b) Convolution SFG II.

The systolized SFG is shown in Figure 4.20(a), and α is two for this array. The cut-set delay transfers for Figure 4.19(b) are straightforward; all edges are in one direction of the cut-sets, and hence all edges in the cut-set are given one additional delay. The resulting systolized SFG arrays are shown in Figure 4.20(b). There is no time-scaling needed; therefore, $\alpha = 1$. Note that the same systolic designs for both arrays may be directly obtained by using a systolic schedule $\vec{s}^T = [1 \ 2]$ in DG as shown in Figure 4.21.

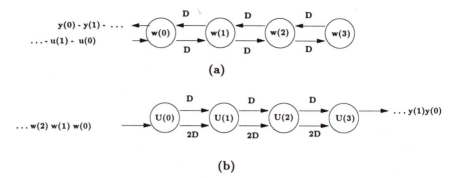

(a)

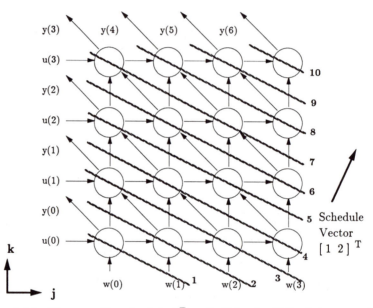

(b)

Figure 4.20: (a) Convolution systolic array I. (b) Convolution systolic array II.

Figure 4.21: The systolic schedule $\vec{s}^T = [1\ 2]$ in the DG of the convolution algorithm.

4.3.4   Bit Level Systolic Arrays

Here a somewhat different perspective of mapping is considered. The DGs discussed previously are all at the word level. In the following, we shall consider mapping bit level DGs to systolic arrays. Bit level systolic arrays were proposed to improve the pipelining rate of systolic arrays and to better utilize the current integration level of VLSI technology [McCan82], [McCan84],

[McCan86]. The bit level arrays can be used either as building blocks or processing elements in the construction of word level systolic arrays. Consider the problem of performing an inner product of two vectors $\mathbf{a}$ and $\mathbf{b}$, each of length n. To design a bit level systolic array for this problem, there are two approaches. One approach is to use *two level pipelining* [HTKun84], in which the first level refers to pipelining at the PE level and the second refers to pipelining within the PE. For this case, the first level is to design a word level systolic array and the second is the bit level pipelining inside each PE. Another approach is to treat the entire problem at the outset as a bit level problem (i.e., each node in the DG performs bit operations instead of word operations). The latter approach is further detailed below.

The inner product c of two vectors $\mathbf{a}$ and $\mathbf{b}$ is computed as:

$$c = \sum_{k=1}^{n} a_k \times b_k$$

Assume that elements of $\mathbf{a}$ or $\mathbf{b}$ are m-bit integers. For example, $a_k = (a_{k,m-1} \, a_{k,m-2} \, \cdots \, a_{k,0})$. An expression for the jth bit of the result c_j takes the form:

$$c_j = 2^j \sum_{k=1}^{n} \sum_{i=0}^{m-1} a_{k,i} b_{k,j-i} + \text{carries}$$

One way to form the bit c_j is to sum over k index first (i.e., sum partial products of the form $a_{k,i} b_{k,j-i}$ first) and then accumulate all those with the same significance by summing over i. Using this approach, the partial products can be formed in the following recurrence:

$$c_{j-i,i}^k = c_{j-i,i}^{k-1} + 2^j a_{k,i} b_{k,j-i} + \text{carries} \tag{4.3}$$

where $c_{j-i,i}$ is the sum over k of all partial products involving bits $a_{k,i}$ and $b_{k,j-i}$.

A DG which represents the above computation is shown in Figure 4.22. This DG is embedded in a three-dimensional index space represented by (i, j, k). Each horizontal plane of this graph shows the interaction of bits within each word a_k and b_k and is such that each bit of one word interacts with each bit of the other word as is required. The accumulating sum bits are connected in the vertical direction so that the equivalent products of the form $a_{k,i} b_{k,j-i}$ are accumulated as indicated by Eq. 4.3. In order to form the final bit in the result, all partial products with the same significance must be added together. This is represented by the bottom plane in the DG.

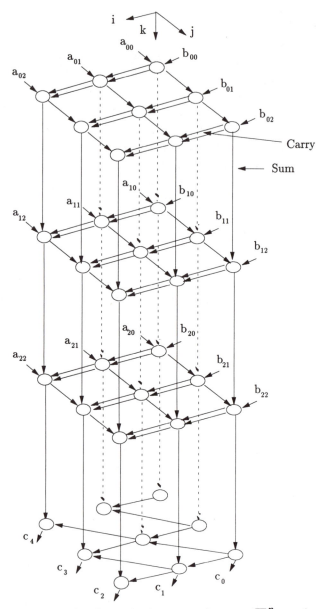

Figure 4.22: Bit level DG for inner product $c = \sum_{k=1}^{n} a_k b_k$.

We can project this DG in several directions to obtain different bit level systolic designs. For example, if we project the DG in the direction of $[1\ 1\ 0]^T$ and let the schedule direction be $[1\ 1\ 1]^T$, the bit level systolic array design shown in Figure 4.23(a) will be obtained. This systolic array in which the bits of a_k and b_k move in a contraflow direction is the same as

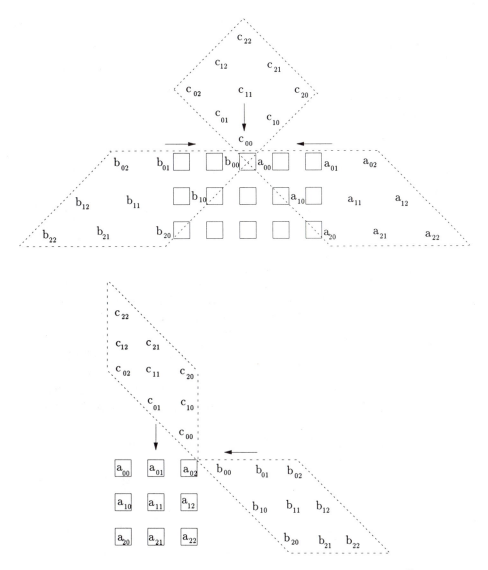

Figure 4.23: (a) Bit level systolic array obtained by $[1\ 1\ 0]^T$ projection. (b) Bit level systolic array obtained by $[0\ 1\ 0]^T$ projection.

the array proposed by [McCan84]. Not shown in this figure is the linear accumulator circuit at the bottom of the array which is required to complete the computation.

One can also obtain systolic arrays in which the bits of one set of words remains fixed by projecting along either the $[0\ 1\ 0]^T$ (for fixed $a's$) or along the $[1\ 0\ 0]^T$ axes (for fixed $b's$). An example of such a circuit is shown in

Figure 4.23(b), which corresponds to the design reported by Urquhart and Wood [Urquhar].

4.4 Performance Analysis and Design Optimization

In this section, we are concerned with the performance and the optimality of systolic arrays. The most suitable optimality criteria are hard to pinpoint and optimizing one factor may sacrifice other factors. However, there are several instrumental formula and useful observations for the searching of an optimal design.

4.4.1 Optimality Criteria and Basic Formula

There are many factors in determining the optimization criteria for the design of systolic arrays. The final choice of optimality criteria will have to be application dependent. Some typical factors are: computation time, pipelining period, block pipelining period, array size, and I/O channels. They are explained below.

Computation time (T): The time interval between starting the first computation and finishing the last computation of a problem instance by the processor array, denoted by T.

Pipelining period (α): The time interval between two successive computations in a processor is denoted by α. In other words, the processor is busy for one out of every α time intervals. (Note: α is the reciprocal of the pipelining rate.)

Block pipelining period (β): The time interval between the initiations of two successive problem instances by the processor array, denoted by β.

Array Size: The number of processors in the array. The array size obviously determines the basic hardware cost.

I/O channels: the number of I/O lines to communicate with the outside world (the host computer). Input/output channels are directly tied to hardware cost in terms of I/O pins of a VLSI chip or I/O wires of a PC board.

Sometimes, combinations of two or more of the above factors are of special interest. For example, product of the array size and the computation time provides a useful measure for the hardware cost-effectiveness. Another useful criterion is *processor utilization*, which depends on the pipelining period, the block pipelining period, and data input schemes. The following case studies should help demonstrate how various design criteria may naturally arise in certain application environments.

Case 1: A Single Problem Instance with Finite Input Data In the case that the DG is finite, the computation time is perhaps a more important criterion than the pipelining period. There have been several works on how to minimize the computation time of a systolic array, such as [Moldo83], [Li85], [Wong85],and [O'Kee86]. A simple formula will be discussed in this section.

Case 2: A Single Problem Instance with Indefinite Input Data
In many DSP applications, such as filtering, input data lengths are usually very long or indefinite, which will result in a DG with a very *slim* shape. In this case, minimizing the computation time is equivalent to minimizing the pipelining period. For example, in convolution, if the input sequence is indefinite, the pipelining period will become the focal point of optimization. If the pipelining period is small, so is the computation time.

Case 3: Many Problem Instances In the case that many problem instances are to be processed by the same systolic array, then we should be concerned with the block pipelining period, β. If there are M problem instances to be computed, then the total processing time is $M \times \beta$. If we ignore the effects caused by the I/O scheme for the systolic array, it is rather straightforward to calculate the block pipelining period. For example, we can use the reservation table, which specifies for all PEs in the array the time steps a certain PE is busy during the computation of one problem instance. Based on this table, we can calculate the *time span* for each PE, which indicates the difference between the last time step and the first time step a PE is busy in the reservation table. Now, it is easy to calculate the block pipelining period, since it is simply the largest time span of any PE in the array. An example of the reservation table is shown in Figure 4.24.

4.4.1.1   Minimal Computation Time Designs

Computation Time We call a schedule vector $\vec{s} = [s_1 \; s_2 \; \cdots]^T$ *coprime* if its components are coprime (i.e., the greatest common divisor of all $\{s_i\} =$

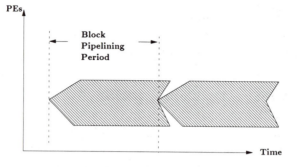

Figure 4.24: Calculating block pipelining period from the reservation table.

1). Similar definition also applies to $\vec{d}$. Given a coprime schedule vector $\vec{s}$, the computation time of a systolic array is computed as [Wong85]:

$$T = \max_{\vec{p},\vec{q}\in L}\{\vec{s}^{T}(\vec{p}-\vec{q})\} + 1 \qquad (4.4)$$

where L is the index set of the nodes in the DG.

This formulation has a simple justification: Since components of $\vec{s}$ are coprime, the range of time steps generated by points $(\vec{p})$ in the index set L, $\vec{s}^{T}\vec{p}$, should be a set of consecutive integers. [4] Therefore, the computation time is equal to the difference between the largest and the smallest time steps in the range set.

Min-Max Formulation   To compute Eq. 4.4, integer programming can be used. To search for a minimal computation time schedule, we can formulate it as a min-max problem:

$$\min_{\vec{s}}\ \max_{\vec{p},\vec{q}\in L}\{\vec{s}^{T}(\vec{p}-\vec{q})\} + 1 \qquad (4.5)$$

The minimal computation time schedule $\vec{s}$ can be found by solving the min-max problem. In most cases, it is also possible to enumerate the allowable schedules and find the minimal computation time solution.

---

[4]Rigorously speaking, this statement is valid only with the assumption that the size of the index space is larger than certain finite size determined by the coprime components in the schedule vector. However, this condition is practically always true.

4.4.1.2   Minimal Pipelining Period Designs

The *pipelining period (α)* is often a good optimality criterion for the case that the input data is either very long or indefinite, as frequently incurred in signal processing applications. Given a systolic schedule $\vec{s}$ and a projection direction $\vec{d}$ of a DG, with both $\vec{s}$ and $\vec{d}$ in coprime form, the pipelining period α can be calculated as:

$$\alpha = \vec{s}^T \vec{d} \tag{4.6}$$

To verify Equation 4.6, we note that the pipelining period equals the time between two consecutive computations for a processor [Rao85]. If the (coprime) projection vector is $\vec{d}$ and $\vec{p}$ is a point in the index space, then $\vec{p}$ and $\vec{p}+\vec{d}$ represent the indices of two consecutive nodes that are projected to the same processor. Therefore, the pipeline period α is equal to the separation between the computation times of these two nodes, i.e.,

$$\alpha = \vec{s}^T(\vec{p}+\vec{d}) - \vec{s}^T \vec{p} = \vec{s}^T \vec{d} \tag{4.7}$$

Integer Programming for Minimizing α   Given the projection vector $\vec{d}$, to find the schedule vector $\vec{s}$ which minimizes α, we can formulate it as an integer programming problem as follows:

$$\text{Minimize } \vec{s}^T \vec{d} \tag{4.8}$$

under the constraints

$$\vec{s}^T \vec{d} > 0 \text{ and } \vec{s}^T \vec{e} > 0 \text{ for any dependence vector } \vec{e}$$

Example 1: Calculation of T and α
    With reference to the DG in Figure 4.25, suppose that the schedule vector $\vec{s}^T = [2\ 3]$. Then the computation time, T, for this schedule vector is determined as:

$$T = \max_{\vec{p},\vec{q}}\{[2\ 3](\vec{p}-\vec{q})\} + 1$$

Note that from the DG, the maximum difference of any two points is $[4\ 4] - [1\ 1] = [3\ 3]$. Therefore, T is found to be

$$T = [2\ 3]\begin{bmatrix} 3 \\ 3 \end{bmatrix} + 1 = 6 + 9 + 1 = 16.$$

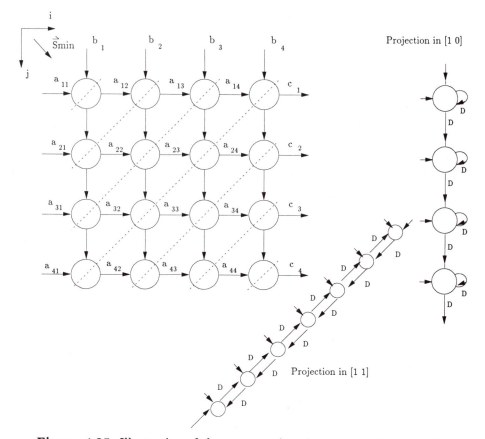

Figure 4.25: Illustration of the computation time and the fact that the projection directions bear no effect on the computation time.

In this example, if any two points $\vec{q}$ and $\vec{p}$ in the DG are on two successive equi-temporal hyperplanes separately, then

$$[2\ 3](\vec{p} - \vec{q}) = 1$$

The solution is $(\vec{p} - \vec{q}) = [2\ -1]^T + k[3\ -2]^T$, for any integer k. Therefore, for example, if $\vec{q} = [2\ 2]^T$, then all the points on the next hyperplane have the form $\vec{p} = [2\ 2]^T + [2\ -1]^T + k[3\ -2]^T = [4\ 1]^T + k[3\ -2]^T$. In fact, the one (and only one) such point inside the index space is $[4\ 1]^T$, i.e. setting $k = 0$.

If the projection vector $\vec{d} = [1\ 1]^T$, the pipelining period can be calculated as $\alpha = [2\ 3][1\ 1]^T = 5$. Given the fixed projection direction, then the optimal schedule vector, according to Eq. 4.8, can be found to be $\vec{s} = [1\ 1]^T$ and the optimal pipelining period is 2.

In the following, we discuss the options available for optimizing T and α in each of the three design stages previously mentioned.

4.4.2   Optimization in the DG Design Stage

Many design rules potentially useful for optimizing the DG design are provided in Section 4.3. Here we use one example to show how to make good use of those rules and the formulations introduced above.

Example 2: DG and Systolic Designs for Multiplication of a Band Matrix and a Vector

In order to find a design with the optimal pipelining period, it is necessary that a *good* DG be derived in the first place. We explain this point by the example of multiplication of a band matrix $\mathbf{A}$ of size $N \times N$, with bandwidth P, and a vector $\mathbf{b}$ of size $N \times 1$, i.e., $\mathbf{c} = \mathbf{A}\,\mathbf{b}$. An example with $N = 7$ and $P = 4$ is shown below.

$$
\mathbf{A} =
\begin{bmatrix}
x & x & 0 & 0 & 0 & 0 & 0 \\
x & x & x & 0 & 0 & 0 & 0 \\
x & x & x & x & 0 & 0 & 0 \\
0 & x & x & x & x & 0 & 0 \\
0 & 0 & x & x & x & x & 0 \\
0 & 0 & 0 & x & x & x & x \\
0 & 0 & 0 & 0 & x & x & x
\end{bmatrix}
\quad
\mathbf{b} =
\begin{bmatrix}
x \\ x \\ x \\ x \\ x \\ x \\ x
\end{bmatrix}
$$

Systolic Design 1:   As discussed in Section 3.4, a DG for this algorithm can be derived, as shown in Figure 4.26(a) (for $P = 4$ case). In most applications, $N \gg P$; therefore, *it is very uneconomical to use a size N linear array for computing $\mathbf{c} = \mathbf{A} \times \mathbf{b}$*. In order to reduce the size, a diagonal projection direction $\vec{d}^T = [1\ 1]$ can be adopted to derive an array of size P (as opposed to an array of size N). For this projection direction, the schedule vector $\vec{s}$ which optimizes the pipelining period can be derived by the following integer programming formulation (cf. Equation 4.8):

$$
\text{Minimize } \vec{s}^T \begin{bmatrix} 1 \\ 1 \end{bmatrix}
$$

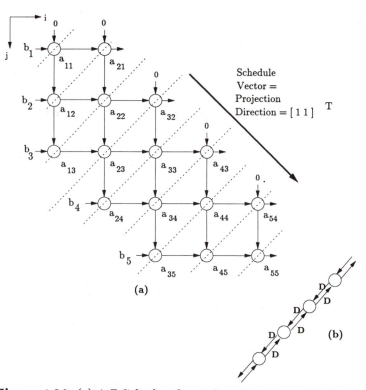

Figure 4.26: (a) A DG for band-matrix and vector multiplication; (b) The systolic design as a result of [1 1] projection; From the top of the array, the band matrix **A** is input (in its band-direction), and the vector **b** is loaded from the left. The final result **c** is output from the leftmost PE.

under the constraints

$$\vec{s}^T \begin{bmatrix} 1 \\ 0 \end{bmatrix} > 0; \quad \vec{s}^T \begin{bmatrix} 0 \\ 1 \end{bmatrix} > 0$$

and

$$\vec{s}^T \begin{bmatrix} 1 \\ 1 \end{bmatrix} > 0$$

The optimal solution is $\vec{s}^T = [1\ 1]$ and $\alpha = 2$. Based on this schedule vector, the resulting systolic array is shown in Figure 4.26(b).

Systolic Design 2: To improve the pipelining rate, modification to the DG in Figure 4.11 is necessary. Note that in the DG we can let **b** data be sent from right to left, instead of from left to right. Or, equivalently, in the single assignment formulation, we adopt

$$b(i-1, j) \leftarrow b(i, j) \quad \text{instead of} \quad b(i+1, j) \leftarrow b(i, j).$$

The new DG with **b** data reversed is shown in Figure 4.27(a). Note that the dependence edges $[1\ 0]^T$ in the original DG is now changed to be $[-1\ 0]^T$ in the new DG. It is a valid change since $b(i, j)$ is actually broadcast data; thus it does not matter whether the (same) data are loaded from left to right or vice versa.

Such a simple modification in fact brings about a very different pipelin-

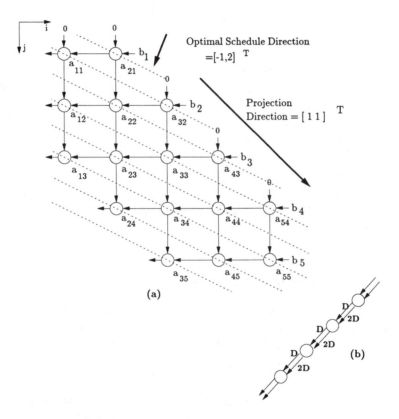

(a)

(b)

Figure 4.27: (a) Another DG for band-matrix and vector multiplication and the equi-temporal hyperplanes corresponding to the optimal schedule; (b) The systolic design with $\alpha = 1$. The major modification here is that the input **b** vector is now loaded from the right and travels leftward.

ing period. For the projection vector $\vec{d}^T = [1\ 1]$, the optimal schedule vector $\vec{s}$ can now be derived by a new formulation (cf. Equation 4.8):

$$\text{Minimize } \vec{s}^T \begin{bmatrix} 1 \\ 1 \end{bmatrix}$$

under the constraints

$$\vec{s}^T \begin{bmatrix} -1 \\ 0 \end{bmatrix} > 0; \quad \vec{s}^T \begin{bmatrix} 0 \\ 1 \end{bmatrix} > 0$$

and

$$\vec{s}^T \begin{bmatrix} 1 \\ 1 \end{bmatrix} > 0$$

Solving the above, we have the optimal schedule vector $\vec{s}^T = [-1\ 2]$ as shown in Figure 4.27(a), and the optimal pipelining period $\alpha = 1$. (It can be easily verified that the schedule vector is also optimal with respect to computation time T.) With this schedule vector, the resulting systolic array is shown in Figure 4.27(b). The new version offers a throughput-rate of one data per unit time for each channel, *twice as fast as* the previous systolic design.

Similar improvements (i.e., reduction of α to one) can be made to the design for multiplication of one band matrix and a full matrix, or to the hexagonal array design for multiplication of two band matrices.

4.4.3 Optimization in the SFG Design Stage

Now, we discuss how to select the projection and the schedule vectors. As shown earlier, the array size and I/O structures of the resulting SFG depend on the projection vector, while the other factors, T, α, and β are also affected by the chosen schedule vector.

Fixed Projection Direction It is often desirable to project along the direction of data arrival, as such a projection is likely to yield minimal hardware. In this sense, the projection direction is almost fixed. The pipelining period is then determined by the selection of the schedule. In this case if the pipelining rate is higher, then the processor utilization is also higher, and the total computation time is also likely to become shorter.

Effects on α and β by the Projection Direction   In many instances, the projection vector chosen have significant effects on the optimal α and β obtainable. This will be illustrated in detail in Section 4.5 when we consider the transitive closure systolic design example.

No Effects on T by the Projection Direction   Note that the computation time is independent of the projection vector chosen. This may be illustrated by a matrix-vector multiplication example $\mathbf{Ab} = \mathbf{c}$. The DG and the minimal computation time schedule direction $\bar{s}^T_{\min} = [1\ 1]$, as obtained through Eq. 4.5, are shown in Figure 4.25. Note that the design corresponding to the projection in $[1\ 0]^T$ direction has 4 PEs and $\alpha = 1$, whereas the design corresponding to the projection in $[1\ 1]^T$ direction has 7 PEs and $\alpha = 2$. However, both designs have the same computation time. Usually, the projection direction can be selected to minimize the array size or the pipelining period.

4.4.4   Optimization in the Systolization Stage

As discussed in Section 4.3, the cut-set systolization can be used to convert an SFG to a systolic array. Here, we address the issues of minimizing the pipelining period and the total number of delays in the systolic array. We also discuss how a multirate systolic design may be used to yield a nearly optimal pipelining period.

4.4.4.1   Systolization Procedure to Achieve Optimal Pipelining Rate

We assert that the array has a linear-rate pipelinability *if and only if* α remains constant with respect to M, where M is the number of PEs. It is clear that if the total time-scaling factor is α, (i.e., $\mathbf{D} = \alpha\mathbf{D}'$), then the data input rate is slowed down by α. Consequently, on average only one of α PEs can be active. This implies that the full processing rate speedup M reduces to $\alpha^{-1}M$. (If α remains constant with respect to M, then the array delivers a linear rate speedup.) Therefore, minimum α is very desirable.

Cycle Time and Pipelining Period   The minimum pipelining period (scaling factor),α, of an SFG is decided by the cycle time and the total number of delays in the cycle. This can be best explained by an example. As shown in Figure 4.28(a), an SFG, consisting of 5 nodes in a cycle, is to be

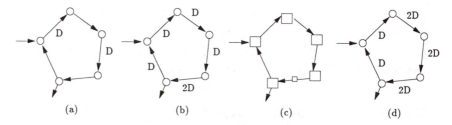

Figure 4.28: Optimal and nonoptimal α systolized SFGs. (a) An SFG with 5 nodes. (b) An optimal-α systolized SFG with $\alpha = 3$. (c) An optimal-α systolic array, where the large box stands for a processor with one unit of processing time, and a small box on an edge denotes a pure delay (D). (d) A systolized array with $\alpha = 4$.

systolized. First, we note that the minimum time for the data x_1 to travel around the cycle is 5, since there are 5 nodes in the cycle, and each node (and edge) takes 1 time unit to compute and pass data. The data x_1', (which is the data x_1 after traveling around the cycle,) should meet input data x_3 at the input node, since there are two delays in the SFG that separate the recursions. Suppose that data is piped into the SFG at the period of α time units. The above requirement can be stated as:

$$2\alpha \geq 5 \quad \text{or}$$

(Total delay in the cycle)$\times \alpha \geq$ total number of nodes in the cycle

In this example, the minimum integer α that satisfies the above inequality is 3, which is therefore the optimal pipelining period of the SFG. The systolized optimal-α SFG is shown in Figure 4.28(b). Note that an extra pure delay is added to the systolic array (Figure 4.28(c)) to ensure the correct sequencing of the computations. It is clear that adding extra pure delays into a systolic array increases hardware cost. Therefore, it is desirable to minimize the number of pure delays in the systolic arrays.

Note that if one arbitrarily scale the time, α can be made unnecessarily large. For example, the systolic array in Figure 4.28(d) is a systolized array of Figure 4.28(a); however, $\alpha = 4$ for this array, which is not optimal.

If we generalize the result of this example, we observe that every cycle C in an SFG will decide a minimal pipelining period of its own, i.e.,

$$\alpha_C = \lceil \tfrac{T_C}{D_C} \rceil;$$

where T_C is the total number of nodes in cycle C, and D_C is the total number of delays in C, and $\lceil x \rceil$ is the ceiling function for a number x, i.e., the smallest integer greater than or equal to x.

The optimal pipelining period of the entire SFG is then decided as:

$$\alpha = \max\{all\ \alpha_C\} \qquad (4.9)$$

And thus the overall systolic array is synchronized by the *slowest* cycle in the graph. Since all cycles must satisfy the pipelining constraint, α from the preceding Eq. 4.9 is the smallest that can meet the requirements for all cycles.

Optimal α in the Systolization Procedure   We have so far discussed Theorem 4.1, which asserts the temporal localizability of SFGs. Now let us address the optimal α cut-set procedure [Kung84b].

To ensure optimality of the cut-set procedure, the only required modification is to confine the application of the delay-transfer operation to a restrictive class of good cut-sets, namely, **nonrescaling** (NR) cut-sets. An NR cut-set is defined as a cut-set in which all the edges in the opposite direction to the target edge have at least two (as opposed to one) delays. Therefore, the (optimal) cut-set procedure can be regarded as a simple modification of the cut-set systolization rules.

Theorem 4.2:   *By following the two rules outlined next, the cut-set systolization procedure suffices to convert an SFG into a systolic array with optimal throughput rate (i.e., minimum α).*

1. **Delay-transfer** *On identifying an appropriate NR cut-set, we apply the delay-transfer operation along the cut-set and localize the target edge(s). Note that no additional time-scaling is incurred.*

2. **Time-scaling** *If no NR cut-set exists, it implies that the current rate is too fast and needs to be slowed. Therefore, α is incremented (each time by one) until an NR cut-set can be found.*

Proof: To prove the theorem, we need to show that if an NR cut-set does not exist, then an increment of the scaling factor is necessary. Suppose that no NR cut-sets exist for a given target edge. According to the colored arc lemma, there exists a cycle which contains the target edge (with zero delay) and the other edges with zero or one delay. Suppose that the node computation times are uniformly equal to one time unit. Then completing the computation around the cycle (so that the resultant is available for the next operation at the beginning node), will take as many time units as the number of edges in the cycle. This means that the present time-delay assigned will be short by at least one time unit. (This is because of the zero delay in the target edge.) Thus a time rescaling is necessary.

On the other hand, a proper time rescaling can certainly settle the timing problem of the cycle. Again, according to the colored arc lemma, if there are no more *bad* cycles containing the target edge, then an NR cut-set should now exist and *Rule 1*(delay transfer) can then be implemented.

Optimal α Cut-set Algorithm   Theorem 4.2 suggests that the algorithm should search for an NR cut-set. When such a cut-set does not exist, then a cycle containing the target edge will be identified. Starting with a target edge, we redefine the meaning of the bad predecessor edges. The *bad predecessor edges* are now defined to be those edges with *one or zero delays* (instead of zero delays as defined earlier). The NR cut-sets should not contain any bad predecessor edges; therefore, in the procedure to find the NR cut-set, the bad edges should be grouped together and avoided. More specifically, we introduce the notion of a *supernode*, which is a clustering of predecessor bad edges and nodes connecting them. For an example of a supernode, see Figure 4.29(a). It is clear that the NR cut-sets should not

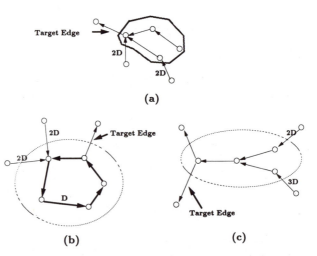

Figure 4.29: Illustration of the optimal-α cut-set localization: (a) An example of a supernode circled by the dashed line. (b) A cycle is found (indicated by bold arcs). (c) An NR cut-set indicated by the dashed circle is found.

cut through the supernode cluster. The supernode is formed according to the following search procedure:

Starting with a target edge, the supernode is expanded by tracing all of the bad predecessor edges (cf. the procedure in Figure 4.7) until either one of the two following criteria is true:

1. The supernode is surrounded by eligible edges only, or

2. The supernode cluster ends up to and includes the terminating vertex of the target edge.

For case 1, these edges which surround the supernode form an NR cut-set. Then a proper delay transfer can be used to systolize the target edge. For case 2, a cycle containing only bad edges and the target edge is found. A proper time rescaling can now be applied. We then restart the NR cut-set search procedure for the same target edge.

The above procedure is repeated until there are no more target edges. For a pictorial example, corresponding to case 2, it is possible that the search procedure ends up with a cycle shown in Figure 4.29(b). On the other hand, corresponding to case 1, a possible supernode (and NR cut-set) will be the one corresponding to the cluster encircled by the cut (dashed line) as shown in Figure 4.29(c).

The preceding rules are the basis of the optimal α cut-set procedure. Due to its simplicity, the algorithm may become preferable when the network or the cycles in the network are of small scale. For an efficient algorithm, α is computed according to the delay distribution along the cycle (instead of being incremented only by one each time). Another possible improvement is to localize the potential target edges inside the supernode in addition to those surrounding it. The preceding method, which is easy to understand, is not necessarily an efficient approach.

4.4.4.2   Leiserson's Procedure for Minimizing the Number of Delay Elements

Let us show an example to illustrate the nonuniqueness of the total number of delays in a systolized array. In Figure 4.30(a), we show an SFG consisting of 3 nodes. In order to systolize this SFG, a scaling factor of 2 is required, resulting in Figure 4.30(b). In Figure 4.30(c), a cut-set delay transfer is shown, and in Figure 4.30(d), another cut-set delay transfer is shown. Both

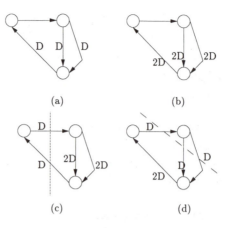

Figure 4.30: A simple example of delay assignment: (a) an SFG of 3 nodes; (b) scaling of delays ($\alpha = 2$); (c) an optimal α systolic array; (d) another optimal α systolic array.

systolic arrays in Figure 4.30(c) and (d) have the same $\alpha = 2$. Note, however, that the total number of delays in Figure 4.30(c) is one more than the total number of delays in Figure 4.30(d).

As just mentioned, it is desirable to minimize hardware cost by using minimal delays in a systolic array, while preserving optimal α. Now we shall introduce a procedure to optimize the total number of delays in a systolic array.

Leiserson et al. [Leis83a] proposed a now-popular systolization scheme (or the retiming of a synchronous circuit, according to the notation in [Leis83a]), which is based on delay-transfer through nodes. [5]

An example of such a delay transfer is shown in Figure 4.31(a), in which two delays are transferred from the incoming edges to the outgoing edges of a node v. Obviously, we can assign any number of delay transfers between the set of incoming edges and the set of outgoing edges of a node, since they cross the node cut in opposite directions. Therefore, the delay transfers of the whole SFG can be specified as a function r from the set of nodes of the SFG to integers. The function r of a node v, denoted by $r(v)$ means that r delays are taken from the incoming edges of v and $r(v)$ delays are added to the outgoing edges of v. For example, $r(v) = 2$ for the node v in Figure 4.31(a).

---

[5]The equivalence of using delay-transfer through nodes and the cut-set delay-transfers for systolization is given as a problem at the end of this chapter (see Problem 11).

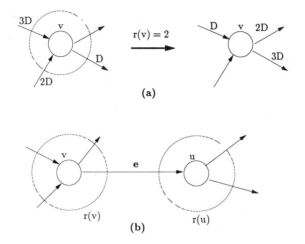

Figure 4.31: (a) Transfer of delays through nodes with $r(v) = 2$. (b) The change of delays on an edge e from node v to node u.

If we define two functions D and D_n for all edges in the graph, where $D(e)$ denotes the number of delays on edge e before delay transfers and $D_n(e)$ denotes the number of delays on edge e after the delay transfer, then for an edge e from node v to node u, $D_n(e)$ is related to $D(e)$ by: (see Figure 4.31(b).)

$$D_n(e) = D(e) + r(v) - r(u) \tag{4.10}$$

The requirement of a systolic design means that

$$D_n(e) \geq 1, \text{ for all edges } e \text{ in the SFG.}$$

The total number of delays can be expressed as:

$$
\begin{aligned}
S(r) &= \sum_{e \in E} D_n(e) \\
&= \sum_{e \in E} [D(e) + r(v) - r(u)] \\
&= K + \sum_{v \in V} r(v)[\text{outdeg}(v) - \text{indeg}(v)] \tag{4.11}
\end{aligned}
$$

where the indeg(v) and the outdeg(v) denote the number of edges pointing into and the number of edges pointing out of node v respectively, and the term K denotes the constant term in the summation, which is independent of $r(v)$. Therefore, the problem of minimizing $S(r)$ can be reformulated as:

$$\textbf{Minimize} \sum_{v \in V} r(v)[\text{outdeg}(v) - \text{indeg}(v)] \qquad (4.12)$$

under the constraints

$$r(u) - r(v) \le D(e) - 1$$

The optimization problem, cast in this manner, is the linear programming dual of a minimum-cost flow problem and is solvable in polynomial time.

4.4.4.3   Trade-off Between Minimal Pipelining Period and Minimal Delay Elements

The above two optimizing factors of the cut-set retiming for an SFG, namely, the pipelining period and total delay elements, cannot in general be minimized at the same time. A trade-off between these two optimizing functions has to be carefully assessed. This may be explained by the following example of the linear phase filter.

A Case Study   As discussed in Section 2.3.5, the linear phase filter is frequently used for its property of retaining linear phase. Figure 4.32(a) details the SFG for the linear phase filter. In this SFG, nodes correspond either to an adder or to a MAC function. Since there are no cycles in the SFG, the minimal-α cut-set procedure discussed earlier may be applied to derive a minimal pipeline period, i.e., $\alpha = 1$. The cut-sets used and the resultant systolized array are shown in Figure 4.32(b). Note that the number of delays corresponding to the 1st, 2nd , ... Nth stages are, respectively, $(N)D$, $(N-1)D$, ..., $1D$. Therefore, the total number of delay elements in this systolic array amounts to $O(N^2)$, which is certainly not desirable. However, it can be shown that (see Problem 9) there exist no cut-sets that would yield the minimal α and require fewer delays (in the sense of orders).

It is possible to realize a systolic design with $O(N)$ delay elements at the expense of a slower pipeline rate. Such a systolic array and the corresponding cut-sets are shown in Figure 4.32(c). Note that in the new array, although there are only $O(N)$ total delay elements, the pipeline rate is reduced by a factor of two, i.e., $\alpha = 2$. This shows that in general the pipeline period α and

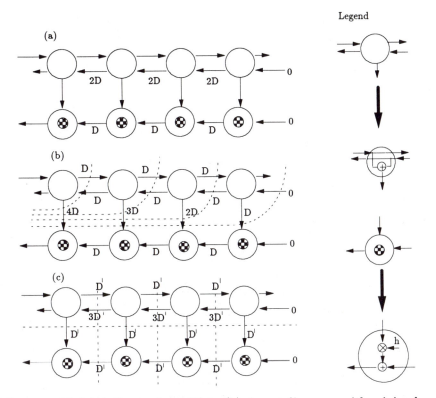

Figure 4.32 (a) A linear phase filter. (b) A systolic array with minimal-α. (c) A systolic array with minimal total delay elements.

delay elements cannot be optimized at the same time. A trade-off between speed and hardware has to be made, depending on the design requirement. For example, we note that by a pipeline interleaving method (discussed later in this section) the processor utilization of the second design may be doubled.

4.4.4.4   Multirate Systolic Design

In a systolic array, the processing or communication times of different operations may vary a great deal. As a result, the global clock period has to be the *maximum* of these operation times, plus some safety margin. This is clearly undesirable if we want to achieve maximum array throughput. One solution is to allow different operations in the array take different time periods by the use of a finer clocking period. This type of systolic array is called a *multirate* systolic array. A multirate systolic array is a generalized systolic array, allowing different operations to consume different time units. Let us look at the following example of a multirate systolic array.

Example 4: Multirate Systolic Array for the Linear Phase Filter

From our earlier discussion of the systolic array for the linear phase filter, we have the dilemma that we cannot simultaneously minimize α and the total number of delay elements. However, if a multirate systolic array is used, this problem can be solved. First, note that the data moving from left to right or from right to left in the upper part of the SFG in Figure 4.32(a) do not change value, i.e., they are transmittent data. Therefore, it is not necessary to assign an equal amount of time for this data transfer to that for the multiplication or addition. Let us define Δ as the time for data transfer of the transmittent data and T as the time for MAC or addition plus a data transferring activity. We further assume that $T \gg \Delta$. Now, if we let the delay $\mathbf{D}$ in Figure 4.32(a) be replaced by $\Delta + T$, which is shown in Figure 4.33(a), this is equivalent to saying that $\mathbf{D}$ is scaled to be $\Delta + T$. Now, we transfer delay of T from the right-going edges to the left-going edges along the vertical cut-sets and transfer delay of T to the down-going edges along the horizontal cut-set, as shown in Figure 4.33(a). The resulting SFG array, shown in Figure 4.33(b), represents a multirate systolic array. To show the final multi-rate array, we can group the delays with the nodes to form basic modules as indicated by the dark *circles* in Figure 4.33(b). The resulting multirate systolic array is shown in Figure 4.33(c). Note that a pure delay of $2T$ (indicated by a small square) is needed on the upper left-going edges to ensure the timing.

The pipelining period of this multirate systolic array is $\Delta + T$. If the original clock period of the (uniform-rate) systolic array for the linear phase filter is T, then effectively, $\alpha = \frac{(\Delta+T)}{T} \approx 1$ for the multirate systolic array. Moreover, the total number of delays in this array is still of $O(N)$.

This example shows that by using a multirate systolic array for the linear phase filter, almost minimal pipelining period and minimal total delays can be achieved. In Section 5.2, we will show that the wavefront array for the linear phase filter also shares this desirable result, since the wavefront array is the asynchronous version of the multirate systolic array.

Reverse SFG edges for optimal design   Note that in the above example, the tradeoff between minimum α and minimum total delays is valid for

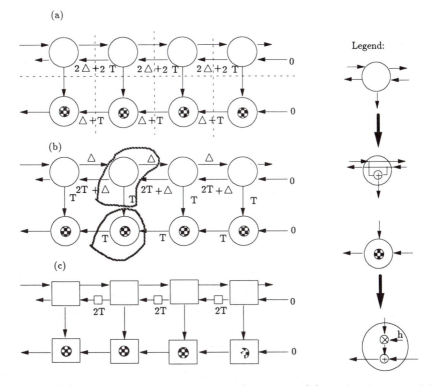

Figure 4.33: (a) time scaling: $\mathbf{D} \to \Delta + T$ and cut-sets. (b) result of cut-set delay transfers and the grouping of delays with nodes to form modules. (c) multirate systolic array for the linear phase filter.

a fixed SFG as the starting point. However, the dilemma we were facing for the linear phase filter can be solved by modifying the SFG (or equivalently the DG). This is illustrated in Figure 4.34.

First, note that the upper right-going edges in Figure 4.32(a) carries the same data, i.e., these edges form a broadcast line. Therefore, we can *reverse these edges* and send data from the right to the left, since the broadcasting direction does not affect the computation. This modified SFG is shown in Figure 4.34(a). Now, there are no cycles in the modified SFG. By transferring one delay along the cuts in Figure 4.34(a), we obtain a systolized SFG with $\alpha = 1$ and total delays of $O(N)$, which is shown in Figure 4.34(b).

From these discussions, we can see that the optimal systolic design depends on careful considerations in all three stages. This issue will be further discussed in Section 4.5.

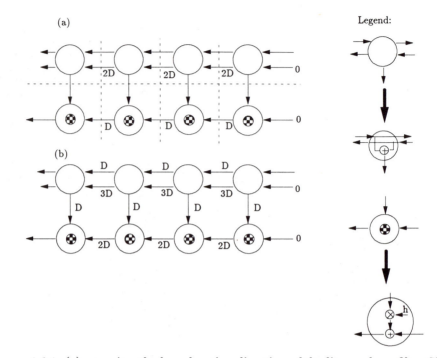

Figure 4.34: (a) reversing the broadcasting direction of the linear phase filter SFG and showing the cut-sets. (b) the systolic array with $\alpha = 1$ and $O(N)$ delays for the linear phase filter.

4.4.5   Improving PE Utilization Efficiency

From a consideration of the systolic arrays discussed earlier, we note that the pipelining period α is usually 1, 2, or 3. When α is larger than 1 for a designed systolic array, the input data rate is slowed down by a factor α; therefore the utilization of the array is only $1/\alpha$. In general, there are two ways to improve the processor utilization. One is pipeline interleaving, and the other is processor sharing. Another factor causing inefficient PE utilization is the case when different operations take various times. Besides using the multirate systolic arrays, we can make operation times more compatible by enhancing hardware for some time-consuming operations.

4.4.5.1   Pipeline Interleaving in Systolic Arrays

If the pipelining period α of a systolic array is larger than 1, each PE in the array is doing one useful computation every α time steps. We can partition the global time steps in the systolic array into α equivalent classes by a

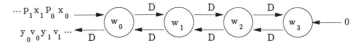

Figure 4.35: Pipeline interleaving for convolving two input sequences: Note that $y = x * w$ and $v = p * w$.

modulo α relation, i.e., time t_1 is in the same class as time t_2 if $(t_1 - t_2) \bmod \alpha = 0$. Since the pipeline period of the systolic array is α, the computations in each of the α time classes are independent of each other. Hence, the systolic array can compute α independent problems in each of the α time classes by interleaving α independent data and pipelining them into the array [Leise81]. By this method, the utilization of the PE becomes maximal (100%).

An example of convolving two input sequences $x(n)$ and $p(n)$ with a window $w(n)$ by the pipeline interleaving method is shown in Figure 4.35.

4.4.5.2 Processor Sharing in Systolic Arrays

In general, it is possible to improve the processor utilization rate in a systolic array by as much as α times, where α is the pipelining period. The scheme is straightforward, noting that the data will have to be α units apart, and therefore only one of α consecutive processor modules will be active at any instant. Therefore, a group of α consecutive PEs can *share a common arithmetic unit* without compromising the throughput rates [Kung84a]. Now let us use an example for a better illustration. Note that, according to the snapshots for the Lattice Systolic Array (Type B) (see Section 4.3.1.3) as depicted in Figure 4.36, *only one upper MAC module is active* in every three PEs at any time instant, and the same is true for the lower MAC module. Therefore, as shown in Figure 4.36, the three PEs can be combined into a (macro) PE and share the two common MA modules (one upper and one lower). A special-purpose ring register (with *period* $= \alpha$) can be designed to handle the resource scheduling. For a formal mathematical treatment on the subject, please refer to Section 6.4.

4.4.5.3 Forcing Uniform PE Processing Times

We have noticed that a uniform rate systolic array may not be efficient since different operations may take different time periods to complete and that a multirate systolic array can be used to improve α. On the other hand,

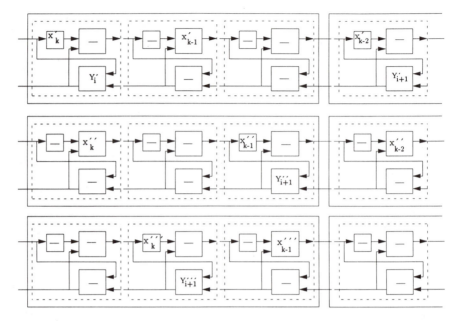

Figure 4.36: Snapshots and time-sharing of three PEs within a (macro) PE for a lattice systolic array (Type B).

the processing speed of the critical PEs (most likely to be the boundary PEs) often dictates the overall speed performance. Therefore, a hardware enhancement focused on these critical PEs (to equalize the processing times of all the PEs in the array) is often cost-effective and desirable. For example, in the QR decomposition discussed in Section 3.4, the bottleneck of the computation is the PEs for performing Givens Rotation, therefore a fast square-rooter, or a square-root-free scheme becomes appealing in order to speed up the rotation.

4.5   Systolic Arrays for the Transitive Closure and Dynamic Programming Problems

4.5.1   Dynamic Programming Method

Dynamic programming is an effective recursive approach to problem solving just as mathematical induction is to proving theorems. It is a general technique and can be applied in many situations. Dynamic programming is in essence a bottom-up procedure in which solutions to all subproblems are first calculated and the results are used to solve the overall problem.

The computation is divided into stages, proceeding from small subproblems to larger ones. The intermediate results are stored and utilized in latter stages. The advantage of the method lies in the fact that once a subproblem is solved, the answer is stored and never recalculated. This is in contrast to top down divide and conquer techniques, in which the same subproblem may be repeatedly solved in different partitions of the problem.

Dynamic programming techniques are widely used in optimization problems, in which an objective function must be either minimized or maximized subject to a set of constraints. Dynamic programming can be generally applied to problems of this class if they adhere to the principle of optimality. Simply put, this means that the overall optimal solution can be decomposed into recursively solving optimal solutions to a number of associated subproblems. The basic principles of dynamic programming are the following [Gondr84]:

1. The problem is embedded in a family of problems of the same nature.

2. The optimal solutions of these problems are linked by a recurrence relation.

A number of algorithms developed in this book employ dynamic programming techniques. In Floyd's algorithm for the shortest path problem, the solutions are found by starting with the direct connections that utilize no intermediate nodes (in other words, the initial permissible intermediate node set is an empty set). Nodes are then added, one at a time, to the set of permissible intermediate nodes. At each stage the shortest paths utilizing the permissible intermediate node set are calculated. The final result is achieved when all nodes have been added to the set. Dynamic time warping algorithms (see Section 8.5) are another instance. They correspond to a single source shortest path problem [Hu82], [Aho74].

4.5.2   Optimal Systolic Design for the Transitive Closure and Shortest Path Problems

In this section, we consider a very important type of dynamic programming problem, i.e., the shortest path problems (or the algebraic path problems, which will be explained in Section 4.5.3). The most familiar problems in this class are the transitive closure and the shortest path problems in graph theory, and Gauss-Jordan elimination in linear algebra. In the following, we first present the design of several systolic arrays, including an optimal one

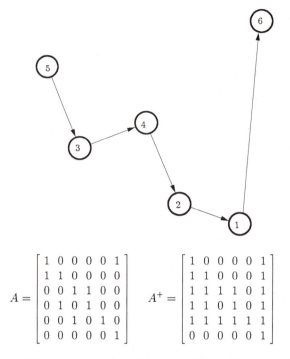

$$A = \begin{bmatrix} 1 & 0 & 0 & 0 & 0 & 1 \\ 1 & 1 & 0 & 0 & 0 & 0 \\ 0 & 0 & 1 & 1 & 0 & 0 \\ 0 & 1 & 0 & 1 & 0 & 0 \\ 0 & 0 & 1 & 0 & 1 & 0 \\ 0 & 0 & 0 & 0 & 0 & 1 \end{bmatrix} \qquad A^+ = \begin{bmatrix} 1 & 0 & 0 & 0 & 0 & 1 \\ 1 & 1 & 0 & 0 & 0 & 1 \\ 1 & 1 & 1 & 1 & 0 & 1 \\ 1 & 1 & 0 & 1 & 0 & 1 \\ 1 & 1 & 1 & 1 & 1 & 1 \\ 0 & 0 & 0 & 0 & 0 & 1 \end{bmatrix}$$

Figure 4.37: An example of the transitive closure problem. (a) A directed graph. (b) Its adjacency matrix **A**. (c) The transitive closure matrix **A**$^+$.

in terms of pipelining period and block pipelining period for the transitive closure problem, and then explain the extension to other problems in this class.

The Transitive Closure Problem A directed graph G can be represented as a tuple $G(V, E)$, where V is the set of vertices and E is the set of edges in the graph. The graph $G^+(V, E^+)$, which has the same vertex set V as G but has an edge from v to w if and only if there is a path (length zero or more) from v to w in G, is called the reflexive and transitive closure of G, or simply the *transitive closure* of G [Aho74].[6]

As shown in Figure 4.37, given a directed graph, its *adjacency matrix* **A** has element $a_{ij} = 1$ if there is an edge from vertex i to vertex j or $i = j$ and $a_{ij} = 0$ otherwise. The transitive closure problem is to compute the

---

[6] A relation R is called *reflexive* when $a\,Ra$ for every a in V. A relation R is *transitive* when it has the property $a\,Rb$ and $b\,Rc$ imply $a\,Rc$ [Ore62].

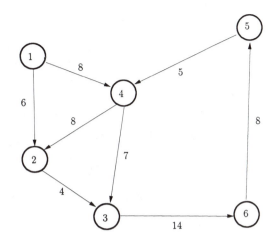

$$A = \begin{bmatrix} 0 & 6 & \infty & 8 & \infty & \infty \\ \infty & 0 & 4 & \infty & \infty & \infty \\ \infty & \infty & 0 & \infty & \infty & 14 \\ \infty & 8 & 7 & 0 & \infty & \infty \\ \infty & \infty & \infty & 5 & 0 & \infty \\ \infty & \infty & \infty & \infty & 8 & 0 \end{bmatrix} \quad A^+ = \begin{bmatrix} 0 & 6 & 10 & 8 & 32 & 24 \\ \infty & 0 & 4 & 31 & 26 & 18 \\ \infty & 35 & 0 & 27 & 22 & 14 \\ \infty & 8 & 7 & 0 & 29 & 21 \\ \infty & 13 & 12 & 5 & 0 & 26 \\ \infty & 21 & 20 & 13 & 8 & 0 \end{bmatrix}$$

Figure 4.38: An example of the shortest path problem. (a) A directed graph. (b) Its distance matrix **A**. (c) The shortest path matrix $\mathbf{A}^+$.

transitive closure matrix $\mathbf{A}^+$, whose element $a_{ij}^+ = 1$ if there is a path of length zero or more from vertex i to vertex j; $a_{ij}^+ = 0$ otherwise.

The Shortest Path Problem The shortest path problem for a directed graph is to determine the lengths of the shortest paths between all pairs of nodes in the graph. For example, in Figure 4.38, given a sample graph of six nodes, then the elements of its distance matrix **A** are defined as follows:

1. a_{ij} = the distance from vertex i to vertex j if there is an edge from i to j.

2. $a_{ij} = \infty$ if there is no edge connecting i and j.

3. $a_{ij} = 0$ if $i = j$.

The shortest path problem is to compute the shortest path matrix $\mathbf{A}^+$, whose element a_{ij}^+ is the shortest path length from vertex i to j.

4.5.2.1   The Warshall-Floyd Algorithm

Given the input matrix $\mathbf{A}$, the sequential Warshall-Floyd algorithm to find $\mathbf{A}^+$ of the graph is quite well known and is as follows [Aho74]:

For k from 1 to N

For i from 1 to N

For j from 1 to N

$$x_{ij}^k \longleftarrow x_{ij}^{k-1} + x_{ik}^{k-1} \times x_{kj}^{k-1};$$

where the initial $\mathbf{X}$ matrix, $\mathbf{X}^0$, is equal to $\mathbf{A}$; and the output matrix $\mathbf{A}^+$ equals $\mathbf{X}^N$.

This algorithm solves both the transitive closure and the shortest path problems. The difference between the two problems can be stated as follows:

1. For the *transitive closure* problem, the input $\mathbf{A}$ is the adjacency matrix. The + operation is the boolean **OR** operation; the × operation is the boolean **AND** operation. The output $\mathbf{A}^+$ is the transitive closure matrix.

2. For the *shortest path* problem, the input $\mathbf{A}$ is the distance matrix. The + operation is the minimum operation; the × operation is the addition operation. The output $\mathbf{A}^+$ is the shortest path matrix.

The transitive closure problem and the shortest path problem actually belong to a more general class of problems, the *algebraic path problem (APP)*, which also includes the Gauss-Jordan elimination procedure for matrix inversion. The definition of APP and examples of it and an algorithm to solve it are discussed in the next subsection.

Dynamic Programming Formulation   The Warshall-Floyd algorithm is based on a dynamic programming formulation. We take the Warshall algorithm for the transitive closure problem for the explanation. First note that the computation of a_{ij}^+ is embedded in the family of the following N problems: Find if there are paths of length zero or more from vertex i to vertex j, with intermediate node indices less than or equal to k, $1 \leq k \leq N$. Secondly, if x_{ij}^k is the value of the transitive closure matrix in the kth subproblem, then that value is directly linked to the value in the $(k-1)^{\text{st}}$ subproblem by the equation:

$$x_{ij}^{k} \leftarrow x_{ij}^{k-1} + x_{ik}^{k-1} \times x_{kj}^{k-1}$$

which defines the dynamic programming formulation for the Warshall algorithm.

Snapshots   To illustrate how the Warshall-Floyd algorithm works, let us first show the snapshots of the recursions of the $\mathbf{X}^{k}$ matrix. Initially, the input matrix $\mathbf{A} = \mathbf{X}^{0}$ is assumed to be distributed in the $N \times N$ square matrix. For the transitive closure problem, the graph and its adjacency matrix $\mathbf{A}$ are shown in Figure 4.37. The snapshots of the Warshall algorithm for finding the transitive closure of this graph are shown in Figure 4.39. The snapshots of the Floyd algorithm for the shortest path problem in Figure 4.38 are shown in Figure 4.40. Note that, at the kth recursion, data on the kth

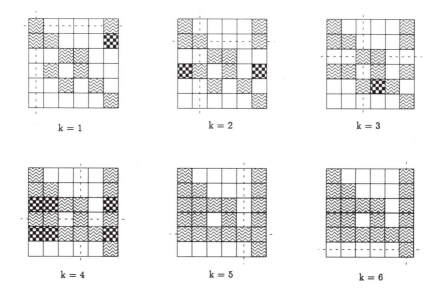

Figure 4.39: Snapshots of the Warshall algorithm: A *blank* node at cell (i, j) denotes that no connection (from i to j) is made yet. On the other hand, a node whose connection is made prior to the current recursion is drawn as a *shaded* one, and a node that is connected at a current recursion as a *checkered* one.

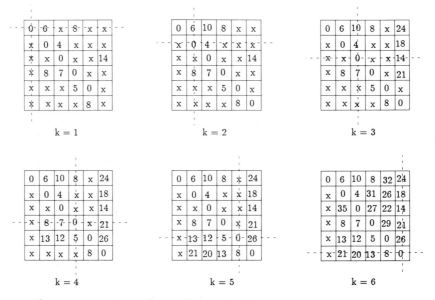

Figure 4.40: Snapshots of the Floyd algorithm for the shortest path problem: An x indicates no path exists.

column and kth row, which are marked with dashed lines in both snapshots, are used in the computation of all matrix elements.

Dependence Graph In the following, we derive systolic arrays for the Warshall-Floyd algorithm in the context of the transitive closure problem only. The readers should note that the results are directly applicable to the shortest path problem.

The sequential Warshall-Floyd algorithm can be easily converted to a single assignment form by introducing an iteration dimension k.

For i, j, k from 1 to N

$$x(i,j,k) \longleftarrow x(i,j,k-1) \; + \; x(i,k,k-1) \times x(k,j,k-1)$$

These dependencies can be localized by adding propagating variables for the row variable r and column variable c at each level k.

For i, j, k from 1 to N

$$c(i,j,k) \longleftarrow \begin{cases} x(i,j,k-1) & if \quad j = k \quad \text{\#distribute column } k \\ c(i,j+1,k) & if \quad j < k \quad \text{\#over row } i \\ c(i,j-1,k) & if \quad j > k \end{cases}$$

$$r(i,j,k) \longleftarrow \begin{cases} x(i,j,k-1) & if \quad i=k \quad \#\text{distribute row } k \\ r(i+1,j,k) & if \quad i<k \quad \#\text{over column } j \\ r(i-1,j,k) & if \quad i>k \end{cases}$$

$$x(i,j,k) \longleftarrow x(i,j,k-1) + r(i,j,k) \times c(i,j,k)$$
$$\#\text{form } kth \text{ connections}$$

where the input is $x(i,j,0) \longleftarrow a_{ij}$ and output is $a_{ij}^+ \longleftarrow x(i,j,N)$.

The cubic DG is shown for the $N = 4$ case in Figure 4.41, with each ij-plane drawn separately. What should have been included but not drawn are the additional dependence lines in the k-direction that run from points

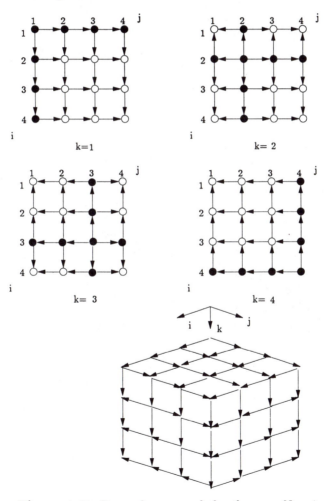

Figure 4.41: Dependence graph for the case $N = 4$.

(i, j, k) to $(i, j, k+1)$. In each plane the points *which* serve as the "source" of the row and column values are marked as dark nodes.

In the following, we project the DG in several directions to find the optimal systolic array.

4.5.2.2 Spiral and Hexagonal Systolic Arrays

Reindexing the DG The DG in Figure 4.41 is not totally regular. The source nodes in the DG move from row (column) k to row (column) $k+1$ from the kth ij-plane to the next one, and the transmittent data streams in the ij planes propagate in two opposite directions. *Because of these irregularities, there is no possible systolic schedule for this DG. One way to get around this problem is by reindexing the nodes in the DG, thereby transforming it to a more regular DG, with possible systolic schedules.* For this purpose, we would like to reindex the nodes in the DG such that all source nodes are located in the first row or the first column in each ij-plane and the transmittent data propagate in only one direction. This can be achieved by the following reindexing:

$$\text{node } (i, j, k) \rightarrow \text{node } ((i-k)_{\text{mod } N} + 1, \ (j-k)_{\text{mod } N} + 1, \ k).$$

Let us denote the original DG by DG-1 and the reindexed DG by DG-2. In DG-2, the preceding reindexing rearranges nodes in any ij-plane such that the source nodes are always on the first column or the first row. *Since the data propagating in the ij-planes are transmittent data, i.e., they do not change their values, the dependence arcs in the ij-planes may still remain as local arcs.* Therefore, all four ij-planes should look the same in DG-2 (see Figure 4.42). However, the dependence arcs between the ij-planes will become somewhat complicated. In the original DG-1, the dependence arcs from the kth ij-plane to the $(k+1)^{\text{st}}$ ij plane are represented by the vector $[0\ 0\ 1]^T$. In DG-2, these dependence vectors are transformed to:

$$[(i-k-1)_{\text{mod } N} - (i-k)_{\text{mod } N}, \ (j-k-1)_{\text{mod } N} - (j-k)_{\text{mod } N}, \ 1]^T.$$

The modulo operation will lead to two types of dependence arcs:

1. One is represented by the dependence vector

$$[-1, \ -1, \ 1]^T,$$

when

$$(i-k)_{\text{mod } N} \neq 0 \ \text{ and } \ (j-k)_{\text{mod } N} \neq 0.$$

The dependence arcs of this type are local.

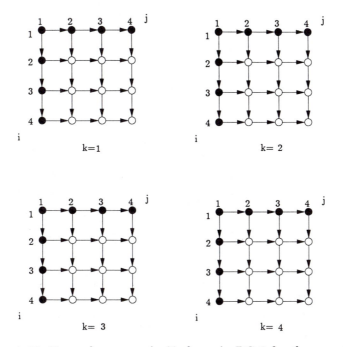

Figure 4.42: Dependence arcs in ij-planes in DG-2 for the case $N = 4$.

2. Another is represented by the vectors

$$[(N-1), -1, 1]^T \quad \text{or} \quad [-1, (N-1), 1]^T \quad \text{or} \quad [(N-1), (N-1), 1]^T,$$

when

$$(i - k)_{\text{mod } N} = 0, \quad \text{or} (j - k)_{\text{mod } N} = 0.$$

Note that dependence arcs of this type are global. They are termed as *spiral* arcs.

Now, we are able to draw a more complete DG-2. The dependence arcs in the ij-planes are shown in Figure 4.42. The two types of dependence arcs between any two adjacent ij-planes (in this case, the first two planes) in DG-2 are shown in Figure 4.43.

The Reindexed Single Assignment Algorithm   We have shown the effect of the modulo reindexing on the DG. The corresponding effect on the

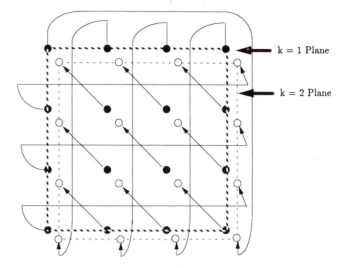

Figure 4.43: Local and spiral dependence arcs between the first two ij-planes in DG-2. Note that the shaded nodes denote nodes in the $k = 1$ plane and the unshaded ones denote nodes in the $k = 2$ plane.

single assignment algorithm can be similarly derived. The reindexed single assignment algorithm is shown in the following:

For i, j, k **from** 1 **to** N

$$c(i,j,k) \longleftarrow \begin{cases} x(i_{\mathrm{mod}\,N} + 1, \ j_{\mathrm{mod}\,N} + 1, \ k - 1) & if \ \ j = 1 \\ c(i, j - 1, k) & if \ \ j \neq k \end{cases}$$

$$r(i,j,k) \longleftarrow \begin{cases} x(i_{\mathrm{mod}\,N} + 1, \ j_{\mathrm{mod}\,N} + 1, \ k - 1) & if \ \ i = 1 \\ r(i - 1, j, k) & if \ \ i \neq k \end{cases}$$

$$x(i,j,k) \longleftarrow x(i_{\mathrm{mod}\,N} + 1, \ j_{\mathrm{mod}\,N} + 1, \ k - 1) + r(i,j,k) \times c(i,j,k)$$

where the input and output have to be also adjusted according to the reindexing.

Spiral SFG Array: Projection in the k-direction Note that the dependence arcs are invariant from one ij-plane to another in DG-2, i.e., they are DSI in the k-direction. Thus, if we project DG-2 in the k-direction (the $[0\ 0\ 1]^T$ direction) and use a linear schedule vector in the k-direction, (i.e., all the nodes in an ij-plane are computed at the same time), then we can derive a regularly structured SFG. According to the projection, the input matrix **A** should reside inside the array, i.e., a_{ij} should be at PE (i, j) to be ready for the execution. Practically, we have to load (input) the matrix **A** from the outside (i.e., the host) to the processor array in the *loading phase*. In our case, we choose to input the matrix **A** from the lower right corner of the SFG array along the diagonal (northwest) direction, as shown in Figure 4.44. The *loading phase* takes N time steps so that all the input a_{ij} arrives at PE (i, j). The SFG array is then ready for the computation, i.e., the *execution phase*, which also takes N time steps. After the computation is done, the output matrix $\mathbf{A}^+$ is sent out from the left and upper side of the array (i.e., the *unloading phase*).

Elimination of the Spiral Arcs   Due to the spiral arcs, the SFG in Figure 4.44 is called a *spiral* SFG array. The spiral interconnections imply potentially expensive global wiring in VLSI technology. Fortunately, there is a simple way to circumvent this problem based on the following observation:

The x variables in the kth row or kth column of the kth ij-plane in the original DG-1 do not change values. Or, equivalently, the x variables in the first column or first row in all ij-planes in DG-2 do not change values.

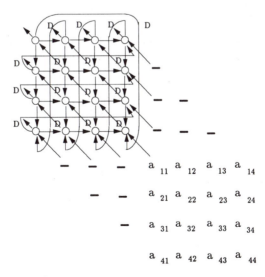

Figure 4.44: The spiral SFG array derived by projection in the k-direction of the DG-2.

For example, let us examine the computation of the first two ij-planes in DG-2. From the original Warshall-Floyd algorithm, we observe that

$$x(i,1,1) = x(i,1,0) + x(i,1,0) \times x(1,1,0)$$

since nodes are connected to themselves; therefore

$$x(1,1,0) = 1$$

Thus

$$x(i,1,1) = x(i,1,0)$$

This shows that no change is effected for the x variables in the first column of the $k = 1$ ij-plane. Similarly, no change is effected for the x variables in the first row of the $k = 1$ ij-plane. Further, notice that $c(i,1,1) = x(i,1,0)$ are the transmittent data in the $k = 1$ ij-plane. Then

$$x(i,1,0) = c(i,1,1) = c(i,N,1) = x(i,1,1) \qquad (4.13)$$

Therefore, the original global arc in DG-2 from node $(i,1,1)$ to node $(i-1,N,2)$, carrying data $x(i,1,1)$, is now replaced by a local arc from node $(i,N,1)$ to node $(i-1,N,2)$, carrying data $c(i,N,1)$ (see Eq. 4.13). Similarly, the original global arc from node $(1,j,1)$ to node $(N,j-1,2)$ carrying data $x(1,j,1)$, is now replaced by a local arc from node $(N,j,1)$ to node $(N,j-1,2)$, carrying data $r(N,j,1)$ (note that $r(N,j,1) = r(1,j,1) = x(1,j,0) = x(1,j,1)$). We further note that the datum $x(1,1,1)$ is a constant 1 in the computation. Therefore, it is not necessary to send the constant 1 from node $(1,1,1)$ to node $(N,N,2)$ by the long global arc. Instead, it can be locally generated by node $(N,N,2)$ and therefore this global arc can be removed. These local dependence arcs between the first two ij planes are shown in Figure 4.45.

The above observation is true for other ij-planes, i.e., the x variables in the first row or first column of the kth ij plane in DG-2 (or the kth row or kth column of the kth ij-plane in DG-1) do not change their values. This can be seen from the original Warshall-Floyd algorithm:

$$x(i,k,k) = x(i,k,k-1) + x(i,k,k-1) \times x(k,k,k-1) = x(i,k,k-1)$$
$$x(k,j,k) = x(k,j,k-1) + x(k,k,k-1) \times x(k,j,k-1) = x(k,j,k-1)$$

since $x(k,k,k-1) = 1$.

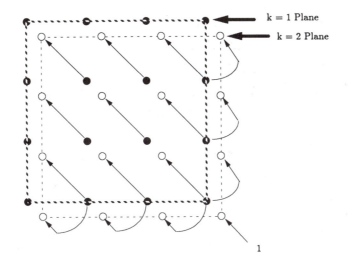

Figure 4.45: Removing spiral arcs between the first two ij planes of DG-2.

Local and Regular Dependence Graph (DG-3)   By the same argument, all global arcs in DG-2 can be replaced by local arcs for all ij plane computations. This localized DG is denoted by DG-3 and is shown in Figure 4.46. Note that DG-3 is also DSI in the k-direction.

Hexagonal SFG Array   Now if we project DG-3 again in the k-direction, the result will be the locally connected hexagonal SFG array, shown in Figure 4.47. Note that the delays associated with the original spiral arcs in Figure 4.44 are the same as the delays on the corresponding local arcs in Figure 4.47 to ensure the correct scheduling. We can further simplify this hexagonal SFG array by deleting the first row and first column PEs, since they do not perform any computation. This leads to an array of size $(N-1)^2$.

Hexagonal Systolic Array Design   The hexagonal systolic array is derived by applying the cut-set systolization [Kung84a] to the hexagonal SFG array in Figure 4.47. The result is shown in Figure 4.48. Note that the *pipelining period* $\alpha = 3$ for this systolic array. As is shown in Figure 4.48, the input data are loaded from the southeast end of the array. Once the data are loaded into their respective PE positions, execution of the Warshall-Floyd algorithm can proceed.

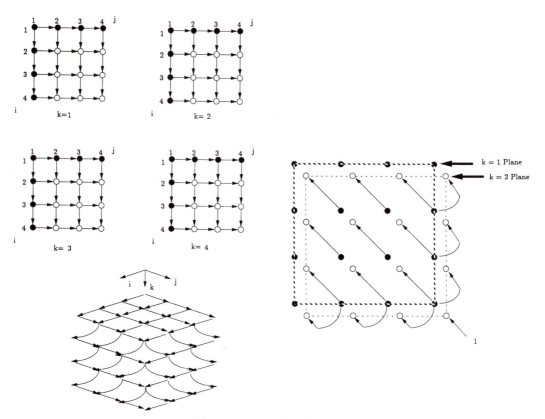

Figure 4.46: The localized DG-3.

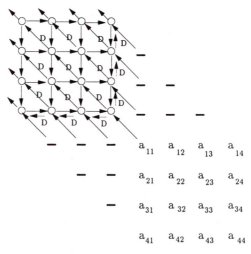

Figure 4.47: Locally connected hexagonal SFG.

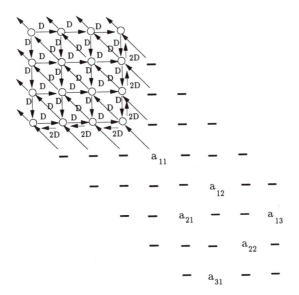

Figure 4.48: The hexagonal systolic array.

A short historical note is in order here. The fact that the spiral connections can be avoided was pointed out by Lin and Wu [Lin85]. However, their scheme actually includes an extra PE column (on the right-hand side) and an extra PE row (at the bottom), which are not needed in the design specified above. Rote [Rote85] also developed the hexagonal systolic array independently.

4.5.2.3   Optimal Orthogonal Systolic Array

Although the hexagonal systolic array has only local connections, the pipelining period $\alpha = 3$ and the number of I/O pins of this array is $4N$. This is not very desirable. To obtain an SFG with $\alpha = 1$ and with fewer I/O pins, we have to opt for an alternative projection direction. After a close investigation, we find that projection in either the i- or j-direction will serve the purpose. Note that projections in both the i-direction and j-direction are similar due to the symmetry in DG-3. Consider, now, projecting DG-3 along the i-direction.

Orthogonal Systolic Array   Since DG-3 is also DSI in the i-direction, we can project it directly. However, a permissible schedule vector for DG-3 is difficult to be determined by simple observation. In the earlier discussion (see Section 4.3.3), we proposed the option of combining both Stage 2 and Stage 3 for systolic designs. This option is now adopted here. We will use the algebraic approach (see Section 3.3) to determine the systolic schedule vector $\vec{s}$ and the projected systolic array directly.

Note that the projection vector $\vec{d} = [1\ 0\ 0]^T$. First, we list all the dependence vectors in DG-3. There are five of them, $\vec{e_1}, \ldots, \vec{e_5}$, as displayed below.

$$\begin{bmatrix} \vec{e_1} & \vec{e_2} & \vec{e_3} & \vec{e_4} & \vec{e_5} \end{bmatrix} = \begin{bmatrix} 1 & 0 & -1 & -1 & 0 \\ 0 & 1 & -1 & 0 & -1 \\ 0 & 0 & 1 & 1 & 1 \end{bmatrix}$$

A valid systolic schedule vector $\vec{s}$ should satisfy the following two constraints:

$$\vec{s}^T \vec{e_i} > 0 \quad \text{for } i = 1, 2, 3, 4, 5; \quad \text{and} \quad \vec{s}^T \vec{d} > 0$$

The minimum computation time solution (see Eq. 4.5) for $\vec{s}$ under these constraints is

$$\vec{s}^T = [1\ 1\ 3].$$

Note that in this case, the minimum computation time T is

$$T = \max_{\vec{p},\vec{q}} \{\vec{s}^T (\vec{p} - \vec{q})\} + 1 = [1\ 1\ 3] \begin{bmatrix} N-1 \\ N-1 \\ N-1 \end{bmatrix} + 1 = 5N - 4$$

With $\vec{s}$ defined, we can find the transformation matrix $\mathbf{T}$ as

$$\mathbf{T} = \begin{bmatrix} \vec{s}^T \\ P^T \end{bmatrix} = \begin{bmatrix} 1 & 1 & 3 \\ 0 & 1 & 0 \\ 0 & 0 & 1 \end{bmatrix}$$

where P, orthogonal to $\vec{d}$, determines the projection. Now we can find, for the edge $\vec{e_i}$, the projected edge $\vec{e_i}' = P^T \vec{e_i}$ in the systolic array are:

$$\begin{bmatrix} \vec{e_1}' & \vec{e_2}' & \vec{e_3}' & \vec{e_4}' & \vec{e_5}' \end{bmatrix} = \begin{bmatrix} 0 & 1 & -1 & 0 & -1 \\ 0 & 0 & 1 & 1 & 1 \end{bmatrix}$$

And the delays on the edges, $D(\vec{e_i}') = \vec{s}^T \vec{e_i}$:

$$\begin{bmatrix} D(\vec{e_1}') & D(\vec{e_2}') & D(\vec{e_3}') & D(\vec{e_4}') & D(\vec{e_5}') \end{bmatrix} = \begin{bmatrix} 1 & 1 & 1 & 2 & 2 \end{bmatrix}$$

With nodes determined by the projection in the i-direction and the edges and delays determined above, we are able to draw the systolic array,

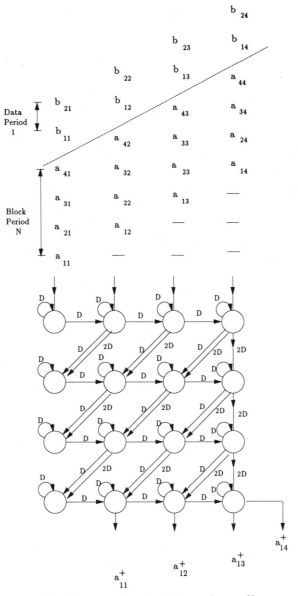

Figure 4.49: The optimal orthogonal systolic array.

which is shown in Figure 4.49. Note that in order to make sure that data are first loaded in the nodes and then reused for a period of N cycles, certain control is required in the systolic array. The details are left as an exercise for the reader (see Problem 24).

Optimality of the Design   This orthogonal systolic array is an optimal one for the transitive closure problem. The array uses N^2 processors and $2N$ I/O ports. It supports a noninterleaved block pipeline period of N, so successive computations can be repeated every N steps. (Note that in Figure 4.49, the $\{b_{ij}\}$ data block follows immediately after the $\{a_{ij}\}$ data block.) Due to the systolization process, the total computation time is $5N - 4$. We can see from the original "parallelized" Warshall-Floyd algorithm dependence graph (DG-1) that this is indeed optimal. Since N^3 computations must be performed with N^2 processors, N is the minimal period. This minimal period insures 100% processor utilization for series of pipelined computations. In DG-1 we find that there exists a directed path of length $5N - 4$ along the path $(1,1,1) \rightarrow (N,1,1) \rightarrow (N,N,1) \rightarrow (N,N,N) \rightarrow (1,N,N) \rightarrow (1,1,N)$. This path is shown in Figure 4.50 by bold-faced arcs. Hence, any systolic implementation, in which all inputs must be calculated at least one step before they are used, have a computation time of at least $5N - 4$. Finally, we note that for a period of N we need to input N data and output N data at each step, and hence $2N$ is the minimal number of I/O connections required.

4.5.3   The Algebraic Path Problem

The basic structures developed in this section for the transitive closure problem can be applied to a wide range of other problems. This is because mathematically transitive closure is a special case of a general problem known as

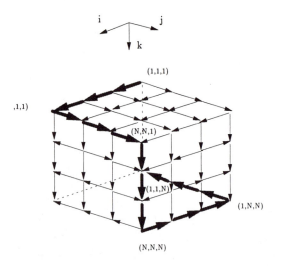

Figure 4.50: The longest path in DG-1 of the Warshall-Floyd algorithm.

the APP. Other problems in this class are the *shortest-path* problem from graph theory, the *matrix inversion* problem from linear algebra, and the generation of *regular languages* from automata theory [Gondr84]. The basic structures developed here can be applied to any instance of the APP.

In this section we briefly summarize the APP and a set of recursive equations to solve it, along with several examples. More detailed discussion for the systolic arrays for the general APP can be found in [Lewis86], [Rote85], and [Rober86]. Thorough derivations of the mathematics associated with the APP can be found in [Lehma77], [Mahr84], and [Gondr84].

4.5.3.1   Problem Definition

We are given a weighted directed graph $G = (V, E, w(e))$, with V the set of vertices, E the set of directed edges, and $w(e)$ a weighting of the edges. The edge weights, $w(e)$, are drawn from a semiring (or dioid) $S = \{H, +, \times, 0, 1\}$. In S, addition $(+)$ and multiplication $(\times)$ are closed binary associative operations over the set of elements H, with 0 and 1 the respective identity (or neutral) elements. Addition is commutative, multiplication distributes over addition, and 0 is absorptive with respect to multiplication. (That is, $a \times 0 = 0 \times a = 0$.) In addition, a unary operation of closure, denoted by a^*, is defined. (That is, $a^* \equiv 1 + a + (a \times a) + (a \times a \times a) + \ldots.$) The weight of a directed path, $w(p)$, is defined as the product of the edge weights in the path p.

$$w(p) \equiv \prod_{\forall e \in p} w(e)$$

If we number the graph vertices from 1 to N, then the APP is to find, for each pair of vertices i and j $(1 \leq i, j \leq N)$, the sum of the weights of all distinct paths from i to j. Denoting this sum as d_{ij} we have

$$d_{ij} \equiv \sum_{\forall p} w(p) \quad \text{for } p: \text{ path from } i \text{ to } j$$

The algorithm to find the d_{ij}'s is defined in terms of a matrix formulation of the problem. Here we define $\mathbf{A} = \{a_{ij}\}$ as the *adjacency matrix* of the graph, where

$$a_{ij} = \begin{cases} w(e) & \text{if } e \text{ is the edge from } i \text{ to } j \\ 0 & \text{if there is no edge from } i \text{ to } j \end{cases}$$

We can then define the result of the algebraic path calculation as a matrix $\mathbf{D} = \{d_{ij}\}$. For a graph with N nodes, the following recurrence equations in $\mathbf{C}^k$ allow us to calculate $\mathbf{D} = \mathbf{C}^N$, starting with $\mathbf{C}^0 = \mathbf{A}$ [Gondr84].

$$
\begin{aligned}
c_{ij}^k &= c_{ij}^{k-1} + \{c_{ik}^{k-1} \times (c_{kk}^{k-1})^* \times c_{kj}^{k-1}\} && \text{for } i \neq k, j \neq k \\
c_{ij}^i &= (c_{ii}^{i-1})^* \times c_{ij}^{i-1} && \text{for } i \neq j \\
c_{ij}^j &= c_{ij}^{j-1} \times (c_{jj}^{j-1})^* && \text{for } i \neq j \\
c_{ii}^i &= (c_{ii}^{i-1})^*
\end{aligned}
$$

4.5.3.2 Examples of the APP

Transitive Closure Problem   Finding the transitive closure $G^+ = (V, E^+)$ of a directed graph $G = (V, E)$ is is a special case of the APP where the edge weights of the graph belong to the semiring $S1 = \{H1, \vee, \wedge, 0, 1\}$, where $H1 = \{0, 1\}$. For each edge in G, $w(e) = 1$, so the adjacency matrix $\mathbf{A}$ contains just zeros and ones. In this case, the unary operation $*$ is equivalent to the constant operation 1. (That is, $a^* \equiv 1 \vee (a \wedge a) \vee (a \wedge a \wedge a) \vee \cdots = 1$.) The preceding recurrence equations, along with a serial ordering, simplify in this case to Warshall's algorithm for transitive closure [Warsh62].

Shortest Path Problem   Determining the lengths of the shortest paths between all pairs of nodes in a directed graph is another instance of the APP. In this case the edge weightings, corresponding to distances, belong to the semiring $S2 = \{H2, \min, +, \infty, 0\}$, where $H2$ is the set of nonnegative real numbers augmented with ∞, $(H2 = \{[0, \infty), \infty\})$. In this case the unary $*$ operation is equivalent to the constant operation 0. (That is, $a^* \equiv \min(0, a, a + a, a + a + a, \cdots) = 0$.) These recurrences, along with a serial ordering, simplify to the shortest path algorithm of Floyd [Floyd62].

Gauss-Jordan Elimination (Matrix Inversion)   Finding the inverse of a real matrix, if it exists, can also be cast as a special case of the APP. The matrix to be inverted is related the adjacency matrix $\mathbf{A}$. Its elements, corresponding to the edge weights, belong to the semiring $S3 = \{H3, +, \times, 0, 1\}$, where $H3$ is the set of real numbers and $+$ and $\times$ correspond to ordinary addition and multiplication. An extended closure is defined as:

$$
c^* \equiv \begin{cases} 1/(1 - c) & \text{for } c \neq 1 \\ \text{undefined} & \text{for } c = 1 \end{cases}
$$

Solving the APP yields $(\mathbf{I} - \mathbf{A})^{-1}$. Minor changes in the algorithm allow $\mathbf{A}^{-1}$ to be computed directly. With these changes, the recurrence equations, along with a serial ordering, correspond to the Gauss-Jordan algorithm (without pivoting) of linear algebra [Rote85].

4.5.3.3 Systolic Array Design for the APP

The two systolic arrays presented in this section can be directly transformed to the APP by taking care of the extra closure computation. For details, the readers are referred to [Lewis86].

4.6 Systolic Design for Artificial Neural Network

Neuroscientists have revealed that the massive parallel processing power in the human brain lies in the global and dense interconnections among a large number of identical logic elements or *neurons*. These neurons are connected to each other with variable strengths by a network of *synapses* [Taked86]. The original discrete-state Hopfield model [Hopfi82] (see Section 2.5), and the continuous-state Hopfield-Tank model (including the proposed analog neural circuits) [Hopfi84],[Hopfi85] have recently become popular in the realization of artificial neural networks (ANNs). They can be programmed to perform computational networks for associative retrieval or for optimization problems. Several optical computing approaches, exploiting the global interconnectivity of optical signal flow, have been proposed [Farha85], [Mada85], [Eichm85].

Both the analog neural circuits and the optical neural networks suffer from the disadvantages of low precision, difficulty of modifying the synaptic strengths, convergence to local optima and global interconnectivity. In order to overcome the disadvantages mentioned above, a locally interconnected systolic architecture for an ANN is proposed [Kung87b].

4.6.1 Hopfield and Hopfield-Tank Models

Original Hopfield Model   As we have discussed in Section 2.5, the original Hopfield model used two-state threshold *neurons*. (The states are usually taken to be 0 and 1). In this discrete model, each neuron i receives input currents of $T_{ij}V_j$ from neuron j, and a bias current input I_i (see Figure 4.51(a)) [Hopfi82], [Taked86]. The discrete-time transition of neuron i can thus be formulated as:

$$U_i(k) = \sum_{j}^{N} T_{ij}V_j(k) + I_i$$

$$V_i(k+1) = step[U_i(k)] \tag{4.14}$$

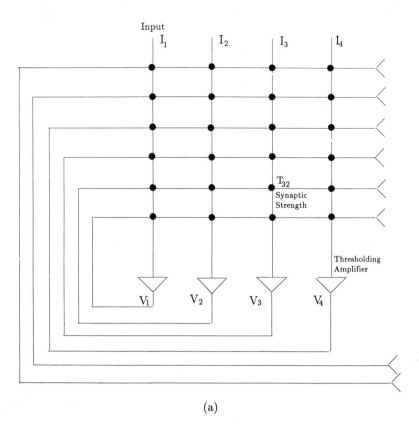

Input

I_1 I_2 I_3 I_4

T_{32}
Synaptic
Strength

Thresholding
Amplifier

V_1 V_2 V_3 V_4

(a)

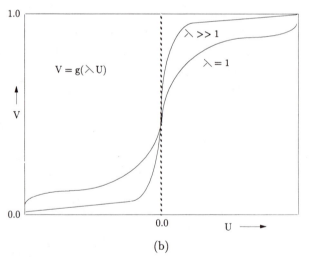

1.0

$V = g(\lambda U)$

$\lambda \gg 1$

$\lambda = 1$

V

0.0

0.0   U

(b)

Figure 4.51: (a) A simplified schematic diagram of the Hopfield model. In the discrete-state model, the amplifiers act like a hard limiter, while in the continuous-state model, the amplifiers follow the input-output relationship $g_i[\lambda u]$ with dynamic behavior. (b) The typical input-output relationship, the λ value determines the gain of the thresholding function.

270

where $step[x]$ is a unit step function, which is 1 for $x \geq 0$ and 0 for $x < 0$. N is the number of neurons and T_{ij} are elements of an interconnection matrix representing the synaptic strengths of neural connections (with $T_{ii} = 0$). The inputs U_i are thresholded by the amplifier (neuron), and the output V_i can then be fed back to the input of any other amplifier to modify their states. The state space transitions will gradually converge to a set of stable fixed points (see Section 2.5).

Hopfield and Tank Model   In order to imitate the continuous input-output relationship of real neurons, and also to simulate the integrative time delay due to the capacitance of real neuron, the continuous-state model proposed by Hopfield and Tank can be approximated by the following dynamic equations [Hopfi84], [Hopfi85], [Taked86]:

$$U_i(k) - U_i(k - 1) \;\; = \;\; \sum_{j}^{N} T_{ij} V_j(k) + I_i \qquad (4.15)$$

$$V_i(k + 1) \;\; = \;\; g[U_i(k)] \qquad (4.16)$$

where $g[x]$ is a nonlinear function, e.g.,

$$g[x] = (1/2)[1 + tanh(x/x_0)] \qquad (4.17)$$

which approaches a unit step function as x_0 tends to zero (see Figure 4.51(b)). The right hand side of Eq. 4.15 can be considered as the new excitation source, which effects "modification" of the states as shown in the left hand side.

For both models, an energy function is defined as [Hopfi82], [Hopfi84]

$$E = -(1/2) \sum_{i=1}^{N} \sum_{j=1}^{N} T_{ij} V_i V_j - \sum_{i=1}^{N} I_i V_i \qquad (4.18)$$

Hopfield has shown that, if $T_{ij} = T_{ji}$, then neurons in the continuous-state model always change their states in such a manner (see Problem 25) that the energy function is reduced [Hopfi84]:

Matrix Formulation   The Hopfield model can be formulated as a consecutive matrix-vector multiplication problem with some prespecified thresholding operations [Farha85]. For example, a matrix-form expression of Eq. 4.15 and Eq. 4.16 can be written as

$$\begin{aligned} \mathbf{u}(k) &= \mathbf{Tv}(k) + \mathbf{i} + \mathbf{u}(k-1) \\ \mathbf{v}(k+1) &= G[\mathbf{u}(k)] \end{aligned} \tag{4.19}$$

where $G[\mathbf{x}]$ function specifies the nonlinear thresholding of each element of the vector $\mathbf{x}$, and the vectors and matrices used are given as:

$$\mathbf{u} = [U_1, \ U_2, \ \cdots, \ U_N]^T$$

$$\mathbf{v} = [V_1, \ V_2, \ \cdots, \ V_N]^T$$

$$\mathbf{i} = [I_1, \ I_2, \ \cdots, \ I_N]^T$$

$$\mathbf{T} = \begin{bmatrix} T_{11} & T_{12} & \cdots & T_{1N} \\ T_{21} & T_{22} & \cdots & T_{2N} \\ \vdots & \vdots & \ddots & \vdots \\ T_{N1} & T_{N2} & \cdots & T_{NN} \end{bmatrix} \tag{4.20}$$

Example 1: Solving Image Restoration Problem

Artificial neural networks have been successfully applied to low level vision processing [Koch86]. Here, an image restoration example is given to illustrate how to map applicational problems onto ANNs. Consider an observed degraded image vector $\mathbf{g}$, which can be mathematically expressed as

$$\mathbf{g} = \mathbf{Hf} + \mathbf{n} \tag{4.21}$$

where $\mathbf{H}$ is a known blurring matrix. The statistical properties of the noise vector $\mathbf{n}$ are also assumed known. Image restoration is the scheme whereby an image vector $\mathbf{f}$ is restored from the linear blurring degradation mechanism and the additive noise. The most straightforward method is to solve the least squares problem as discussed in Chapter 2, i.e., to find an estimate $\hat{\mathbf{f}}$ which minimizes the total estimation error.

$$\min \ (\mathbf{g} - \mathbf{H\hat{f}})^T(\mathbf{g} - \mathbf{H\hat{f}}) \tag{4.22}$$

In many cases, however a priori information about the image properties (e.g., smoothness, intensity distribution) is known. In order that the estimated solution also reflects this information, a modified least squares formulation should be adopted:

$$\min \ (\mathbf{g} - \mathbf{H\hat{f}})^T(\mathbf{g} - \mathbf{H\hat{f}}) + \gamma(\mathbf{W\hat{f}})^T(\mathbf{W\hat{f}}) \tag{4.23}$$

where the $\mathbf{W}$ matrix represents the intensity weighting for the overall smoothness measure of the image, and γ is a proper regularization parameter.

To formulate the regularized image restoration problem in terms of ANN, the key step is to derive an energy function so that the lowest energy state (the most stable state of the network) would correspond to the best restored image. Once the energy function is determined, the synaptic strengths and input can be immediately derived. Let each pixel of image f_i correspond to the neuron state V_i, then the derived energy function is equal to the expression given in Eq. 4.23. By comparing Eq. 4.18 and Eq. 4.23, the corresponding $\mathbf{T}$ matrix and $\mathbf{i}$ vector are found to be (see Problem 26)

$$
\begin{aligned}
\mathbf{T} &= -2(\mathbf{H}^T\mathbf{H} + \gamma\mathbf{W}^T\mathbf{W}) \\
\mathbf{i} &= 2\mathbf{H}^T\mathbf{g}
\end{aligned}
\tag{4.24}
$$

Once $\mathbf{T}$ and $\mathbf{i}$ are determined, the ANN can be programmed accordingly to solve the image restoration problem.

4.6.2   Systolic Design via Cascade DG

Cascade DG Design   As we have discussed in Section 3.5, the consecutive matrix-vector multiplication array architecture design can be derived from a cascaded DG with nonlinear assignment. The DG of Eq. 4.19 is shown in Figure 4.52. By the same nonlinear assignment given in Section 3.5, a locally interconnected systolic array with bidirectional communication links (see Figure 4.53) can be obtained. This systolic architecture requires some smart switches to change the operations of each PE at different time slots (see Problem 27). However, according to Figure 4.53, some T_{ij}'s need to be repetitively stored in multiple PEs, and making the design unsuitable for adaptive modification of T_{ij}.

Modified Cascade DG Design   Examining the DG design shown in Figure 4.52 more closely, we note that many undesirable design aspects are due to the fact that the input direction of neuron states $V_i(k)$ is orthogonal to the output direction of the thresholded neuron states $V_i(k+1)$. It is possible to rearrange the data ordering of the T_{ij} elements, so that the input direction $V_i(k)$ becomes parallel to the output direction of $V_i(k+1)$. Such

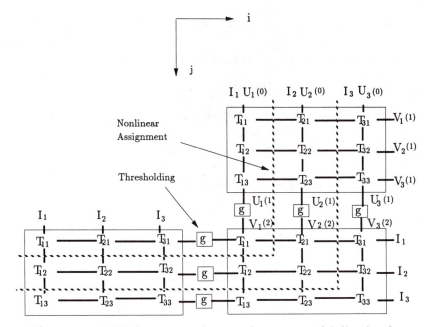

Figure 4.52: DG for consecutive matrix-vector multiplication formulation of Hopfield model.

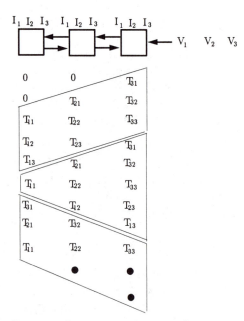

Figure 4.53: Systolic array for consecutive matrix-vector multiplication formulation of Hopfield model.

274

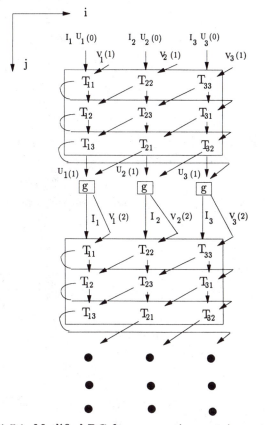

Figure 4.54: Modified DG for consecutive matrix-vector multiplication formulation of Hopfield model.

a modified DG is depicted in Figure 4.54. In this DG, for $i = 1, 2, \ldots, N$, the i-th column of the T_{ij} data array is circularly shifted-up by $i - 1$ positions. This DG is not totally localized due to the presence of the global spiral communication arcs. However, the input direction (from the top) and the output direction (from the bottom) are parallel. The advantage is that when many such DGs are cascaded top-down, the inputs and outputs data can be matched perfectly.

Ring Array for ANN For the top-down cascaded DG, the projection can be taken along the vertical direction, which will result in a ring array architecture as shown in Figure 4.55.[7] In the ANN implementation each

---

[7]The same architecture was previously proposed by Porter [Porte86], for consecutive matrix-vector multiplication in solving discrete state space equations.

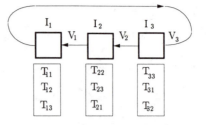

Figure 4.55: Ring systolic array for consecutive matrix-vector multiplication formulation of Hopfield model.

PE, say the i-th PE, is treated as a neuron, and the synaptic strengths $(T_{i1}, T_{i2}, \ldots, T_{iN})$ are stored in it. At the k-th iteration, the operation of the PE is as follows:

1. Each of the neuron outputs $(V_1, V_2, \ldots, V_N)$ is cycling through the ring array, and will pass through the i-th PE once during the N clock cycles.

2. When V_j passes through the i-th PE, it is multiplied with T_{ij}, and the result is added to the sum of $U_i(k-1)$ and I_i (according to the Eq. 4.15).

3. After all N clock cycles, the computation for $U_i(k)$ is completed, and it is ready for the thresholding operation.

4. After the thresholding operations, the PE sends the thresholded neuron output $V_i(k+1)$ to the left-side neighbor PE.

The above procedures is repeated until a convergence is reached. For implementing a large number of neurons, the problem of the long wrap-around line can be solved by a special 2-D arrangement scheme [Kung87b] (see Problem 29).

Advantageous Properties of Systolic ANN   The advantages of this array architecture for ANN are summarized as follows:

- The pipelining period $\alpha = 1$, which implies 100% utilization efficiency during the iteration process.

- Only N synaptic strengths of T_{ij} are stored in each PE. This results in easier modification of the synaptic strengths, making the "learning" capability possible.

- Using this proposed systolic architecture, the gain parameters can be easily updated during the iterations. (It is observed that if the gain control parameter of the sigmoid input-output relation in the neuron can be dynamically changed, then a faster convergence and better performance can be obtained [Hopfi85].)

- Compared to the analog RC neural circuit and the optical neural networks, the digital implementation can achieve higher precision in computation.

4.6.3   Using ANN to Solve Combinatorial Optimization Problems

Hopfield and Tank's continuous-state ANNs have been used to search for the solution of some combinatorial optimization problems. Examples are the travelling salesman problem [Hopfi85], the linear programming problems [Tank86], and the Hitchcock problem [Taked86].

Most combinatorial optimization problems can be formulated as linear programming problems, which involve manipulating a cost function, while satisfying some linear equality and inequality constraints among these real variables [Rumel86]. In some special applications, the variables are further constrained to take on only two discrete values 0 and 1. This is called *zero-one* programming problem. As an example of a zero-one programming problem [Papad83], the perfect matching problem is discussed below.

Example 2: Perfect Matching Problem

In the perfect matching problem, there are two sets of nodes of equal number, S_1 and S_2. All nodes in S_1 are to be matched to one (and only one) node of S_2, and vice versa. The matching can be represented by a connection matrix $[x_{ij}]$, where $x_{ij} = 1$ means node i of S_1 is matched to node j of S_2. For every connection, an associated weight w_{ij} is assigned. The perfect matching problem is: Given a weight matrix $[w_{ij}]$, to find a connection matrix $[x_{ij}]$ ($x_{ij} = 0$ or 1) which minimizes the following cost function:

$$\sum_{i,j} w_{ij} x_{ij} \tag{4.25}$$

under the constraints

$$\sum_j^n x_{ij} = 1 \quad i = 1, 2, \ldots, n \tag{4.26}$$

$$\sum_{i}^{n} x_{ij} = 1 \quad j = 1, 2, \ldots, n \tag{4.27}$$

Note that Eq. 4.26 and Eq. 4.27 imply that $[x_{ij}]$ is a permutation matrix.

In order to use ANNs for the perfect matching problem, the network should be first described by an energy function so that the lowest energy state corresponds to the best match. Let x_{ij} be represented by a (double index) neuron state V_{ij}, i.e., $V_{ij} = x_{ij}$, then the energy function to be minimized has the following form:

$$\begin{aligned}
E = \;& A/2 \sum_{i1,j1} w_{i1,j1} V_{i1,j1} \\
+\;& B/2 \sum_{i1} \sum_{j1} \sum_{j2 \neq j1} V_{i1,j1} V_{i1,j2} \\
+\;& C/2 \sum_{j1} \sum_{i1} \sum_{i2 \neq i1} V_{i1,j1} V_{i2,j1} \\
+\;& D/2 (\sum_{i1} \sum_{j1} V_{i1,j1} - n)^2 \tag{4.28}
\end{aligned}$$

where A, B, C and D are proper (positive) weighting constants to be determined.

The motivation for using the above form in Eq. 4.28 is as follows: The first term corresponds to the minimized cost function given in Eq. 4.25. The second term is zero if and only if at most one entry in each row of the matrix $[x_{ij}]$ is "1". The third term is zero if and only if at most one entry in each column of the matrix $[x_{ij}]$ is "1". The fourth term of Eq. 4.28 is zero if and only if there are exactly n entries in matrix $[x_{ij}]$ equal to "1". The second and fourth terms together constitute the constraint specified in Eq. 4.26. The third and fourth terms together constitute the constraint specified in Eq. 4.27. The "performance" of this energy function is discussed below.

Finding the Synaptic Strengths and Inputs   If the neuron states have double indices, then the energy function in Eq. 4.18 can be rewritten as:

$$E = -(1/2) \sum_{i1=1}^{n} \sum_{j1=1}^{n} \sum_{i2=1}^{n} \sum_{j2=1}^{n} T_{i1,j1,i2,j2} V_{i1,j1} V_{i2,j2} - \sum_{i1=1}^{n} \sum_{j1=1}^{n} I_{i1,j1} V_{i1,j1} \tag{4.29}$$

By comparing Eq. 4.28 and Eq. 4.29, the synaptic strengths $T_{i1,j1,i2,j2}$ and the input strengths $I_{i1,j1}$ can be thus derived

$$
\begin{aligned}
T_{i1,j1,i2,j2} &= -B\delta_{i1,i2}(1 - \delta_{j1,j2}) \\
&\quad -C\delta_{j1,j2}(1 - \delta_{i1,i2}) \\
&\quad -D
\end{aligned}
$$

$$
I_{i1,j1} = Dn - \frac{A}{2}\, w_{i1,j1} \tag{4.30}
$$

where $\delta_{ij} = 1$, if $i = j$ and is 0 otherwise

The ANN can now be programmed using the above values of T_{ij} and I_{ij} to search for the optimal solution. The same constraint optimization problem formulation can be extended to more general graph matching problems, e.g., bipartite graph matching, allowing matching between two sets of nodes of different size [Papad83].

Remark: Although the constraints are incorporated in Eq. 4.28, the solution state corresponding to the global minimum does not necessarily satisfy the constraints. In an attempt to enforce the ANN to converge to a valid state (i.e., discrete V_{ij} and $[x_{ij}]$ is a permutation matrix), B, C, and D should be chosen to be sufficiently large compared to A (see Problem 31). This should ensure that the low energy state of the network does in fact correspond to nearly valid solutions. However, it is still possible when B, C, D are sufficiently large, the ANN is trapped in a stable local minimum. (These factors prompt the use of simulated annealing and hardening annealing techniques discussed below.)

4.6.4   Global Optimization Searching Schemes

As mentioned in the previous examples, two main difficulties exist in using the current ANNs for searching a global optimal and valid solution. One is the potential pitfall of convergence to local minima. The other is that a direct minimizing solution may not satisfy the given hard constraints. There are presently no completely satisfactory solutions and they remain important subjects in open research domain. In the following, two promising, although tentative, approaches to the problems are discussed.

Incorporating Simulated Annealing into ANNs In many optimization problems, the minimization of the given energy function suffices to obtain the correct solution. However, the minimization process in the continuous-state ANN will sometimes converge only to a local minimum instead of the global minimum. In order to overcome this difficulty, the idea of simulated annealing is adopted and incorporated into the ANN model [Ackle85].

By appropriate substitution of the gain control parameter x_0 in Eq. 4.17 of the continuous model with the temperature control parameter T, and confining the neuron response to be discrete valued (0 and 1 only) with a stochastic decision mechanism, this modified version of the Hopfield model is essentially the same form as the Boltzmann machine [Ackle85] as discussed in Section 2.5 (see Problem 30). The deterministic thresholding operation in the continuous-state Hopfield model is now replaced by the on-off (discrete-state) stochastic decision mechanism, allowing the network to gradually realize the global optimum.

Incorporating Hardening into ANNs Many combinatorial optimization problems involve *hard constraints* (that must be satisfied) on the solution state. A solution state which minimizes the cost function is, in general, not a valid solution state which satisfies those hard constraints.

To force a valid solution, a technique called hardening is proposed [Kung87b]. By gradually decreasing the ratio of the weighting parameters between the cost function and the hard constraints during the iteration while the weighting ratio, R, is decreasing, the hard constraints are increasingly emphasized (i.e., hardening).

Systolic Design for Hardening annealing ANNs Both the simulated annealing and the hardening techniques can be properly combined during the search process. This is called hardening annealing (this involves the dynamic changing of the synaptic strengths and the gain control parameters). The proposed systolic ANN can be easily adapted to the hardening annealing technique, thus making this architecture more attractive to solving the ANN problems (see Problem 32). Basically, the hardening can be introduced into the systolic ANN by allowing the PEs to update T_{ij} and I_i values, and the simulated annealing may be incorporated into the ANN by adaptively changing the thresholding function according to the "temperature" drop. Both the adaptions may be performed at the same time when the systolic ANN is "running". This technique will hopefully (although still under in-

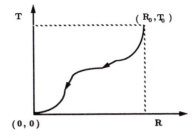

Figure 4.56: Possible annealing schedule for hardening annealing technique, where T_0 represents the initial temperature of the simulated annealing, and R_0 represents the initial weighting ratio of hardening.

vestigation) allow convergence to the global optimal solution state and yet also satisfy the hard constraints embedded in the combinatorial optimization problems when the weighting ratio and temperature reach a very small value (see Figure 4.56).

4.7   Concluding Remarks

The underlying principle of systolic design is to achieve massive parallelism with a minimum communication overhead. Featuring modularity, regularity and local interconnection, systolic arrays are amenable to VLSI implementation. However, their applications are restricted to a special class of algorithms that exhibit regular and local data structure which is naturally compatible with systolic design. From a research perspective, the techniques for mapping algorithms directly to systolic arrays have recently received a lot of attention. Along the same line, this chapter presents a cut-set based retiming procedure for converting SFG computing networks into synchronous systolic arrays. This corresponds to the realization stage (Stage 3) in the systematic mapping methodology of Chapter 3. In order to achieve an optimal systolic design, the design of the DGs and the selection of the projection method will be critical and may greatly affect the performance of the resulting systolic designs.

Although systolic design is generally thought of as being for dedicated systems, it is often appealing to incorporate into the design some degree of flexibilities. For this there are two important designs: (1) programmable systolic design and (2) reconfigurable systolic design. As to programmable systolic design, a high degree of flexibility can be obtained by allowing programmability in PEs. The level of programmability will have to depend on the application. As to reconfigurable systolic design, a very typical approach is to interleave PEs with switch lattices. This scheme has found a great popularity in the design of fault tolerant arrays (see Chapter 6). One prominent example is the CHiP (configurable highly parallel) computer [Snyde82]. More recently, some research has explored the feasibility of enhancing PE arrays with (dynamic) interconnection networks to make the arrays more reconfigurable.

It is worth mentioning that a balanced design must guarantee that the I/O can support sufficiently fast bandwidth so that data may be continuously fed for the high-speed array processing. This very formidable implementation issue, although overlooked by many researchers, cannot be avoided in the final system design. There are many more technical problems to be fully addressed in the system design. They include (but are not limited to) the interface between the host and arrays, clock distribution schemes, partitioning, fault tolerance, and software. (see Chapter 6.) For comparisons, some implementation examples for systolic processor chips and systolic array systems are provided in Chapter 7.

4.8   Problems

1. *Polynomial division*: Let $f(x)$ and $g(x)$ be polynomials, where $f(x) = a_n x^n + a_{n-1} x^{n-1} + \cdots + a_0$, and $g(x) = b_m x^m + b_{m-1} x^{m-1} + \cdots + b_0$. We further assume that $n \geq m$. Then there exists a unique quotient $q(x)$ and remainder $r(x)$, such that $f(x) = g(x)q(x) + r(x)$, where the degree of $r(x)$ is less than the degree of $g(x)$.

   (a) Derive a DG of an algorithm to compute $q(x)$ and $r(x)$ from inputs $f(x)$ and $g(x)$.

   (b) Find at least three systolic arrays from the DG by projections. Try to compare these arrays in terms of pipelining period, computation time, number of PEs, and so on.

2. *Back substitution (BS)*: The linear system equation $\mathbf{A}\mathbf{x} = \mathbf{b}$ can be fast solved if $\mathbf{A}$ is an upper triangular or lower triangular matrix, by the method of back substitution. Assume $\mathbf{A}$ is an upper triangular

matrix of size $n \times n$, and **b** and **x** are $n \times 1$ vectors. The unknown **x** is solved by the following procedure:

$$x_n = \frac{b_n}{a_{nn}}$$

$$x_i = \frac{1}{a_{ii}}(b_i - \sum_{j=i+1}^{n} a_{ij}x_j) \text{ for } i = n-1, n-2, ..., 2, 1.$$

(a) Give the single assignment form of this algorithm.

(b) Draw the DG.

(c) Derive three SFG arrays by projections.

(d) Transform the SFG arrays to systolic arrays by cut-set systolization. Compare these systolic arrays.

3. *Systolic arrays for DFT*: Let $\{x(n), n = 0, 1, ..., N-1\}$ be a finite length sequence. The DFT of $x(n)$ is defined as

$$X(k) = \sum_{n=0}^{N-1} x(n) \cdot W_N^{nk} \qquad (4.31)$$

where $k = 0, 1, 2, ..., N-1$ and $W_N = e^{-j2\pi/N}$.

Complete a three-stage design (DG $\rightarrow$ SFG $\rightarrow$ systolic array) for the DFT. Justify that your systolic design is a "good" one.

4. *Solving linear system*: We have developed parallel algorithms for the LU-decomposition and back-substitution (BS). Now, we have a linear system $\mathbf{Ax} = \mathbf{b}$, where **A** is an $n \times n$ matrix and **x** and **b** are $n \times 1$ vectors. Use LU and BS together to solve this problem. Draw the SFG and systolic array for this linear system solver.

Hint (Step 1a) When **A** is decomposed to be **L** and **U**, (Step 1b) **b** can also be transformed to be $\hat{\mathbf{b}} = \mathbf{L}^{-1}\mathbf{b}$ by adding one more column to the LU array (see Fig. 4.57). The $\mathbf{L}_i$'s are broadcast to both the LU block and $\mathbf{b}\hat{\mathbf{b}}$ block. (Step 2) To solve the back-substitution, you may need some LIFO stacks to convert the data sequence.

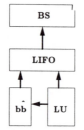

Figure 4.57: Solving linear system using LU and BS.

5. *Band-matrix QR decomposition*: We know that the square LU array can be used for band matrices. If we have a banded matrix with upper bandwidth $p = 4$ and lower bandwidth $q = 4$, but with size 100×100, show that the minimum PE array (i.e., the array with fewest PEs) for QR decomposition of this matrix is of size 4×7.)

6. *Projecting a DG onto a linear array*: In the DG shown in Figure 4.58, assume that the linear array lies in the j axis and the schedule executed is also shown in the figure.

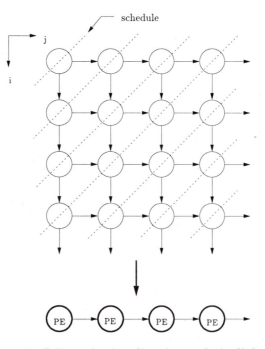

Figure 4.58: DG: projection direction and pipelining period.

(a) If the projection direction is $[1\ 1]^T$, what is the pipelining period, α, for this array?

(b) If the projection direction is $[1\ 2]^T$, what is the new pipelining period α?

(c) Concerning the speed and computation time, is the pipelining period itself a good measure of performance?

7. *Systolic arrays for the LU decomposition*: The SFG projection mappings of the two versions of SFG arrays in the projection directions of $[1\ 1\ 1]^T$ and $[1\ 0\ 0]^T$ are discussed in Section 3.4, and the corresponding systolic arrays were presented in this chapter. We can also directly map the DG to the two versions of systolic arrays by changing the schedule direction. Find the systolic schedule for the LU decomposition and derive the mappings for both systolic arrays.

8. *Two versions of LU decomposition*: We have discussed two different arrays for LU decomposition. The first version is square $(n \times n)$, and the second, triangular (about $n \times n/2$). If we have 10 matrices of the same size $(n \times n)$ (you can assume n to be 5), compare the time needed for each array to perform the decomposition.

9. *Cut-set systolization of the linear phase filter*: For the SFG of the linear phase filter in Figure 4.32(a), show that there exists no cut-set systolization that will obtain a systolic array with $\alpha = 1$ and $O(N)$ total delay elements.

10. *An example of nonpipelinable SFG*: For most practical computational models, the scaling factor α is 1, 2, or 3. For example, in the ARMA and lattice (Type A) systolic arrays, $\alpha = 2$, and in the lattice (Type B) array, $\alpha = 3$, regardless of how large M is. The same is true for most 2-D graphs, e.g. the SFGs for matrix multiplication LU decomposition. However, there are examples of SFGs that, when localized, lead to nonconstant α. The arrays cannot therefore exhibit $O(M)$ execution-time speedups. As an example of a regular but nonpipelinable SFG, consider the SFG shown in Figure 4.59, which is originally due to Dewilde [Dewil83], [Jagad83]. Note that there exists no simple, uniform cuts with which to achieve the conversion.

(a) Systolize this SFG by nonuniform cut-sets and find out α for the systolic array.

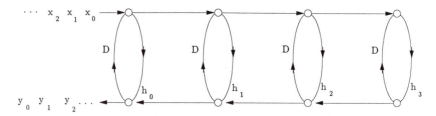

Figure 4.59: A nonpipelinable SFG example.

(b) If the SFG is extended to have M sections ($2M$ nodes), what is α for this SFG after systolization? Is this SFG pipelinable?

11. *Equivalence of node delay-transfer and general cut-sets*: Show that any cut-set delay transfer can be replaced by a series of node delay-transfers. Therefore, any cut-set systolization can be performed by using only node delay-transfers.

    Hint Consider the partition of nodes in a cut-set. Let the two partitions of nodes be A and B. Suppose k delays are subtracted from all edges coming out from A. Start from these edges and trace their predecessor nodes, and apply a node cut around those nodes with delay transfer of k. Repeat the same procedure for the edges coming into these nodes. Eventually, this procedure will terminate. (You have to consider all possible stop conditions.) The same procedure can be applied to the partition B.

12. *Comparing the cut-set systolization with Leiserson's retiming*: We have demonstrated the equivalence between node delay-transfers and general cut-set delay transfers.

    (a) Systolize the SFG in Figure 4.60 without time-scaling by the node delay-transfers only.

    (b) By grouping nodes with the same delay-transfers, find out the corresponding cut-set delay transfers.

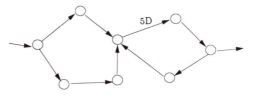

Figure 4.60: An SFG.

13. *Cut-set and fundamental loops*: A tree subgraph, T, of an SFG, G, may be defined such that it (1) contains no cycles; (2) is a connected graph; (3) contains all nodes of G. Arcs in T are referred to as *branches* and the arcs of G not included in T are called *links*. A *fundamental loop (FL)* is any closed path in G that does not pass through any node more than once and that closes a set of branches with one link. Note that a particular link may close more than one defined FL. With each FL, an arbitrary direction should be assigned. A loop delay may be associated with each FL. For a delay edge, if both the edge and the loop directions are the same, then the delay magnitude is added to the cumulative sum for that loop. If the directions are opposite, then the delay magnitude is subtracted from the sum. These concepts are illustrated in Figure 4.61. A *minimum set of FLs* for an SFG should include all edges.

With these notions, we have a useful theorem:

Theorem: Delays may be inserted or removed anywhere in G, in such a way that the loop delay is maintained constant for each loop in a minimum set.

(a) Show that after suitable time scaling, an SFG can be retimed using the above theorem to yield a systolic array.

(b) Define a different tree for the example in Figure 4.61 and repeat the retiming procedure.

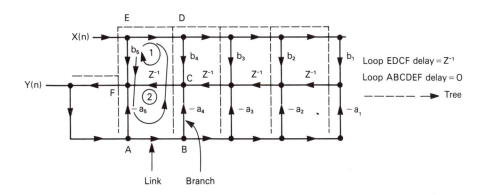

Figure 4.61: Example illustrating branch, link, and fundamental loops.

(The interested reader may note that a duality exists between cut-sets and fundamental loops. Whereby if the dual graph G' is derived, a fundamental loop in G is a fundamental cut-set in G' and vice-versa.)

14. *Nonuniform cut-sets* In Section 4.3.3, we show that uniform cut-set systolization is equivalent to rotation of the hyperplanes in the DG. In fact, it can be shown that the cut-set procedure is actually more flexible. The cut-sets in a systolization for an SFG do not have to be uniform. It is worth pointing out here that: *the nonuniform cut-sets* for deriving the optimal pipeline rate systolic array would not be derivable through other *uniform* mapping methodologies such as those by Moldovan and Quinton. Show that to systolize the SFG shown in Figure 4.62 (which can compute the transitive closure problem) to achieve minimal α, the cut-sets are nonuniform.

15. *Coprime properties*:

    (a) Prove that for the linear schedule function $\vec{s}^T\vec{i}$, where $\vec{i}$ is any point in a index space L, in order to obtain a consecutive set of integers, the components of the schedule vector $\vec{s}$ must be coprime, that is, $GCD\{s_i\} = 1$.

    (b) Prove that for the projection $\vec{d}$ to be valid, i.e., every point $\vec{i}$ in the index space must have a processor indexed by $\vec{d}^T\vec{i}$ to execute it, the components of the $\vec{d}$ vector must be coprime.

16. *Improving α by even-odd splitting*: It has been noted many linear systolic arrays have $\alpha = 2$ due to the bidirectional data flow, such as the systolic convolution array in Figure 4.20(b). There is another way to

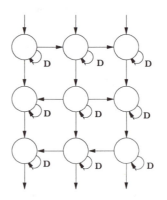

Figure 4.62: An SFG array for illustrating the use of nonuniform cut-sets.

improve the PE utilization, i.e., by grouping the even-numbered PEs on one side and odd-numbered PEs on the other side of the linear array, then connecting the two parts in the center. In this way, α can be reduced to 1. Show how this method works [Cheng86].

17. *Bit level systolic array for matrix multiplication*: Show that the bit level systolic array for inner product computation in Figure 4.23(a) can be also used for matrix-matrix multiplication.

18. *Systolic array for a dynamic programming problem*: [Chen86] Many dynamic programming problems can be stated in the following form, where $C(i,j)$ is some cost function to be minimized.

$$C(i,j) = \begin{cases} s > 0 & \rightarrow \min_{i<k<j}\{F(C(i,k), C(k,j))\} \\ & \text{where F is some function on the costs} \\ s = 0 & \rightarrow C_i \quad \text{for some individual cost} \end{cases}$$

where $s = j-i-1$ and i and j are integers in the range of $0 < i < j \leq n$, for some constant n.

(a) Draw the DG of this algorithm. Note that the DG may not be local, i.e., the dependence arcs are not all local.

(b) By localizing the DG and selecting the proper projection and scheduling, derive a systolic array that is of size $n(n-1)/2$.

19. *Snapshot of Guibas' algorithm*: The first systolic algorithm for the transitive closure problem was proposed by Guibas et al. [Guiba79], although it is far from being optimal in terms of computation time or pipelining period. Guibas' algorithm consists of several passes of Boolean matrix multiplication-like operations. The input matrices are $\mathbf{A}$ and $\mathbf{A}'$, where $\mathbf{A} = \mathbf{A}'$ initially. There are $N \times N$ PEs in the array. We denote the values stored inside the PE (i,j) by b_{ij}. In each pass of the algorithm, the steps performed at processor (i,j) are as follows:

(1) *Matrix Boolean multiplication* $b_{ij} = b_{ij} + (a_{ik} \bullet a'_{kj})$. That is, if both a_{ik} and a'_{kj} are 1, set b_{ij} to 1, otherwise, leave b_{ij} as it is.

(2) *One-sided information updating* At $t = k$ (kth recursion), a_{ik} will be updated at cell (i,k) as

$$a_{ik} \leftarrow a_{ik} + B_{ik};$$

and the new value of a_{ik} will be sent to all the cells to its right, i.e., cells (i, j) for all $j > k$. (That is why it is termed one-sided updating.) Similarly, a'_{kj} will be updated at cell (k, j) at $t = k$.

Three passes The Guibas' systolic algorithm consists of three identical passes. As the data completes one pass and reaches the rightmost (or the bottom) cells, they are immediately fed back to the left stacks (or the top stacks), respectively, to begin (or to be ready for) the next pass. Note also that wraparound connections are then required to support this feedback. At the end of three passes, a_{ij}^+ is just the b_{ij} in the array. Note that global connections are required.

Using the convention of snapshots in Section 4-5, draw the snapshots of the the first pass of the SFG array (in which broadcasting of a value is assumed) of Guibas' algorithm for the graph shown in Figure 4.37.

20. (a) *First pass of Guibas' algorithm*: From the snapshot of the last problem, it should be clear that although node 5 and node 6 are connected after the first pass; the path from node 5 to node 1 is *not* connected after the first pass. Explain why.

   (b) *Second pass of Guibas' algorithm*: Note that one pass cannot complete the computation for Guibas' algorithm from the observation of the snapshot. Continue the snapshots of the 6 node graph of the last problem, and verify that, in this case, two passes are sufficient.

21. *Proof of Guibas' algorithm*: Prove that three passes are required to compute the transitive closure for any directed graph using Guibas' algorithm.

22. *Projection of the DG for transitive closure in the $[1\ 1\ 1]^T$ direction*:

   (a) Derive the SFG by projecting the DG-1 in Figure 4.41 in the $[1\ 1\ 1]^T$ direction.

   (b) How do you modify this SFG such that it becomes the hexagonal SFG in Figure 4.47?

23. *Verification of optimality of transitive closure array*: It was shown earlier that for a graph of N nodes, the optimal systolic array takes $5N - 4$ steps to compute the transitive closure of a given graph. Verify this statement by simulating the operation of the systolic array. Note that the computation time T is equal to

$$T = \mathrm{Gentime}(x(N, N, N)) - \mathrm{Gentime}(x(1,1,1)) + 1$$

where $\mathrm{Gentime}(x(i,j,k))$ is the time at which $x(i,j,k)$ is generated.

24. *Control of data loading*: Referring to the orthogonal systolic array in Figure 4.49, note that the first row of the input matrix are first loaded into the first row of the PEs in the array and stay there to be reused for the next N clocks. Similar phenomenon happens to other PEs (involving other data). Describe a simple way to control the PEs in the array such that they properly execute the computations specified by DG-3.

25. *Convergence proof of Hopfield Model*: If we assume $T_{ij} = T_{ji}$ and $T_{ii} = 0$, show that by using the discrete Hopfield model as shown in Eq. 4.14, the neurons will always change their states in such a manner that they minimize the energy function defined in Eq. 4.18.

26. *Finding the synaptic strengths and inputs*: Show that by appropriate matrix manipulation and comparison of Eq. 4.18 and Eq. 4.23, the corresponding synaptic strengths matrix $\mathbf{T}$ and input vector $\mathbf{i}$ of this problem can be written in the forms as shown in Eq. 4.24.

27. *Switching operations of systolic array*: As we have discussed in Section 4.6.2, the systolic array shown in Figure 4.53), which adopts conventional cascade DG design, requires a smart control strategy to switch the PE among different operations. Draw all the distinct operations executed in this array, and design a control strategy to switch the PE among those different functions.

28. *Cylindrical systolic array*: The idea of rearranging the data array in the DG for consecutive matrix-vector multiplication problems, and obtaining a ring array architecture, can be easily extended to consecutive matrix-matrix multiplication problems in the estimation and control applications [Porte86]. This will result in 2-D cylindrical systolic arrays. Draw the modified cascade DG for the following consecutive matrix-matrix multiplication problem, and the resulting systolic array.

$$\mathbf{d}(k + 1) = \mathbf{A}\ \mathbf{d}(k)$$

where $\mathbf{A}$ is a 4×4 matrix, and $\mathbf{d}$ is a vector with 4 entries.

29. *2-D arrangement of 1-D ring array*: In most cases, an ANN will consist

of many simple PEs, and this will introduce an undesirable long wrap-around wire. In order to overcome this problem, a 2-D arrangement of $N \times N$ PEs (with N being even) can be adopted to avoid the long wrap-around wire. Draw the connection pattern of this 2-D arrangement.

30. *Applying simulated annealing to Hopfield model*:   If the gain control parameter x_0 in Eq. 4.17 is replaced by T, and also change the $\tanh(x)$ function into exponential form, what will the thresholding operation become?

    Compare this result with the stochastic formulation of the Boltzmann machine [Ackle85] as discussed in Section 2.5. Show that this will change the Hopfield model into a Boltzmann machine.

31. *Hardening Annealing Procedures*:   When the weighting ratio between the cost function and the hard constraints is very small, then the solution should be enforced to satisfy the hard constraints. Assume two sets of nodes with size three are to be perfectly matched with weighting matrix $\mathbf{W}$ equal to:

$$\mathbf{W} = \begin{bmatrix} 4 & 6 & 7 \\ 5 & 3 & 6 \\ 2 & 7 & 4 \end{bmatrix}$$

    Find the solution states $\mathbf{v} = [V_{ij}]$, i, $j = 1$, 2, 3, which minimizing the quadratic energy function shown in Eq. 4.29. The $\mathbf{T}$ and $\mathbf{i}$ can be obtained by using Eq. 4.30 for the following given weighting constants.

    (a) $A = 1$, and $B = C = D = 1$.

    (b) $A = 1$, and $B = C = D = 50$.

    (c) $A = 1$, and $B = C = D = 1000$.

    Hint: The energy function expresed in Eq. 4.18 can also be expressed as the following matrix form:

$$E = -1/2 \mathbf{v}^T \mathbf{T} \mathbf{v} - \mathbf{i}^T \mathbf{v}$$

32. *Combining hardening and simulated annealing techniques in the systolic ANN*: To incorporate simulated annealing into the systolic ANN,

the thresholding function $g(\lambda U)$ is changed iteratively. Hardening is introduced into the systolic ANN by allowing the PEs to update the T_{ij} and I_{ij} values.

(a) Show that in the perfect matching example, the updated $T_{i1,j1,i2,j2}(k+1)$ and $I_{i1,j1}$ can be expressed as:

$$T_{i1,j1,i2,j2}(k+1) = T_{i1,j1,i2,j2}(k) + m(k)s_1(k) \triangle T_{i1,j1,i2,j2}$$

$$I_{i1,j1}(k+1) = I_{i1,j1}(k) + m(k)s_2(k) \triangle I_{i1,j1}$$

where the masking function $m(k) = 1$ or 0, and $s_1(k)$, $s_2(k)$ are proper scaling factors, e.g.,

$$s(k) = 2^{-\sigma(k)}$$

where $\sigma(k)$ is a non-decreasing sequence of integers, and

$$\triangle T_{i1,j1,i2,j2} = -\triangle B\delta_{i1,i2}(1 - \delta_{j1,j2})$$
$$-\triangle C\delta_{j1,j2}(1 - \delta_{i1,i2})$$
$$-\triangle D$$

$$\triangle I_{i1,j1} = \triangle Dn - \frac{\triangle A}{2} w_{i1,j1}$$

(b) Both the hardening and annealing operations can be performed while the systolic ANN is "running". Design a strategy to implement these modifications which will not slow down the processing speed of the systolic array.

33. *Permissible Cut-sets*: The graph shown in Figure 4.63 is an SFG with four nodes. Define all possible cut-sets and characterize them as *permissible* or *not permissible*. Find three different α − optimal systolic arrays that result from three cut-set systolization procedures based on

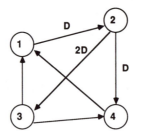

Figure 4.63: An SFG array for illustrating the use of permissible cut-sets.

three different sequences of cut-sets. What is the overall optimum systolized graph? Explain. What is the pipelining period and the pipeline rate of this graph?

34. *Shortest route problem*: In many applications, it is not only necessary to know the shortest path distance between any pair of nodes in the graph, but also the actual shortest route (the path itself) between any pair of nodes in the graph. Modify the algorithm and the systolic array designs in Section 4.5.2 and derive an optimal systolic array for computing the shortest distance and route between any pair of nodes in a directed graph.

Chapter 5

WAVEFRONT ARRAY PROCESSORS

5.1 Introduction

In Chapter 4 we noted that systolic arrays are very amenable to VLSI implementation of computation-bound signal/image processing algorithms. They have the critical advantages of modularity, regularity, local interconnection, highly pipelined multiprocessing, and continuous flow of data between PEs. The disadvantages of systolic arrays, however, lie in the fact that the activities are controlled by global timing-reference beats. From a hardware perspective, this global synchronization incurs problems of *clock skew, fault-tolerance, and peak power.* A simple solution to these problems is to adopt the principle of *data flow computing* in array processors, which is the main theme of this chapter.

Unlike the Von Neumann or conventional stored program computer, which is based on control flow (i.e., instructions are executed by the flow of control through the program), dataflow computers use the flow of data to initiate the execution of an instruction. Thus in dataflow computers an instruction is queued for execution only after all the operands to the instruction have arrived and activated it. The result of the instruction is in turn forwarded and consumed by other instructions. This approach eliminates the need for global control and global synchronization. It motivates us to adopt a data-driven, self-timed approach to array processing. It also neatly handles data dependencies, because the execution of instructions depends on the availability of their operands. More importantly, this approach replaces the requirement of correct timing by correct sequencing.

A general purpose dataflow computer often involves a large amount of data and resource management and requires a powerful supervising system. For this reason, current working dataflow computers have not demonstrated convincing effectiveness. As shown in the previous chapters, most algorithms have the inherent properties of *modularity* and *locality* which are readily exploitable. A proper incorporation of these features into dataflow multiprocessors may effectively alleviate the interconnection and memory conflict problems in a general-purpose dataflow computer. This leads to the development of the wavefront array processor.

5.2   Wavefront Array Processors

5.2.1   Evolution from Synchronous to Asynchronous Arrays

Array processor structures may be broken into two major groups: synchronous and asynchronous. Due to increasing array size and clock speed, synchronous design based on a global clock is becoming less suitable. This motivates us to look at other possible clocking schemes.

Several clocking schemes (representing an evolution from synchronous to asynchronous design) are illustrated here for a ring array [Dolec84]. A ring array has a sequential connection between processors, with the output of processor j connected to the input of processor $j + 1$, and so on, until the input of processor n is connected to the input of the first ring node, as shown in Figure 5.1(a). In the usual configuration, one node processor is used for I/O to the ring and a central clock is used for synchronous transfer of data around the ring (e.g., a systolic ring array). As mentioned earlier, for large (long) arrays, clock skew can be significant. One scheme for reducing the effect of clock skew is to use the relayed clock scheme shown in Figure 5.1(b). In this network each processing element uses the clock to its left for communication with the element to its left and generates a new clock for use in communication to the right. Thus, data transfers are resynchronized with the clock at each PE. Unfortunately, this scheme is not applicable to more general arrays, such as a 2-D array processor.

The addition of command/data message packets to a ring array (Figure 5.1(c)) enables each PE to communicate with all other PEs in the network

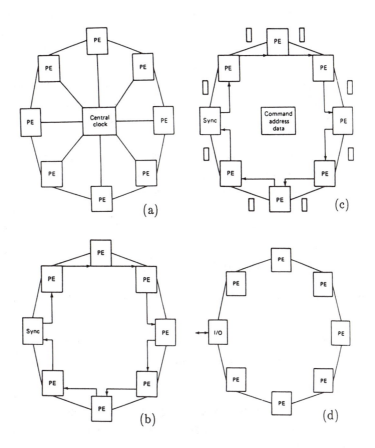

Figure 5.1: (a) Synchronous ring array (b) Relayed clock ring array (c) Command/data-packet-driven relayed clock ring array (d) Asynchronous ring array (adapted from [Dolec84]).

without necessitating direct communication links. The command is a high-level processor operation, such as "load a program" or "read data from bus," with the address being either for a particular PE or for a subsystem in a PE. Each processor reads input messages, retransmitting only those with destination addresses other than its own. Thus, as messages are used they are stripped off the bus. If each processor sends messages only to its adjacent processor, communication bandwidth does not accumulate along the bus.

The ring array can also be used with an asynchronous clocking scheme, this yields an asynchronous data-transfer structure (Figure 5.1(d)). In this structure, each PE has a bidirectional buffer and independent status and control flags for handshaking with adjacent elements. This structure obviates

clocking between elements, permits local synchronous PEs at different clock
rates, and allows different VLSI chip families to be used for the various PEs.
The only requirement is a common interface structure to the ring bus. This
structure can be regarded as a wavefront ring array.

The dataflow approach (i.e., asynchronous data transfer approach) has
recently been advocated in the use of large regular arrays of interconnected
processors. This leads to the design of wavefront array processors (WAPs)
[Kung82a]. In this approach, *the arrival of data from neighboring processors
is interpreted as a signal to change state and to activate new actions.* Such
a data-driven phenomena in wavefront arrays is reminiscent of the action
of wavefront propagation. It envisages a distributed and (globally) asynchronous array processing system. The wavefront array approach is very
suitable for a full exploitation of VLSI technology.

5.2.2   Definition of Wavefront Arrays

Definition: A **Wavefront Array** is a computing network with the following features:

1. *Self-timed, data-driven computation:* No global clock is needed, since
   the computation is self-timed.

2. *Regularity, modularity and local interconnection:* The array should
   consist of modular processing units with regular and (spatially) local
   interconnections. Moreover, the computing network may be extended
   indefinitely.

3. *Programmability in wavefront language or data flow graph (DFG):*
   Wavefront arrays can be implemented as dedicated or programmable
   processors, but the latter are usually preferred. Tracing computation
   wavefronts or adopting may facilitate programming wavefront arrays.

4. *Pipelinability with linear-rate speed-up:* A wavefront array should exhibit a linear-rate speed-up, i.e., it should achieve an $O(M)$ speedup,
   in terms of processing rates, where M is the number of PEs.

Note that the major feature distinguishing the wavefront array from
the systolic array is the data-driven property, i.e., there is no global timing
reference in the wavefront array. In the wavefront architecture, the information transfer is by mutual convenience between a PE and its immediate
neighbors. Whenever the data are available, the transmitting PE informs
the receiver, which accepts the data whenever required. It then conveys to

the sender the acknowledge that the data have been consumed. This scheme can be implemented by means of a simple handshaking protocol [Kung82a], [Kung82b], which ensures that the computational wavefronts follow in an orderly manner instead of crushing into other fronts.

A simple way to compare the wavefront array and its systolic array counterpart is

Wavefront array = systolic array + data flow computing

In other words, wavefront processing utilizes the localities of both data flow **and** control flow inherent in many signal processing algorithms. Since there is no need to synchronize the entire array, a wavefront array is truly architecturally scalable. The wavefront array processor possesses most of the advantages of the systolic array processor, such as extensive pipelining and multiprocessing, regularity and modularity. More significantly, it also possesses the asynchronous data-driven capability of dataflow machines and can, therefore, accommodate the critical problem of timing uncertainty in VLSI array systems.

5.2.3   Comparison with Other Array Architectures

In this section, we compare the wavefront array processors with the systolic arrays, SIMD arrays and MIMD multiprocessors. A complete and objective comparison is very involved as there exists complex trade-off between numerous criteria, such as programmability, modularity, synchronization, and interprocessor communication. To highlight the characteristic differences among these architectures, we propose a classification as shown in Figure 5.2. Note that a systolic array has local instruction codes and external data are piped into the array concurrently with the processing. SIMD and wavefront arrays are somewhat more complex classes. An SIMD array has control (instruction) buses and data buses (in lieu of the local instruction codes adopted in the systolic arrays). Wavefront arrays, on the other hand, provide data-driven processing capability. MIMD multiprocessors in general offer all the features just mentioned with an additional feature of shared memories. More detailed comparisons are provided below.

5.2.3.1   Comparison with MIMD and Dataflow Multiprocessors

Asynchronous distributed control and localized communication (for both data and control) are more attractive in VLSI. In contrast to MIMD arrays, wavefront arrays employ local communication, local instruction storage, and

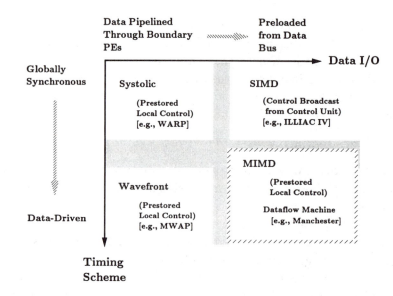

Figure 5.2: Classifications of SIMD, MIMD machines, systolic arrays, and wavefront arrays.

data flow based control, making them very appealing for VLSI and yet applicable to a large number of algorithms with inherent locality and recursiveness. The advantages gained by restricting applications to a special class are that of programming ease, functional modularity, and consequent reduction in design cost. However, many trade-offs exist between the simplicity of the wavefront array versus the flexibility of the MIMD design. Current research is directed at how to optimally compromise between specialized applications of the wavefront array and the general purpose MIMD approach.

The dataflow multiprocessor and the wavefront array share the same property of self-timed, data-driven computing. The basic principle of the wavefront architecture is that each PE waits for the arrival of a primary wavefront and then executes the required computation and acts as a secondary source of new wavefronts. This is equivalent to the key concept in dataflow machines, i.e., the arrival of data fires each PE which subsequently sends relevant data to the next PE. Hence the wavefront array can be regarded as an array of homogeneous dataflow processing elements.

Both the wavefront array and dataflow multiprocessor share the key common feature of being globally asynchronous, i.e., scheduling and synchronization are distributed, instead of centralized. This allows for efficient, fine-grained parallelism. Because they are language-based architectures, they support the mapping of application algorithms directly onto the multiprocessor in a way that achieves high performance.

5.2.3.2   Comparison with Systolic Arrays

Systolic and wavefront arrays share the important common feature of using a large number of modular and locally interconnected processors for massive pipelined and parallel processing. The differences are in hardware design, e.g., on clock and buffer arrangements, architectural expandability, pipeline efficiency, programmability in a high-level language, and capability to cope with time uncertainties in fault-tolerant designs.

Synchronization   The clocking scheme is a critical factor for large-scale array systems, and global synchronization often incurs severe hardware design burdens in terms of clock skew. The synchronization time delay in systolic arrays is primarily due to the clock skew which can vary drastically depending on the size of the array. On the other hand, in the data-driven wavefront array, a global timing reference is not required, and thus local synchronization suffices. The asynchronous data-driven model, however, incurs a fixed time delay and hardware overhead due to handshaking.

Processing Speed and Pipelining Rate   The data-driven computing in the wavefront array may improve the pipelinability. This becomes especially helpful in the case where variable processing times are used in individual PEs. As shown in Figure 4.28, the optimal systolic design requires an average pipeline period $\alpha = 3$. Because of the asynchronous processing feature of wavefront processing, the *average* pipeline period for the same network is only $\alpha = 2.5$. Furthermore, in many computations, the computing time is data-dependent. In a synchronous processing array, the "worst" time period has to be adopted, whereas asynchronous processing permits some computation to be performed in shorter time. This feature becomes very important in, for example, many sparse matrix operations, where an abundance of zero entries are encountered in many multiplication operations. In a data-dependent processing environment, such a trivial multiplication consumes much less time than a regular multiplication. Wavefront processing techniques can therefore be adopted to speed up the processing time [Melhe86]. A simulation study on a recursive least squares minimization computation also reports a speedup by a factor of almost two, in favor of the wavefront array over a globally clocked systolic array [Broom85].

Programmability *Programming a wavefront array* means specifying the sequence of operations for each PE. Each operation includes the following specifications:

- the type of computation (addition, multiplication, division, and so on).

- the input data link (e.g., north, south, east, west or internal register).

- the output data link.

Note that an additional specification on the schedule of the operations is required when programming systolic arrays with fixed interconnections, since the correct timing is necessary. This is, however, not required in the wavefront array programming due to the data driven nature. In this sense, programming a wavefront array is easier than a systolic array. Programming a wavefront array can be easily based on *data flow graph* descriptions of the computing structures and suitable languages such as the Matrix Data Flow Language (MDFL) and Occam.

Fault Tolerance While many fault-tolerance schemes are known for linear systolic arrays [HTKun83], 2-D systolic arrays are in general not amenable to run-time fault tolerant design because they require a global interrupt of PEs when any failure occurs. It is known that certain fault tolerance issues (rollback, suspension of execution and so on) are simpler to handle in dataflow architectures than in other multiprocessors [Denni80]. Since wavefront arrays incorporate the data-driven feature, they provide similar advantages in dealing with time uncertainties in the fault tolerant environment, where the actual communication paths fluctuate due to the necessary rerouting around faulty PEs. Once a fault is detected, further propagation of the wavefront can be automatically suspended, due to its data-driven nature. More details are discussed in Chapter 6.

Peak Power and Quality of Power Supply In a large array processor system, it is often inadvisable to synchronize the transfer of data between a large number of PEs so that transfer occurs simultaneously between these PEs. This can lead to power supply noise-induced problems due to large current surges as the components (e.g., buffers) are simultaneously energized or change state. From this point of view, wavefront arrays are more suitable for large scale system implementation.

When Wavefront Arrays? In summary, the choice between a systolic and a wavefront array depends on several important factors, such as (global) synchrony, programmability, hardware complexity, extendibility, fault tolerance, and testability. The final choice between the two array processors hinges upon the specific applications intended. In general, a systolic array is useful when the PEs are simple primitive modules, since the handshaking hardware in a wavefront array would represent a nonnegligible overhead for such applications. On the other hand, a wavefront array is more applicable when the modules of the PEs are more complex (such as floating-point multiply-and-add), when synchronization of a large array becomes impractical or when a reliable computing environment (such as fault tolerance) is essential.

5.3   Mapping Algorithms to Wavefront Arrays

In general, there are three approaches to deriving wavefront arrays:

1. Trace the computational wavefronts and pipeline the fronts through the processor array.

2. Map the DG directly to a wavefront array (or DFG).

3. Convert an SFG array into a wavefront (DFG) array, by properly imposing several key data flow hardware elements.

5.3.1   Notion of Computational Wavefronts

The notion of computational wavefronts offers a very simple way to design wavefront computing, which consists of three steps:

1. Decompose an algorithm into an orderly sequence of recursions;

2. Map the recursions onto corresponding computational wavefronts in the array;

3. Pipeline the wavefronts successively through the processor array.

Example 1: Wavefront Processing for Matrix Multiplication
    The notion of computational wavefronts may be better illustrated by the example of the matrix multiplication algorithm, $C = A \times B$. The topol-

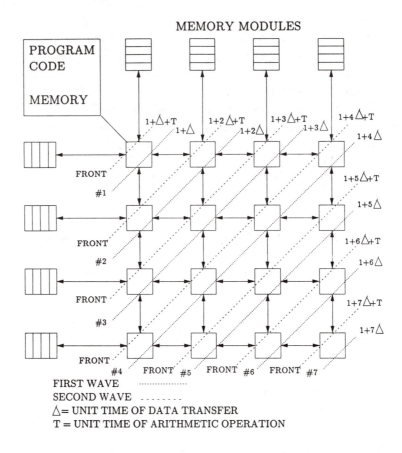

Figure 5.3: Wavefront processing for matrix multiplication. In this example, the wavefront array consists of $N \times N$ processing elements with regular and local interconnections. The figure shows the first 4×4 processing elements of the array. The computing network serves as a (data) wave propagating medium. Hence the hardware has to support pipelining the computational wavefronts as fast as resource and data availability allow, which can often be accomplished simply by means of a handshaking protocol (cf. [Kung82a]). The (average) time interval **T** between two separate wavefronts is determined by the availability of the *operands and operators.*

ogy of the matrix multiplication algorithm can be mapped naturally onto the square, orthogonal $N \times N$ matrix array of the WAP, as in Figure 5.3. The computing network serves as a (data) wave-propagating medium.

More elaborately, the computational wavefront for the first recursion in matrix multiplication is now examined. Suppose that the registers of all the PEs are initially set to zero:

$$C_{ij}^{(0)} = 0 \quad for\ all\ (i, j)$$

The elements of **A** are stored in the memory modules to the left (in columns) and those of **B** in the memory modules on the top (in rows). The process starts with PE $(1, 1)$ which computes:

$$C_{11}^{(1)} = C_{11}^{(0)} + a_{11} * b_{11} \tag{5.1}$$

The computational activity then propagates to the neighboring PEs $(1, 2)$ and $(2, 1)$, which execute:

$$C_{12}^{(1)} = C_{12}^{(0)} + a_{11} * b_{12} \tag{5.2}$$

and

$$C_{21}^{(1)} = C_{21}^{(0)} + a_{21} * b_{11} \tag{5.3}$$

The next front of activity will be at PEs $(3, 1)$, $(2, 2)$, and $(1, 3)$, thus creating a computation wavefront traveling down the processor array. This computational wavefront is similar to optical wavefronts (they both obey Huygens' principle),[1] since each processor acts as a secondary source and is responsible for the propagation of the wavefront. It may be noted that wave propagation implies localized data flow. Once the wavefront sweeps through all the cells, the first recursion is over. As the first wave propagates, we can execute an *identical* second recursion in parallel by *pipelining* a second

---

[1] Huygens' principle states that each point on the wavefront of a wave can be considered to be a new source of a "secondary" spherical disturbance, then the wavefront at any later instant could be found by constructing the "envelope" of the secondary wavelets [Goodm68]. Consequently, the two fronts never overlap.

wavefront *immediately after* the first one. For example, the $(1, 1)$ processor executes

$$C_{11}^{(2)} = C_{11}^{(1)} + a_{12} * b_{21} = a_{11} * b_{11} + a_{12} * b_{21} \qquad (5.4)$$

Likewise each processor (i, j) will execute (from $k = 1$ to N)

$$C_{ij}^{(k)} = C_{ij}^{(k-1)} + a_{ik} * b_{kj} = a_{i1} * b_{1j} + a_{i2} * b_{2j} + \ldots + a_{ik} * b_{kj} \qquad (5.5)$$

and so on.

In the wavefront processing, the pipelining technique is feasible because the wavefronts of two successive recursions never intersect. The processors executing the recursions at any given instant are different, thus any contention problems are avoided.

Note that the successive pipelining of the wavefronts furnishes an additional dimension of concurrency. The separated roles of pipeline and parallel processing also become evident when we carefully inspect how parallel processing computational wavefronts are pipelined successively through the processor arrays. Generally speaking, parallel processing activities always occur at the PEs on the same front, whereas pipelining activities are perpendicular to the fronts. With reference to the wavefront processing example in Figure 5.3, PEs on the antidiagonals execute in parallel, since each of the PEs processes information independently. On the other hand, pipeline processing takes place along the diagonal direction, in which the computational wavefronts are piped.

5.3.2 Mapping DGs to Wavefront Arrays via the DFG Model

It is important to formalize and systemize the design of the wavefront arrays directly from algorithm descriptions. In this subsection, we propose the *data flow graph* (DFG) as a formal abstract model for WAPs and discuss how to map a DG to a DFG.

5.3.2.1   The DFG Model

A DFG is a weighted and directed graph[2]

$$DFG \equiv [N, A, D(a), Q(a), \tau(n)] \qquad (5.6)$$

in which nodes in N model computation and arcs in A model asynchronous communication. Each node n has an associated nonnegative real weight $\tau(n)$ representing its computation time. Each arc a has an FIFO queue capacity, represented by a *positive integer* weight $Q(a)$.[3]

Each arc is also associated with a nonnegative integer weight, $D(a)$, representing the number of initial data tokens on the arc (initial state). *The state of a DFG is represented by the distribution of tokens on its arcs.* These tokens represent data values, either initial values or intermediate results. In analyzing the computational behavior of the DFG, we do not need to consider the actual data values. (The actual values determine the overall result of the computation but not how it is executed.)

Each arc may contain a nonnegative number of tokens that is less than or equal to its queue capacity. These tokens may be thought of as filling in a part of the arc queue. This leaves the remainder of the queue empty. Each empty queue slot is called a space. Therefore, each arc is also associated with a nonnegative integer weight $S(a)$, representing the number of initial spaces on the arc. Obviously,

$$Q(a) = D(a) + S(a) \qquad (5.7)$$

hence, the initial state of a DFG may be represented by either the distribution of tokens or spaces on its arcs.

Figure 5.4 shows an example of a DFG. Nodes are represented by circles and arcs by arrows. The computation time of a node, $\tau(n)$, is written inside of the circle. The size of the queue on an arc is represented by $Q(a)$ perpendicular dashs across the arc. The number of tokens on an arc is represented by $D(a)$ dots drawn on top of the queue dashs. The spaces on an arc then correspond to the number of queue dashes without token dots.

---

[2]For the definition of the graph terminologies, see Section 3.2

[3]Using queues is a way to implement asynchronous communication in the wavefront array. A queue is a mechanism to store and retrieve data, which can be implemented by software or hardware.

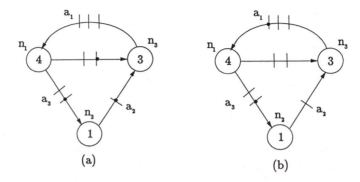

Figure 5.4: A DFG: (a) before n_3 fires; (b) after n_3 fires.

The state of the DFG at any time is represented by the distribution of tokens (or spaces) on its arcs. Since the queue size is fixed, Equation 5.7 must hold not only for the initial state at $t = 0$ but also at any time t.[4]

$$Q(a) = D^t(a) + S^t(a) \tag{5.8}$$

State changes occur through the "firing" of enabled nodes. A node is *enabled* when all input arcs contain a positive number of tokens and all output arcs contain a positive number of spaces. A node *fires* after it has been enabled for its computation time. This means that the tokens on the input arcs and spaces on the output arcs are locked in place during the computation period. The new state resulting from the firing of a node is obtained by subtracting one token (and adding one space) from each input arc of the fired node and adding one token (and subtracting one space) from each output arc. Because of the node enable condition, Equation 5.8 holds for all states. In addition, each node is assumed to be nonpipelined, so that a new computation period cannot begin until the previous one is complete. (In other words, a node cannot be re-enabled until it is first fired.)

This firing rule is illustrated in Figure 5.4. In (a) node n_1 is not enabled because arc a_1 contains no tokens. Node n_2 is not enabled because arc a_2 contains no spaces. Node n_3 is enabled and can fire after 3 time units. Figure 5.4(b) shows the state of the DFG after node n_3 fires. Note that the time a token spends on an arc is not a direct function of the queue length of the arc. It depends on the overall data flow pattern.

---

[4]We use $D^t(a)$ and $S^t(a)$ to refer to the number of tokens and spaces on arc a at time t. The initial values are then the values at $t = 0$. They are abbreviated as $D(a) \equiv D^0(a)$ and $S(a) \equiv S^0(a)$.

5.3.2.2   Mapping DGs to DFGs

We can derive the wavefront array for an algorithm by directly mapping the DG to a DFG. Here we assume that a DFG represents a programmable wavefront array, and that each arc has a queue (FIFO buffer) capacity which is large enough to accommodate the target algorithms.

For a shift-invariant DG, some boundary nodes of the DG may appear to have a different dependency structure (e.g., fewer dependency arcs) than that of the internal nodes. For our mapping, it is necessary to enforce a uniform appearance by assigning some initializing data (usually a constant such as zero) to the boundary nodes so that *all* the nodes exhibit the same dependency arcs (see Figure 5.5(b)). In the following discussion, all the data input to boundary nodes are called *input data tokens.*

For the shift-invariant DG and a given projection direction $\vec{d}$, we can derive the DFG in a manner similar to the SFG mapping, except that the delay elements are now replaced by initial tokens ($D(a)$). Each input data token in the DG is mapped to an initial token on the corresponding arc in the DFG. An example of mapping a DG to a DFG (with its initial tokens) is shown in Figure 5.5(b).

In contrast to the systolic mapping as shown in Figure 5.5(a), *the DFG mapping does not need any schedule vector* $\vec{s}$, since the data-driven computing nature of the wavefront array obviates the need of specifying the exact

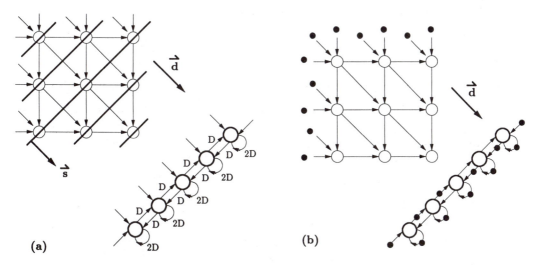

Figure 5.5: (a) Mapping a DG to a systolic array, and (b) mapping a DG to a DFG (wavefront array).

timing. In other words, the operation of a wavefront array is only dictated by the data dependency structure and the availability of initial data tokens. Furthermore, based on the dataflow principle, we claim that *an optimal schedule (as permitted by the DG) will be automatically carried out.* This is explained below.

Assume that each DG node is assigned to one PE, then minimum computation time can be achieved. Suppose that the projection direction $\vec{d}$ is chosen so that *there is a strict dependency among the nodes which are mapped to the same PE.* Thus the sequential processing among these nodes by the single PE should not in any way impose extra slow-down in the execution time, so the resulting DFG can perform the same computations in the minimum time. This provides a simple guideline for the selection of the projection direction $\vec{d}$. (In fact, this rule is also useful for the systolic mapping.)

If the *pipeline period* is the objective function to be minimized, then the selection of $\vec{d}$ requires a different guideline based on minimizing the time difference between two successive DG nodes assigned to the PE.

In the above discussions, we have assumed that the queues in the wavefront arrays (or DFGs) are large enough for the target algorithms. Insufficient queue size usually results in additional slow-down of the computation. Therefore, it is natural to ask the following question: How is the *minimal required queue size* for each DFG arc determined so that the minimum computation time can still be achieved? A more detailed discussion will be provided in Section 5.4 and Problem 6 and 7.

5.3.3 Deriving DFGs from SFGs

As we discussed earlier, SFGs provide a popular description for recursive parallel algorithms used in digital signal processing. In this subsection, we address the issue of transforming an SFG into a DFG by an equivalence transformation from SFG to DFG. The following theorem defines the *equivalence transformation* between SFGs and DFGs.

Theorem 5-1: *Barring deadlock situations (which means that no node in the DFG can ever fire), the computation of any SFG can be equivalently executed by a self-timed, data-driven machine with a topologically identical DFG. The number of initial tokens assigned on each DFG edge is equal to the number of delays on the corresponding SFG edge.*

Proof: It is necessary to verify that the global timing in the SFG can be replaced by the corresponding sequencing of the data tokens in the DFG. Note that the transfer of the data tokens is now "timed" by the processing node. This ensures that the relative time between data tokens received at the node is the same as it was in the SFG, as far as that individual node is concerned. By induction, this can be extended to show the correctness of the sequencing in the entire network.   Q.E.D.

Remark: In the preceding transformation, any queue capacity greater than or equal to the number of initial tokens on an edge is acceptable, if the only concern is that of the functional correctness. The discussion on the throughput of the derived DFG is given in Section 5.4.

For convenience, we call this transformation the *SFG/DFG equivalence transformation.*   It helps establish a theoretical foundation for the wavefront array as well as providing an insight towards programming techniques.

The **D**s in the SFG locate the proper setting of the initial conditions in the corresponding wavefront array. The initial data token distribution plays a vital role in assuring the correct sequencing in a data-driven computing network. The initial state assignment is straightforward: for each delay in the SFG there is an initial data token (regarded as an initial value) assigned to the corresponding DFG edge. We also note that a queue of arbitrary length can be inserted on any edge without affecting the validity of the equivalence transformation. It may however affect the deadlock situation and the throughput rate.

Example 2: DFG Design for Linear Phase Filter   To illustrate the role of the initial tokens and the correctness of data sequencing as guaranteed by the equivalence relationship, let us discuss the SFG/DFG equivalence transformation via a linear phase filter example. Linear phase filters have two key features: They have a symmetrical impulse response function, i.e., $h(n) = h(N - 1 - n)$, and they do not add phase distortion to the signal. Figure 5.6(a) shows an SFG that takes advantage of the symmetry property and reduces the number of multipliers by one half. By the SFG/DFG equivalence transformation, the DFG is derived as in Figure 5.6(b). In order to ensure the correct sequencing of data, the **W** data should propagate twice as slowly as the **Y** data.

Let us now explain the initial conditions in the linear phase filter example shown in Figure 5.6(c). Note that one initial zero-valued token is

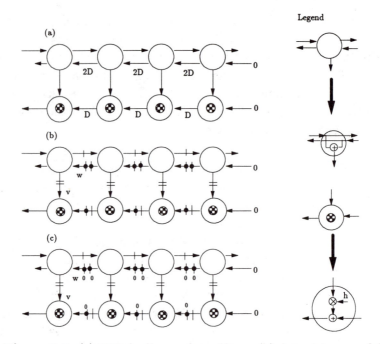

Figure 5.6: (a) SFG for linear phase filters, (b) data-driven model (DFG) for the linear phase filter, (c) initial condition of linear phase filter. An empty buffer is denoted by a bar on an edge and a full buffer is denoted by a bar with a dot on it, the allocations and distribution of initial data token are extremely important in ensuring a correct sequencing.

placed on each **Y**-data edge; and two initial zero-valued tokens are placed on each **W**-data edge. The first zero-valued token of the **Y**-data edge, when requested by the **Y**-summing node, is passed to meet the **V**-data token arriving from the upper node. When the operation is done, the way is cleared for sending the next **Y**-data token from the right-hand side PE. The situation is similar for the **W** summing node, but only one zero-valued token is used and the **W**-data is still one token away from meeting an **X**-data in the summing node. It will have to wait until the **Y**-data and the second zero meet in the lower summing node. This explains why the propagation of **W** is slower than **Y**. (This is precisely the requirement to ensure a correct sequencing of data transfers.)

In this example, we see that the wavefront array implementation of the linear phase filter needs only $O(N)$ queues and runs at the fastest allowable rate (again we assume that the data transfer time Δ is much shorter than arithmetic computation time T). This result is in accordance with the specified performance of the multirate systolic array for a linear phase filter (see Section 4.3). The total number of queues in this wavefront array is minimal, although this is not immediately obvious. The question then is how we decide the assignment of minimal queues to achieve the fastest pipelining rate for other even more complicated structures. This is addressed in Section 5.4.

5.4   Timing Analysis and Optimal Queue Assignment

In this section, we use the DFG model for the purpose of analyzing and predicting the performance of asynchronous wavefront arrays. We will demonstrate that the optimal performance and minimal queue assignments are determined by the structure of the DFG and its initial token state. These problems are first worked out for the general case and then for the case of regular wavefront arrays. For comparison purpose, another approach to the optimal queue assignment problem based on a timing analysis on DGs will also be presented.

Computation Graphs and Petri Nets   Timing analyses of parallel computations have been studied in the context of a wide variety of models. Here we compare the DFG model with other important computation graph models. The earliest model used was the *computation graph (CG)* proposed by Karp and Miller [Karp66]. Like a DFG, a CG is a directed graph in which the nodes represent computations and the arcs communication. Each arc has a FIFO queue, but no restrictions are placed on its length. Using this model, Karp and Miller[Karp66] proved that the the results of the CG are independent of the operation times, derived upper bounds on the queue lengths needed for each arc, and studied the deadlock characteristics of CGs.

The basic CG does not model computation time and so does not support performance analysis. Largely based on the CG model, Reiter [Reite68] incorporated computation times for the *arcs* into the original model, which is called the *timed computation graph (TCG)*. Murata and Onaga [Murat77] added queue-length constraints or, in their terminology, capacitance constraints, to another version of the CG model. Tani and Murata [Tani78] further incorporated Reiter's timed model to form the so-called *timed capacitated computation graphs (TCCG)*, which is very similar to the DFG model.

The DFG modeling of decision-free computation may be viewed as a subclass of *Petri nets* [Peter81], called marked Petri nets or simply *marked graphs (MGs)*. Commoner et al. [Commo71] used marked graphs to study space requirements of systems. They demonstrated the conservation of tokens in a cycle and showed that deadlock can be avoided only if there are no empty cycles. Ramchandani [Ramch73] provided an independent treatment of *timed Petri nets*.

5.4.1 DFG Timing Analysis

For motivation, let us first look into a simple DFG example. Figure 5.7 shows two DFGs differing only in queue lengths on the lower arc, a_1. The DFG in (a), with only one space on the lower branch, can fire at times 1, 6, 11, ... This yields an average throughput rate of 1/5. (After firing once it must wait for a_1 to clear.) However, the DFG in (b), with two spaces on

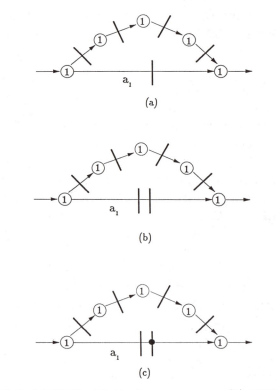

Figure 5.7: (a) DFG with single space on a_1. (b) DFG with two spaces on a_1. (c) DFG with a space and a token on a_1.

a_1, can fire at times 1, 3, 6, 8, 11, 13, ... , for an average throughput rate of 2/5. In Figure 5.7(c), we show a similar DFG in which there is one space and one token on the lower path. This DFG can fire at times 1, 6, 11, ... , yielding the same average performance as (a).

It is clear from this example that the queue capacities and token distribution on the arcs play an important role in determining the average firing rate or its inverse, the *pipelining period*, of the DFG nodes. The *pipelining period* of a DFG node is *the average time between its consecutive firings*. In this section we provide a complete timing analysis for general (acyclic and cyclic) DFG networks. Our objectives are twofold:

1. **Analysis:** Given a DFG with defined initial tokens and spaces on the arcs: Is it deadlock free? What is the average pipelining period?

2. **Synthesis** Given a desired (achievable) pipelining period: How can minimal queues on the arcs of a DFG be assigned to support this speed?

The DFG was developed to provide a convenient model that would approximately represent hardware implementations. The drawback of this approach is the added complexity of the various modeling terms. In order to investigate and develop properties of the DFG, it is easier to consider a simpler model, the timed marked graph (TMG). Development of general properties for TMGs is simpler. In essence, DFGs are models for describing data flow hardware, whereas TMGs are models more suitable for theoretical analysis. As we will show, any DFG can be equivalently represented by a larger (and hence topologically more complex) TMG. This correspondence allows us to develop useful properties in the simpler TMG domain and then transport them to DFGs. To begin and motivate this transition we examine the duality of tokens and spaces in a DFG.

5.4.1.1   The Duality of Token and Space

Figure 5.8(a) shows a cycle with many spaces but only one token. Since any firing will both create and destroy a token in this cycle, it is easy to see that there will always be only one token in this cycle. This is easily generalized to the following properties:

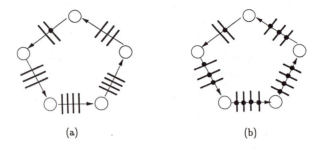

(a)   (b)

Figure 5.8: (a) A cycle with only one token. (b) A cycle with only one space.

Property 1:   The number of tokens in every cycle of a DFG remains constant.

The token can traverse the cycle repeatedly by a series of firing of the nodes in the cycle. Each trip around the cycle takes $\tau(c)$ time, which is the sum of all node operation times in the cycle. The pipelining period, α, is $\tau(c)$, and the average firing rate of each node in the cycle is $1/\tau(c)$. However, if we put one more token in the cycle, then by the time one token traverses around the cycle, the other one also finishes its own trip around the cycle. So, the period is only half as long, or $\tau(c)/2$, assuming that $\tau(c)/2 \geq \max_{n \in c} \tau(n)$. Generalizing this we have Property 2:

Property 2:   For any DFG cycle c, the pipelining period, $\alpha(c)$, will satisfy

$$\alpha(c) \geq \frac{\tau(c)}{D(c)} \qquad (5.9)$$

where $D(c)$ is the number of tokens in the cycle.

If we kept adding tokens to the cycle we would observe that the period decreases for a while, and then increases. To understand this we must examine the role played by spaces.

In Figure 5.8(b) we have a cycle with many tokens but only one space. Since the firing of a node requires at least one space on all output arcs, we see that only one node can fire at any time. Each time a node is fired, one space is created in the cycle and one is destroyed. The situation is analogous to that of the token, leading us to the third property:

Property 3:   The number of spaces in every cycle of a DFG remains constant.

So, in our example, the space traverses (in reverse direction) around the cycle in $\tau(c)$ time. The pipelining period of this cycle is again $\tau(c)$. By the same analogy, if we put in one more space in the cycle, the period becomes $\tau(c)/2$. Hence we can see that the next property holds:

Property 4:   For any DFG cycle c, the pipelining period, $\alpha(c)$, satisfies:

$$\alpha(c) \geq \frac{\tau(c)}{S(c)} \tag{5.10}$$

where $S(c)$ is the number of spaces in the cycle.

Figure 5.9 shows a DFG consisting of a single elementary undirected cycle. Note that whenever node n_1 fires, it both destroys a token on the clockwise arc a_C and creates a space on the counterclockwise arc a_{CC}. Likewise, it creates a space on a_C and destroys a token on a_{CC}. Generalizing this we have Property 5:

Property 5:   For any undirected cycle y of a DFG.

$$D_C^t(y) + S_{CC}^t(y) = D_C(y) + S_{CC}(y)$$
$$D_{CC}^t(y) + S_C^t(y) = D_{CC}(y) + S_C(y) \tag{5.11}$$

where $D_C^t(y)$ is the number of tokens and $S_C^t(y)$ is the number of spaces on clockwise arcs of y at time t. $D_{CC}^t(y)$ and $S_{CC}^t(y)$ are defined likewise for counterclockwise arcs. (And consistent with our previous notation, $D_C(y)$, $D_{CC}(y)$, $S_C(y)$, and $S_{CC}(y)$ are the initial values at $t = 0$.)

This means that for an undirected cycle, the sum of the tokens on

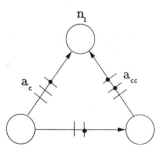

Figure 5.9: An elementary DFG undirected cycle.

arcs in one direction and spaces on arcs in the opposite direction remains constant. Notationally we indicate this by a single constant quantity:

$$DS_C(y) \equiv D_C(y) + S_{CC}(y)$$
$$DS_{CC}(y) \equiv D_{CC}(y) + S_C(y) \tag{5.12}$$

From this example we see clearly that tokens and spaces play a very similar role in firing nodes. They are both the "resources" to be used by the nodes in order to fire. They display a duality relationship which is analogous to the duality of electrons and holes in semiconductors. More precisely, a space in one direction plays the same role as a token in the reverse direction.

5.4.1.2   Timed Marked Graphs (TMG)

Motivated by the duality of tokens and spaces, we can express our DFG in a simpler form, that of a TMG.

The TMG Model   A TMG is a (weighted, directed) graph

$$TMG \equiv [N, \ A, \ D(a), \ \tau(n)] \tag{5.13}$$

which is equivalent to a DFG in which all the arc queues are of infinite length ($\forall a \in A : Q(a) = \infty$). To be enabled, a TMG node requires only a positive number of tokens on each of its input edges. Enabling is not affected by the condition of the output arcs. As with a DFG, a node n fires after it has been enabled for its computation time $\tau(n)$. The new state after firing a node is obtained by subtracting a token from each input arc and adding a token to each output arc, just like the DFG case. As in the DFG case, it is straightforward to verify Property 6:

Property 6:   The number of tokens in every cycle of a TMG remains constant.

Figure 5.10 shows an example of a TMG. Dots on the arcs indicate the token count. In (a) nodes n_2 and n_3 are enabled, whereas node n_1 is not. Firing of n_2 results in the new state shown in (b).

TMG Model of a DFG   Given a $DFG \equiv [N, \ A, \ D(a), \ Q(a), \ \tau(n)]$, an equivalent TMG is:

$$TMG \equiv [N, \ \hat{A}, \ K(a), \ \tau(n)] \tag{5.14}$$

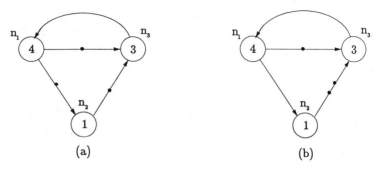

Figure 5.10: A TMG: (a) before firing n_2; (b) after firing n_2.

If we define the *reverse* of arc $a = i \xrightarrow{a} j$ to be $\bar{a} = j \xrightarrow{\bar{a}} i$, and $\bar{A} \equiv \{\bar{a}\}$ to be the set of reversed arcs, then:

$$\hat{A} \equiv A \cup \bar{A} \tag{5.15}$$

For every arc in the DFG there will be two in the equivalent TMG, one in the original direction and one in the reverse direction.

The initial tokens in the equivalent TMG are defined as follows: for each arc $a \in \hat{A}$,

$$
\begin{aligned}
&\text{If} \quad a \in A, \quad \text{then} \quad K(a) \equiv D(a) \\
&\text{If} \quad a \in \bar{A}, \quad \text{then} \quad K(a) \equiv Q(\bar{a}) - D(\bar{a})
\end{aligned}
\tag{5.16}
$$

In essence, the initial token count on TMG arcs corresponding to forward or original DFG arcs, is equal to the DFG token count. TMG arcs corresponding to added or reversed arcs is equal to the corresponding DFG space counts. If a TMG arc corresponds to both a forward and a reversed arc, then its token count is the minimum of the two.

Figure 5.11 shows a DFG and its equivalent TMG. The additional reverse arcs in the TMG are drawn as dashed lines. We can see that spaces

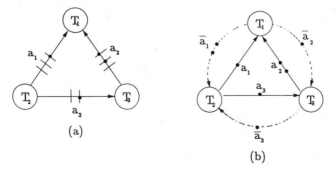

Figure 5.11: DFG/TMG equivalence: (a) a DFG; (b) its equivalent TMG.

in a DFG are modeled by additional tokens in the TMG. The TMG is more easily analyzed precisely because it has only one class of resources rather than two. It is straightforward to show that the operation of this TMG is the same as the original DFG.

Property 7: The pipelining period, α, (or average throughput rate) of a DFG and its equivalent TMG are the same.

5.4.1.3   TMG Pipelining Period

Property 8: The pipelining period, α, of a strongly connected TMG is

$$\alpha = \max_{n,\ c} \left(\tau(n), \frac{\tau(c)}{K(c)} \right) \tag{5.17}$$

where $n \in N$, c is an elementary cycle of the TMG, $\tau(c)$ is the sum of the node times in the cycle, $(\tau(c) \equiv \sum_{n' \in c} \tau(n'))$, and $K(c)$ is the number of tokens in the cycle $(K(c) \equiv \sum_{a' \in c} K(a'))$.

Formally, this property can be derived from a number of proofs given in varying contexts. The earliest version of this result can be found in [Reite68] in terms of computation graphs. Another can be found in [Ramam80] in terms of Petri net marked graphs. In order to give an intuitive feel for why this result is reasonable, we make the following observations.

Consider a single, elementary cycle c of the TMG operating independently of the rest of the graph. If this cycle contains only one token, then its pipelining period, $\alpha(c)$, is the sum of the node times, $\tau(c)$. If a second token is added to the cycle, then the period is halved. The period can be further decreased by adding more tokens until a limit is reached. This limit is equal to the longest node time of the nodes in the cycle. All the nodes in the cycle fire at the same average rate and hence all operate at the same pipeline period. So for this isolated cycle;

$$\alpha(c) = \max_{n \in c} \left(\tau(n), \frac{\tau(c)}{K(c)} \right) \tag{5.18}$$

If the elementary cycle just discussed is no longer isolated from the rest of the TMG, then its nodes may also be a part of additional cycles. The nodes may now not fire at the average rate indicated by Equation 5.18, but *they cannot fire at a faster rate.* Hence the pipeline period of a given node cannot be less than the pipeline periods of the cycles of which it is a member:

$$\alpha(n) \geq \max_{c\,:\,n \in c} \alpha(c) \tag{5.19}$$

All nodes in a strongly connected TMG are in a common cycle. Consequently all nodes in a TMG fire at the same average rate, so all nodes operate at the same pipelining period. Hence the *slowest cycle prevails.*

To understand why only elementary cycles are considered in Equation 5.17, we examine Figure 5.12. Here there are two elementary cycles, denoted as $c_1 = (n_1\, a_1\, n_2\, a_3\, n_1)$ and $c_2 = (n_2\, a_2\, n_3\, a_4\, n_2)$. A nonelementary cycle containing c_1 and c_2 is $c_3 = (n_1\, a_1\, n_2\, a_2\, n_3\, a_4\, n_2\, a_3\, n_1)$. We see that:

$$\begin{aligned}
\tau(c_1) = \tau(n_1) + \tau(n_2), \quad &K(c_1) = K(a_1) + K(a_3)\\
\tau(c_2) = \tau(n_2) + \tau(n_3), \quad &K(c_2) = K(a_2) + K(a_4)
\end{aligned} \tag{5.20}$$

$$\begin{aligned}
\tau(c_3) &= \tau(n_1) + \tau(n_2) + \tau(n_3) + \tau(n_4) &= \tau(c_1) + \tau(c_2)\\
K(c_3) &= K(a_1) + K(a_2) + K(a_3 + K(a_4) &= K(c_1) + K(c_2)
\end{aligned} \tag{5.21}$$

$$\begin{aligned}
\text{if } \frac{\tau(c_1)}{K(c_1)} \geq \frac{\tau(c_2)}{K(c_2)}, \quad \text{then} \quad \frac{\tau(c_1)}{K(c_1)} \geq \frac{\tau(c_1)+\tau(c_2)}{K(c_1)+K(c_2)}\\[2mm]
\text{if } \frac{\tau(c_2)}{K(c_2)} \geq \frac{\tau(c_1)}{K(c_1)}, \quad \text{then} \quad \frac{\tau(c_2)}{K(c_2)} \geq \frac{\tau(c_1)+\tau(c_2)}{K(c_1)+K(c_2)}
\end{aligned} \tag{5.22}$$

In either case:

$$\max[\alpha(c_1), \alpha(c_2)] \geq \alpha(c_3) \tag{5.23}$$

Remark Property 8 has been derived previously by several researchers. Reiter demonstrated a bound on the maximum average throughput rate α using the TCG [Reite68]:

$$\alpha \geq \max_c \frac{\tau(c)}{K(c)} \tag{5.24}$$

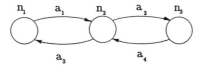

Figure 5.12: Elementary versus complex cycles in a TMG.

where c is a cycle in the TCG, $\tau(c)$ is the sum of the computation times on the arcs in c, and $K(c)$ is the number of tokens in c. In addition, Reiter also demonstrated that a *periodic schedule*, in which every node fires at period α, could be found to meet this bound.

In a manner similar to our use of TMGs, Tani and Murata [Tani78] showed that any TCCG could be modeled by a TCG constructed by adding additional reverse arcs. Using the TCG equivalent of a TCCG, they apply Reiter's earlier result to calculate the maximum performance rate. Ramamoorthy and Ho studied a Petri net model equivalent to the TMG model. Using this model they rederived Reiter's result in Equation 5.24 with a different proof based on graph theory [Ramam80].

5.4.1.4 DFG Pipelining Period

Property 9:   Given a connected DFG with preassigned initial data tokens and spaces, the pipelining period α of the DFG is

$$\alpha = \max_{n,a,y} \left(\tau(n), \frac{\tau(i)+\tau(j)}{Q(a)}, \frac{\tau(y)}{DS_C(y)}, \frac{\tau(y)}{DS_{CC}(y)} \right) \qquad (5.25)$$

where $n \in N$, $a = i \xrightarrow{a} j \in A$, and y is an elementary undirected cycle of the DFG.

Recall from Equation 5.12 that $DS_C(y)$ is the sum of the clockwise tokens and counterclockwise spaces, whereas $DS_{CC}(y)$ is the sum of the counterclockwise tokens and clockwise spaces.

Property 9 can be derived from the DFGs equivalent TMG. Every undirected cycle y in the DFG generates two cycles, c_C and c_{CC}, in the equivalent TMG over the same set of nodes, one in each direction. Clearly,

$$\begin{aligned} K(c_C) &= DS_C(y) \\ K(c_{CC}) &= DS_{CC}(y) \end{aligned} \qquad (5.26)$$

All elementary cycles of the TMG are present as elementary undirected cycles of the DFG except those cycles of length two consisting of an arc and its reverse, hence the addition of the $\{\tau(i)+\tau(j)\}/Q(a)$ term.

5.4.1.5 Deadlock Analysis

The issue of deadlock can be regarded as a special application of Property 9. From this it follows that a DFG is deadlock free if and only if there exists

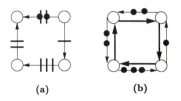

(a)   (b)

Figure 5.13: (a) A deadlocked DFG; (b) its corresponding TMG.

a finite solution α for Eq. 5.25. A finite solution for α exists if and only if for every undirected cycle y,

$$DS_C(y) \neq 0 \ \text{ and } \ DS_{CC}(y) \neq 0 \tag{5.27}$$

Or, to express it in another way, the equivalent TMG must contain no empty cycles. An example of deadlock is shown in Figure 5.13. Here the arcs in the empty cycle are drawn as dashed lines.

5.4.2   DFG Performance Optimization

5.4.2.1   General Issues

To understand the issues involved in the optimization of throughput for a DFG, it is useful to emphasize several points. First, initial token distribution plays a critical part in both the functionality and performance of a DFG. *Two topologically equivalent DFGs with different initial token distributions represent different algorithms.* Because of this, the initial token distribution cannot be arbitrarily altered to improve performance. In fact, the initial token distribution provides a fundamental limitation on the performance of the DFG.

The distribution of spaces, however, does not affect the functionality of the DFG. *The distribution of spaces affects only the performance of the DFG.* Hence, in the optimization of a DFG the initial token distribution must be considered as a property of the algorithm, but the space distribution can be treated as a design variable.

In analysis, where queue size and initial conditions are fixed, tokens and spaces can be treated alike, as is done in the TMG equivalent of the DFG. However in synthesis, where we wish to determine the queue sizes given the initial conditions and computation times, the distinction between tokens and spaces must be made. Tokens are fixed constraints representing the initial conditions of the algorithm, whereas spaces are free variables representing the initial excess queue capacity to be designed into the system.

Property 9 specifies the pipelining period of a DFG when the initial token assignments and number of queues on arcs are given. The dual problem of determining the assignment of minimum queue lengths on arcs of the DFG while still maintaining maximal throughput (minimum period or α^*) is treated in this section.

5.4.3   Optimal Pipelining Period (α^*)

Property 10:   Given a DFG with initial token assignment and unspecified queue lengths, the minimum (or optimal) pipelining period, α^*, that can be achieved is:

$$\alpha^* = \max_{n,c} \left(\tau(n), \frac{\tau(c)}{D(c)} \right) \tag{5.28}$$

where $n \in N$, c is an elementary cycle of the DFG, and $D(c)$ is the number of tokens in the cycle.

This can be derived by setting $Q(a) = \infty$ for all arcs. The DFG then becomes a TMG and Property 10 follows from Property 8.

Property 11:   To support a given pipelining rate, α^*, the DFG queue assignments must satisfy

$$
\begin{aligned}
\forall y: \quad & S_C(y) \geq \frac{\tau(y)}{\alpha^*} - D_{CC}(y) \\[2mm]
\forall y: \quad & S_{CC}(y) \geq \frac{\tau(y)}{\alpha^*} - D_C(y) \\[2mm]
\forall a: \quad & S(a) \geq \frac{\tau(i)+\tau(j)}{\alpha^*} - D(a) \\[2mm]
\forall a: \quad & S(a) \geq \max(1 - D(a), 0)
\end{aligned}
\tag{5.29}
$$

where $a = i \xrightarrow{a} j \in A$ and y is an elementary undirected cycle of the DFG.

These inequalities follow directly from Property 9 and Equation 5.7. They are, in fact, valid for any desired pipelining period $\alpha \geq \alpha^*$.

Example 1: Optimal Pipeline Period for a Lattice Filter

A DFG implementing a lattice filter is shown in Figure 5.14(a). The initial data tokens, queue capacities, and node operation times are also displayed. We first apply Property 8 to determine the pipelining period of its

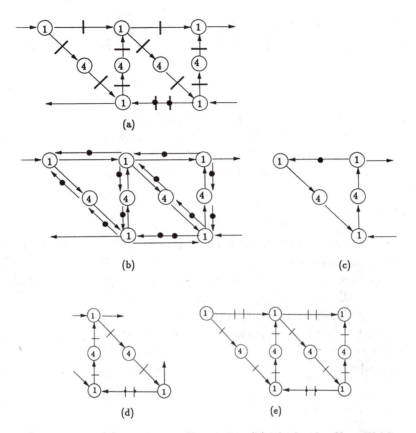

Figure 5.14: (a) The lattice filter DFG. (b) The lattice filter TMG. (c) The slowest TMG cycle. (d) The limiting DFG cycle. (e) A DFG to support α^*.

equivalent TMG, shown in Figure 5.14(b). The TMG cycle that has the maximum period α is shown in Figure 5.14(c). From this cycle we obtain $\alpha = 11$.

Applying Property 10 we find that the the optimal pipelining period α^* is 5.5. This can be seen by examining the limiting DFG cycle shown in Figure 5.14(d). Figure 5.14(e) shows a queue assignment that meets all of the constraints in Property 11 and hence will support the optimal period of 5.5.

Simple Minimal Queue Assignment   The assignment of the minimal queue capacities to arcs to support the optimal pipeline period can be formulated as an integer programming problem. Essentially we want to minimize the cost function $\sum Q(a)$. Since the initial tokens are fixed, we can rewrite the cost function in terms of initial spaces. So the problem formulation becomes

$$\text{Minimize} \sum_{a \in A} S(a) \tag{5.30}$$

subject to the constraints in Equation 5.29

The cost function and constraint equations are all linear combinations of the initial space distribution. However, the number of initial spaces on each arc must be an integer, so the overall problem corresponds to one of integer programming. Approaches and algorithms for solving this class of problems can be found in [Leis83a].

Linear Programming Approximation for Simple Minimal Queue Assignment   If we do not wish to solve the integer programming problem just presented, then a linear formulation can be used that will provide sufficient, although perhaps not minimal, queue assignments. Simply remove the restriction that the $S(a)$ must be integers and solve for $S_{\text{real}}(a)$, using the cost function in Equation 5.30, under the constraints in Equation 5.29. This corresponds to a linear programming problem. Actual integer values to be used for $S(a)$ are then found by rounding $S_{\text{real}}(a)$ up to the next integer.

$$S(a) = \lceil S_{\text{real}}(a) \rceil \tag{5.31}$$

5.4.4   Timing Analysis for Regular Wavefront Arrays

5.4.4.1   Regular DFGs

The preceding methods all deal with the analysis and optimization of an arbitrary DFG. If the DFG is a regular graph, then the effort needed to apply these results can be greatly reduced. In this section we show how to apply the preceding results to the case where the DFG is a *completely regular* graph, or wavefront array.

To characterize a completely regular graph, we imbed it in a finite-dimensional index space such that each node in the graph resides at an index point. For an n-dimensional index space we define an index vector as an ordered set of index coordinates.

$$\mathbf{i} \equiv \{i_1, i_2, \ldots, i_n\} \tag{5.32}$$

A node can then be described by its location in the index space. A *regular DFG (RDFG)* has the following properties:

1. It is defined over a finite, contiguous region of the index space.

2. It has functionally identical nodes (with the same $\tau(n)$) at every index point in this region.

3. For every node, the set of arcs for which the node is a terminal endpoint have their initial points at the same relative offsets. This means that if there exists an arc from a to b, then there will be an arc from every point $a + x$ to every point $b + x$.

4. For every node corresponding arcs have corresponding properties (both $D(a)$ and $Q(a)$).

An RDFG can be fully described by a matrix and two vectors. The offset matrix $\mathbf{A}$ specifies the (relative) initial points of the arcs coming into a node. Its entries are all integers. For an n-dimensional index space, each arc is specified by an n-dimensional vector.

$$\mathbf{a}_i^T = \{a_{i1}, a_{i2}, \ldots, a_{in}\} \tag{5.33}$$

If each node has m incoming arcs then $\mathbf{A}$ is an $n \times m$ matrix with integer entries:

$$\mathbf{A} \equiv [\mathbf{a}_1, \mathbf{a}_2, \ldots, \mathbf{a}_m] \tag{5.34}$$

Since the RDFG is assumed to be connected, the rank of $\mathbf{A}$ cannot be smaller than n.

The tokens, spaces, and queue lengths on each arc can be specified by m-dimensional vectors with nonnegative integer entries:

$$\begin{aligned}
\mathbf{d}^T &\equiv [D(\mathbf{a}_1), D(\mathbf{a}_2), \ldots, D(\mathbf{a}_m)] \\
\mathbf{s}^T &\equiv [S(\mathbf{a}_1), S(\mathbf{a}_2), \ldots, S(\mathbf{a}_m)] \\
\mathbf{q}^T &\equiv [Q(\mathbf{a}_1), Q(\mathbf{a}_2), \ldots, Q(\mathbf{a}_m)]
\end{aligned} \tag{5.35}$$

5.4.4.2   RDFG Optimal Pipelining Period (α^*)

Any path can be represented as an m-dimensional vector with nonnegative integer entries

$$\mathbf{p}^T \equiv [p_1, p_2, \ldots, p_m] \tag{5.36}$$

where p_i is the number of times an instance of $\mathbf{a}_i$ is traversed. The endpoint of path $\mathbf{p}$ relative to the initial point is

$$\mathbf{i}_{end} = \mathbf{A}\mathbf{p} \tag{5.37}$$

Since any cycle must end where it began, it can be represented by a path vector $\mathbf{c}$, which is a right null vector of $\mathbf{A}$ with nonnegative integer entries.

$$\mathbf{A}\mathbf{c} = \mathbf{0} \tag{5.38}$$

The sums of the node times, tokens, and spaces in the cycle are:

$$
\begin{aligned}
\tau(\mathbf{c}) &= \mathbf{c}^T\mathbf{1} \\
D(\mathbf{c}) &= \mathbf{c}^T\mathbf{d} \\
S(\mathbf{c}) &= \mathbf{c}^T\mathbf{s}
\end{aligned} \tag{5.39}
$$

Finding the maximal pipelining rate can now be expressed as

$$\alpha^* = \max_{\mathbf{c}} \left(1, \frac{\mathbf{c}^T\mathbf{1}}{\mathbf{c}^T\mathbf{d}}\right) \tag{5.40}$$

subject to the constraints that $\mathbf{c}$ is a right null vector of $\mathbf{A}$, i.e., $\mathbf{A}\mathbf{c} = \mathbf{0}$, $\mathbf{c} \neq \mathbf{0}$, and $\mathbf{c}$ has only nonnegative integer entries.

The ratio to be maximized is independent of the magnitude of $\mathbf{c}$. Ignoring the integer conditions, the constraints form a convex set consisting of the intersection of the right null space of $\mathbf{A}$ and the closed half-spaces of the nonnegativity constraints. The cost function is convex (over the constraint set), so its maximum must lie on a boundary of the constraint set. This problem is an instance of the class of nonlinear convex programming problems with linear fractional objective functions [Mital76]. These problems can be solved by variations of the simplex method. Adding the integer constraints to the problem make its solution similar to integer programming.

Note that the dimension of this problem formulation is much smaller than that of the general formulation in Property 10. Here the size of the problem is a constant, independent of the size of the graph, whereas the size of the general formulation is proportional to the number of elementary cycles in the graph.

5.4.4.3   Fixed RDFG Pipelining Period (α)

The pipelining rate of a DFG given fixed tokens and spaces (or queues) can be found in a similar manner. First we define the equivalent TMG of the RDFG. The regular TMG (RTMG) can be represented by a connection matrix representing the arcs:

$$\hat{\mathbf{A}} \equiv [\mathbf{A} \mid -\mathbf{A}] \tag{5.41}$$

and by a vector representing the tokens;

$$\mathbf{k}^T \equiv [\mathbf{d}^T \mid \mathbf{s}^T] \tag{5.42}$$

The pipeline period is found by

$$\alpha = \max_{\mathbf{c}} \left(1, \frac{\mathbf{c}^T \mathbf{1}}{\mathbf{c}^T \mathbf{k}}\right) \tag{5.43}$$

subject to the constraints that $\mathbf{c}$ is nonzero with nonnegative integer entries and $\hat{\mathbf{A}}\mathbf{c} = \mathbf{0}$.

5.4.5   Minimal Queue RDFG

Expanding the above we note that:

$$\alpha \geq \frac{[\mathbf{c}_d^T \mid \mathbf{c}_s^T]\mathbf{1}}{[\mathbf{c}_d^T \mid \mathbf{c}_s^T][\mathbf{d}^T \mid \mathbf{s}^T]^T} \tag{5.44}$$

In order to find the distribution of the minimal number of spaces to support α^* (without token redistribution) we set up the following problem:

$$\text{Minimize } \mathbf{s}^T \mathbf{1} \tag{5.45}$$

subject to the constraints that $\mathbf{s}$ has nonnegative integer entries and

$$\mathbf{c}_s^T \mathbf{s} \geq \frac{\mathbf{c}^T \mathbf{1}}{\alpha^*} - \mathbf{c}_d^T \mathbf{d} \tag{5.46}$$

for every $\mathbf{c}^T = [\mathbf{c}_d^T \mid \mathbf{c}_s^T] \neq \mathbf{0}$ with nonnegative integer entries such that $\hat{\mathbf{A}}\mathbf{c} = \mathbf{0}$.

5.5   Programming Languages for Wavefront Arrays

5.5.1   Concurrency and Communication

Two major factors in expressing wavefront processing are *concurrency* and *communication*. Concurrent process concepts and channel concepts are included in Hoare's communicating sequential processes as the parallel command and input and output commands [Hoare78]. Early studies of modules and processes can be found in [Parna72] and [Horni73], respectively. A formal model of both are abstract machines [Dijks68] for which a mathematical foundation has been given in [Creme78]. Algebraic specification attempts to express modules in terms of parameterized abstract data types. Aspects of buffer-free communication and data flow are discussed in [Hoare78] and [Denni79], respectively. An important notion of state has given rise to the computational concepts of data spaces [Creme76] and applicative state transition systems [Backu78], respectively. Another VLSI notation based on a state transition model is CRYSTAL [Chen85a].

Concurrency (Parallelism and Pipelining)   Concurrency is usually achieved by decomposing a problem into independent subproblems or into pipelined subtasks. As shown in Section 1.1 (cf. Figure 1.4) the concurrency in the systolic/wavefront arrays is derived from pipeline processing or parallel processing or both.

Conventional programming languages for multiprocessors describe only the global parallelism in data executions. They cannot describe the parallel data movements that occur in a pipeline processing environment. There is an enormous number of PEs in a VLSI system, and direct data transfer between PEs increases the speed and alleviates the storage problem by eliminating the fetching and storing operations. How to express the pipeline or data flow properties in a VLSI language is a critical issue. Therefore, adoption of repetitive or recursive representations and functional expression is preferred.

Communication   For fine-grained operations, dataflow (i.e., data-driven) communication appears to be an appropriate model for I/O of processing elements of locally clocked algorithms; a synchronized port of a PE at any time has a status of either *ready to read* or *ready to write*. In the former, if data are to be written into a port, the transition is delayed until the communication partner in one of its transitions has read the present contents. The opposite occurs in the latter. As a consequence, a PE makes a transition only when all of its input and output ports are ready. Note that in this mode of exchange there is no queuing of data sent. Note also that this simple protocol is easily implemented in hardware.

We deliberately choose to ignore the exact timing of occurrences of events. The advantage of this is that designs and reasoning about events are simplified and furthermore can be applied to physical and computing systems of any speed and performance. In cases where timing of responses is critical, these concerns can be treated independently of the logical correctness of the design. Independence of timing is a critical condition of the success of high-level programming languages. Therefore, for programming large-scale systems, the flexible asynchronous scheme is perhaps more preferable [Kung82a], [Kung82b].

5.5.2 Wavefront Programming Techniques

New programming languages are needed to allow systems consisting of many interconnected microcomputers to be designed and programmed. The wavefront notion can facilitate the description of parallel and pipelined algorithms and drastically reduce the complexity of parallel programming. Based on the ideas of tracing computational wavefronts and DFG models, high-level languages can be developed to support array processors. They should describe parallel data movements as well as parallel program executions, with the following features: (1) computational wavefronts, (2) asynchronous timing capability, (3) high level notation, and (4) unification of algorithm expression, simulation, and design.

In this subsection, we use the MDFL (Matrix Data Flow Language) – an experimental language proposed in [Kung82b] – as an example to illustrate the basic ideas of wavefront programming. In Section 5.5.3, we describe INMOS's Occam language which may be directly applied to wavefront arrays constructed from the commercially available *transputer* chips. Their main features are discussed below.

Global and Local Wavefront Programming The advantage of the wavefront language is that it allows the programmer to address an entire front of processors, instead of having to deal with each processor individually. There are actually two approaches to programming the WAP: a global and a local approach. A global wavefront program describes the algorithm from the viewpoint of a wavefront passing across all the processors. A local wavefront program describes the operations of an individual processor using the perspective of one processor encountering a series of wavefronts. Implementing a program on such a system requires transforming the high-level global wavefront description by a preprocessor into a set of lower level programs (or microinstructions) for the individual processors.

Space and Time Invariance Based on the assumption of regularity and recursivity, we observe two typical types of programming features.

1. *Space Invariance*: The task performed by the wavefront in a particular kind of processor must be identical or similar at all fronts. For example, a space repetitive construct in global MDFL is shown here.[5]

WHILE WAVEFRONT IN ARRAY DO
BEGIN < TASK T > END

Task T is repeated at all fronts.

2. *Time Invariance*: Recursions for the successive wavefronts are identical or similar. A time-repetitive construct is

REPEAT
< ONE RECURSION >
UNTIL TERMINATED

Since wavefronts do not intersect at any time, pipelining of the successive wavefronts can be incorporated in this form.

Interprocessor Communication   The connection of PEs in an array is such that they communicate with their adjacent orthogonal processors to simulate a wavefront. Since the array is data-driven, the software must ensure that each processor performs a FETCH instruction from a buffer between each successive set of computations. Each processor can FLOW data to the orthogonally connected neighbor only when the previous value in the connection buffer has been read. The wavefront principle ensures that no two wavefronts intersect at any time. This in turn offers an asynchronous waiting facility, which is critical for solving the problems of global clocking, random communication delays, and other fluctuations in computing times.

Example 1: Wavefront Programming for Matrix Multiplication
An example of a global MDFL program for matrix multiplication on a WAP (see Figure 5.3) is given below, which illustrates the simplicity of the description.

---

[5]Part of the global MDFL instruction repertoire is provided in Table 5.1.

Array size: $N \times N$
Computation: $\mathbf{C} = \mathbf{A} \times \mathbf{B}$

$$k^{\text{th}} \; wavefront : \; C_{ij}^{(k)} = C_{ij}^{(k-1)} + a_{ik}b_{kj}$$

$k = 1, 2, \cdots, N$

Initial: Matrix $\mathbf{A}$ is stored in the Memory Module (MM) on the left (stored row by row). Matrix $\mathbf{B}$ is in MM on the top and is stored column by column (see Figure 5.3).

Final: The result will be in the C registers of the PEs.

```
BEGIN
SET COUNT N;
REPEAT;
     WHILE WAVEFRONT IN ARRAY DO
        BEGIN
           FETCH A, LEFT;
           FETCH B, UP;
           FLOW A, RIGHT;
           FLOW B, DOWN;
              (Now form C := C + A × B)
           MULT A, B, D;
           ADD C, D, C;
        END;
     DECREMENT COUNT;
UNTIL TERMINATED;
ENDPROGRAM.
```

To demonstrate the flexibility of wavefront-type programmability, let us look at the multiplication of a band matrix $\mathbf{A}$, $N \times N$, with bandwidth P and a rectangular matrix $\mathbf{B}$, $N \times Q$ (see Section 4.4.2). Only a slight modification to the above program is needed. First, the data storage in the memory modules for matrix $\mathbf{A}$ are skewed in the band direction. The major modification on the wavefront propagation is that, between the recursions of outer products, there should be an upward shift of the partial sums. (This is because the input matrix **A** is loaded in a skewed fashion.) Therefore, the program remains almost the same, except for the two additional *bracketed* instructions to shift the partial sum upwards. The modified program is given below.

```
BEGIN
SET COUNT N;
REPEAT;
    WHILE WAVEFRONT IN ARRAY DO
        BEGIN
            [ FETCH C, DOWN; ]
            FETCH A, LEFT;
            FETCH B, UP;
            FLOW A, RIGHT;
            FLOW B, DOWN;
                (Now form C := C + A × B)
            MULT A, B, D;
            ADD C, D, C;
            [ FLOW C, UP; ]
        END;
    DECREMENT COUNT;
UNTIL TERMINATED;
ENDPROGRAM.
```

Example 2: Mapping DFG to Wavefront Programming

Another approach to programming wavefront arrays is based on the SFG/DFG descriptions. Many 1-D or 2-D digital filters are initially given in SFG/DFG form. Therefore, it is desirable to have a simple mechanism to convert its SFG/DFG representation into array processing programming codes. In order to demonstrate the simplicity of programming based on the DFG representation, an MDFL program implementing the linear phase filter as shown in Figure 5.6 is given below.[6]

---

[6]Note that the separators in the DFG are implemented simply by adding three lines of (internal register transfer) code to the program, as opposed to adding a separate buffer register external to the PE.

```
BEGIN
SET COUNT N;
REPEAT;
     WHILE WAVEFRONT IN ARRAY DO
        BEGIN
           FETCH X, LEFT;
           FLOW X, RIGHT;
           TRANSFER W2 TO W1;
           FLOW W1, LEFT;
           FETCH W2, RIGHT;
              (now compute V := (W1 + X) × h(k))
           ADD W1, X, U;
           MULT U, h(k), V;
           FETCH Y, RIGHT;
              (now compute Y := Y + V)
           ADD Y, V, Y;
           FLOW Y, LEFT;
        END;
     DECREMENT COUNT;
UNTIL TERMINATED;
ENDPROGRAM.
```

An MDFL program for a linear phase filter.

5.5.3   The Occam Programming Language

MDFL may be considered to be a good experimental language for demonstrating wavefront programming. For practical development of wavefront type arrays, it is very desirable to have a programming language supported by a commercial company and equipped with widely distributed programming system software. The programming language Occam enjoys such advantages. In many ways, Occam is very similar to MDFL, and therefore a direct translation between the two languages is possible.

The Occam programming language [Wils83] [May84] and the *transputer* microprocessors developed by Inmos Ltd. are based on the process model of computation, which is sufficiently general to include both sequential and concurrent processing in a natural manner. The Occam model is a dual of CSP (communicating sequential processes) designed by Hoare [Hoare78]. The Occam language is based on the models of communication and concurrency. Occam describes the structure of a system of connected microcomputers. It can also be used to program individual computers. In addition, Occam is a design formalism. Its formal semantics allow a program to be read either as a set of commands or predicates in an extension of the predicate calculus. Its semantics provide a set of rules for transforming programs [Pount86].

5.5.3.1 Basic Occam Constructs

Occam has a role to play in describing both *transputer* based systems and those constructed from dedicated hardware. It can be used as a behavioral specification language for digital hardware in general. A signal processing algorithm may be expressed as a program in Occam. The program can be executed and debugged on a single processor. When the algorithm works, the program is further developed to work efficiently in its intended implementation.

Occam introduces two fundamental concepts: process (PROC) and channel (CHAN). Process is a statement or a group of statements or even a group of processes. For an Occam compiler, a group of contiguous lines of code indented at the same level share the same context and are termed as a process. A channel is a basic communication element that enables two processes to communicate with each other. In contrast to the fine granularity of concurrency in data flow machines, Occam's concurrency is at the procedural level. This appears to be a very desirable feature to system engineers.

A process is taken to mean an independent computation, autonomous in the sense that it has its own program and data but is able to communicate with other current processes by message passing via explicitly defined channels. In general, a process may itself consist of a number of processes. These are grouped into sets, called constructs.

At the root of this hierarchical structure are three primitive processes:

- Assignment, which changes the value of a variable, e.g., $a := b + c$.

- Input, which receives a value from a channel and assigns it to a variable, e.g., chan1 ? a.

- Output, which sends a value to a channel, e.g., chan2 ! $b + c$.

The basic constructs are:

- *Sequential construct (SEQ)*: The component processes are executed one after another.

- *Parallel construct (PAR)*: All component processes are executed together.

- *Alternative construct (ALT)*: The component process which is ready to communicate first is executed.

- *Conditional construct*: The maximum (and only) one able to proceed.

In keeping with the philosophy underlying the development of the *transputer*, Occam naturally describes a concurrent system as a set of independent processes that use locally defined variables and can communicate only via declared channels. Channel communication in Occam is exactly mirrored by communication via *transputer* links. In particular, there is a handshake protocol that allows data to pass only when the sender and receiver have both signified their readiness to each other. This imposes order, although there is no concept of global time in Occam, and all concurrent processes run asynchronously.

For example, reading from and writing to a channel are represented in Occam by

$$chan1 \ ? \ x$$
$$chan2 \ ! \ x$$

Thus a process that reads from a channel and writes the result to a second channel, using an intermediate variable x, is written as:

```
VAR x:
SEQ
    chan.in ? x
    chan.out ! x
```

All the following processes that are indented to the right of SEQ are to be executed sequentially in the order that they appear. An example for PAR, which indicates that the indented processes are to be executed in parallel, is the following "vote counting" process, which continues to read (in parallel) the inputs from the channels *for* and *against*.

VAR *for-vote, against-vote*:
PAR
        For ? *for-vote*
        Against ? *against-vote*

In addition to SEQ and PAR, another useful construct to implement *replications* is ALT. An example follows:

ALT i = [0 for 100]
    inp [i] ? x
    oup ! x:

This process will read the first available value from the array of channels, inp[i], into variable x, and output this to channel oup.

5.5.3.2 Properties of Occam

Occam has the following useful properties:

1. It is simple and concise. A few primitive processes are combined in sequence (SEQ), parallel (PAR), and/or multiplexed (ALT) ways to create new, powerful abstractions.

2. The synchronization of parallel processes is built in the language by the support of the concept of the channel. An Occam channel provides a direct substitute of the kernel of a real-time executive. Therefore the programmer does not require to implement explicitly semaphore operations, process queuing mechanisms, scheduler and interrupt servers, as is often required in other concurrent languages.

3. Occam can be used as a system description language for systems built from a large number of concurrently operating components.

4. Occam supports a hierarchical structure and is applicable to different levels. It allows the user to define named processes which can be used in a program as any other process. Named processes can be compiled as separate entities and down loaded onto specified nodes of a *transputer* array.

5. From the point of view of system design, Occam is designed to run in the same logical fashion on a single processor as on several. Thus designs can be tested by running the Occam program on a conventional Von Neumann machine – the Occam compiler implements time sharing so that all concurrent processes gain access to the processor. After this initial development stage, the program can then be augmented with the necessary constructors to define the way in which it loads onto a given network and locations of the named processes in the local node memories.

6. The Occam programming system provides: a supportive and responsive environment; a folding screen editor, which facilitates the compiler/checker; performance, storage, and speed estimation (for the *transputer*); structured editing; cross-compilation facilities.

7. Occam is very suitable for programming wavefront arrays and simulating all kinds of wavefront processing.

A new version of Occam (Occam-II) will provide some useful features not available in the initial version, such as data types and the capability for multidimensional arrays and user-defined data structures. It could be desirable if some common processes are included as predefined macros in the language.

5.5.3.3   Occam for Wavefront Array Programming

Note that the channel communications defined in Occam is basically the same as the communication protocol envisaged for wavefront array processors. *Transputer* networks programmed in Occam may, quite naturally, be regarded as wavefront processors, given that they have a sufficiently simple data flow. The communication requirements in array programming often vary among various applications and involve issues such as buffering strategies and flow control. Since the channel in Occam is unbuffered it may not be possible to achieve maximum throughput because of the mismatch between the sending and receiving time instants. A solution to this problem is to insert a FIFO buffer between the two linked processors, which will presumably filter out the spikes in the data rate. Some strategies are proposed in [Kung86b].

Note also that Occam is very suitable for programming wavefront arrays and simulating all kinds of wavefront processing. The Occam programming

Figure 5.15: FIFO buffers: all the memory cells read the data from their respective left channels and output the previous data stored onto their right channels.

system may be adopted to develop a complete software library of all algorithms suitable for systolic/wavefront-type parallel processing.

5.5.3.4   Examples of Occam Programming

In this section, the concepts of pipelinability, concurrency, and locality inherent in the structure of Occam are highlighted through simple examples and then used to generate programs for more complicated signal processing algorithms [Taylo84]. Some useful keywords for Occam programming are provided in Table 5.3 at the end of this chapter.

Example 3: Occam Programming for FIFO Buffer   An n-element FIFO buffer can be visualized as a chain of memory cells linked together, as shown in Figure 5.15:

A program in Occam for such a structure will look like this:

```
CHAN c[n + 1]:
    PAR i = [0 FOR n]
        WHILE TRUE
            VAR x:
            SEQ
                c[i] ? x
                c[i + 1] ! x:
```

Every time there is a data input request, the whole process is to be duplicated. We can give a name to this process and then call it as and when required. The previous process must be modified as follows:

```
PROC buffer(CHAN c.in, c.out) =
    CHAN c[n − 2]:
        WHILE TRUE
            PAR
                VAR x :
                SEQ
                    c.in ? x
                    c[0] ! x
                PAR i = [0 FOR n − 2]
                    VAR x:

                    SEQ
                        VAR x:
                        SEQ
                            c[i] ? x
                            c[i + 1] ! x
                VAR x :
                SEQ
                    c[n − 2] ? x
                    c.out ! x:
```

To call such a process we just need to write the following statement:

$$buffer(c1, c2)$$

where c1 and c2 are pre-declared channels.

Example 4: Reading Synchronization Signals   In certain situations it is necessary to know only when the input can be read and not its value. Consider a situation where it is required to copy the data from one channel to another until a stop control signal arrives. Let there be three channels, c1, c2, and stop. The data are read from channel c1 and output to c2 until the stop channel is full, in which case the stop channel can be read.

A simple program to implement this is:

```
VAR running:
SEQ
    running := TRUE
    WHILE running
        VAR x :
        ALT
            stop ? ANY (check if control signal has arrived)
                running := FALSE
            c1 ? x (otherwise read in variable x and output)
            c2 ! x :
```

Example 5: Occam Programming for Lattice Processor

In this example [Chapm85] it will be shown how to come up with systolic and wavefront program representations for the same algorithm. An SFG representation of an n-th order all zero (MA) filter is depicted in Figure 5.16.

An Occam program that implements a single lattice module is as follows:

```
PROC latt (VALUE k, CHAN IN.1, IN.2, OUT.1, OUT.2) =
    VAR w, x, y, z:
    WHILE TRUE
        SEQ
            in.1 ? x
            y := x + (w * k)
            out.1 ! y
            z := w + (x * k)
            PAR
                in.2 ? w
                out.2 ! z :
```

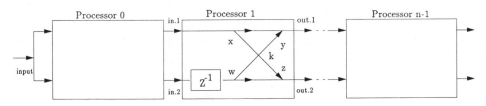

Figure 5.16: n-th order MA lattice filter.

Note that after the old value of w is used in the computation of y and z a new value of w is input via channel in.2. In this way we are able to simulate a delay. The main program where this process is called will look like this:

```
VAR input, output.a, output.b :
CHAN top[n + 1], bottom[n + 1] :
PAR
      WHILE TRUE
      PAR
            top[0] ! input
            bottom[0] ! input
            top[n] ? output.a
            bottom[n] ? output.b
      PAR i = [0 FOR n]
      latt (k[i], top[i], bottom[i], top[i + 1], bottom[i + 1] ):
```

The process *latt* will be executed n times in parallel. Since a handshaking protocol is adopted between the communicating channels this program actually implements a WAP. This can also be seen in the program. First the data are read in by the channel in.1. They then have to be modified before they can be retransmitted to the next lattice via the channel out.1. Hence there is an asynchronous ripple through the process.

Example 6: Occam Programming for Matrix Multiplication

Here Occam is used to design and simulate a 2-D wavefront array for matrix multiplication. Refer to Figure 5.3: The computing network serves as a (data) wave-propagating medium, which can be implemented by using the channels in Occam.

One matrix enters the array of processors from the left (column-wise), while the other matrix enters from the top (row-wise). As the data values move right and down, they are multiplied and accumulated. Finally, when the whole matrix has passed through the array, each processor has the elements of the final matrix. Again, all the PEs perform the same tasks of reading the data and multiplying, accumulating and transmitting the data further right and down.

An Occam program describing such a PE will be a process (for instance, mult).

```
; PE NODE PROGRAM
PROC mult (CHAN up, down, left, right) =
    VAR acc, a, b :
    SEQ
        acc := 0
        SEQ i = [0 FOR n]
            SEQ
                PAR
                    up ? a
                    left ? b
                acc := acc + a * b
                PAR
                    down ! a
                    right ! b :
```

Now the links have to be established between the adjacent PEs, and then mult is called $n \times n$ number of times. The main program describing the array is:

```
; MAIN PROGRAM
CHAN vertical[n * (n + 1)]:
CHAN horizontal[n * (n + 1)]:
PAR i = [0 FOR n]
    PAR j = [0 FOR n]
        mult (vertical[(n * i) + j],
              vertical[(n * i) + j + 1],
              horizontal[(n * i) + j],
              horizontal[(n * (i + 1)) + j)]):
```

5.6 Hardware Design

In this section, we concentrate on the hardware implementation of PE design and handshaking protocols of the WAP.

5.6.1   PE Design for Wavefront Array Processor

A programmable wavefront array must be able to execute the entire range of instructions of the wavefront language repertoire. Each PE includes the full complement of hardware necessary to support the language. However, within the class of algorithms for WAP applications, there is a finer division of applications, which may lead to significant simplification in architectural structures and great savings in hardware implementation. It is clearly beneficial, when the scope of applications of the wavefront arrays is limited, to streamline the PE design to the need at hand by deleting unnecessary hardware and enhancing certain other hardware components of the PE. The benefits involved include decreased PE area (i.e., increased number of PEs that can be assembled on a chip) and an increase in the speed and throughput rate of the PEs and the wavefront arrays.

The key characteristics for DSP applications, for example, are adequate word length, fast multiply and accumulate, high-speed RAM and fast coefficient table addressing. The functionality of the PE should be designed to support the above operations. As shown in Figure 5.17, there are four main components in the PE, i.e., the arithmetic and logic unit (ALU), the memory unit, the control unit, and the I/O unit.

1. *Arithmetic and Logic Unit (ALU)*   Since high throughput is usually demanded for the wavefront array processor, the ALU must com-

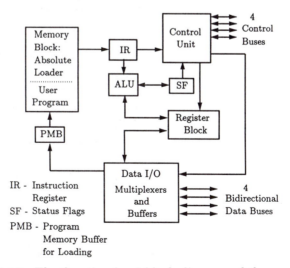

Figure 5.17: The function level block diagram of the wavefront array PE.

pute any frequently encountered operations fast. Fixed-point ALUs are cheaper to build while floating-point hardware gives higher precision and dynamic range, which is often required by many DSP applications.

2. *Memory Unit*   A design with separate program memory and data memory is now becoming popular. Although on-chip memory has limited capacity as opposed to off-chip memory, it does allow faster processing speed.

3. *Control Unit*   There are two approaches to the control unit design. The first is the reduced instruction set computer (RISC) approach, which uses a small set of simple instructions and obtains a simple control unit with faster clocking rate. The second is the complex instruction set computer (CISC) approach, which uses a large and complex instruction set and allows complicated tasks to be completed with fewer instructions. The current trend of VLSI implementations appears to favor the RISC approach.

4. *I/O Unit*   The PE should be able to simultaneously perform data transfers in four directions in a mesh array concurrently with the processing. The transfer of data is controlled by the I/O controller, one for each of the four directions, which handles the two-way handshaking functions.

5.6.2   Asynchronous Communication Protocols

One of the most important features that distinguish the wavefront array processor from other array processors is the data-driven operation of each PE. To ensure the correct sequencing and data transfers between adjacent PEs, handshaking protocols must be adopted to synchronize the operations. In general, there are two types of asynchronous communication schemes: the one-way control and the two-way control schemes. In the one-way control scheme, the sender will send data without waiting for the acknowledgement signal of the receiver. It is suitable only when large buffers are provided. The two-way control scheme, usually known as *handshaking*, is more preferable for the wavefront array processor. A proposed handshaking circuit is shown in Figure 5.18(a). This circuit can be considered as an improved version of a previous design [Kung82a]. This new design is more robust due to the fact that the flip-flops are driven by internal clocks, and are less sensitive to the glitch noise encountered in the communication links (see Problem

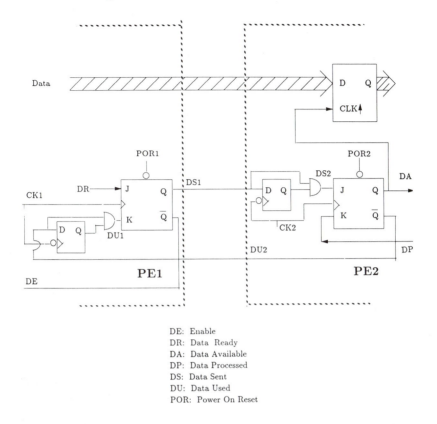

DE: Enable
DR: Data Ready
DA: Data Available
DP: Data Processed
DS: Data Sent
DU: Data Used
POR: Power On Reset

Figure 5.18: (a) The proposed handshaking circuit with glitch protection ability.

15). The timing diagram of this circuit is shown in Figure 5.18(b), two rising-edge triggered JK flip-flops and two falling-edge triggered D flip-flops plus two AND gates are used to implement this handshaking circuit. The basic operations and protocols during one handshaking process are briefly explained below:

1. When $DE = 1$, $CK1 =$ falling; then $DR = 1$ (for one clock cycle), *Data* are on the bus.

2. When $DR = 1$, $DU1 = 1$, $CK1 =$ rising; then $DS1 = 1$, $DE = 0$.

3. When $DS1 = 1$, $Q_b = 1$, $CK2 =$ falling; then $DS2 = 1$.

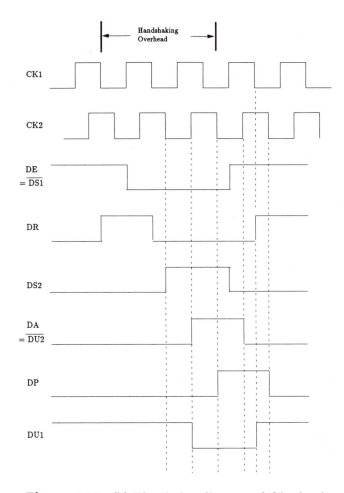

Figure 5.18: (b) The timing diagram of this circuit.

4. When $DS2 = 1$, $DP = 0$, $CK2$ = rising; then $DA = 1$, $DU2 = 0$, *Data* latched on PE2.

5. When $DA = 1$, $CK2$ = falling; then $DP = 1$ (for one clock cycle), *Data* are used.

6. When $DS2 = 1$, $DP = 1$, $CK2$ = rising; then $DA = 0$, $DU2 = 1$.

7. When $DU2 = 1$, $CK1$ = falling; then $DU1 = 1$.

8. When $DU1 = 1$, $DR = 0$, $CK1$ = rising; $DS1 = 0$, $DE = 1$.

If we define the handshaking communication overhead to be equal to the time interval between the rising edges of the DR flag and that of the DP

flag, this time interval is determined by several delay factors: the flip-flop time delay d_{ff}; the propagation time delay d_{pr}; the phase difference between $CK1$ and $CK2$; and the delays introduced by the DP flag response after the DA is set high. On average, this overhead is shorter than three clock periods, and in most cases during this handshaking time interval, both PEs still continue their computations simultaneously. When the granularity of PEs is large, this handshaking time penalty is relatively tolerable.

Block Handshaking Scheme   One method to reduce the handshaking time overhead is to use a *block handshaking* scheme, in which a block of data can be transmitted and received with only one handshaking. The success of the block handshaking scheme relies on the assumption that the clock frequency and phase of the sending PE remain stable during the period of the block data transfer. This is suitable for communication between systems operating with almost identical clock frequencies.

Two-level Pipelining and Handshaking   In many cases, to further explore speed enhancement, internal pipelining within the ALU of the PE is required. In this case, the handshaking scheme of participating PEs is the same as before. However, due to the uncertainty of the continuous supply of data into the pipe, the PE should record the time when a data enters the pipe and then it can retrieve the processed data from the pipe later. If there are no data coming into the pipe, then the corresponding output is considered as garbage. A simple way to implement this scheme is to employ a one-bit shift register of the same length as the pipeline in the PE. When a valid (invalid) data comes into the pipe, a 1 (0) is entered into the shift register at the same time. This 1 or 0 is shifted in the shift register as the data moves down the pipe. When the data comes out from the pipe, the 1 or 0 bit simultaneously exits from the shift register and gates the output data. In this way, the PE can retrieve (only) the valid processed data.

5.7   Concluding Remarks

This chapter introduces the wavefront architecture which eliminates the need for global control and global synchronization incurred in a systolic design. It permits a data-driven, self-timed approach to neatly handle data dependencies in array processing. The power and flexibility of the wavefront type arrays are best demonstrated by its very broad application domain, includ-

Table 5.1: The (Global) MDFL Instruction Set

Data-transfer instructions

    FLOW < SOURCE REGISTER >, < DIRECTION >;
    FETCH < DESTINATION REGISTER >, < DIRECTION >;
    READ;

Recursion-oriented instructions

    REPEAT ... UNTIL TERMINATED;
    WHILE WAVEFRONT IN ARRAY DO
        BEGIN .... END;
    SET COUNT < NUMBER OF WAVEFRONTS >;
    DECREMENT COUNT;
    ENDPROGRAM.

Conditional instructions

    IF EQUAL THEN < STATEMENT >;
    IF NOT-EQUAL THEN < STATEMENT >;
    IF GREATER THEN < STATEMENT >;
    IF LESS-THAN THEN < STATEMENT >;
    IF < DIRECTION > DISABLED THEN < STATEMENT >;
        CASE KIND =
            (1, 1) : < STATEMENT >;
            (1, *) : < STATEMENT >;
            (*, 1) : < STATEMENT >;
            INT : < STATEMENT >;
        ENDCASE;

Internal processor instructions

    TSR < SOURCE >, < DESTINATION >;
    ADD < SOURCE #1 >, < SOURCE #2 >, < DESTINATION >;
    SUB < SOURCE #1 >, < SOURCE #2 >, < DESTINATION >;
    MULT < SOURCE #1 >, < SOURCE #2 >, < DESTINATION >;
    DIV < SOURCE #1 >, < SOURCE #2 >, < DESTINATION >;
    SORT < SOURCE >, < DESTINATION >;

Table 5.1: cont'd

```
CMP < SOURCE #1 >, < SOURCE #2 >;
TST < SOURCE >;
STORE;
NOP;
RESET;
DISABLE SELF;
BEGIN ..... END;
```

ing spectrum analysis, adaptive array processing, image/vision processing, seismic and medical signal processing, PDE solutions, and many others. Furthermore, at the expense of extra communication overhead, the wavefront processing technique (as opposed to the systolic technique) may cope better with global communication algorithms, or computing networks with somewhat irregular interconnection.

We have proposed a DFG as a formal abstract model for wavefront computing networks. A DFG may be derived from a DG by assigning a projection vector and letting the data-driven computing dictate the schedule. Deriving a DFG from a SFG is also straightforward following an equivalence transformation technique. We have investigated timing and retiming analysis based on the DFG, which leads to several algorithms for queue assignment for optimal throughput. In fact, our analysis technique follows very closely the approach adopted in the analysis of asynchronous computation systems by *timed Petri nets*. The timed Petri net, in our opinion, is a methodology which has not been as well appreciated as is warranted by its merit. This chapter has demonstrated its potential application to wavefront-type parallel processing.

Some guidelines will be useful for selecting between a synchronous design and an asynchronous one. A synchronous design (e.g., systolic array) is preferred when clock distribution for the integration level is feasible and practical and the PE primitives are of small granularity. As the integration level grows, however, the synchronization of a large array will become formidable. On the other hand, an asynchronous design (e.g., wavefront array) suffers from the hardware and time overhead of requiring a handshaking interface between the PEs. However, for PEs with more complex functionality, the overhead penalty will become relatively negligible. A popular scheme to reduce the time penalty is to use block data transfer per handshaking.

Table 5.2: Keywords for Occam Programming

KEYWORD	*DEFINITION*	*COMMENTS*
SEQ	SEQ {base FOR count} p1 p2 . .	Executes processes p1, p2, ... in sequence (count-base) number of times.
PAR	PAR {base FOR count} p1 p2 . .	Executes processes p1, p2, ... simultaneously with the index varying from base to count.
ALT	ALT {base FOR count} p1 p2 . .	All the processes are checked simultaneously to see which one can be executed first.
WHILE	WHILE < condition > process	
IF	IF < condition > process	
VAR	VAR name1,name2,...: VAR name[dimension]:	A variable may be a single variable or an array of variables.
CHAN	CHAN name1,name2,...: CHAN name[dimension]	
PROC	PROC name (< parameters >) subprocess p1 subprocess p2	

To further decrease the overhead, it might be useful to adopt a computing network, partitioned into many blocks of PEs, which is synchronized within a block but is globally asynchronous.

5.8   Problems

1. *Comparison of systolic and wavefront array*: Give a general comparison between systolic and wavefront arrays from the following perspectives.

    (a) hardware complexity;

    (b) operating speed;

    (c) program flexibility;

    (d) fault tolerant capability.

2. *Some properties of the marked graph*: The TMG model is very similar to the marked graph model in Petri net theory; the only difference is that usually the marked graph does not specify the node operation times. Prove the following results for the marked graph [Commo71]:

    (a) The sum of tokens in a directed loop does not change after firings.

    (b) A marked graph is not deadlocked if and only if the token count (sum of tokens) of every directed loop is positive.

    (c) A marked graph is called *safe* if no edge is assigned more than one token and if no sequence of firings can bring two tokens or more to one edge. Prove that a marked graph is not deadlocked and safe if and only if every edge in the graph is in a directed loop with token count 1.

3. *Queue assignment for the linear phase filter*: Referring to Figure 5.6(c), we have said that the queue assignment for this linear phase filter is minimal if Δ is much smaller than T (compare with the multi-rate linear phase systolic array in Chapter 4). Assuming that $\Delta = \frac{1}{5}T$, show that the queue assignment is indeed a minimal one.

4. *Minimal queue assignment with initial token redistribution*: In some instances we find that part of the queue capacity of an edge needed

to accommodate initial tokens is never used in the rest of the computation after the initialization stage. In these cases a smaller edge
queue capacity can support the computation after initialization. Use
of this smaller capacity necessitates the redistribution of initial tokens
so that they can all fit into the reduced edge capacities. Show that this
redistribution can be formulated along the same line as we proposed
for the integer programming formulation and can be integrated into
a larger size integer programming problem to solve the real minimal
queue assignment problem [Kung86b].

5. *DG boundary conditions:* For the DG shown in Figure 5.19, derive the
   SFG and DFG arrays by applying the systolic and DFG mappings,
   respectively, along the given projection direction. Show that to ensure
   the correctness of the computation some attention has to be paid to
   the boundary condition of the DG. That is, some PEs should remain
   inactive until certain time.

6. *Minimum queue assignment for SI DG:* Assume that a DG is shift-
   invariant with the same and deterministic node computation time.
   Then the DG can be scheduled a priori and the *minimum computation
   time* can be determined.

   (a) Show that all DG nodes are on the critical path, i.e., no node
       can be rescheduled without increasing the minimum computation
       time.

   (b) Suppose that a DG arc a is projected along $\vec{d}$ onto a DFG arc a'.
       To determine the *minimal required queue size* for a', we note the
       following: (1) The scheduled completion time, t_1, for the initiating
       node of a indicates when the output data of the node is *produced*
       (or *put*) on a'. (2) The scheduled completion time, t_2, for the

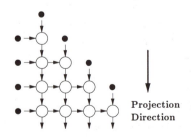

Figure 5.19: A shift-invariant DG with a projection direction.

terminating node of a represents when the data is *consumed* from a'. (3) The node computation time is τ. Show that $t_2 - t_1 + \tau$ represents *the length of time* a data token stays in a' and its two end nodes.

(c) Show that the queue size for a', Q, can be calculated as

$$Q = \lceil \frac{t_2 - t_1 + \tau}{\alpha} \rceil$$

where $\lceil \cdot \rceil$ denotes the ceiling function and α is the pipelining period, i.e., the time period between two consecutive data being put on a'.

If the queue size of a wavefront array is less than the minimal required one, then the overall speed of the array will be slowed down.

(d) Suppose that the node computation time τ is normalized (i.e., $\tau = 1$). If the DG is mapped to an SFG along $\vec{d}$, show that the number of delays on the arc a' will be $D = t_2 - t_1$. This can lead to the conclusion that the queue size on a DFG can be determined by its corresponding SFG. Verify that the queue size for a' can be determined as

$$Q = \lceil \frac{D + 1}{\alpha} \rceil$$

7. *Mapping DSI DG to DFG*: A DSI DG is shown in Figure 5.20, where the number in each DG node shows the required computation time of the node.

   (a) Derive a structurally time-invariant DFG from the DG. For the time being, the queue size in the DFG is assumed to be infinite.

   (b) What is the schedule executed by the DFG?

   (c) What is the minimum queue size to achieve the minimum computation time? (**Hint:** The latest completion time (c.f., Sec. 3.4.6), instead of the earliest completion time, can be used for some DG nodes.)

   (d) (Optional) How to systematically derive the minimum queue assignment for a general 2-D DSI DG? How about N-dimensional DSI DGs?

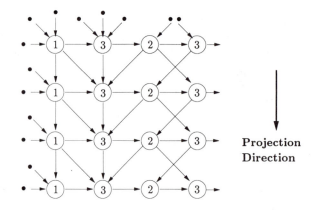

Figure 5.20: A DSI DG showing the computation time of each node.

8. *Programming Parallel Algorithm*:

   (a) Program the transitive closure problem in Chapter 4 in MDFL. Assume matrix **A** comes from both the left and top directions.

   (b) Redo part (a) if matrix **A** comes from the diagonal direction.

   (c) Redo part (a) using Occam.

   (d) Redo part (b) using Occam.

9. *Architecture-oriented languages*: The operation in each systolic array processor element is synchronized by a global clock, whereas wavefront array processors are asynchronous. The wavefront concept also means the data can be piled up if queue or memory is available. Propose appropriate communication mechanisms using languages suitable for each architecture. Discuss the related programming issues, such as the race problems, the cost of negotiation, deadlock, and so on.

10. *Shortest-path problem*: Given a shortest-path problem, an undirected graph G and a nonnegative cost $c(e)$ associated with each edge e, devise an array processor and write a program in Occam to solve it.

11. *LU decomposition*: Given the LU decomposition problem, program a VLSI array processor in MDFL.

    Hint: Two versions of LU decomposition can be discussed; one is the mesh connected array version, and the other is the triangular array version.

12. *Occam for systolic and wavefront arrays*: Discuss and compare the programming issues of Occam for systolic (synchronous) and wavefront (asynchronous) array processors.

13. *Toeplitz solver programming*:

    (a) Program the Levinson algorithm in Occam.
    (b) Program the Schur algorithm in Occam.
    (c) Compare and discuss the results of (a) and (b).

14. *Handshaking protocols*: There are two kinds of schemes of asynchronous communications [Hayes78]:

    - *One-way control*: Timing signals are supplied exclusively by one of the two communicating devices.
    - *Two-way, or interlocked, control*: Both devices generate timing signals.

    Source-initiated timing diagrams of the two schemes are shown in Figure 5.21.

    (a) Compare the pros and cons of the two schemes.
    (b) Which scheme is more suitable for the WAP? Why?

15. *Handshaking circuit without glitch protection:* Compared to the handshaking circuit given in Figure 5.18, a simpler circuit is given in Figure

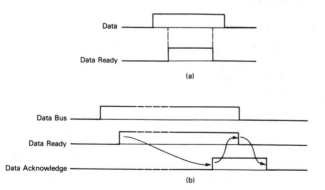

Figure 5.21: (a) One-way control protocol. (b) Two-way control handshaking protocol.

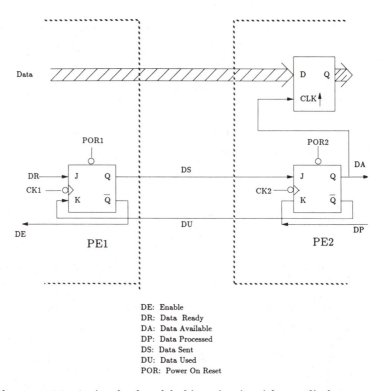

DE: Enable
DR: Data Ready
DA: Data Available
DP: Data Processed
DS: Data Sent
DU: Data Used
POR: Power On Reset

Figure 5.22: A simpler handshaking circuit without glitch protection ability.

5.22. This circuit is sensitive to glitch noise, which occurs frequently in a large array processors when all the PEs are active. Draw the timing diagram of this circuit, and find out in what situation this circuit will fail to work properly.

Hint: The simplified operations can be summarized by the following steps.

(a) When $DE = 1$, $CK1 = $ rising; then $DR = 1$, $Data$ are on the bus.

(b) When $DR = 1$, $DU = 1$, $CK1 = $ falling; then $DS = 1$, $DE = 0$, $Data$ are latched in PE2.

(c) When $DS = 1$, $DP = 0$; $CK2 = $ falling; then $DA = 1$, $DU = 0$.

(d) When $DA = 1$, $CK2 = $ rising; then $DP = 1$, $Data$ are used.

(e) When $DS = 1$, $DP = 1$, $CK2 = $ falling; then $DA = 0$, $DU = 1$.

(f) When $DU = 1$, $CK1 = $ falling; then $DS = 0$, $DE = 1$.

16. *Different signaling schemes*: The most elementary signal event that can be used to compare self-timed signal conversions is transition. Two types of transitions are generally used [Seitz80]:

- Two-cycle signaling (non-return-to-zero, NRZ)
- Four-cycle signaling (return-to-zero, RZ)

Based on the two-way control handshaking in the previous problem, the two types of signaling are shown in Figure 5.23.

Compare the relative advantages and the disadvantages of these two types of self-timed signaling.

17. *Block handshaking*: One argument against the use of one handshaking per data transfer is that the communication time is usually about the same as the computation time, which is not very efficient. A solution to this problem is to transfer a block of words per handshaking instead of just one word. Is this a good scheme? What are the advantages and disadvantages?

18. *Irregular wavefront arrays for sparse matrix manipulations*: As discussed earlier, the sparse matrix operations involve many useless computations with zeros. One way to improve this is to use the content-addressable systolic array, in which only non-zero elements are used for the computation [Charn86]. Another way is to use a wavefront array which allows irregular wavefronts, corresponding to the computations of non-zero elements, to be processed [Melhe86]. Compare these two methods.

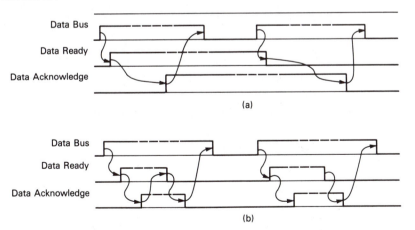

Figure 5.23: (a) Two-cycle signaling, (b) Four-cycle signaling.

Chapter 6

SYSTEM AND SOFTWARE DESIGN

6.1 Introduction

In the previous chapters, we have treated the design and analysis of PE arrays. An important subject naturally follows: *How to incorporate these arrays into an overall computing system.*

An array processor may be used either as an accelerator attached to a compatible host machine or as a stand-alone machine equipped with a global control processor. Generally speaking, in an overall array processor system, one seeks to maximize the following performance indicators: effective system organization, flexibility on problem partitioning, fault-tolerance to improve system reliability, and programmability for adequate software support. More elaborately, the desirable features for an array processor system are:

1. *High-speed performance*: The overall system should provide a speed performance which at least matches the real-time processing requirement.

2. *Cost-effectiveness*: The hardware cost depends not only on the number of hardware components used but also on the number of module types

used. Therefore, the array processor should use as few types of PE as possible and tailor the program within each PE to suit the application.

3. *Flexibility*: The array processor system should be sufficiently programmable and reconfigurable so that it may satisfy the need for a broad range of applications. For example, it is important to provide hardware and/or software support for an efficient *partitioning* scheme, which allows large problems to be decomposed into smaller subproblems to be solved on the array.

4. *Reliability*: Actual array processor hardware will be very large scale and additional attention should be taken to achieve an acceptable system reliability. It is inevitable that temporary or permanent failure of one or more PEs in a large scale array processor will be encountered. Therefore, *fault-tolerance* is an essential survival attribute of a highly parallel array processor system.

5. *Software support*: To facilitate the use of the array processor system, it is useful to develop coherent software techniques for programming or design of the system, as well as a set of applicational software packages.

To support the construction of array processors, the implementational issues in the PE and system levels should also be dealt with. At the system level, the major components of an array processor system are: host computer and/or array control unit (ACU) (with optional I/O units); interface unit; interconnection networks (at the level of PE-to-PE network and/or at the level of PE-to-memory network); and PE arrays.

An important issue of array processing system design is that of *matching* algorithms to arrays. Basically, given a specific array network, the *matching* problem is *to program a set of algorithms so that they may be efficiently executed on the array*. If the topology of the interconnection network is fixed, some conversion methods must be adopted to match the structure of algorithms to the networks. When the size of array or the size of local memory is fixed, then the center of the matching problem is the issue of partitioning. A partitioning scheme permits decomposing a large problem into several smaller subproblems, solving each of them on the given array, and finally combining the solutions of the subproblems to yield an overall solution. A systematic partitioning methodology is discussed in Section 6.3.

To improve the reliability, a fault-tolerant array processor design based on a distributed reconfiguration algorithm is proposed in Section 6.4. An algorithm-based fault-tolerant approach, where the input data to the algorithm are encoded at the system-level in the form of some error-correcting or error-detecting code is also addressed. The dependence graph representation is found very useful in the fault-tolerance analysis.

Due to frequently changing application specifications, programmable array processor systems are often required. Parallel programming is significantly more complicated than the conventional programming techniques. It is advisable to explore new notations or languages which are more appropriate to array processors, instead of committing the choice to conventional languages. Along a similar line, a hierarchical design methodology may be developed to pave the way for the concept of the *array compiler*, which maps a high-level specification to a certain architectural design.

6.2   System Organization

A possible overall system configuration is depicted in Figure 6.1, which consists of the following four major components:

- Host computer and/or array control unit (ACU).

- Interface unit.

- PE array(s).

- Interconnection network(s).

The general considerations for the four major components are discussed in this section. For practical design examples and a more elaborate discussion on interconnection networks, the reader is referred to Chapter 7.

6.2.1   Host Computer and Array Control Unit

The host computer should provide system monitoring, data storage, management, and formatting, determine the schedule program that controls all the system units, and generate global control codes and object codes of PEs.

The host machine may be selected out of a broad range which covers very different levels of computing power. It can be a microcomputer, workstation, minicomputer, main-frame, or supercomputer. The selection will depend on the desired applications. It is important for the system designer to identify in advance a suitable host machine for the usually very high-speed array processor units.

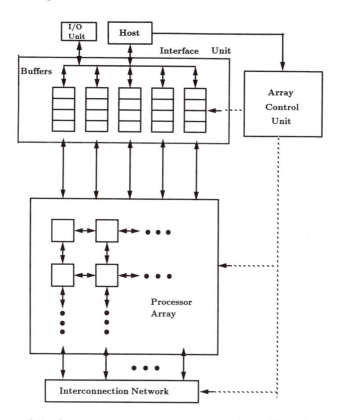

Figure 6.1: An array processor system consists of host/ACU, interface system, interconnection networks, and processor array.

Array Control Unit: The host generates control codes to coordinate all system units. The ACU follows the schedule commands to perform data rearrangement and direct data-transfer traffic, so that parallel processing tasks are timely assigned to the appropriate array processor modules. Sequence control guides the sequence of operations.

Data Formatting/Storage Management: The data may have many formats, such as floating-point/fixed-point numbers, bit serial/bit parallel forms, or non-numerical data. The data formats may be different among the different array processor units,; therefore, conversions between different formats are often required, which is often performed by the host.

Storage management may be specified either by the programmer or by the system controller. System controlled management is often safe and free from undesirable interference, but its efficiency of allocation is somewhat limited. A stack based (run-time) storage management is a simple and useful technique.

Operating System: In order to provide a complete programming environment for the user, there is need for disk I/O support, terminal I/O support, resource management (CPU, arrays, memory, I/O devices), protection from unauthorized accesses, multiuser and multitasking capabilities and virtual memory management. These features are already adequately supported by most conventional operating systems (e.g., Unix).

The array system is in general *asymmetric* (master-slave), that is the array processors are controlled by the host and are not expected to perform any of the host's tasks. This greatly simplifies the operating system design since the host can treat the arrays just as a special resource. The main extension to the conventional operating system is a driver for the arrays which must be able to treat arrays both as a whole (to allocate tasks to arrays) and as collections of processors (to load programs and data to individual PEs). The information to perform this task is provided by the language compiler which performs the dependence analysis.

In the case of a real-time special purpose system, a general purpose operating system is not necessary since the system will perform one function for a long time and any changing of the function (reprogramming) can be carried out *off-line*. However, an *operating system kernel* may be very useful. In other words, there is a need for some system programs running on the host or ACU which will handle external interrupts and I/O equipment, manage array-host communications, perform recovery from errors, and provide some means of system programmability to facilitate reprogramming (i.e., interconnection network reprogramming, code loading and I/O control). The program development is done on a separate *development system* which provides the facilities of a general purpose operating system and has additional capabilities for off-line simulation and debugging and subsequent loading of the finished program to the array processor.

In the environment of a special purpose system (e.g., a guidance control system), the following software tools are needed:

1. *Software Development:* Basic tools provided by a conventional operating system (i.e., editors, file management), and additional tools to assist in mapping/matching algorithms to arrays (i.e., high level languages, projection programs).

2. *Test and Debugging:* Target array architecture simulator.

3. *Code Downloading:* Tools to program ROMs for existing systems, or interface to drive CAD systems for custom/semicustom implementation.

4. *Run Time Support:* Library of routines to support inter-PE, host-PE communications, basic control functions, and global resource management such as memory and external I/O.

In this case we see that the functionality of system programs at run time is limited but since the system will be real-time, the speed of their execution is very important which leads to the selection of a very fast host and the need for optimized coding of the system programs.

6.2.2   Interface Unit

Since an array processor is to be used as an attached processor, the design of the interface unit is important. The interface unit is monitored by the system controller based on the schedule program. *The interface unit, connected to the host via the host bus, or DMA, has the function of down-loading, up-loading, and buffering array data (e.g., cache memory, or special purpose stacks), handling interrupts and data formatting.*

Direct Memory Access: To support high-bandwidth communication (accompanying high-speed processing), direct memory access (DMA) channels may be utilized for both instruction and data communication between the host and the arrays. In a DMA transfer, the interface unit communicates directly with memory without disturbing the internal registers of the computer. The array processor machine codes and voluminous raw and processed data are loaded to and returned from the arrays through high-speed DMA transfers. For certain applications, optional *I/O units* must be included to handle the high bandwidth I/O requirements [Horii86].

Memory/Storage Unit: The interface unit should furnish adequate hardware support for many common data management operations. Other challenging tasks for the system designer are managing blocks of data and making sure the memory (buffer) unit is able to balance the low bandwidth of the system I/O and the high bandwidth of array processors. To satisfy the high bandwidth of the processor array, memory/storage units should be carefully designed. The adoption of cache memory should be considered. In a parallel processing environment, the shared main memory is partitioned into several independent memory modules and the addresses distributed across these modules. This scheme, called *interleaving*, resolves some of the interference by allowing concurrent accesses to more than one module. For matrix data accessibility, it is desirable to have memory accessible via both columns and rows. It has many useful applications such as matrix transposition, sorting (e.g., a generalized bitonic sorting), or for PDE solutions. For many image processing applications, certain data reordering capabilities provided by the interface unit are desirable. For example, the staging memory in the Goodyear's massively parallel processor (MPP) provides such a facility [Batch80].

Partitioning: It is often necessary to decompose a large task into smaller subtasks and assign the subtasks to appropriate arrays, and recombine all the results. Programmable interfaces play a key role in providing temporary storage space and matching the memory bandwidth and the array processing load.

6.2.3   PE Arrays

A PE array comprises a number of processor elements with local memory. Most existing parallel array processor systems (e.g., systolic, wavefront, and SIMD arrays) emphasize both the fact that the PEs execute the same instruction, and that the PEs are interconnected in a regular and expandable manner. They are very amenable to a large class of regular and recursive algorithms, such as signal and image processing with local/global operations.

An important factor in the design of the PE arrays is the local memory available to each PE. Local memory is desirable to give each PE maximum flexibility. Although on chip memory is much less dense than off-chip memory, it allows more extensive simultaneous I/O and processing. When there is sufficient local memory space in the PE, it can be effectively utilized in the following fashions:

1. *Multiple program storage*: This allows the array processor to perform various functions without having to up-load/down-load the programs.

2. *Temporary data storage*: Whenever possible, it is advantageous to retain certain data within the PE for some subsequent processing. This can save communication time and avoid tying up the interface buses.

For simplicity, only one processor array is depicted in Figure 6.1. (In many applications, one programmable processor array suffices the need.) However, the concept of networking multiple arrays offers many more potential applications, especially where multiple arrays must act in unison to solve complex problems at high speed.

6.2.4   Interconnection Networks

Interconnection networks between PEs may significantly broaden the application domain and enhance the speed performance of processor arrays.

Intra-Array Communication   Within a PE array, certain structured (intra-array) interconnection networks may be incorporated to provide direct, global and high-speed communications. There are two suitable ways to adopt such an interconnection network. The first configuration as shown in Figure 6.2(a), in which a network is used to support direct communication among PEs, is appropriate when the number of PEs is equal to that of the memory blocks. Each PE is permanently linked to one (and only one) memory block. When several computations in different PEs are successively performed using the same set of data, this structure is very efficient since the memory operations are not involved in the transfer of data between PEs. This design is chosen for most array computers currently in use or under development [Lenfa78].

The second configuration is often used when the number of memory blocks and PEs is not equal. Usually, the number of memory blocks will be greater than the number of PEs (e.g., in matrix manipulation algorithms). Figure 6.2(b) illustrates this configuration where an interconnection network is used to connect memory blocks to the PEs.

Inter-Array Communication   For the case where there are multiple arrays (or multiple clusters of PEs) in the array processor system, a bottleneck is often created if the host machine is asked to handle all of the *inter-array* data transactions. Therefore, a global bus or interconnection network is added between the global memory blocks and the local memory blocks for the inter-array communication, while in each array a local interconnection network is provided for the intra-array communication (see Figure 6.2(c)).

The configuration shown in Figure 6.2(c) is a hierarchical version of that shown in Figure 6.2(b). In each sub-array (cluster), a local interconnection network is provided between the faster local memory blocks and PE array

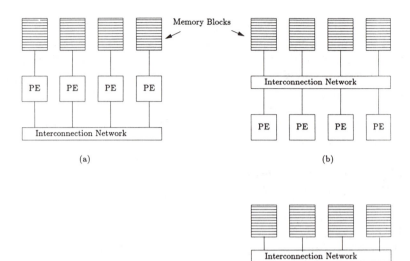

Figure 6.2: Three different configuration of the interconnection networks (INs) in a array processor system: (a) intra-array communication – among PEs; (b) intra-array communication – between PEs and memory blocks; (c) inter-array communication – between array and global memory blocks (ACU is not shown here).

for the intra-array communication. A global interconnection network, which communicates with local memory through the local control unit, is provided for the inter-array communication. This also allows access to the globally shared data [Gajsk83]. It may be desirable to use the global interconnection network for both intra-array and inter-array communication. For that, the interconnection network must have some kind of partition capability to allow it to be used as a local (intra-array) or global (inter-array) network [Feng81].

6.3   Matching Algorithms to Arrays

Chapter 3 concentrates on *mapping* an algorithm onto a dedicated array. The main theme addressed there is: *How to design a (dedicated) array processor so as to best reflect the data structure of the algorithm under consideration.*

Now let us address a closely related question: *How is a set of algorithms best programmed to fit in the architecture of a specific array processor?* This is called the *matching* problem, in which the key parameters of the array such as the number of PEs, communication links, or memory size are assumed to be constrained. In other words, *matching* is basically a *constrained mapping*. If constraints are on the communication links, some conversion methods will be necessary to match the algorithm to the architecture. When the constraints are on the number of PEs or the size of memory, the matching problem becomes the issue of *partitioning*.

6.3.1   Matching Algorithms to Fixed Array Structures

To help illustrate the general methodology of matching algorithms to certain structures, let us study two examples. One matches regular arrays to mesh arrays; and the other matches mesh algorithms to hypercube structures.

6.3.1.1   Matching Regular Arrays to Mesh Arrays

We first consider the case that the target array processor is a linear or a mesh array with a bidirectional single communication channel between neighboring PEs. This target array can directly support a single communication link in either direction between neighboring PEs. However, many algorithms, in the form of the SFG arrays, have more than one communication link for each node. In order to implement such SFGs by the linear or mesh array, the communication channels have to be *time-shared* in order to accommodate all the data links in the SFG array. Therefore, the main problem is to specify a proper time-division scheme for the communication channels. Let us first use several examples to illustrate the time-division scheme.

Example 1: Time Sharing of Channels

Suppose that the target architecture is a simple linear array with one unidirectional channel, and the SFG to be implemented is a linear array with two arcs, as shown in Figure 6.3(a). Note that since there are two pieces of data to be sent through a single channel in every recursion period (D) of the SFG, we can divide the recursion period D into two minor cycles, i.e., $D = 2\tau$ (where each minor cycle is used for one data transfer). Then we can redistribute the delay time 2τ on the lower arc of the SFG, as shown in Figure 6.3(b). Note that the upper arc of the array has zero delay and the lower arc has delay τ (denoted by a small box) at the initiating node of the arc and delay τ at the terminating node of the arc. The time-division

Figure 6.3: (a) An SFG with two arcs. (b) The time-division of the SFG in (a). A box on an arc denotes a delay of τ.

for this example is straightforward. Data on the upper arc is sent at $t = 0$ and data on the lower arc is delayed by τ before being transmitted through the channel. This means that data on the upper arc can be sent in the first minor cycle and data on the lower arc, in the second minor cycle; therefore, there is no time conflict between these two events.

Example 2: From Hexagonal Arrays to Mesh Arrays

The hexagonal SFG array shown in Figure 6.4(a) has neighboring east and south arcs (represented by [1 0] and [0 1]) along with a southeast (diagonal) arc (represented by [1 1]). To map this SFG to a target mesh array, the *southeast* arc has to be decomposed into an *east* arc and a *south* arc, as shown in Figure 6.4(b). Therefore, each south or east arc has to use time division to handle the two data transfers in one recursion time D. This will require that $D = 2\tau$. As shown in Figure 6.4(c), the time-division of the

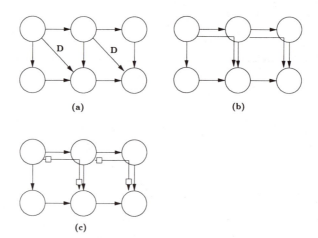

Figure 6.4: (a) An SFG with two neighboring arcs and a diagonal arc. (b) Rerouting the diagonal arc by a east and a south arc. (c) The time-division of the SFG in (a).

SFG array can be derived similarly as in Example 1. (Verify that there is no time conflict in this scheme.)

Example 3: A More General Case Study

A more complicated example of an SFG array is shown in Figure 6.5, where an arc represented by [2 1] is present in addition to the [1 0] and [0 1] arcs. Decomposing the [2 1] arc into *two* east arcs and *one* south arc, we see that the east channel has to handle $2 + 1 = 3$ data transfers and the south channel, $1 + 1 = 2$ data transfers in one recursion time. This will require $D = 3\tau$. In order to redistribute the delay time 3τ on the arcs, we first draw all decomposed arcs in the SFG as shown in Figure 6.5(b). The mesh arcs, such as arcs 6, 7, and 8 have no delays. Arc 1 is an initial arc of the [2 1] link. We can assign one τ at its initiating node. Since arc 4 is also an initial arc for the [2 1] link, which has one τ at its initiating node, a delay of τ is required at the initiating node of arc 2 so that the accumulated delay at the

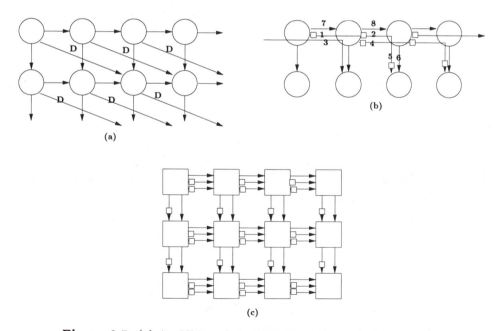

Figure 6.5: (a) An SFG with two neighboring arcs and a [2 1] arc. (b) The time-division of the SFG in (a). (c) The systolic array of (b), where the large box denotes a PE with any computation time, e.g., 3τ.

beginning of arc 2 is 2τ. In this way, there is neither time conflict with arc 8 (which has zero delay), nor with arc 4 (which has τ delay). Another τ delay is put at the end of arc 5 so that the time conflict with arc 6 is averted.

Still More General Cases   Now we can describe the time-division for a general SFG array to match the target mesh array. Assume there are N data arcs, represented by $[x_i \ y_i]$ in the SFG array. For ease of discussion, let us also assume that $x_i \geq 0$ and $y_i \geq 0$ for all i. (The case of reverse directions ($x_i < 0$ or $y_i < 0$) can be similarly treated.) First, we require that any arc in the SFG except the $[1 \ 0]$ and $[0 \ 1]$ arcs has at least one delay (D) in order to specify the time-division. (This condition can always be satisfied by proper cut-set delay transfers on the SFG.) After rerouting these arcs by decomposing them into east ($[1 \ 0]$) or south ($[0 \ 1]$) arcs, a PE has to take care of $X = \sum_i x_i$ data arcs in the east channel and $Y = \sum_i y_i$ data arcs in the south channel in one recursion of the SFG. Therefore, a scaling of $D = Z\tau$ for all delays is required to route all the data, where $Z = \max\{X, Y\}$.

The distribution of the delay time $Z\tau$ on each of the non-mesh arcs can be done using a method similar to that in Example 3. The basic strategy is to avoid time-conflicts and to anticipate a regular distribution of the delays. The details are left as an exercise in Problem 10.

Systolization of Time-divided SFG   Now let us address the issue of systolization of a time-divided SFG. Note first that the speed of the SFG array is now dominated by the communication, which takes $Z\tau$ for each recursion. The computation time for the PEs can actually take any amount of time, say $\tau, 2\tau, \cdots$, at the designer's option. However, since communication time is $Z\tau$, it is reasonable to let the computation time be the same as $Z\tau = D$. In this case, in the cut-set systolization of the time-divided SFG, there should be delay transfer of $Z\tau$ to be absorbed by each node as its computation time.

Example 4: Systolization of a Time-divided SFG

To systolize the time-divided SFG in Example 3 above, we apply cut-set delay transfer of 3τ to all the arcs in the SFG. Now the computation nodes absorb the added delay of 3τ, and the resulting systolic array is shown in Figure 6.5(c), where a large box denotes the PE with computation time of 3τ, and a small box denotes the communication delay τ.

6.3.1.2   Matching Mesh Algorithms to Hypercube Structures

An n-cube is a hypercube consisting of 2^n nodes. The nodes of an n-cube can be labeled by n-bit binary indices, from 0 to $2^n - 1$. By the definition of *hyper-network, two processors are directly linked if and only if their binary index representations differ only by one bit.* For example, if $n = 3$, then the 8 nodes can be represented as the vertices of a three-dimensional cube as shown in Figure 6.6. To see how to embed a mesh in a hypercube, let us consider a two dimensional mesh of size 8 × 4 as shown in Figure 6.7 and examine how it might be embedded into a 5-cube. (Note that $2^5 = 32$.) By using a 3-bit Gray code $(b_1 b_2 b_3)$ for labeling the horizontal dimension and a 2-bit Gray code $(c_1 c_2)$ for the vertical dimension, each node in the mesh is labeled as a 5-bit representation $(b_1 b_2 b_3 c_1 c_2)$ [Chan86]. Due to the Gray coding, the corresponding indices between any two neighboring nodes in the mesh differ by exactly one bit. By the definition of *hyper-network*, their corresponding nodes in the hypercube are thus directly connected. Thus, the embedding of this 8 × 4 mesh into a 5-cube is completed. In a similar

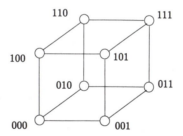

Figure 6.6: 3-dimensional view of the 3-cube.

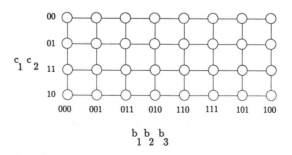

Figure 6.7: 2-dimensional Gray code for an 8 × 4 mesh.

manner, any algorithm executable in a mesh can be easily matched to (and executed by) a hypercube computer.

We have demonstrated that a mesh algorithm (of any dimension) may be embedded in a hypercube [Chan86]. Combining this with the previous observation that any local SFG (or DG) can be matched to a mesh array, it is concluded that all local SFG (or DGs) may be matched to a hypercube array.

6.3.2   Partitioning

The partitioning problem is basically *mapping computations of a larger size problem to an array processor of a smaller size.* It is a basic requirement in many practical system designs, since no matter what special-purpose computing hardware is available, there is a computation too large for it [Helle84]. There have been several works on this problem [Hwang82], [Helle84], [Rao85], [Moldo86], [Kisha86], and [Horik87]. In general, the mapping scheme (including both the node assignment and scheduling) will be much more complicated than the regular projection methods discussed in the previous chapters. To design the partitioning scheme, the following factors should be taken into account [Moldo86]:

1. Minimum overall computation time.

2. Minimum control overheads.

3. Balanced tradeoff between external communication and local memory.

It is necessary to specify *at what time and in which PE the computation of each DG node takes place.* For a systematic mapping from the DG onto a systolic/wavefront array, the DG is regularly partitioned into many *blocks*, each consisting of a cluster of nodes in the DG. For convenience of presentation, we adopt the following mathematical notations. Suppose that an N-dimensional DG is (linearly) projected to an $(N-1)$-dimensional SFG array of size $L_1 \times L_2 \times \cdots \times L_{N-1}$. The SFG is partitioned into $M_1 \times M_2 \times \cdots \times M_{N-1}$ blocks, where each block is of size $Z_1 \times Z_2 \times \cdots \times Z_{N-1}$. Obviously, for $i = 1, 2, \ldots, N-1$, $Z_i = L_i/M_i$. As a simple example, we look at the computation of a matrix-vector multiplication $\mathbf{Ab} = \mathbf{c}$, where $\mathbf{A}$ is 6×6 and $\mathbf{b}$ and $\mathbf{c}$ are 6×1, with a linear array of 3 PEs. The DG of this problem is shown in Figure 6.8(a). Two ways to partition of blocks are also shown in in Figure 6.8(b), $L_1 = 6$, $M_1 = 3$, and $Z_1 = 2$. In Figure 6.8(c), $L_1 = 6$, $M_1 = 2$, and $Z_1 = 3$.

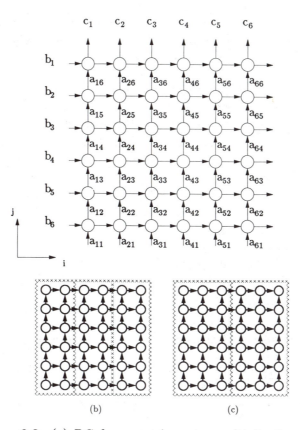

Figure 6.8: (a) DG for a matrix-vector multiplication; (b) DG being partitioned into three blocks; (c) DG being partitioned into two blocks.

There are two methods for mapping the partitioned DG to an array: *locally sequential globally parallel* (LSGP) method and *locally parallel globally sequential* (LPGS) method.

1. *LSGP scheme:* In the LSGP scheme, one block is mapped to one PE. Thus, the number of blocks is equal to the number of PEs in the array, i.e., the array size equals to the product $M_1 \times M_2 \times \cdots \times M_{N-1}$. Each PE sequentially executes the nodes of the corresponding block. In order to store the node data in the block, local memory within each PE is needed, i.e., local memory size should be $c \times Z_1 \times Z_2 \times \cdots \times Z_{N-1}$, where c is a small constant (most often, 1,2, or 3) depending on the data dependencies. As long as the local memory is large enough for the computation under consideration, the LSGP approach is quite appeal-

ing. The LSGP scheme for the matrix-vector multiplication example is shown in Figure 6.8(b). The number of blocks M_1 is equal to the array size 3. The local memory size is $Z_1 = 2$.

2. *LPGS scheme:* In the LPGS scheme, the block size is chosen to match the array size, i.e., one block can be mapped to one array. All nodes within one block are processed concurrently, i.e., *locally parallel.* One block after another block of node data are loaded into the array and processed in a sequential manner, i.e., *globally sequential.* Hence the name LPGS. In this scheme, local memory size in the PE can be kept constant, independent of the size of computation. All intermediate data can be stored in certain buffers outside the processor array. Usually, simple FIFO buffers are adequate for storing and recirculating the intermediate data efficiently. In the above matrix-vector multiplication example, for the LPGS scheme, the DG is partitioned into two blocks as shown in Figure 6.8(c), where the block size matches the array size, i.e., $Z_1 = 3$.

6.3.2.1   Schedule Strategies for LSGP Scheme

The preceding discussion specifies only the node assignment scheme, i.e., the spatial mapping of a DG to the processor array. The scheduling part, i.e., at what time the computation takes place, is yet to be specified.

Linear Schedule Approach   Given an N-dimensional DG and a projection direction $\vec{d}$, according to the algebraic mapping method discussed in Chapter 3, there is a corresponding matrix **P**. Let $\mathbf{P}^T = [\vec{P}_1 \vec{P}_2 \cdots \vec{P}_{N-1}]$ be a set of independent vectors, each orthogonal to $\vec{d}$. This set of vectors specifies the $(N-1)$-dimensional SFG space. From the *processor sharing* perspective (c.f., Section 4.4), there are $Z_1 \times Z_2 \times \cdots \times Z_{N-1}$ nodes in each block in the SFG, which share one PE. An *acceptable (i.e., sufficiently slow)* linear schedule is chosen so that at any time instant, there is at most one active PE in each block. This problem can be formally stated as follows:

Find a scheduling vector $\vec{s}$ which minimizes

$$\alpha \left(= \vec{s}^T \vec{d} \right)$$

and satisfies the following constraints:

1. $\vec{s}^T \vec{e} > 0$,   for all dependency arcs $\vec{e}$.

2. For all integer vector $\vec{m}$, which satisfies $\vec{s}^T \vec{m} = 0$ (i.e., $\vec{m}$ is on an equitemporal hyperplane), there exists i, such that $|\vec{P}_i^T \vec{m}| \geq Z_i$.

Note that in the LSGP scheme, *the minimum α is very often determined by the number of processor sharing nodes in the SFG*, i.e.,

$$\alpha = Z_1 \times Z_2 \times \cdots \times Z_{N-1}.$$

Since the solution $\vec{s}$ should have its elements reasonably small, it is usually possible to obtain a solution by *enumeration* on $\vec{s}$ and $\vec{m}$. That is, different $\vec{s}$ can be checked to see if the constraints are satisfied. Let us now look at two examples:

Example 5: LSGP Schedule for 1-D Array

In the previous matrix-vector multiplication problem,

$$\vec{d} = \begin{bmatrix} 0 \\ 1 \end{bmatrix}, \quad \vec{e}_1 = \begin{bmatrix} 1 \\ 0 \end{bmatrix}, \quad \vec{e}_2 = \begin{bmatrix} 0 \\ 1 \end{bmatrix},$$

$$\text{and} \quad \vec{P}_1 = \begin{bmatrix} 1 \\ 0 \end{bmatrix}.$$

The lower bound of α is obviously $Z_1 = 2$. Thus if $[1\ 2]$ is used for $\vec{s}^T$, we may find that it satisfies all the constraints and the resulting α is 2 which is minimal. The result for LSGP partitioning using the linear scheduling is shown in Figure 6.9.

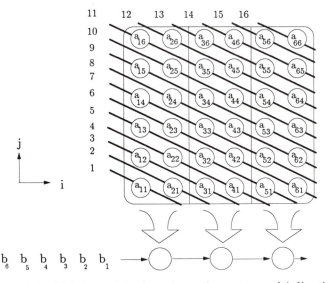

Figure 6.9: LSGP partitioning of matrix-vector multiplication.

Example 6: LSGP Schedule for 2-D Array

The 3-D DG for a 9 × 9 matrix multiplication consists of three dependency arcs,

$$\vec{e}_1 = \begin{bmatrix} 1 \\ 0 \\ 0 \end{bmatrix}, \quad \vec{e}_2 = \begin{bmatrix} 0 \\ 1 \\ 0 \end{bmatrix}, \quad \vec{e}_3 = \begin{bmatrix} 0 \\ 0 \\ 1 \end{bmatrix},$$

If the projection direction is chosen to be $[0\ 0\ 1]^T$, then the resulting
array size is 9 × 9 and the processor space can be spanned by two vectors

$$\vec{P}_1 = \begin{bmatrix} 1 \\ 0 \\ 0 \end{bmatrix}, \text{ and } \quad \vec{P}_2 = \begin{bmatrix} 0 \\ 1 \\ 0 \end{bmatrix}.$$

If a 3 × 3 array is to be used, then $Z_1 = Z_2 = 3$. Apparently, the lower
bound of α is $Z_1 \times Z_2 = 9$. Thus different scheduling vectors $\vec{s}$ with the
form $[s_1\ s_2\ 9]^T$ ($s_1 > 0$ and $s_2 > 0$) can be enumerated and tested for the
second constraint. It can be shown that $\vec{s} = [1\ 3\ 9]^T$ is a good choice. This
LSGP scheduling and the input data scheme for the 3 × 3 array is shown in
Figure 6.10.

6.3.2.2   Schedule Considerations for LPGS Scheme

As to the scheduling scheme for the LPGS method, a general rule is to select
a (global) scheduling vector $\vec{s}$ which does not violate the data dependencies.
As a simple example, a LPGS design for matrix-vector multiplication is
shown in Figure 6.11, where a FIFO buffer is used to store intermediate
data.

Note that the above LPGS design example has the advantage that
blocks can be executed one after another in a *natural* order. However, this
simple ordering is valid only when there is no *reverse data dependence* for
the chosen blocks. In general, such a property may not exist. For example,
the blocks shown in Figure 6.12 exhibit a reverse data dependence. Thus the
choice of blocks should take the data dependencies into account. A procedure
to decide the blocks can be found in [Moldo86], in which an N-dimensional

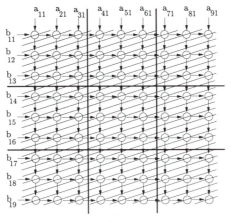

(a)

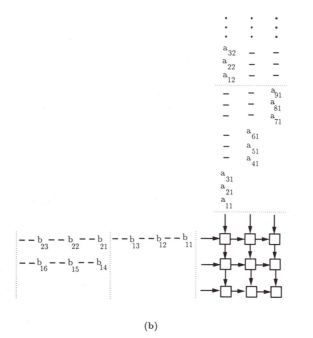

(b)

Figure 6.10: (a) LSGP scheduling on the 3-D DG for a 9 × 9 matrix multiplication (only one DG layer is shown), (b) data input scheme for the 3 × 3 array.

379

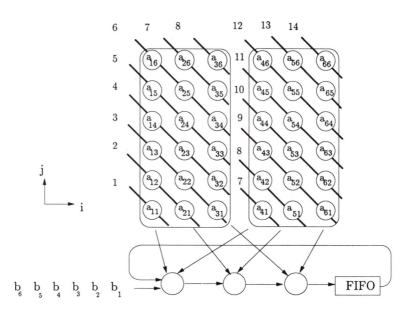

Figure 6.11: LPGS partitioning of matrix-vector multiplication.

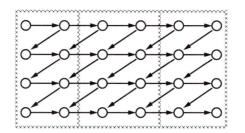

Figure 6.12: A partition of DG with reverse data dependence.

DG is partitioned into blocks by $N - 1$ sets of parallel hyperplanes. Each set of hyperplanes can be represented by a vector normal to the hyperplanes, say $\vec{P_i}$ ($i = 1, 2, \ldots, N - 1$). Thus the node assignment problem is to find $N - 1$ linearly independent vectors, $\{\vec{P_i}\}$, such that there is no reverse data dependence. Or equivalently, for $i = 1, 2, \ldots, N - 1$

$$\vec{P_i}^{\mathrm{T}} \cdot \vec{e} > 0, \quad \text{for all dependence arcs } \vec{e}$$

Since these $N - 1$ vectors span the processor space, they are orthogonal to the projection direction $\vec{d}$. The projection direction can thus be determined from the $N-1$ processor space vectors. Furthermore, $\{\vec{s}, \vec{P_i}, i = 1, 2, \ldots, N-1\}$ should be linearly independent. (Otherwise, some nodes on an equi-temporal hyperplane will be unacceptably assigned to the same PE).

In summary, the LPGS scheme is to find N linearly independent vectors, $\{\vec{s}, \vec{P_i}, i = 1, 2, \ldots, N-1\}$, which optimize a desired object function, e.g., α, and at the same time satisfy the following conditions.

1. No reverse data dependency,

$$\vec{P_i}^{\mathrm{T}} \cdot \vec{e} > 0.$$

2. The causality constraint

$$\vec{s}^{\mathrm{T}} \vec{e} > 0.$$

When using the LPGS scheme, some FIFOs are required to feed the *output* of the array back to the array. By noting that some *output* of the array may have the same value as the input (e.g., when the variable is transmittent), some FIFOs are actually not necessary. Thus, *the property of transmittent variables should be exploited to reduce the number of FIFOs in LPGS schemes*. For example, a 3 × 3 array for 9 × 9 matrix multiplication without using FIFOs is shown in Figure 6.13.

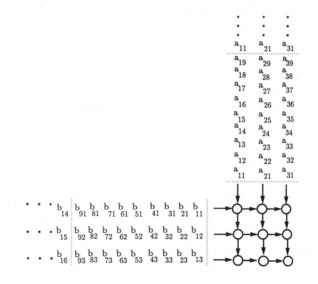

Figure 6.13: A 3 × 3 array for 9 × 9 matrix multiplication *without* using FIFOs.

6.3.2.3   Other Partitioning Approaches

In the above, the partitioning is viewed as a mapping of a DG to a processor array. In many circumstances, the SFG (or systolic) array is derived first, and the partitioning has to be applied directly to that array. Fortunately, the LSGP and LPGS schemes are still applicable at this stage. Again, a proper scheduling strategy can be determined. The detailed derivation is left to the reader.

There are some other partitioning methods for different problems. Here, we mention only one example based on *algebraic partitioning*. For example, a 2-D convolution can be expressed by sums of products of several pairs of 1-D convolutions. The basic idea is first to decompose the 2-D window (matrix) into a sum of unit-rank matrices by the SVD algorithm. A unit-rank 2-D convolution window can then be decomposed into two consecutive 1-D convolutions. Therefore, the original 2-D convolution can be replaced by a sequence of 1-D convolutions, and fewer PEs are required. If the 2-D window has a small rank, then fewer computations are needed. The details of this approach are further discussed in Chapter 8.

6.4   Fault-Tolerance on VLSI Array Processors

Actual systolic/wavefront array processing hardware is very space consuming and complex. Failure of PEs in such a large-scale processor array is basically inevitable. Therefore, design of array processors must take *fault-tolerance* (FT) into account, especially for real-time applications.

6.4.1   Fabrication-Time, Compile-Time, and Run-Time FT

In the literature, various kinds of fault-tolerance are being considered. Firstly, in the production stage of a VLSI array, some form of fault tolerance must be included to allow cost-effective yield of parts in the presence of fabrication defects. This is called the *fabrication-time FT* problem. Secondly, if a fault occurs after the array is installed in a system, then given adequate time, the array may be recompiled (reconfigured) to function properly. This is called *compile-time FT* problem. Thirdly, some form of FT is necessary to overcome faults that occur during the processor's mission and to allow the system to continue to function. This is called the *run-time FT* problem. Several practical approaches to FT include: hardware redundancy, temporal redundancy, and check-sum-based algorithmic designs [HTKun83], [KHHua84].

A three-level model (see Figure 6.14) can be used to describe the three FT problems. The first level is the *real array*. The real array is the one

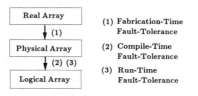

Figure 6.14: The three-level model for general array FT schemes.

obtained on the fabricated chip (or wafer) which may have some manufacturing defects. The second level is the *physical array*, which is assumed to be free from manufacturing defects and contain some spare PEs. The third level is the *logical array* which represents the desired array structure directly corresponding to an algorithm.

With reference to Figure 6.14, the fabrication-time FT is basically a technique to construct the physical array from the real array; while the compile-time and run-time FT are techniques to construct the logical array from the physical array. The run-time FT scheme proposed in this section is shown in Figure 6.15 in terms of the three-level model. Note that this scheme, built on the physical array, can be combined with most fabrication time FT schemes. The details are explained in a moment.

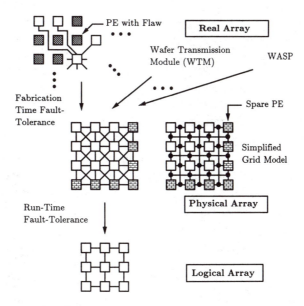

Figure 6.15: The logical array, physical array, and real array in our scheme.

6.4.1.1   Fabrication-Time Fault-Tolerance

Fabrication-time FT is to recover from circuit faults in the *after-fabrication* stages. The goal is to enhance the yield to a satisfactory rate. The basic technique is to reconstruct a working physical array of a certain size from a real array with redundant components. If the number of working PEs is less than the array size, the wafer (or chip) should be rejected. If there exists a sufficient number of healthy PEs, then they may be reconfigured to form a desired structure.

Wafer Scale Integration (WSI)   With today's technologies, it is realistic to develop fault-tolerant systolic/wavefront array processors using WSI. Regularity and modularity characteristics for systolic/wavefront architectures should be exploited in devising suitable fault-tolerant techniques. The strength and weakness of systolic/wavefront architectures and WSI technology with FT design are summarized in Table 6.1.

Restructuring Methods   The major drawback of WSI is the very low yield; thus redundancy and reconfiguration will be necessary to enhance the yield. The following methods have been proposed to restructure the array processors in WSI [Moore86]:

	Strength	Weakness
Systolic/ Wavefront Architectures	• Homogeneous and modular structure • Inclusion of spare PEs is economically feasible	• Utilization of spare PEs will be limited due to local interconnection
Wafer-Scale Integration Fabrication	• Higher speed • Less noise interference • Mixing switches and wires for dynamic interconnection • Shorter inter-PE connection • Lower power required • Smaller driving pad areas	• Difficulty in power dissipation • Low yield and difficulty in testing (for monolithic approach)

Table 6.1: The compatibility between Systolic/Wavefront Arrays and WSI Technology.

- *Metallization techniques* To probe test the wafer before the circuits are completely connected and then to tailor a metallization pattern to connect up the working cells.

- *Laser surgery* To use a laser to blow fusible links or to make connections.

- *Electrical surgery* To make or break connections by passing high currents through the circuit.

- *Reprogrammable techniques* To design a circuit that can select those parts of the circuit that are to be used.

- *Switch lattice techniques* To select the suitable connection topology from the switching functions of embedded switches.

- *Hybrid packaging techniques* To make individual connections during chip packaging.

The use of programmable switches is for flexible data routing. The switches allow the good processors to be connected and the faulty ones to be avoided. The flexibility inherent in the switches not only allows faults occurring during the machine lifetime to be tolerated but also provides a wide variety of interconnection patterns for multiprocessor reconfiguration. An example of extensive use of switch lattice is the configurable highly parallel computer (CHiP) [Hedlu82].

6.4.1.2   Compile-Time Fault-Tolerance

The compile-time FT reconfiguration scheme is similar to software supported fabrication-time FT, except with different testing and technology methods. A run-time FT may be applied to the (less demanding) compile-time FT, but not vice-versa. In addition, most real-time applications demands real-time FT. Therefore, a major emphasis of our discussion is placed on run-time FT.

6.4.1.3   Run-Time Fault-Tolerance

The main objective here is to guarantee that a large scale array processor will continue to function when it is in the operation (run-time) mode. Basically, the FT design in this stage includes fault-testing, reconfiguration, and restart. Our objectives in the run-time FT design are (1) to exploit the regularity and locality of the arrays and the data-driven feature of WAPs, (2) to explore the structural properties of array algorithms, and (3) to minimize the potential time overhead due to FT consideration.

Architectural versus Algorithmic Approaches　There are two categories of run-time FT: architectural approaches and algorithmic approaches. Architectural approaches are algorithm-independent. Here we emphasize fixed array size cases. For example with PEs and switches, an array may be reconfigured so that the same array size may be supported even with some faulty PEs. Another architectural approach is fault-masking. Fault-masking stands for immediate recovery without noticing the occurrence of faults. It is an effective technique for run-time FT and can be realized by triple module redundancy (TMR) or, in general, N-module redundancy (NMR). (In Problem 15, a TMR fault-masking scheme is discussed.)

Algorithmic FT approaches, on the other hand, exploit certain algebraic properties inherent in many run-time DSP applications. For example, a popular approach is to employ a data encoding scheme, such as a checksum, to recover the correct data of a faulty processor. It is possible to simultaneously reconfigure the array structure and the algorithm such that the problem can be executed on a reduced size array. This design may lead to a gracefully degradable design [Fort85b].

Transient Faults　An important issue in run-time FT is the coverage of transient faults. It is reported that the occurrence of transient faults, mainly due to temporary environmental changes, is ten times more frequent than that of permanent faults [Siewi82]. Since the length of transient duration is random, some transient faults may be treated as permanent with a conventional retry process. However, the reconfiguration scheme proposed in this section is especially suitable for the handling of the transient faults. In general, recovery from transient faults improves significantly the overall system reliability.

When a PE has a transient fault, it becomes a connecting element and enters a dormant state. In the dormant state, the PE tests itself to check its status repeatedly. When the transient fault is removed, the dormant PE reactivates and generates an interrupt immediately. The reactivation scheme recovers the failed PE and turns it into an active PE.

6.4.2   Architectural Approach to Run-Time FT

Typically, a FT scheme consists of fault detection, reconfiguration and recovery. The fault detection (good/fault test) is concurrently performed in every PE by means of some self-testing circuits, e.g., via duplication of arithmetic and logic units with matching circuits. Reconfiguration allows the failed elements to be replaced by fault-free standby ones and the circuits to be routed so that a new array may be formed. There is no rollback problem (i.e., retracing previously executed steps), since concurrent error detection is adopted. For systolic/wavefront arrays, it is preferred to have FT distributively processed in individual PEs. Several different reconfiguration approaches can be found in [Koren81], [Rosbe83], [Sami83], and [Hedlu82].

Wavefront versus Systolic Arrays   Some advantages of using a wavefront array instead of a systolic array for run-time FT are (1) Transient faults can be recovered since the reassignment of computations is negotiated between two neighboring PEs; and (2) The potential difficulty of having to accommodate the adaptively changed schedule can be easily solved in a wavefront array, since it requires only correct *sequencing* instead of exact *timing*. For a more detailed discussion on systolic versus wavefront design, please see Problem 22.

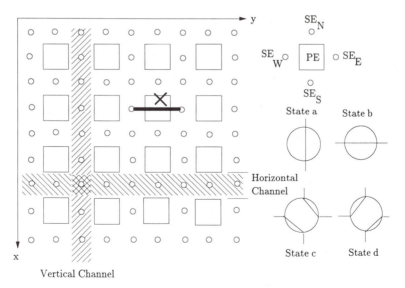

Figure 6.16: VLSI array processor grid model and the functions of switching elements.

Generally speaking, global control may be more suitable for a systolic design, while the local control is more suitable for a wavefront design. This is illustrated in the following example.

Example 1: Fault-Tolerance for 1-D Arrays

To illustrate the reconfiguration process, let us look into 1-D array examples. A linear array consisting of 6 PEs with the rightmost PE as a spare one is shown in Figure 6.17. Two possible reconfiguration control strategies may be implemented on this array, one is with *global control* and the other is with *local control*. When a fault occurs, say at PE_2, the global control instantly notifies the other PEs, and temporarily halts the executions.

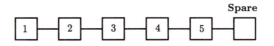

Figure 6.17: A linear array with one spare PE.

The array is then reconfigured so that PE_{j+1} is assigned the job originally assigned to PE_j for $j = 2, \ldots, 5$. Note that PE_2 will be simply treated as a connecting element either with or without delay.[1] A space-time activity diagram for a synchronous and reconfigurable linear array with *global control* is shown in Figure 6.18(a). The space-time diagrams may be different if other scheduling schemes are used. In contrast, for the design with local control, the information about the failure of PE_2 is transmitted one PE by one PE and the job reassignments occur at different clock periods for different PEs. This is illustrated in Figure 6.18(b) which shows a space-time diagram for the design with *local control*.

Switch Model   To discuss FT for 2-D arrays, a switch grid model is useful. Embedding switch lattices into the array structure was proposed by, and applied successfully to CHiP [Hedlu82]. The flexibility inherent in the switches not only allows faults to be more tolerable but also supports a wide variety of interconnection patterns for multiprocessor reconfiguration. The

---

[1]Without a delay in PE_1, the introduction of the longer data communication link between PE_1 and PE_3 will result in slow-down of the system clock rate. Whereas with a suitable delay inserted in PE_2, the system clock rate can be kept the same as that in a system with only nearest neighbor communication [HTKun84].

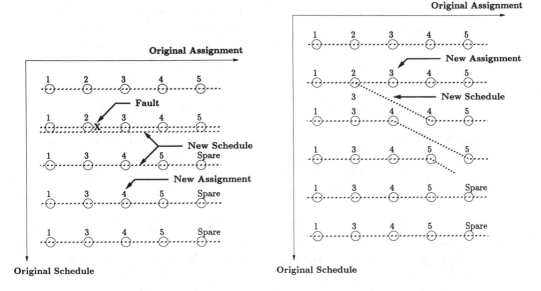

Figure 6.18: Space-time diagrams for a synchronous and reconfigurable linear array (a) with global control; (b) with local control.

basic grid model consists of processing elements and switches as shown in Figure 6.16, with the following assumptions:

1. The switching elements and the interconnection wirings are fault-free. This assumption is adopted by many researchers [Fort85b], [Sami86] and is justifiable by the fact that these elements have much lower hardware complexity as compared to the PEs.

2. Each PE can be decomposed into a computation part, a communication part, and a reconfiguration part. The computation part will be responsible for all computation functions. The communication part will be responsible for the communications between PEs. And the reconfiguration part is used for the fault detection, the reconfiguration, and the switch control. Both the communication and reconfiguration parts are assumed to be fault-free. Therefore, the faulty PEs can provide the connection functions, i.e., PEs can be converted into connecting elements [Sami86].

Single-track Switch   A track means one communication path in each direction of switching. Reconfiguration capability varies with the number of tracks provided by the switch situated between two adjacent PEs. The switching functions provided by a single-track switch are shown in Figure 6.16. The simplicity of the single-track switch not only saves area but also makes the assumption of fault-free switching more realistic. Since the PE failure rate is assumed to be small, a sufficient coverage can be provided by a single-track switching model.

6.4.2.1   Reconfigurability and Compensation Path Theorems

Reconfiguration in array processors comprises two tasks (1) *placement* and (2) *routing*.

1. *Placement*: A *placement* is denoted by a mapping $f(\cdot)$, which maps all the PEs on the logical array L (size $M \times M$) to the good PEs on the physical array P (size $N \times N$), (see Figure 6.19). Mathematically, for every $PE(i,j) \in L$, there exists a good PE on the physical plane location $(x(i,j),\, y(i,j)) \in P$, so that the logic $PE(i,\, j)$ is mapped to the physical $PE(x(i,j),\, y(i,j))$.

2. *Routing*: A *routing* is determined by all the *horizontal links* and *vertical links* on the physical array. The path linking $PE(x(i,j),\, y(i,j))$ and $PE(x(i,j+1),\, y(i,j+1))$ is called a *horizontal link*. The path linking $PE(x(i,j),\, y(i,j))$ and $PE(x(i+1,j),\, y(i+1,j))$ is called a *vertical link*.

Conversely, the pattern of horizontal and vertical links is a routing scheme which dictates the number of tracks required for the reconfiguration. An example of placement and routing, for the case $N = 5$, is depicted in Figure 6.20, which shows that the reconfiguration scheme requires two tracks.

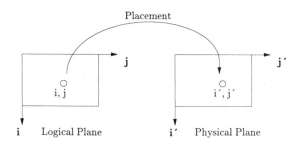

Figure 6.19: Illustration of the relationship between logical array and physical array.

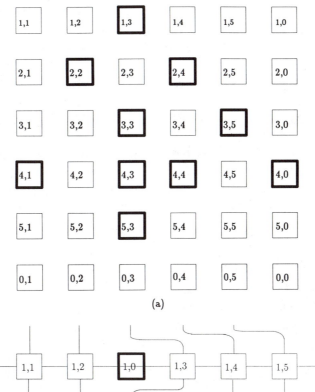

(a)

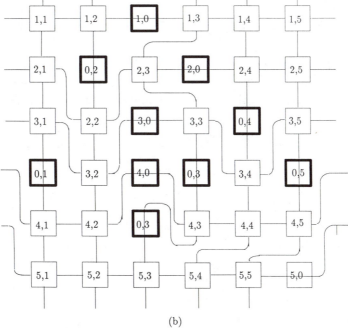

(b)

Figure 6.20: Example of direct reconfiguration, (a) fault distribution, (b) final placement (renaming) and reconfiguration with 2-track routing (adapted from [Sami83]).

Our run-time FT discussion concentrates on the single-track switch model. Due to the limited routing capability of the single-track switch, the placement has to satisfy certain conditions in order to permit a successful reconfiguration. These conditions are stated in the following two theorems: the *reconfigurability theorem* and the *compensation path theorem*. The reconfigurability theorem gives the necessary and sufficient conditions of a placement which bears a successful reconfiguration using single track.

Reconfigurability Theorem: *Given an N × N physical array and the logical array of size $(N-1) \times (N-1)$, a successful reconfiguration may be achieved by using single track if and only if the following placement conditions are all satisfied:*

1. $x(i, j) < x(i+1, j)$.

2. $y(i, j) < y(i, j+1)$.

3. *If* $x(i_1, j_1) = x(i_2, j_2)$ *and* $|j_2 - j_1| > 1$,
    then $|y(i_2, j_2) - y(i_1, j_1)| > 1$.

4. *If* $y(i_1, j_1) = y(i_2, j_2)$ *and* $|i_2 - i_1| > 1$,
    then $|x(i_2, j_2) - x(i_1, j_1)| > 1$.

for all $(i, j), (i, j+1), (i+1, j), (i_1, j_1), (i_2, j_2) \in L$.

Proof: The relationship between logical arrays and physical arrays for Conditions 1 and 3 are illustrated in Figures 6.21. Conditions 2 and 4 are just the symmetrical versions of Conditions 1 and 3.

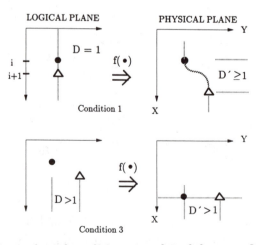

Figure 6.21: Illustration of conditions 1 and 3 of the reconfigurability theorem.

	$x(i,\ j+1)$ $= x(i,\ j)$	$x(i,\ j+1)$ $= x(i,\ j)-1$	$x(i,\ j+1)$ $= x(i,\ j)+1$
$y(i,\ j+1) = y(i,\ j)+1$	A	B	C
$y(i,\ j+1) = y(i,\ j)+2$	D	E,F	G,H

Table 6.2: All possible situations that the four conditions in the reconfigurability theorem are satisfied; $A,\ B,\cdots,\ H$ refer to Figure 6.22.

The necessity part of the proof can be found in [Kung86d]. The sufficiency part of the proof is largely based on the following constructive procedures for routing.

1. *Construction of horizontal links* Table 6.2 lists all the possible situations that the four conditions in the theorem are satisfied (see the proof in [Kung86d] and Problem 19). All interconnection patterns of horizontal links are shown in Figure 6.22.

2. *Construction of vertical links* Similarly, the vertical links may be constructed in a symmetrical manner as in the horizontal case.

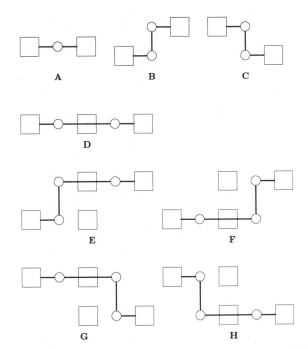

Figure 6.22: All interconnection patterns of horizontal links.

3. From inspection of Figure 6.22 and noting the order-preserving property of the row paths, we conclude that *no two horizontal links can pass through the same vertical channel segment.* Similarly, no two vertical links can overlap. This establishes sufficient conditions for single track reconfiguration.

Compensation Path: Under the local communication link constraint, it is simpler to replace a faulty PE by a neighboring healthy PE. The neighboring PE will in turn be replaced by the next neighbor PE, and so on. Finally, the spare PE will be used. In other words, the replacement process will involve a sequence of "local replacements". This defines the *compensation path.* An example is shown in Figure 6.23. The following theorem is useful for selecting the compensation paths (which determine the corresponding placement), so that single-track routing is feasible.

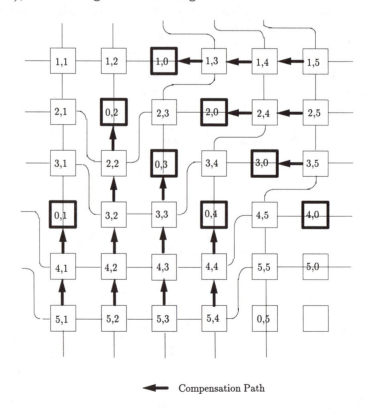

Figure 6.23: Illustration of nonintersecting compensation paths and the corresponding 1-track reconfiguration routing scheme.

Compensation Path Theorem: *Given an $N \times N$ physical array with an $(N-1) \times (N-1)$ logical array embedded, if the placement is based on compensation paths, which are straight segments with no intersection, then a single-track routing scheme is feasible.*

Proof of the compensation path theorem is left to the reader as an exercise (see Problem 20). Since the switches on the "corners" of PEs are not necessary in the reconfiguration scheme, a simplified grid model (shown in Figure 6.24) may be adopted. In this model, there are four switches, labeled as SE_N, SE_S, SE_E and SE_W, surrounding each PE. Note that both the *reconfigurability theorem* and the *compensation path theorem* are still valid for the simplified grid model. The following reconfiguration algorithm will be based on such a simplified model.

6.4.2.2   Reconfiguration Algorithm (With Global Control)

The compensation path theorem provides a basis for the following reconfiguration algorithm, namely, a *COPE* (*CO*mpensation *P*ath *E*ncode) scheme. The COPE scheme consists of a COPE placement algorithm and a COPE routing algorithm. They are discussed below.

COPE Placement Algorithm The COPE placement scheme consists of both the deactivating scheme and reactivating scheme. The *deactivating*

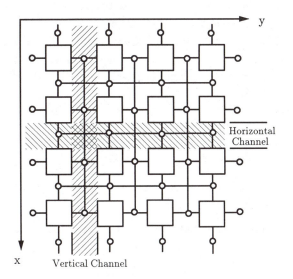

Figure 6.24: Simplified VLSI array processor grid model.

program will be initiated when a PE is found faulty, while the *reactivating* program will be executed when the failed PE is recovered.

1. *Deactivation Scheme:*

   - The faulty PE will perform the following:
     (1) Issue an interrupt request and broadcast its *physical location* and the *direction* of the chosen compensation path.
     (2) Pass the unprocessed input data to the assigned neighboring PE.
     (3) Convert itself into a connecting element and enter into a *dormant* state.

   - Other PEs will perform the following:
     (1) Receive the interrupt signal and the information broadcast by the faulty PE, update the present *placement state* (cf., Figure 6.26(a)), and execute the routing algorithm based on the *routing state* (explained below).
     (2) Every PE on the *new* compensation path will discard its outputs and transfer its unprocessed inputs to the next PE on this path.

2. *Reactivation Scheme:* In the *dormant* state, the faulty PE checks its own status periodically. Once the dormant PE is reactivated, it will reverse the above deactivation process and eventually cancel the compensation path originally due to the deactivation of the said PE. The elimination of a compensation path will put a PE back to the spare row or spare column. For transient faults reactivation, the transition diagram of PE placement states is provided (cf. Figure 6.26(b)).

Updating Placement State: Our placement strategy is based on the compensation path theorem. To guarantee the requirement that the compensation paths do not intersect, every PE has its *placement state*. There are totally five placement states:

1. **HV**: Both horizontal and vertical compensation paths are allowed to pass through this PE.[2]

2. **H$\overline{\text{V}}$**: Only horizontal compensation path is allowed to pass through this PE.

---

[2]When a PE in this state fails, the shorter compensation path will be chosen to decrease the possibility of intersection with the other future compensation paths.

3. $\overline{H}V$: Only vertical compensation path is allowed to pass through this PE.

4. $\overline{HV}$: No more compensation path is allowed to pass through this PE.

5. **FAIL**: (The system is declared to be failed if any) PE enters this state.

Originally, all PEs have the same initial placement state **HV** and all the logical indices of PEs are the same as their physical indices. Once a fault occurs, the reconfiguration part of the faulty PE initiates an interrupt and determines a new compensation path. It also broadcasts to all the PEs its location and the direction of the chosen compensation path, which is the (necessary and sufficient) information for the other PEs to update their *placement states*. Of course, the updating scheme also depends on the PE's location as described in Figure 6.25. The placement state *transition diagram* is shown in Figure 6.26(a).

Routing Algorithm   The routing is determined by the switching states in the *switching elements* (SE). There are two kinds of switches: The first kind, denoted as SW1, located between two vertical neighboring PEs (i.e., SE_N

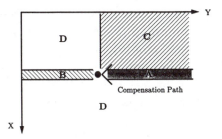

Figure 6.25: PE updating conditions in terms of its relative position to a new horizontal compensation path.

1. Condition A: when the PE is on the compensation path.
2. Condition B: when the PE is located from $(x(i,j),1)$ to $(x(i,j),y(i,j)-1)$.
3. Condition C: when the PE is in the rectangular region of $(1,y(i,j))$, $(1,N-1)$, $(x(i,j)-1,y(i,j))$, $(x(i,j)-1,N-1)$.
4. Condition D: all the other locations.

Similar rules can be applied to vertical compensation paths.

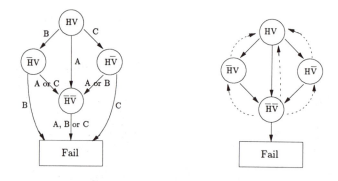

Figure 6.26: (a) Transition diagram of single PE states. (b) Modified transition diagram of PEs states with reactivation capability.

and SE_S). The other kind, denoted as SW2, located between two horizontal neighboring PEs (i.e., SE_E and SE_W). For SW1 type switches, there are only three types of switching patterns, i.e., states a, c, and d. For SW2 type switches, the three switching patterns are b, c, and d (cf., Figure 6.16).

In the following discussion, we shall first concentrate on the switch control design for SW1 type switches which should be adjusted according to any new creation of horizontal compensation path. It will be shown shortly that such a switch control is determined by the so-called *routing states* of its two vertical neighboring PEs.[3] This state, represented by two bits as defined in Table 6.3, will be called the *horizontal routing state* (HRS). Note that, because of the imposed non-intersecting property of compensation paths, the HRS_N and HRS_S (cf., Figure 6.27(a)) provide sufficient information about the relative shift of logical indices to determine the state of the switch. Therefore, the state of the switch is a simple function of HRS_N and HRS_S

00	Good PE, not on a horizontal compensation path
01	Good PE, on a horizontal compensation path
10	Faulty PE, on a horizontal compensation path
11	Faulty PE, on a vertical compensation path

Table 6.3: The definition of the horizontal routing states (HRS). (A similar definition holds for the vertical routing states (VRS).)

---

[3]Since only the PEs on a new compensation path will change their logical-physical relations, and therefore only those switches situated on two sides of the new compensation path need to change states.

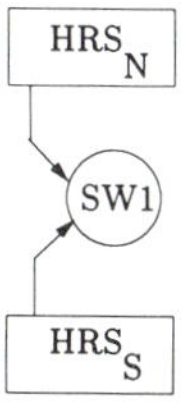

HRS_N	HRS_S	SW1	HRS_N	HRS_S	SW1
00	00	a	10	00	d
00	01	c	10	01	c
00	10	c	10	10	x
00	11	a	10	11	d
01	00	d	11	00	a
01	01	a	11	01	c
01	10	d	11	10	c
01	11	d	11	11	x

Figure 6.27: (a) The switch element state SW1 is a function of HRS_N and HRS_S. Here SW1 = SE_N of the south PE = SE_S of the north PE. Note that, similarly, the switch element state SW2 is a function of VRS_E and VRS_W; (b) The corresponding truth table, where "x" stands for "don't care".

as shown in Figure 6.27(b). Similarly, SW2 type switches may be controlled by the vertical routing states (VRS) of its two horizontal neighboring PEs (see Problem 21). Note that a similar switching scheme may be applied to the reactivation of PEs.

6.4.2.3   Reconfiguration Algorithm (With Local Control)

The reconfiguration algorithm mentioned above requires global communication links for instantly broadcasting the interrupt signal and the location of the faulty PE. This is sometimes difficult to implement. Therefore, a design based on only local control is proposed. In this new design, only local communication is required to transmit all the data and the control signals. The design is based on a typical asynchronous communication protocol (see Section 5.6.2).[4]

The new scheme is presented as follows. We assume that, without loss of generality, the spare PEs are on the rightmost column or the bottom row. When a fault occurs, choose a horizontal compensation path or a vertical compensation path based on the current placement state of the faulty PE. For convenience of discussion, let us denote the PEs on the (horizontal) compensation path as PE_i, PE_{i+1}, PE_{i+2}, PE_{i+3}, and so on (see Figure 6.28), where PE_i is the faulty PE. Two situations may arise:

---

[4]The same local control scheme can be applied to a synchronous system with additional adjustment on the clock cycles (see Problem 22).

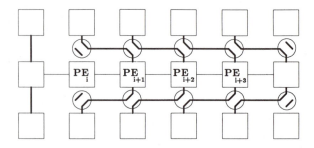

Figure 6.28 Single faulty PE with horizontal compensation path.

1. For $j = i, i+1, i+2,\ldots$, if no other compensation path is encountered, PE_j initiates an interrupt to PE_{j+1} and reassigns the job originally belonging to PE_j to PE_{j+1}, and changes its VRS and HRS states (cf., Table 6.3) and thus changes the state of its neighboring switches (cf., Figure 6.27).

2. If another compensation path also passes through PE_j ($j = i+1, i+2, \ldots$), i.e., there are two overlapping compensation paths, one of these two compensation paths must be "cancelled" (and "replaced" by another kind of compensation path whenever possible). This process is termed *path correction* and depends on the physical location of PE_j. There are two possibilities: (1) if PE_j is on the upper triangular part of the logical array, the path coming from the left should be cancelled and replaced by a vertical compensation path; (2) if PE_j is on the lower triangular part of the logical array, the path coming from the upper part should be cancelled and replaced by a horizontal compensation path. Due to the absence of global links, the *cancellation* and *replacement* must be performed locally. Thus if the horizontal compensation path is to be replaced by a vertical one, the cancellation is initiated at PE_j, then PE_{j-1}, PE_{j-2}, and so on. After PE_{i+1} is reached, PE_{i+1} will initialize a vertical compensation path starting from PE_i and change the states (HRS and VRS) of PE_i.

The flowchart of the reconfiguration algorithm with local control is shown in Figure 6.29. The flowchart guarantees the following two properties: (1) the routing is correct; (2) data will be communicated only when the correct routing is established. Based on these observations, we claim that the PEs which are not on any compensation path will execute the correct

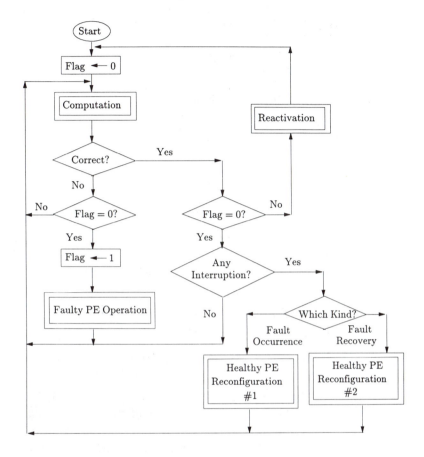

Figure 6.29 The flowchart of the reconfiguration algorithm with local control.

data when they are available. By a similar argument, we claim that those PEs on a compensation path will also yield correct computation results. They involve, however, an additional task of transfering of jobs between PEs.

Note that the reactivation process can also be included in the new scheme (cf., the flowchart in Figure 6.29). Briefly speaking, once a faulty PE is recovered, it initiates an interrupt to its neighboring PE, takes back the job originally assigned to it, changes the states (including placement state, HRS, and VRS), and then continues its unfinished work.

Compared with the global control based design, the local control based design has the following advantages and disadvantages:

- *Advantages:*

    1. The removal of the global link requirement will significantly re-
       duce the implementation cost.

    2. The reconfiguration algorithm with global control has difficulties
       in handling multiple faults which occur at the same time. The
       algorithm with local control can, however, easily handle this prob-
       lem.

- *Disadvantages:*

    1. The control process is more complicated.

    2. Each PE must be able to store two sets of data/processes (depend-
       ing on the homogeneity of the array) at the same time. Thus a
       larger local memory is required.

6.4.2.4   Combination of Single-track Switch and Partitioning Scheme

For very large arrays, system reliability drops rapidly due to the limited reconfiguring capability of single-track switching. However, the reliability can
be drastically improved by partitioning the large array into several smaller
arrays (e.g., 4 × 4) and distributively incorporates single-track switches and
spare PEs in partitioned subarrays at the expense of using more spare PEs.
The reliability improvements are demonstrated in Figure 6.30.

6.4.3   Algorithmic Approach: Weighted Checksum Coding

Recently, an algorithm-based FT has been proposed [KHHua84], based on
a data-encoding approach. The input data to the algorithm are encoded
at the system level in the form of some error-correcting or error-detecting
code. The computation tasks within an algorithm must be appropriately
distributed among every PE, so that any malfunction in the PE will affect
only a portion of the data, which can be detected or corrected by the appropriate code. Since the major computation requirements for many important
run-time signal/image processing tasks can be reduced to a common set of
basic matrix operations, i.e., addition, multiplication, LU decomposition,
transpose, and product of a matrix with a scalar, the weighted checksum
code (WCC) is an appropriate low-cost code to achieve fault-tolerant matrix

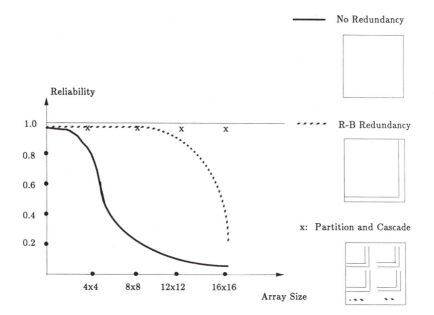

Figure 6.30: Reliability versus array size with no redundancy, the proposed redundancy, and 4 × 4 partition cases: reliability of single PE = 0.99 and DMR in PE module.

operations because the weighted checksum property is preserved in these five matrix operations.

6.4.3.1   The Weighted Checksum Scheme

In an array processor, it is a reasonable assumption that there is only one faulty PE in a very short period of computation. Therefore, if we can recover the error caused by the faulty PE before it propagates to other computations by periodically checking the results of the computation, we can still get the correct output. The weighted checksum scheme [Jou86] is used to make certain matrix-type computations on array processors fault-tolerant, i.e., if we first encode the input data to the computation, and examine the result of the computation later, we should be able to detect if there are errors in the computation and, if so, correct the errors. Simple checksum properties can be used for this purpose. This is illustrated by the following matrix-vector multiplication example.

Design Example – Matrix-Vector Multiplication   Assume we have a matrix-vector multiplication $\mathbf{Ax} = \mathbf{b}$, where $\mathbf{A}$ is a $n \times n$ matrix and $\mathbf{x}$ and $\mathbf{b}$ are $n \times 1$ vectors. This computation can be made fault-tolerant using the WCC.

We first define two row weight vectors $\mathbf{W1}$ and $\mathbf{W2}$ as:

$$\mathbf{W1} = (1\,1\cdots1)$$
$$\mathbf{W2} = (1\,2\,2^2\cdots2^{n-1})$$

Note that if we let $\mathbf{WS1} = \mathbf{W1} * \mathbf{A}$ and $\mathbf{WS2} = \mathbf{W2} * \mathbf{A}$, where $*$ denotes matrix multiplication, and augment the $\mathbf{A}$ matrix by adding two more rows $\mathbf{WS1}$ and $\mathbf{WS2}$, then

$$\begin{bmatrix} \mathbf{A} \\ --- \\ \mathbf{WS1} \\ \mathbf{WS2} \end{bmatrix} * \mathbf{x} = \begin{bmatrix} \mathbf{b} \\ --- \\ R1 \\ R2 \end{bmatrix}$$

Checksum Preservation Property   The purpose of adding two weighted checksum rows $\mathbf{WS1}$ and $\mathbf{WS2}$ to matrix $\mathbf{A}$ is to detect and correct if there exists a single erroneous element in the computed vector $\mathbf{b}$. This scheme works because the checksum property is preserved under the multiplication operation. More precisely, we note that

$$R1 = \mathbf{WS1} * \mathbf{x} = (\mathbf{W1} * \mathbf{A}) * \mathbf{x} = \mathbf{W1} * (\mathbf{A} * \mathbf{x}) = \mathbf{W1} * \mathbf{b} = C1$$

where the two checksums of $\mathbf{b}$ may be computed as $C1 = \mathbf{W1} * \mathbf{b}$.

Therefore, $R1$ is supposed to be equal to $C1$. Similarly, $R2$ is supposed to be equal to $C2 \, (= \mathbf{W2} * \mathbf{b})$.

In general, the operation $*$ may be replaced by certain other linear operators and still preserve the checksum property. It can be shown that five matrix operations exist that preserve the weighted checksum property: addition, multiplication, scaling, transpose, and LU decomposition. Their proofs are left as exercises.

6.4.3.2   The Error-Correction Procedure for Single Error

Let the checksum weighted vector $(b_1\ b_2\ \dots\ b_n\ WS1\ WS2)^T$, where $WS1$ and $WS2$ are the checksums for vector $\mathbf{b} = (b_1\ b_2...b_n)^T$, be the output

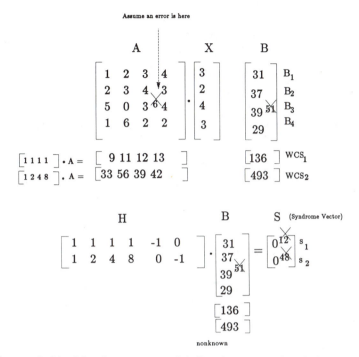

Figure 6.31: Matrix-vector multiplication example of the WCC scheme.

of some computation that preserves the weighted checksum property, such as the output vector of the preceding matrix-vector multiplication example. A single erroneous element in the vector can be corrected. We define two syndromes as follows:

$$S1 = \sum_{i=1}^{n} b_i - WS1$$

$$S2 = \sum_{i=1}^{n} (2^{i-1} b_i) - WS2$$

The correction procedure for a single error is described as follows:

If $(S1 \neq 0)$ AND $(S2 \neq 0)$,[5] then $S2/S1 = 2^{j-1}$, which implies that b_j is erroneous. The corrected value of b_j should be $b_j - S1$.

A numerical example is shown in Figure 6.31.

---

[5] Otherwise, it is assumed that no error occurs.

Systolic Array Design I   The preceding checksum fault-tolerant scheme
can be applied to VLSI array processors, e.g., systolic arrays. Consider the
matrix-vector multiplication example. The DG for the matrix-vector multiplication and the systolic array derived by the projection in the horizontal
direction are shown in Figure 6.32. If there is only one erroneous element
in the resultant vector, the weighted checksum scheme can detect and correct the error. Moreover, note that all computations to calculate one output
vector element (i.e., the inner product of two vectors) are mapped to a single PE. Thus if there are multiple computation errors produced by one PE
(marked by shaded nodes in the DG in Figure 6.32) in the matrix-vector
multiplication, the output vector will have only a single erroneous element.
Therefore, the weighted checksum scheme will work without any modification in this case. After the result vector is obtained, two n-operand adders
can be used to calculate the weighted checksum. The multiplication of two
matrices can be treated as several independent matrix-vector multiplications
and can be performed by the same fault-tolerant scheme.

   Although weighted checksum matrix operations produce code outputs
that provide some degree of error detecting or correcting capability, a faulty

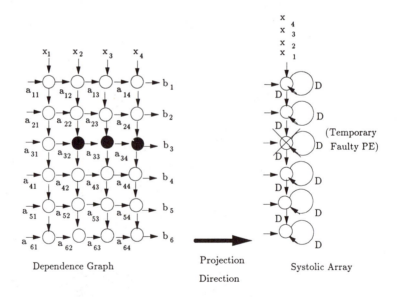

Figure 6.32: DG and systolic array I for matrix-vector multiplication; the shaded nodes indicate the faulty nodes. The vertically
propagating data are *transmittent* (see Section 3.2) and are therefore not affected by the faulty nodes.

PE module may cause more than one data element of the result to be erroneous. From a practical standpoint, the transient error duration must be very short compared with the processor cycle time in order to confine the errors caused by a faulty processor to only one data element. However, this may not always be the case in errors caused by, say, gamma ray radiation. Though the linear systolic array in Figure 6.32 can cope with this problem, in other cases, multiple errors produced by a PE can be disastrous. One such example is described next.

Systolic Array Design II   In practice, the matrix **A** used in many applications is often rectangular with the number of rows much greater than the number of columns. Typical examples are matrices used in filtering, convolution, and correlation. With reference to the DG in Figure 6.32, the horizontal projection will require a large number of processors.

On the other hand, a vertical projection will result in a much smaller array size. Under a single transient fault assumption, this systolic design will still be useful. The error detection and correction procedure will remain the same. With reference to Figure 6.33, note that if the transient fault covers two processor cycles, then a PE will produce two consecutive errors in the computation. This is drawn as shaded nodes in the DG. Correspondingly, the resulting output vector will have two *consecutive* erroneous elements. In

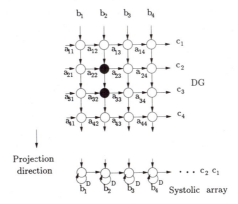

Figure 6.33: DG and systolic array II for matrix-vector multiplication. The shaded nodes indicate the faulty nodes. The vertically propagating data are the so-called *transmittent data* and are therefore not affected by the faulty nodes.

this case the weighted checksum scheme is no longer valid. In order to detect and correct this kind of error, a new decoding scheme is proposed.

6.4.3.3   Weighted Checksum Scheme for Double (Consecutive) Faults

Let us define another row vector **W3** as:

$$\mathbf{W3} = (1 \ (-2)^1 \ (-2)^2 \cdots (-2)^{n-1})$$

and for a column vector **a** define $WS3 = \mathbf{W3} * \mathbf{a}$.

The Error-Correction Procedure   Let $(b_1 \ b_2...b_n \ WS1 \ WS2 \ WS3)$ be the output of some computation that preserves the weighted checksum property, where $WS1$, $WS2$, and $WS3$ are the weighted checksums for vector **b** $= (b_1 \ b_2...b_n)^T$. Two consecutive erroneous elements in the vector can be corrected. We define three syndromes as follows:

$$S1 \ = \ \sum_{i=1}^{n} b_i - WS1$$

$$S2 \ = \ \sum_{i=1}^{n}(2^{i-1}b_i) - WS2$$

$$S3 \ = \ \sum_{i=1}^{n}[(-2)^{i-1}b_i] - WS3$$

Let the two errors δx and δy be located at positions j and $j+1$ in the vector.

The correction procedure for two consecutive errors is described as follows: If $S1 \neq 0$, then it is determined that errors have occurred.[6] We can solve the location j and δx, δy by the following equations:[7]

$$j = \log_2(\frac{3S2 + S3}{S1}) - 1, \ \ (\text{if } j \text{ is odd})$$

---

[6]If $S1 = 0$ then it is considered as no error – regardless what values $S2$ and $S3$ might be. In a very unlikely instance, it may also be that $\delta x = -\delta y$. If this is the case, our error-correction procedure cannot recover the errors.

[7]If the result of taking logarithms in the above procedure is not an integer, most likely it is the case that there are more than two errors, or the errors are not consecutive.

or

$$j = \log_2(\frac{3S2 - S3}{S1}) - 1, \quad (\text{if } j \text{ is even})$$

and

$$\delta x = 2S1 - 2^{-j+1}S2,$$
$$\delta y = 2^{-j+1}S2 - S1.$$

Once the error locations are determined, the errors δx and δy may be properly compensated. The derivation of these results is given as an exercise. The error-checking and error-correction procedure requires summing a column of data, which may be computed in a pipelined fashion.

6.4.4   Time-Redundancy Approach to Fault-Tolerance

Instead of space-redundancy (hardware), time-redundancy can also be used to handle array FT. Two applications using time-redundancy are discussed in this subsection: (1) Fault detection and correction based on interleaved DGs. (2) Graceful degradation based on LSGP or LPGS algorithm partitioning schemes.

6.4.4.1   DG Interleaving for Fault-Tolerance

One of the basic assumptions made in the architectural approach is that every PE is equipped with some means of self-checking. So when a fault occurs during some computation it is diagnosed before the faulty PE sends out its probably incorrect results. Under this assumption there is no need for roll-back since the error latency is supposed to be kept minimal. It is possible to implement such a scheme by duplicating the processing part of every PE. Each computation is performed simultaneously by the two independent processing parts of the PE and the results are compared. A disagreement means that a fault has occurred. The main disadvantage of such a scheme is that it essentially doubles the hardware required.

An alternative solution can be the use of *time-redundancy* instead of PE duplication for the purpose of error detection and correction. The idea is to perform the same computation twice in adjacent PEs at two different but close enough time periods and then compare the results. If they match there

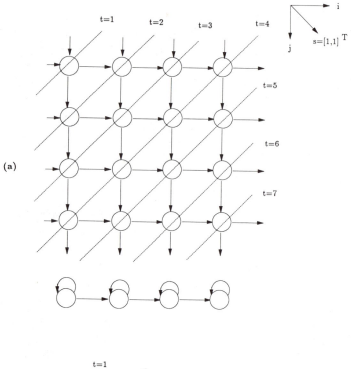

(a)

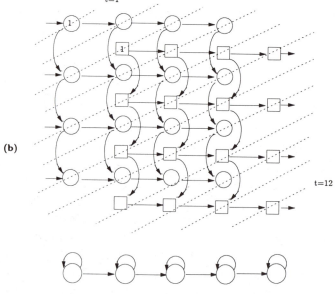

(b)

Figure 6.34: (a) DG for matrix-vector multiplication. (b) The TRIFT version of (a).

is no fault. Otherwise a roll-back is necessary to correct the fault. The most interesting feature of time redundancy versus PE duplication is the limited increase in chip area which in turn leads to lower cost and higher production yields. An obvious drawback is the time overhead due to the increasing total problem latency and the necessary roll-back due to possible faults.

We illustrate this idea by an example in Figure 6.34. In part (a) we give the DG for the matrix-vector multiplication problem (assume a 4×4 matrix). The total problem latency without faults is going to be 7 schedule time units ($2n - 1$ in the case of $n \times n$ matrix), if we use $\vec{s} = [1, 1]^T$ as the schedule vector. Actually this is the minimum latency schedule we can have for the given DG. In Figure 6.34 part (b) we interleave the original DG with a copy of the same DG shifted one position to the right (in space). With reference to the same figure, computation 1 that is performed at the leftmost PE at schedule time $t = 1$ is repeated at its right neighbor PE and the results are compared at time $t = 3$. The hardware overhead is minimal (only one extra PE) but the new total problem latency without faults is going to be 12 schedule time units ($3n$ for an $n \times n$ matrix). Because of the space and time interleaving of identical DGs we call this approach *time-redundancy with interleaving for fault-tolerance (TRIFT)*. It should be noted that the way of interleaving different copies of the same DG is not unique. This fact may be exploited for optimality.

Consider now the computation x performed at PE_i during the time period $t = t_0$, as shown in Figure 6.35. The same computation will be repeated at PE_{i+1} two time units later at $t = t_0 + 2$ and the results are compared. Assuming that there is a mismatch we conclude that a fault has occured at either PE_i or PE_{i+1}. In the wavefront array case PE_{i+1} is interrupting both its neighbors and a localized roll-back is started. No output is sent out of the neighbors until the roll-back has been successfully completed. The roll-back path is shown in Figure 6.35 by the dashed lines. It can be implemented by local communication between the faulty PE and its neighbors. However in the systolic array case PE_{i+1} should halt all the non-neighboring PEs before the roll-back and restart them after it. So global communication is required. A small amount of local memory to hold the input data used for the last computation is necessary for every PE to support this mechanism.

If the fault persists a new roll-back using the same path can be initiated. The minimum time overhead is 3 schedule time units per roll-back (c.f., the dashed lines in Figure 6.35). A fault can be classified as *permanent* if it is not corrected after a prespecified number of roll-backs. In that case we conclude that there exists a permanent fault in PE_i (see Problem 28). Then

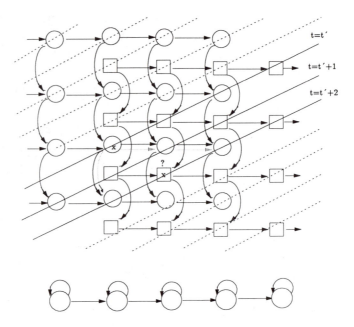

Figure 6.35: Localized roll-back in TRIFT.

a reconfiguration algorithm such as those proposed in Section 6.4.2 can be used to replace the permanently faulty PE. Multiple simultaneous faults can also be handled.

In short, the TRIFT scheme can provide a cost-effective solution to transient fault recovery and it can also be easily integrated with architectural FT schemes for permanent fault recovery.

6.4.4.2   Graceful Degradation: Algorithm Partitioning for Fault-Tolerance

An important idea for reconfigurable arrays is *graceful degradation*, which uses *temporal redundancy* instead of *spatial redundancy* (or hardware redundancy) to achieve FT. The faulty PEs should be disconnected, once they are detected, and the remaining healthy PEs form a new array of a smaller size. The algorithm, executable on the original array, should thus be remapped (i.e., partitioned) to fit into the new array. This explains why algorithm partition techniques may be used for gracefully degradable FT.

Note that there are two algorithm partition schemes: LSGP and LPGS, so there should be two corresponding gracefully degradable FT techniques.

Recall that the LPGS partition scheme leads to an array with wrap-around FIFOs. The architecture proposed in [Fort85b] is to support the LPGS scheme by using a switch lattice such that an entire row and/or an entire column may be disconnected once a fault occurs. For the LPGS scheme, resuming is possible only if the original projection direction (or the original array algorithm) has no reverse data dependency. If the original assignment does not satisfy this condition, then the LPGS scheme needs to find an acceptable projection direction, which according to [Fort85b] may accumulate a much longer computation time. In short, if there is some reverse data dependency, then LSGP may be a better alternative. For more details, see [Fort85b].

When an LSGP FT scheme is implemented, larger local memory will be needed to accommodate the additional DG nodes assigned to the PE when faults occur. In the LSGP scheme, it is always possible (regardless of the presence of reversed data dependency) to proceed after reconfiguration of the array since the computation for the faulty PE is reassigned to its neighboring PE. Since wavefront arrays allow local rearrangement of faulty PEs, they have an advantage over systolic arrays when local based control (COPE) is used.

The gracefully degradable FT may also be viewed as a "constrained" DG node assignment with the constraint on the number of available PEs. In [Moore87], Fortes propose an "optimal" FT technique via a *nonlinear assignment* scheme, in which each healthy PE is assigned one node at each time step. It is optimal in the sense that all the healthy PEs are kept usefully busy. The difficulty of the scheme lies in the high control complexity.

6.5 Programming Languages for Array Processors

The actual implementation of VLSI array processors can be either dedicated or programmable. Programmable arrays are often more attractive due to constantly changing application specifications. However, array programming requires keeping track of several simultaneously occurring events. It is significantly more complicated than sequential programming. For a complete software tool, one needs to develop:

1. New notations or languages which are specifically for array processors (instead of using conventional languages).

2. Coherent software techniques for *programming* or *design* of array processors.

3. A set of applicational software packages for the programmable array processors.

6.5.1   Software versus Hardware Design

In VLSI the boundary between software and hardware has become increasingly vague. Therefore, software and hardware should be coordinated to achieve better performance. We note that for array processors the programming languages (for software development) and the design languages (for hardware development) have similarities in terms of their theoretical basis. Figure 6.36 is provided to help demonstrate this point.

An architecture is the specification of an interface between components maintained at different organization levels. A computer system typically contains many levels and types of architectures. In this section we will discuss the design decisions for the implementation of the proposed architectures in the array system. The different architectural levels shown in Figure 6.36 are defined as follows:

1. *System architecture level* defines how the system appears to users and application programmers and includes the characteristics of languages, user interface and operating system.

2. *Array architecture level* defines the interconnections between different arrays and the functional capabilities of the processors comprising the arrays.

3. *PE architecture level* defines the hardware modules for the PE nodes. It includes both instruction implementation (fetch, decode, execute mechanism) and the interfaces between individual building blocks (e.g., bus widths, queue lengths).

4. *Building block architecture level* defines the structure of individual building blocks (e.g., multiplier, adder, or memory).

It should be noted here that in all architectural levels, except the building block level, the *model of execution* should be defined (e.g., control flow or data flow model). All levels do not necessarily share the same model of

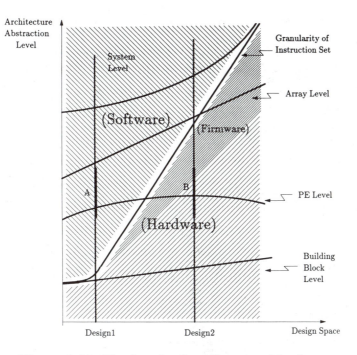

Figure 6.36: The domains for software and hardware.

execution. For example, we may have a data flow model at the system level, allowing the user to see the system as a collection of processes and processors that operate on data availability, while the underlying levels may adopt a control flow model, that is for example the PEs are conventional Von Neumann machines (i.e, Motorola 68000). On the other hand the system level may have a control flow model of execution (i.e., programmed in FORTRAN) and the underlying levels may adopt a data flow model of computation, for example the PEs may be the NEC μPD7281 data flow chips.

In Figure 6.36, the system flexibility relative to these architectural levels is illustrated. The vertical axis represents the complexity of information processed in the nodes. Here a node may represent the total system, a single array, one individual PE, or a single basic building block (BBB). In a way it also indicates the level of abstraction, from high-level (or very abstract) to low level (or very concrete). The four curves, indicating the complexity of the system, array, PE, and BBB respectively, are shown only on a relative scale. The horizontal axis, on the other hand, represents different system designs (e.g., design1, design2, and so on).

An additional virtual architectural level – the *instruction set architec-*

ture – may be introduced. Recall that a programming language basically defines a virtual machine whose (virtual) building blocks depend on the instruction set adopted. The flexibility of programming may be measured in terms of the *granularity of the instruction set*. For convenience, we shall refer to this as the *instruction level*. It defines the instruction set of each processing element, the types of data that can be manipulated and the communication primitives between processors. Accordingly, there are different degrees of flexibilities corresponding to different system designs, although it is impossible to give a strict ordering among them. Roughly speaking, as illustrated in Figure 6.36, the system's flexibility (or programmability) decreases from the left-hand side to the right-hand side of the design space. More precisely, the corresponding granularity of the instruction set grows from BBB, to PE, to array, and ultimately to system nodes. (The coarser the instruction set is, the less flexible the system will be.) Here we give two (contrasting) examples. One extreme, which is at the leftmost and represents a fully custom VLSI design, can satisfy almost any requirement specified by the user but takes much more effort and time to implement. Another is at the rightmost extreme in Figure 6.36, which represents a turn key system design. It is very easy to use but offers very little flexibility.

With reference to Figure 6.36, we note that the design/programming space can be divided into the software and hardware zones. The first zone, above the instruction set level, represents the software space. (Note that the software space can be further divided into three parts: system software, array programming, and intra-PE (local) programming parts, separated respectively by the array and PE lines.) The second zone, below the instruction level, represents the hardware space. Within the hardware space, there is a special zone representing the firmware space. Here the codes defining the instruction set in terms of BBBs are assumed to be semi-permanent, e.g., microprogramming codes in a ROM.

It is possible for the designer to build a system with a variety of options, trading software for hardware and vice versa. This means that if we have the tools that translate one architecture level to another with well defined architecture description tools, hardware design and programming techniques for most of the implementation process are to an extent very much interchangeable. For example, the algorithm analysis for the software (and programming language) development along line *A* in Figure 6.36 can be reapplied to the firmware/hardware development along line *B*. This observation leads naturally to the powerful notion of hierarchical design methodology based on a systematic mapping between all architectural levels. This will be clarified when we discuss CAD tools for array processors in Section 6.6.

6.5.2 Language Design Factors

Software development requires language tools for expressing the algorithm in a high-level language and compiling the language into machine executable object code on the target machine [Hockn83]. The key factors affecting language and compiler design are the following [Hoare78]:

1. *Application Range:* It should cover the desired range of interesting areas, e.g., signal/image processing and scientific computing applications.

2. *Ease of Programming:* A problem solution must be easy to express in the programming language adopted, which leads to the concept of high-level languages. The programming language must *facilitate the expression of parallelism* and *extract it from the available representation.* To achieve this, the compiler must perform a data dependence analysis to identify which parts of the algorithm can be executed independently of others. It should provide clear assistance to the programmer in his tasks specification, design, implementation, verification and validation of complex computer systems.

3. *Execution Efficiency:* To implement an algorithm efficiently on an array processor, maximum use of the available parallelism must be attempted. It should be time-efficient in translation, execution, and processor/memory utilization. Hence the hardware features should be reflected in the software design.

Trade-off Practically, trade-offs between the above three factors will have to be properly weighted to reach an optimal compromise. For example, if high speed processing efficiency in a real-time image processing system is an absolute necessity, then the *efficiency* factor might be accommodated (by hard-wired or microcoded dedicated processors) at the expense of the other two remaining factors (application domain and coding simplicity).

From the perspective of ease of programming, on the other hand, a convenient language should let the programmer concentrate on the solution of the problem without having to worry about the underlying machine architecture and its internal representation. This means that the demand for control information needed by the language (which is not actually a part of the original algorithm) should be kept to a minimum. In some cases however, the compiler intelligence may not be enough to produce the optimal mapping of the algorithm to the underlying architecture so the programmer will have to intervene and explicitly control the allocation of processes to processors. This requirement is conflicting with the previous requirements of programming ease since to make optimal use of the architecture, the programmer must be aware of its details and use this knowledge to write a program that explicitly controls the hardware. At the same time, memory allocation must be performed by the programmer to minimize data movement delays which also conflicts with the previous requirements.

To resolve these conflicts we introduce the notion of an *intermediate representation* which is the target language of the high level language compiler. By introducing this representation, we define an *interface* between the programmer and the underlying architecture which isolates the programmer from the tedious details of controlling the hardware.

The intermediate representation is a much simpler language than the high level language and provides primitives for explicit expression of concurrency and communication control. The existence of these primitives permits the programmer or an optimizing compiler to tailor the algorithm representation to the underlying architecture.

An Integrated Approach to VLSI Algorithmic Notations   A successful array language and compiler design will hinge upon a highly interdisciplinary approach involving algorithm analysis, parallelism extractions, array architectures, functional and structural primitives. The search for an algorithmic notation adequate for VLSI algorithms represents a very challenging field of programming language research. Can we write in a language that permits complete (explicit or implicit) specifications of the space-time activities on the VLSI array processors?

6.5.3   Classes of High Level Languages

Complete presentation of the languages that can be used for array processor programming is beyond the scope of this discussion. We will only briefly discuss the advantages and disadvantages of some classes of languages. For the purpose of selecting a language for an array processor, the languages can be classified in three classes:

1. *Class 1.*   Traditional Von Neumann procedural languages (possibly with a powerful optimizing and vectorizing compiler).

2. *Class 2.*   Von Neumann type languages augmented with synchronization and communication primitives.

3. *Class 3.*   Parallel languages that facilitate the expression of parallel algorithms and perform dependence analysis automatically.

Class 1 Languages: Conventional procedural languages (e.g., Fortran) are based on the Von Neumann model of computation which centers around the assignment statement and the associated processor state. Parallelism is very difficult to express in such languages since they were not designed for this purpose. One way to aid the optimizing compiler to perform the dependence analysis is to represent the algorithms in a *single assignment* form of the language. This however means that the programmer is responsible for transforming an inherently parallel algorithm to a serial form thus making programming very difficult. Since this class of languages does not aid the programmer in expressing a parallel algorithm, it does not meet the ease of programming requirements. In addition, vectorization and optimization are extremely difficult since the original expression of the program does not convey much information for the uncovering of parallelism. The advantage of the conventional languages is that a very large number of programs have been written already and the effort of rewriting them in another form may be prohibiting.

   With the advent of the successful pipelined processors (e.g., Cray-1, Cray X-MP, and the FPS AP120B), several *vectorizers*, optimizing compilers or preprocessors for high-speed processors, have been developed. They include: The MCA vectorizer developed by Massachusetts Computer Associates, the Cray Research Fortran compiler, the Fortran compiler for the TI ASC, the vectorizer for the Burroughs BSP, the Fortran compiler for the Cyber 205, the vectorizers for the Fujitsu and Hitachi vector supercomputers, and Parafrase compiler for the Cedar project [Kuck80]. *A vectorizing compiler processes a source code written in a sequential language and, where possible, generates parallel machine instructions.* Sometimes, a more compact intermediate form, such as the directed graph representation of the source code for the Texas Instruments Advanced Scientific Computer (ASC NX), may also be compilable.

Example 1: Matrix Multiplication in a Fortran-Like Language

   The standard way of expressing a matrix multiplication in a sequential (Class 1) language is:

$$\text{For } i \text{ from } 1 \text{ to } N_1$$

$$\text{For } j \text{ from } 1 \text{ to } N_2$$

$$\text{For } k \text{ from } 1 \text{ to } N_3$$

$$c(i,j,k) \;=\; c(i,j,k-1) \;+\; a(i,k)b(k,j)$$

while in single assignment form it would be:

$$\text{For } i \text{ from } 1 \text{ to } N_1$$

$$\text{For } j \text{ from } 1 \text{ to } N_2$$

$$\text{For } k \text{ from } 1 \text{ to } N_3$$

$$a(i,j,k) \;=\; a(i,j-1,k)$$

$$b(i,j,k) \;=\; b(i-1,j,k)$$

$$c(i,j,k) \;=\; c(i,j,k-1) \;+\; a(i,j,k)b(i,j,k)$$

The conversion is not trivial since it involves a dependence analysis in order to specify the indices of each expression.

Class 2 Languages: This class contains those languages that permit the expression of algorithms as a collection of *communicating sequential processes (CSP)* (e.g., Ada, Occam). The languages of this class inherit some of the advantages and disadvantages of the previous class. Expressing an algorithm in such a language requires the programmer to identify those parts of the algorithm that can be executed in parallel and then use the language primitives to organize the communication and synchronization of the cooperating processes that implement the algorithm procedures. This enhances the efficiency of the machine greatly since the programmer directly allocates processes to processors and organizes their interactions. On the other hand however, this may be an unacceptable burden for the programmer especially in large applications.

Example 2: Matrix Multiplication in Occam

Occam is a simple but powerful language based on the ideas developed in CSP [Hoare78]. We will use the matrix multiplication algorithm as an example. (The primitives of the Occam language were discussed in Section 5.5.)

The matrix multiplication program in Occam is divided into two parts. The first is the node program and the second is the main program that interconnects the node processes to form the array. The program list is given in Section 5.5, Example 6, and is omitted here.

It is obvious from this example that in order to express the algorithm in Occam, the dependence analysis must have already been done by the programmer.

Class 3 Languages   This class of languages is fairly recent and much less developed than the previous classes. Research results however have produced languages that show considerable promise for array processor programming. The presentation of all languages of this type is beyond the scope of this book. We will concentrate on the subclass of *applicative* or *functional* languages which have features very close to our requirements. (e.g., Lisp variations, SISAL [Sked85b], FP, FFP, Paralfl [Hud86]). Functional languages stem from the realization that the assignment statement is an artifact of the Von Neumann model of computation and that the mathematical expressions of algorithms are in essence sets of functions to be evaluated. By eliminating the assignment statement and by designing the languages to facilitate writing of *functions* and describing their evaluation, very simple and elegant languages have been produced with great expressive power. Since there is no concept of processor state in these languages, the *Church-Rosser* [O'Don85] property holds. This property in essence states that no matter what the sequence of execution of functions is (provided that dependence constraints are observed), the program will always produce the same results (if it terminates). This property enables the compiler to produce code for the program, the execution of which is subject only to dependence constraints. Thus we see that functional languages satisfy our ease of programming and application requirements.

There are possible problems with the function evaluation only nature of these languages. For example, sometimes too much of the underlying architecture is hidden from the programmer so that he cannot optimize his algorithm to the array processor. This problem has been alleviated by the introduction of single assignment statements (e.g., SISAL [Sked85b]) or communication primitives (e.g., Paralfl [Hud86]). Another way to permit the programmer to intervene, is to have an intermediate language that the high level compiler is targeted for and then use an optimizer to generate the actual machine code. The most convenient intermediate form for these languages is a representation of the DG in a graph language (IF1 for SISAL [Sked85a]) or a Class 2 language (e.g., Occam). In this way we retain the advantages of both classes, that is we retain the high level, functional form of the functional language for initial algorithm description and the detailed communication and synchronization description of the Class 2 languages.

Example 3: Matrix Multiplication in SISAL

SISAL stands for *Streams and Iterations in a Single Assignment Language* and is an *applicative* language that compiles to IF1 which is its intermediate form. Before presenting our example we will first define the features of the language that will be used.

SISAL has two forms of *for* loops, the *product form* and the *non product form*. The latter is equivalent to the conventional *for* loop regarding the range of indices generated. For the product form we have two cases, the *cross* form and the *dot* form. The former produces all pairs of indices (Cartesian product) while the latter produces the pairs of indices that would be generated in an inner product calculation. Another feature that must be mentioned is the reduction operation *value of sum* that produces a scalar value, equal to the sum of all variable instances following the keyword *sum*.

```
type OneDim = array [ integer ];
type TwoDim = array [ OneDim ];
function MxM  (A,B: TwoDim; M,N,L: integer
returns TwoDim)
     for i  in 1,M cross j in 1,L
     S:=
               for K in 1,N
               R:=A[i,k]*B[k,j]
               returns value of sum R
               end for
     returns array of S
     end for
     end function %MxM
```

This program generates all pairs of elements along the i-direction and j-directions, performs the multiplications and adds the partial products along the k-direction. From this example we see that we can easily express our algorithm in SISAL. In Figure 6.37 the output of the compiler is shown in graphical form. (The actual IF1 code is not shown because of space limitations.)

The *RangeGenerator* nodes along with the *AElement* nodes generate the array elements that are multiplied and then added to form the final product. The IF1 graph consists of two graphs, the program structure graph (PSG) which is very useful for allocation and partitioning decisions and the actual DG which consists of the leaves of the PSG along with their joining arcs. The graph of Figure 6.37 does not correspond directly to our form of DFG but with a little further processing, that is expanding the *AElement* and *RangeGenerator* nodes we can arrive at our representation.

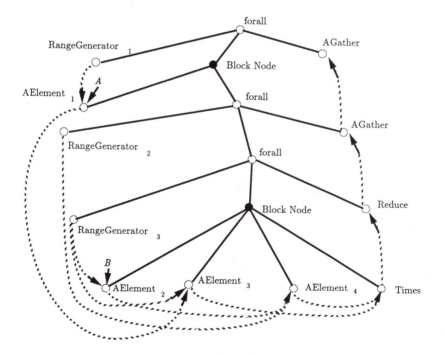

Figure 6.37: Matrix-matrix multiplication in IF1.

Example 4: Matrix Multiplication in FP

The matrix multiplication algorithm can be very concisely and easily written in the FP language [Backu78]. Before we present the program, we must define some of the primitives and the notation used.

In the following, f, g, h denote *functions*, $< x >$ is an *argument* which may be a scalar or a vector denoted by $< x_1, x_2, \ldots, x_n >$ or a matrix denoted as a vector of row vectors such as for a matrix of dimension 2: $<< x_{11}, x_{12} >< x_{21}, x_{22} >>$. The symbols $f :< x >$ mean that function f is applied to argument $< x >$. The following functions are some of the language primitives:

$$distr :<< x_1, x_2, \ldots, x_n >, y > \rightarrow << x_1, y >, < x_2, y >, \ldots, < x_n, y >>$$

$$distl :< x, < y_1, y_2, \ldots, y_n >> \rightarrow << x, y_1 >, \ldots, < x, y_n >>$$

$$trans :<< x_1, \ldots, x_n >, < y_1, \ldots, y_n >> \rightarrow << x_1, y_1 >, \ldots < x_n, y_n >>$$

$$/f :< x_1, x_2, \ldots, x_n > \rightarrow f :< x_1, f :< x_2, \ldots, f :< x_n > \ldots >$$

$$af :< x_1, x_2, \ldots, x_n > \rightarrow < f : x_1, f : x_2, \ldots, f : x_n >$$

$$[f, g] :< x > \rightarrow < f : x, g : x >$$

$$f \circ g :< x > \rightarrow f : (g :< x >)$$

$$1 :< x_1, x_2, \ldots, x_n > \rightarrow x_1$$

$$k :< x_1, x_2, \ldots, x_k, \ldots, x_n > \rightarrow x_k$$

Using the above primitives we can now proceed to write the program in FP. First we define the *inner product* operation as

$$\textbf{def IP} = (/+) \circ (a\times) \circ trans$$

which means that if IP is applied to $<< x_1, \ldots, x_n >, < y_1, \ldots, y_n >>$ will first execute the *trans* operation forming pairs of elements $<< x_1, y_1 >, \ldots, < x_n, y_n >>$. All pairs are multiplied forming a vector $< z_1, \ldots, z_n >$ of products and then the $+$ operator is applied recursively, that is $+ :<$

$z_1, + :< z_2, \ldots, + :< z_{n-1}, z_n > \ldots >$ giving the inner product. The matrix-matrix multiplication program takes the form:

$$\textbf{def MM} = aaIP \circ adistr \circ distl \circ [1, trans \circ 2]$$

This is equivalent to saying that we first form the transpose of the second matrix, then form the row column pairs and finally apply the inner product operation to all pairs.

The above program can be easily converted into a DG if we interpret the *distr* and *distl* operators as *broadcast* operators. With this interpretation, the *distr* $\circ$ *distl* operation form the Cartesian product of the row column vectors. The DG is shown in Figure 6.38.

6.5.4   The Intermediate Representation

The high level language compiler performs the dependence analysis and produces a list of functions that have to be evaluated in some order to produce the correct result. The order of the computations is described by the DG, the nodes of which represent the functions, and the edges the interdependencies between functions. The next step is to create the SFG by projecting

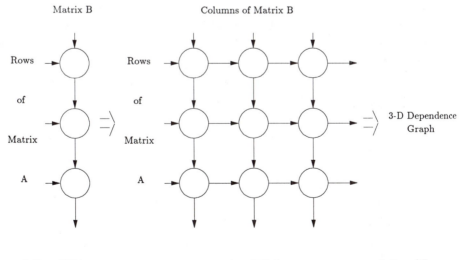

Figure 6.38: DG for matrix-matrix multiplication.

the DG and the scheduling function and then retiming it to obtain the systolic/wavefront array. After this point we can proceed with the actual code generation for the PEs in our physical array and assign code blocks to processing elements. The systolic/wavefront array generation can be done independently of the code generation since it can be obtained by geometrical operations on the DG without any reference to the actual function of the nodes. On the other hand, code generation is only concerned with the function of the nodes in the systolic/wavefront array.

From the above observations we conclude that we need a representation that can adequately represent the location of a node in a graph and the edges that are connected to it and at the same time describe the function of the node, i.e., a representation that has two parts: *graph information* and *functional information.*

In most cases the available computing resources (i.e., the array size) are not enough for the problem at hand and some grouping of operations must be done. Invariably these groups must be assigned to different processing elements. To perform the allocation automatically we need some control information derived from the program control structure. For example all operations of a *for* loop on a data array can be allocated to one processor (if necessary) thus reducing the interprocessor communication need if the processor can store all the necessary data. To facilitate the allocation process we must generate a *Program Structure Graph* which will describe the various blocks our functions can be grouped into. The lowest level nodes of the structure graph with their interconnections are the nodes of the DG. The higher level nodes describe a whole subfunction and the highest level node represents the algorithm. Such a graph enables us to decide the granularity of the algorithm implementation since we can easily see which operations can be grouped together (see Figure 6.37).

Thus our representation must be of the form:

Program structure information

. . .

DG information

. . .

node procedure

. . .

The procedures operating on this representation can use only the appropriate data. Suitable languages for such a representation are IF1 and Occam. In the following we will briefly describe how each representation can be used for our purposes.

IF1   IF1 is the intermediate representation for the language SISAL. This language is used to represent both the structure graph and the DG. The structural information is carried by the *compound nodes* while the dependence information is carried by *simple nodes*. A complete description of the language is beyond the scope of this discussion. We will only give an indicative sample of the semantics of the language.

Graphs in IF1   This language is based on acyclic graphs. There are four components to a graph: *nodes, edges, types* and *graph boundaries*. Nodes denote operations such as multiply and add; edges represent values passed between nodes (dependencies), and types can be attached to either edges or functions to declare the data types required for a particular operation. Graph boundaries surround groups of nodes and edges and convey structural information.

Nodes can be either *simple* or *compound*. Simple nodes have the form: *N label operation* where the operation is an ASCII string that represents an integer associated with the function of the node. For example, the operation $(a + b) \times 5$ would be represented by the sequence of nodes and edges in Table 6.4.

SourceNode	SourcePort	DestNode	DestPort	Type	Comment
N	4	Plus			Node 1 computing $(a + b)$
0	1	4	1	int	name a
0	2	4	2	int	name b
4	1	5	1	int	name
N	5	Multiply			Node 2 computing $\times$ 5
		5	2	int	val 5
5	1	0	1	int	name

Table 6.4: IF1 expression for computing $(a + b) \times$ 5

Compound nodes contain subgraphs and the interconnections between subgraphs. The general form of a compound node is:

{

G ... subgraph 0

. . .

G ... subgraph 1

. . .

G ... subgraph n

. . .

} *label Opcode association-list*

The association list itemizes the interconnections between the subgraphs described by the semantics of the compound node. For example the *for* loop node has four subgraphs: *Initialization, Test, Body* and *Result*. The association list for this compound node is:

$$4 \ < init > < test > < body > < result >$$

The first field is the number of subgraphs and the last four fields are filled with integers that point to the loop subgraphs (i.e., input edges).

Graph boundaries denote functions and can be either *global, local* or *imported*. Local boundaries begin with letter G, global functions begin with X and imported (from other compilation units) begin with I.

Code generation   In the previous discussion we described how IF1 represents the control and DGs. In order for this representation to be useful we must convert it to a machine understandable form which strongly depends on the underlying architecture. In this section we will use two architectures for illustration, the Inmos transputer and the NEC μPD7281 data flow chip.

Example 5: Code Generation for the Transputer

This processor is designed to execute Occam directly. It is thus sufficient to convert our intermediate representation to Occam. This can be easily achieved if we assign an Occam channel to each edge of the DG and a sequential process for the node function. Therefore the previous simple example of the $(a + b) \times 5$ computation will be expressed in Occam as:

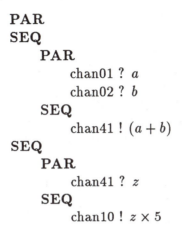

```
PAR
SEQ
    PAR
        chan01 ? a
        chan02 ? b
    SEQ
        chan41 ! (a + b)
    SEQ
        PAR
            chan41 ? z
        SEQ
            chan10 ! z × 5
```

Example 6: Code Generation for the μPD7281

The NEC chip has a data flow architecture as described in Chapter 7. Its assembly language is graph oriented since it has primitives to describe the nodes and edges of a data flow graph. The program for the $(a + b) \times 5$ computation is:

LINK	LABEL		FLABEL
LINK	c01	=	ADDQ
LINK	c02	=	ADDQ
LINK	c41	=	MULQ
LINK	c51	=	OUTQ

FUNCTION	LABEL		OPCODE	FLOW
FUNCTION	ADDQ =		ADD	QUE(?)
FUNCTION	MULQ =		MUL	QUE(QA)
FUNCTION	OUTQ =		OUT1	QUE(?)
MEMORY	QA	=	AREA(1)	

The link entries represent the graph edges and the function entries describe the node functions. The FLOW field describes the queues for each node. The question mark in the QUE primitives means that these queues are not initialized and the QA area in memory, is assumed to contain the number 5.

6.5.5   Software Environment

6.5.5.1   A Structured Programming Environment

To help the programmer tackle the complex issues in programming an array processor, we propose the structure depicted in Figure 6.39 for software systems for array processors. In this software system, the algorithm is expressed in a functional language (e.g., SISAL) and its compiler produces the DG expressed in the intermediate representation which can then be further processed (i.e., projections, retiming) into an SFG or systolic form. Sometimes, it can be directly executed for debugging purposes or mapped to a program that simulates the underlying architecture.

Software Support for Debugging and Simulation Debugging a parallel program at the machine level is impossible since there are so many cooperating processes. It would be convenient to provide either a source level debugger or an intermediate level debugger. The source level debugger is very difficult to implement in Class 1 languages. Class 2 languages can be used however not only to describe the DG but to execute it as well thus

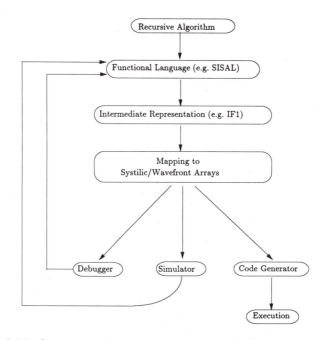

Figure 6.39: Structure of the programming tools for an array processor.

making very good debuggers at the intermediate level. Class 3 languages are too high level to be useful for debugging but can be used for verification of the program since we can execute them on either a sequential or a parallel machine and their results are guaranteed to be the same because of the Church-Rosser property. Thus we conclude that the best choice is a Class 2 language at the intermediate level for both representation and debugging.

Related to the above arguments are the arguments for the same solution for simulation purposes. In this case we want to predict how the algorithm will perform on the actual machine. Instead of running the algorithm on the machine, we simulate the architecture on a sequential machine (host). The most natural way to represent the machine operation is using Class 2 languages since they provide primitives which have one to one correspondence with the machine building blocks (e.g., FIFOs $\rightarrow$ queues, processors $\rightarrow$ processes).

6.5.5.2   Parallelism Extraction and Partitioning

To facilitate the compiler design, it is important to understand the related attributes of the various classes of languages, especially those regarding *parallelism extraction* and *partitioning ease*.

Parallelism Extraction: Parallelizing programs written in Class 1 languages is not easy since there is not enough information in the source program. A lot of effort has been invested however in parallelizing compilers, especially for FORTRAN (e.g., Paraphrase). The problem with Class 1 languages is aggravated by the fact that the vectorizers do not produce a dependence graph from which maximum parallelism may be extracted. Class 2 languages can express all the available parallelism so that they do not need parallelization but they place the burden of the parallelism analysis on the programmer. Class 3 languages are specifically designed to permit parallel processing expression and produce a DG for maximum parallelism.

Partitioning Ease: Partitioning and mapping of algorithms to specific architectures can be greatly facilitated by the generation of a DG. From this point of view, class 2 languages can be used to assign processes to processors but the programmer is responsible for the partitioning. Class 3 languages enjoy a definite advantage in this area since they are specifically designed to generate DG.

6.6   CAD for Array Processors

In order to facilitate mapping/matching algorithms to array processors, a set of integrated design software packages must be developed. This design system accepts inputs such as recursive algorithms and produces VLSI chips as output. This kind of integrated CAD tool may be called an *array compiler*. A complete and high-level array compiler is yet to be developed, though certain special purpose silicon compilers do exist, such as FIRST [Denye85], Lager [Rabae86], and CATHEDRAL [Jain86]. These silicon compilers all aim for a special application domain and adopt a special architecture, and most importantly, they accept only structural input. With a structural compiler, the user directs the formation and connection of tangible hardware blocks. The array compiler we are interested in must transcend this level and interpret hardware structure from user-specified behavior [Denye86].

In general, it is hard to compile hardware directly from behavior specification. However, since DSP algorithms are mostly regular and recursive in nature, the canonical and generalized mapping methodology (discussed in Chapter 3) can effectively map recursive algorithms to systolic/wavefront arrays. Therefore, the development of an array compiler for systolic/wavefront arrays is very promising and realistic.

6.6.1   Characteristics of an Array Compiler System

Design/Simulation System Requirement   The requirements for an array compiler for systolic/wavefront arrays will include the following:

1. User-friendly graphic-based interface.

2. High-level behavior input.

3. Interactive operation for users.

4. Hierarchical iterative top-down/bottom-up design methodology.

5. Consistent database for different levels.

6. Multi-level and mixed-level simulation.

7. Verifiable levels and final result.

Hierarchical Design   This array compiler can be divided into three levels, from top down: the array level; the processor level; and the realization level (see Figure 6.40). The input to the array level is the behavior specification,

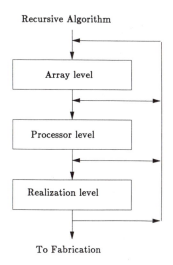

Figure 6.40: Top-down hierarchical design for systolic/wavefront arrays.

such as recursive algorithms, and the output of the array level will be the array processor for the algorithm. The input to the processor level is a functional requirement of the processor element and their interconnections in the array level, and the output is an assembly of components such as data-path and control path, which will be the input to the realization level. The output of the realization level is the layout. Note that there are feedback lines between stages, which allow iterative modifications of a design. Most of the software to be developed will not be "fully automated". Instead, user-friendly "interactive" design approaches are more suitable and effective. Conceivably, expert system techniques may be useful for the purpose.

6.6.1.1   Array Level CAD

In the array level, the design process is to convert an algorithm (a mathematical description of the design) into an architecture (a physical description of the design). This work concentrates on algorithm analysis to architecture mapping, parallelism extraction, and architecture design. The canonical mapping and generalized mapping methodologies discussed in Chapters 3, 4, and 5 provide the theoretical background for the software implementation. A schematic of the design tools at the array level is shown in Figure 6.41, which includes the following key components:

- *Mapping DGs to systolic arrays*: A mapping methodology can be applied to the DG to obtain an "optimal" systolic design (cf., Section 4.4).

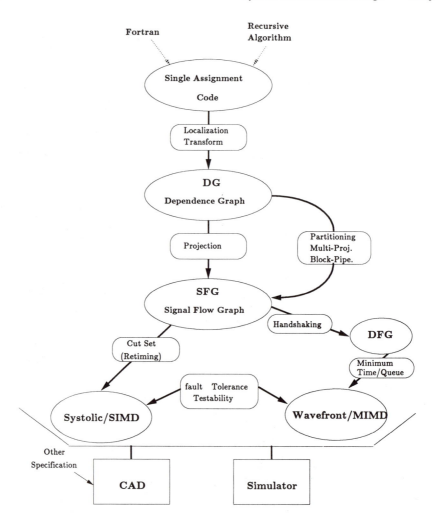

Figure 6.41: CAD packages for the array level design.

- *Mapping DGs to wavefront arrays*: Formal methods for mapping a DG to a wavefront array with the fastest throughput and minimal queues is discussed in Chapter 5.

- *Projection and multiprojection*: For an N-dimensional DG, one can project it to be an $(N-1)$-dimensional array. Further projection can be used to reduce it to a smaller dimensional array. Appropriate linear projections and linear schedules are to be decided by the user.

- *Partitioning*: For the partitioning, the designer selects the LPGS or the

LSGP scheme (see Section 6.3.2), and then determines the processor assignment and scheduling.

- *Other supporting software packages*: Examples are those dealing with communication, I/O, and interconnection networks.

6.6.1.2   Processor Level CAD

Future array compiler systems should support a wide range of processor architectures. One can identify at least the following candidates forms [Denye86]:

- datapath (e.g., AMD 2901 bit-slice design)

- accelerated datapath (e.g., multiplier)

- algorithmic processors (e.g., serial FFT butterfly)

- random logic/finite-state-machine (FSM) (e.g., EXOR and count)

Description of the processor level is shown in Figure 6.42.

6.6.1.3   Realization Level CAD

There are basically three kinds of approaches to realizing a chip: custom designed, standard cell, and gate array. For the array compiler, the standard

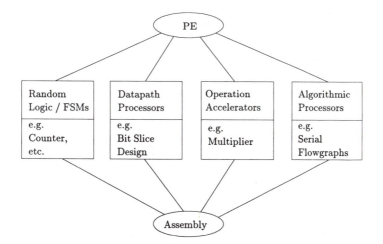

Figure 6.42: Compiler for processor level design (adopted from [Denye86]).

cell approach will be the most appropriate one. With functions properly specified by the higher level descriptions, the realization level consists of the following steps:

1. Logic synthesis: To realize a set of Boolean expressions or representations of an FSM. The standard cell library will provide the basic building blocks.

2. Chip planning, wirability analysis, partition and placement [Dunlo85], [Cheng84], [Kirkp83], [Goto81], [Breue77].

3. Global and detailed routing [Ting83], [Yoshi82], [Burst83], [Hsu83], [Ragha83]. After routing, the complete circuit layout is generated.

4. Design verification: circuit extraction, layout versus schematics verification, and timing analysis.

Though there are a few existing software packages which can support each of the above four steps, it is nevertheless very desirable to have an integrated design environment to do all the work automatically. In fact, the four steps in the realization process are very closely related. Naturally, the result of one step influences the subsequent steps as well as its previous steps. Such an iterative, top down, and bottom up design approach is very crucial [Hu85]. The iterative design scheme should be adopted in all the three levels in the array compiler.

6.6.2   Hadamard Transform Systolic Design Example

The convenience of chip-design support, such as the MOSIS program, allows VLSI chips to be designed in a university environment and then fabricated by manufacturers. It encourages the signal processing designer to use a vertically integrated VLSI design approach. One such design project at the University of Southern California (USC) is described next.

The chip for Hadamard transform designed at USC contains almost 14,000 transistors in a $6900 \times 6800 \ \mu m^2$ area based on 3-μm CMOS technology, fabricated by MOSIS. It is modular and expandable. A linear array of up to $2,000$ PEs can be built by connecting these chips.

The Hadamard transform was introduced in Section 2.3, and is basically a matrix-vector multiplication. Suppose we have N ($N = 2^n$) input data represented in vector form, say $\mathbf{x}$, and a transformed data vector, $\mathbf{y}$. Then, $\mathbf{y} = \mathbf{H}\,\mathbf{x}$ and each element is calculated by

$$y(k) = \sum_{j=0}^{N-1} x(j)(-1)^{p(k,j)} \qquad (6.1)$$

where

$$p(k,j) = \sum_{i=0}^{n-1} k_i j_i \qquad (6.2)$$

The k_i and j_i are the ith bit binary representations of k and j, respectively. Equation 6.1 suggests a recursive way to calculate the transform that is very suitable for systolic arrays, and Equation 6.2 illustrates how to generate the Hadamard matrix. Since the Hadamard matrix is very regular, we can generate the plus or minus ones inside a chip to reduce the communication cost in the sense of reducing input pins.

6.6.2.1   Systolic Array Design

There are two major components in the computational node of the Hadamard transform, i.e., the MAC and Hadamard coefficient generator, as shown in Figure 6.43. The PE is a full adder with an accumulator to hold the intermediate result. The design of the coefficient generator is not obvious. According to this projection, the k-direction is interpreted as the time axis and the j-direction is the spatial axis. From Eq. 6.1, the Hadamard coefficient for PE_j at time k is $(-1)^{p(k,j)}$. If the "-1" is represented by logic 1 and "$+1$" is represented by logic 0, the function of the generator is to generate logic 1 when $p(k,j)$ is odd or generate logic 0 when $p(k,j)$ is even. From

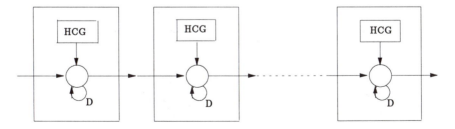

Figure 6.43: Systolic array for the Hadamard transform.

Eq. 6.2, $p(k, j)$ is the summation of a sequence of 1s and 0s; therefore, the generator is simply a parity checker of length n. The n input data are the binary representations of time k and location j ANDing together bitwise.

Besides these two components, I/O is another important issue. Input is simple and data are sent from the boundary. However, output is more complex and results are generated sequentially from the first PE to the last PE. Once the result is generated, it can be sent either to the left or to the right. In our design, we send the result to the right following the input data but slow down the propagation speed to one-half of the original speed. The reason for doing so is to have communication in one direction to save buses and control signals. The first result is produced N time steps after the chip starts working. After another N time steps, the first result moves to the middle of the linear array, and meanwhile the last result is generated. The results are then output from the end of the array and another set of input data can be input to the array at the same time. This I/O scheme provides a high pipelining rate.

6.6.2.2   Realization Procedures

The Hadamard transform chip is realized by the following sequential procedures (see also Figure 6.44). The entire layout is shown in Figure 6.45.

1. *Module Design*   Several modules are defined to perform the required functions. Each single chip contains eight PEs, which are basically accumulators. A Hadamard coefficient generator has also been incorporated into the chip, and is implemented using an XOR tree and counter to implement Eq. 6.2. An address token indicating the physical address of the chip propagates through the array and is increased by one when passing to the next chip. The address token is latched in a register in the chip.

2. *Function Simulation*   Once we have module level descriptions and block diagrams, we have to verify that they do indeed function as expected. Using one of the many available simulation tools (e.g. ISPS), we can describe and simulate the function of each module. Using the ELOGS logic simulator, the logic function correctness can be verified. Examining the simulator outputs, the original design can be corrected until no errors are found.

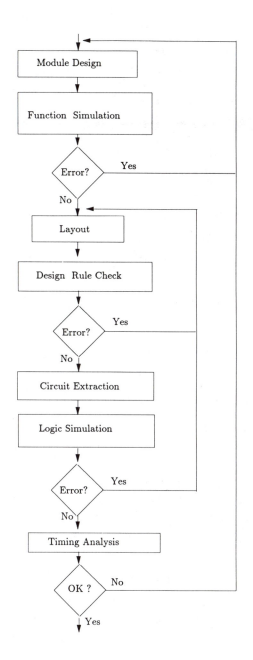

Figure 6.44: Steps of realization procedures.

HADAMARD

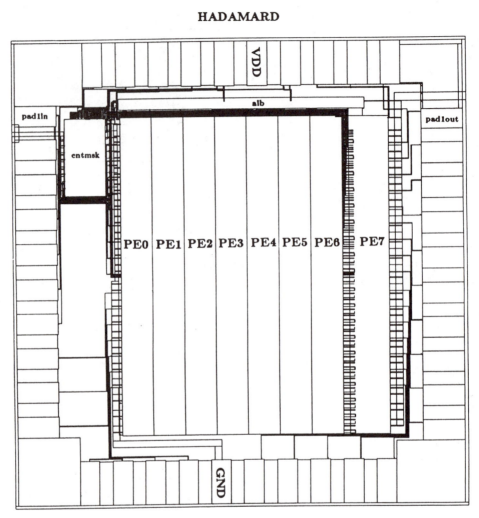

Figure 6.45: Hadamard transform chip.

3. *Circuit Layout* This design is fabricated in 3-μm CMOS technology. A standard cell library containing the required logic gates was utilized. This is obviously a trade-off between design time and chip area. The layout facility is an AED512 graphic terminal running the Caesar Graphics Editor.

4. *Design Rule Check*   This step is used to verify that the geometric patterns generated in Step 3 are correctly following the design rules. If any violation exists, Step 3 must be repeated to modify the layout.

5. *Layout Verification–Circuit Extraction*   Transistors and their interconnections are extracted from the geometric patterns of different material layers. This extracted circuit is used for simulation purposes in later steps.

6. *Layout Verification–Logic Simulation*   This is a transistor-level simulation. If the layout is correct, the extracted circuit will function exactly as verified in Step 5. The ESIM tool is used for this step.

7. *Layout Verification–Timing Analysis*   At this stage it is assumed that the generated layout realizes the Hadamard transform, and we now require to know how fast this chip can function. First, we use the Crystal tool to find the critical path, which is the longest route a signal has to propagate during a state transition. SPICE can then be used to analyze the equivalent circuit of the critical path. For this chip, the critical path is the carry chain in each PE which requires 195 ns to propagate the carry. Therefore, the clock rate is at most 5 MHz, which also turns out to be the throughput rate.

6.7   Concluding Remarks

This chapter considers the issues arising in the overall design of array processor systems. The technical issues addressed in this chapter are overall system organization, software techniques, partitioning problems, and FT.

In the proposed system organization, the basic system components are: host and ACU; interface unit; PE array(s); and interconnection networks. The main objective is to achieve high-speed, but flexible computing by integrating control, communication, and processing functions. The exact organization will depend to a great extent on the intended applications.

A good system should incorporate a partitioning scheme which permits large problems to be decomposed into several smaller subproblems and the solutions of the subproblems to be combined to yield an overall solution. Most signal and image processing algorithms are generally deterministic both in the time domain (with uniform and predictable task execution times) and in the space domain (with very restrictive conditional branching). Therefore, efficient partitioning and scheduling for concurrent processing is achievable for these algorithms.

Fault-tolerant design is of vital importance for enhancing the reliability of array processors. We have explored the issues on redundancy of architectures, run-time diagnosis/test, reconfiguration algorithms, and graceful degradation for FT. We have proposed a reconfiguration algorithm which is *distributively* executed by individual PEs. Several examples of algorithm-based fault-tolerant design are presented, indicating another promising approach to reliable system design.

System software is a very important and integrated part of the overall system development. We have explored the software tools for expressing the algorithm in a high-level language and compiling the program into executable object code on the arrays. The key factors affecting language and compiler design are application range, ease of programming, and execution efficiency. Trade-offs between these factors are weighed to reach an optimal compromise.

The basic features of programming and design languages for array processors have a great deal of commonality. Therefore, it is advisable to develop coherent software packages for programming as well as design tools for the system development.

6.8   Problems

1. *Example of system design*: An FFT based frequency domain filter example which achieves high spectral resolution with moderate levels of computation is presented. This example was developed with the goal of achieving maximum performance by employing a mixture of commercially available and semi-custom VLSI circuits. The basic architecture of the system is shown in Figure 6.46 [Swart86]. Data enters the system through a data acquisition module which blocks the data into packets and "windows" the data. The windowed data passes into the FFT module which computes its Fourier transform. The Fourier transform of the data and the filter kernel (for suppressing the noise) are multiplied in the frequency domain, and the result is transformed into the time domain with an inverse FFT module.

   (a) For this system, the signal bandwidth requires a 40 MHz data rate through the system, while the noise characteristics require filter resolution of 10 KHz. What is the expected transformed length N of the FFT? (i) 1024, (ii) 2048, (iii) 4096, or (iv) 8192.

   (b) It is evident that the best computational density is achieved with the clock rates of the order of 10 MHz. To achieve the desired rate of 40 MHz with circuits operating at a 10 MHz clock rate,

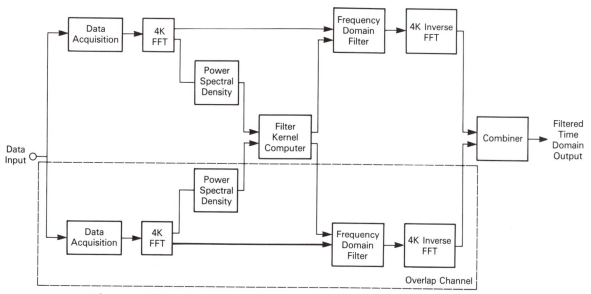

Figure 6.46: System architecture of frequency domain filter (adapted from [Swart86]).

what is the internal parallelism you should use? (i) 1, (ii) 2, (iii) 4, or (iv) 8.

(c) This module interfaces to the data source, produces the four data channels from the single input data stream, and windows the data in the time domain. Complex data is clocked at a 40 MHz rate into four pairs of shift registers (pairs of registers are used to accommodate the two components of each complex data word). The contents of the shift register are then transferred into parallel latches. How many parallel latches are needed? (i) 2, (ii) 4, (iii) 8, or (iv) 16.

(d) If radix 2 FFT is used, how many butterfly computations are required to complete an N point FFT. (i) 6144, (ii) 12288, (iii) 24576, or (iv) 49152.

(e) Each radix 2 butterfly computation consists of four multiplications and six addition operations, and this FFT processor can complete the N point transform in 102.4 microseconds, how many arithmetic operations per second can this processor achieve? (i) 0.6 billion, (ii) 1.2 billion, (iii) 2.4 billion, or (iv) 4.8 billion.

(f) Compare this special purpose chip implementation with the WARP machine (a somewhat more general purpose architecture) in terms of speed performance.

2. *Memory requirements for balanced computer architecture*: A PE can be characterized by its computational bandwidth, I/O bandwidth, and the size of its local memory. In carrying out a computation, a PE is said to be *balanced* if the computation time equals the I/O time [HTKun86]. Consider the problem of multiplying two $N \times N$ matrices, assuming a local memory of size M. We use a decomposition scheme that uses no more than M words of storage at any given time of the computation. The product matrix is computed in $(\frac{N}{\sqrt{M}})^2$ steps, each being the computation of a $\sqrt{M} \times \sqrt{M}$ submatrix of the product matrix. Every step is a multiplication of a $\sqrt{M} \times N$ submatrix of the first input matrix with an $N \times \sqrt{M}$ submatrix of the second.

(a) Show that the computation time $C_{comp} = O(NM)$, arithmetic operations and $C_{io} = O(N\sqrt{M})$ I/O operations. Therefore, $\frac{C_{comp}}{C_{io}} = O(\sqrt{M})$.

(b) Assume that for this computation, the PE is balanced. Now suppose that the computational bandwidth is increased by a factor of γ relative to the I/O bandwidth. Show that the increase for the size of the local memory in order to balance the computation again is:

$$M_{new} = \gamma^2 M_{old}.$$

3. *FP for matrix-matrix multiplication*: Using the FP primitives in this chapter, write a program that calculates a matrix-matrix multiplication using an *outer product* instead of inner product computation.

4. *FP for Convolution*: Using the FP primitives, write a program that performs convolution with the matrix-vector formulation of the algorithm. Is there any difference between this program and that of Problem 3. Why?

5. *FP for sequence manipulation*: Express in FP the following: for sequences x and y, reverse sequence x (that is x_1 to z_n and x_n to z_1, position x such that it overlaps y at the first element (form pair $< z_1, y_1 >$, multiply and repeat, sliding z to the y_n direction.

6. *DG for description in FP*: For problems 3, 4, 5 show how the DG is created from the algorithm description in FP.

7. *Matching to mesh*: An SFG with diagonal links is shown in Figure 6.47.

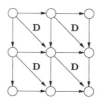

Figure 6.47: An SFG with diagonal links.

(a) Construct a DG from this SFG. (The procedure to construct a DG from an SFG can be found in Section 3.3.)

(b) To avoid the requirement of the diagonal links, extra layers of nodes may be put into the DG and interleaved with the original DG layers so that the data previously requiring diagonal links can be transmitted through these extra layers with no diagonal link requirement. Construct such a DG with extra layers of nodes.

(c) Using the same projection direction as before and assuming the nodes of the modified DG are homogeneous (i.e., the "inserted" DG nodes are treated the same as the original nodes), try to give a scheduling of the nodes of this "modified" DG.

(d) With the same projection direction as before and the scheduling you gave, derive a projected SFG. (Note that the projection direction for the DG with extra layers can be viewed as a PDSI projection direction. Thus the communication links required, i.e., mesh, is time-invariant. The number of buffers (delays) residing in each PE is time varying due to the PDSI projection direction, and the speed is twice as slow as that of the implementation with diagonal links. This is the cost for the reduced number of communication links.)

8. *More matching problems (to mesh)*: Four SFGs are shown in Figure 6.48. Show how these SFGs can be implemented on two-dimensional meshes by modifying the constructed DGs. (Only two dimensional SFGs are shown in Figure 6.48. Higher dimensional SFGs can be transformed to two dimensional SFGs by multiprojection if two-dimensional meshes are going to be used.)

Hint:

(a) For the SFG in Figure 6.48(a), the number of extra layers should be doubled.

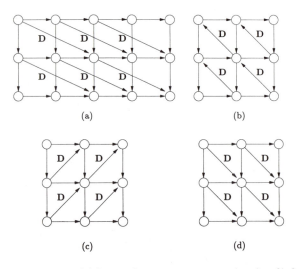

Figure 6.48: Four SFGs which require extra communication links on a mesh.

(b) To deal with the SFG in Figure 6.48(b), the communication links in the mesh are assumed to be bi-directional. Half duplex links on which data can be transmitted in one direction at one time are adequate for this SFG but, in general, full duplex links are preferred.

(c) To deal with the SFG in Figure 6.48(c), extra layers should be carefully inserted to avoid loops in the modified DG.

(d) For the SFG in Figure 6.48(d), extra layers should be inserted in the same way as for the SFG in Figure 6.48(c). If no extra layers are inserted, show that a mesh implementation may still be obtained but with two links between neighboring PEs. That implies the system clock rate may slow down to transmit two sets of data through one link of the mesh.

9. *Time-divided SFG*: Derive a time-divided SFG for the SFG array shown in Figure 6.49 to match a mesh array with a single channel in the east or south link.

10. *General time-divided SFG*: Referring to Section 6.3.1.1, derive a delay distribution scheme for a general time-divided SFG.

11. *Embedding a linear array into a hypercube* [Chan86]: A linear array may be viewed as a one-dimensional mesh and can thus be embedded

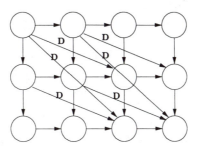

Figure 6.49: An SFG array.

in a hypercube. Draw a figure to show how to embed a linear array with 16 nodes into a 4-cube.

12. *FFT partitioning*: It has been shown that N point FFT can be computed by an array of $N/2$ PEs and a perfect shuffle network of size N (cf., Figure 2.6(a)). If we have an array of four PEs and a perfect shuffle network of size 8, devise a partitioning scheme for computing a 32 point FFT using the above array and interconnection network. What is the extra hardware needed for this partitioning?

13. *Linear schedule for LSGP*: Find a shortest linear schedule vector for the matrix multiplication of size 24×24 to be implemented by a systolic array of size 3×3, assuming the projection direction is $[1, 1, 1]$.

14. *Piecewise linear schedule for LSGP*: In Sec. 6.3.2, a linear schedule $[1\ 3\ 9]^T$ is found to be a good schedule for the 9×9 matrix multiplication using a 3×3 array. Show that this schedule, when implemented on a wavefront array, automatically becomes a *piecewise linear schedule*.

15. *Triple module redundancy (TMR)*: A popular algorithmic fault-masking approach is to use three identical computation modules within each PE. This is termed triple module redundancy (TMR). In TMR, if only one module fails, masking is still possible since the other two modules agree on a consistent (and thus likely to be accurate) result.

   (a) Show that the reliability of the duplicated part of each PE is

$$r_{\text{TMR}} = 3r^2 - 2r^3$$

   where r is the reliability of one module. The reliability of a module is the probability that the module works at time t given that the module works initially.

(b) If $r = 0.99$, what is the reliability of an array processor of 100 PEs without redundancy?

(c) If $r = 0.99$, what is the reliability of an array processor of 100 PEs with TMR?

16. *Redundant PE to improve reliability*: For an $n \times n$ VLSI array processor, incorporating some hardware redundancies can improve reliability. There are two schemes which use the same amount (n^2) of redundant PEs:

   • Scheme A: These n^2 spare PEs form another spare $n \times n$ array processor, which runs in parallel with the original array and executes the same program.

   • Scheme B: The n^2 PEs are distributed to every PE, so that each processor module consists of two identical PEs working in parallel.

   Given that the reliability of an individual PE is r, calculate and compare the reliabilities of the array for the two redundancy Schemes A and B.

17. *Ideal reconfiguration scheme*: Referring to the previous exercise, suppose that the n^2 redundant PEs are not distributed to each processor. If we assume an ideal reconfiguration process exists, i.e., any failed PE in the array can be replaced by a spare PE, what is the reliability of the array?

18. *Maximum wire length*: From the reconfigurability theorem, it can also be derived that for any placement satisfying all four conditions, the physical communication between any two logical adjacent PEs has at most three unit distances. Prove the following corollary based on the statements of the reconfigurability theorem.

   Corollary: In any successful reconfiguration as stated in the reconfigurability theorem, the maximum wire length is $D_f \leq 3$. In other words, lengths of horizontal or vertical links are 1, 2 or 3 units.

19. *Proof of the Reconfigurability Theorem*: Prove that Table 6.2 lists all the possible situations that the four conditions in the reconfigurability theorem are satisfied.

20. *Proof of the compensation path theorem*: The compensation path theorem allows us to select compensation paths so that reconfiguration using 1-track may be successfully accomplished. In order to prove the compensation path theorem, some procedures can be followed [Kung86d]. Assume that a horizontal compensation path affects only $y(i, j)$ and a vertical compensation path affects only $x(i, j)$:

  (a) Prove that any assignment which satisfies the conditions of the reconfigurability theorem will also satisfy the conditions after any nonintersected horizontal compensation path.

  (b) Prove that any assignment which satisfies the conditions of the reconfigurability theorem will also satisfy the conditions after any nonintersected vertical compensation path.

21. *Switch control*: To control a switch located between two horizontal neighboring PEs, a vertical reconfiguration state (VRS) within each PE is defined in Table 6.5.

    Derive a truth table to show how to control a switch by using the vertical reconfiguration states of its two horizontal neighboring PEs.

22. *Fault-tolerance for systolic array with local control*:

  (a) Is it possible to use local control to achieve run-time FT for a 1-D systolic array?

  (b) Is it possible to use local control to achieve run-time FT for a 2-D systolic array?

  (c) What is the overhead compared with a scheme used in a wavefront array with local control (considering both 1-D and 2-D arrays)?

  (d) Does the clock cycle depend on the array size for this scheme (considering both 1-D and 2-D arrays)?

23. *Fault-tolerance using temporal redundancy*: In the compensation path correction process, since two jobs are residing in PE_j (one is originally

00	Good PE, not on a vertical compensation path
01	Good PE, on a vertical compensation path
10	Faulty PE, on a vertical compensation path
11	Faulty PE, on a horizontal compensation path

Table 6.5: The definition of the vertical reconfiguration state (VRS).

assigned to PE_j and the other is originally assigned to PE_{j-1}), PE_j can thus execute these two jobs alternately[8] if it is impossible to do the compensation path correction. That is, the temporal redundancy can be utilized. Compare the usage of space redundancy vs. temporal redundancy in a fault tolerance scheme.

Hint: The system throughput decreases to about half of its original throughput without declaring system failure. This is called *graceful degradation*. Note that although the reduction of throughput may be intolerable for some systems, the detection of the system failure for these systems can be done by checking the system throughput from outside without going into each PE.

24. *Weighted Checksum Property:* Assume that $\mathbf{A}$ is an $n \times m$ matrix. The column, row, and full weighted checksum matrix $\mathbf{A}_c$, $\mathbf{A}_r$, and $\mathbf{A}_f$ of the matrix are defined as:

$$\mathbf{A}_r = \begin{bmatrix} \mathbf{A} & \mathbf{Af} & \mathbf{Af}_w \end{bmatrix} \quad \mathbf{A}_c = \begin{bmatrix} \mathbf{A} \\ \mathbf{e}^T \mathbf{A} \\ \mathbf{e}_w^T \mathbf{A} \end{bmatrix} \quad \mathbf{A}_f = \begin{bmatrix} \mathbf{A} & \mathbf{Af} & \mathbf{Af}_w \\ \mathbf{e}^T \mathbf{A} & \mathbf{e}^T \mathbf{Af} & \mathbf{e}^T \mathbf{Af}_w \\ \mathbf{e}_w^T \mathbf{A} & \mathbf{e}_w^T \mathbf{Af} & \mathbf{e}_w^T \mathbf{Af}_w \end{bmatrix}$$

where

$$\begin{aligned} \mathbf{e}^T &= [1\,1\,\cdots 1] \\ \mathbf{e}_w^T &= [1\,2\,\cdots 2^{n-1}] \\ \mathbf{f}^T &= [1\,1\,\cdots 1] \\ \mathbf{f}_w^T &= [1\,2\,\cdots 2^{m-1}] \end{aligned}$$

Show that for matrices of compatible sizes, the following properties hold:

(a) If $\mathbf{A}_1 \ldots \mathbf{A}_n \mathbf{B} = \mathbf{C}$, then $\mathbf{A}_1 \ldots \mathbf{A}_n \mathbf{B}_r = \mathbf{C}_r$, $\mathbf{B}_c \mathbf{A}_1 \ldots \mathbf{A}_n = \mathbf{C}_c$, and $\mathbf{A}_c \mathbf{B}_r = \mathbf{C}_f$

(b) If $\mathbf{A} + \mathbf{B} = \mathbf{C}$, then $\mathbf{A}_r + \mathbf{B}_r = \mathbf{C}_r$, $\mathbf{A}_c + \mathbf{B}_c = \mathbf{C}_c$, and $\mathbf{A}_f + \mathbf{B}_f = \mathbf{C}_f$

---

[8]Note that the neighboring switches of PE_j is controlled by the working job and thus their states may be time-varying.

(c) If $s\mathbf{A} = \mathbf{C}$, where s is a scalar, then $s\mathbf{A}_r = \mathbf{C}_r$, $s\mathbf{A}_c = \mathbf{C}_c$, and $s\mathbf{A}_f = \mathbf{C}_f$

(d) If $\mathbf{A}^{\mathrm{T}} = \mathbf{C}$, then $\mathbf{A}_r^T = \mathbf{C}_c$, $\mathbf{A}_c^T = \mathbf{C}_r$, and $\mathbf{A}_f^T = \mathbf{C}_f$.

(e) If $\mathbf{A}$ is LU decomposable, then $\mathbf{A}_f = \mathbf{L}_c\mathbf{U}_r$

25. *Single error correction by weighted checksum scheme*: If we use another vector $\mathbf{W}_2 = (1\ 2\ 3\ ...\ n)$ instead of $(1\ 2\ 2^2\ ...\ 2^{n-1})$, we can still detect and correct a single error. Modify the single error correction procedure presented earlier and show how this scheme works.

26. *Double consecutive errors correction*: Referring to the discussion of correcting two consecutive errors by a weighted checksum scheme under the transient faults condition (see Section 6.4.3.3), assume that S_1, S_2, *or* S_3 are not equal to zero. Show that: $j = (\log_2 \frac{3S_2+S_3}{S_1}) - 1$ if j is odd, otherwise $j = (\log_2 \frac{3S_2-S_3}{S_1}) - 1$. $\delta x = 2S_1 - 2^{-j+1}S_2$ and $\delta y = 2^{-j+1}S_2 - S_1$. All variables are defined as before.

27. *Physical Array Design:* The fault-tolerant design based on *simplified grid model* may be a bad design considering the fabrication-time FT issues. In fabrication-time, after testing, some healthy PEs and switches may be connected to form a physical array. Because each switch in the simplified grid model is controlled by its neighboring PEs, it may be difficult to form the (simplified grid model) array.

Show that a mesh of PEs without switches (cf., Figure 6.15), with each PE being connected to its eight neighbors and multiplexers being built in each PE to choose four out of eight neighbors can be adopted to simplify the fabrication-time FT.

Hint: When there is no faulty (physical) PE, the mesh interconnection embedded can be constructed by using multiplexers. When faults occur, the multiplexers are used to reconfigure the array.

28. *DG interleaving time-redundancy FT*: With reference to Figure 6.35 and in the case where there is a permanent fault, rationalize why this fault should be at PE_i and not at PE_{i-1}.

29. *DG interleaving methods*: In Figure 6.34, the DGs are interleaved in a way that there are two time units between same computations in different DGs. Derive a DG interleaving method which has no starting time difference and compare it with the one in Figure 6.34.

Chapter 7

IMPLEMENTATION OF ARRAY PROCESSORS

7.1 Introduction

In the early stages of DSP research, activity focused on the development of DSP algorithms and demonstration of their applications to practical system design. VLSI has made implementation of system hardware or even highly parallel array processors economically feasible and technically realizable. In fact, we have just witnessed a remarkable period in the development of modern VLSI DSP technology. System designers immediately realized the advantages associated with such implementations. VLSI technology acts strongly as an influencing factor, favoring some organizational approaches and penalizing others. For example, the use of on-chip memory is attractive for VLSI, since relatively high densities can be obtained and interchip communication can be eliminated. Array processors implemented with today's VLSI technology can offer a much greater hardware capacity, higher speed, and lower power compared with other existing technologies. For example, based on micro- or submicrometer technology, the VHSIC program in the United States is aimed at increasing the gate count to levels in excess of 30,000 gates per chip and simultaneously increasing the performance of the

452

individual gates. This should allow development of complete systems such as FFT on single chips.

Hardware organization can also dramatically affect performance. This is especially true when the implementation is in VLSI, where the interaction of the architecture and its implementation is more pronounced. Design of VLSI systems should be based on the potential and constraints imposed by the VLSI device technology and design methodology. The influential issues in the VLSI system design include interconnection, communication, functional primitives, modularity, and system clocking, which were discussed in Chapter 1.

The actual implementation of the various signal processing techniques and functions in a specific application area may range from the use of dedicated fixed function hardware to the use of software on a general-purpose digital computer. The choice of implementation method tends to be driven by cost and performance considerations and processing flexibility requirements. The following list indicates the general trends in implementation hardware for several application areas.

- Telecommunications/speech: single chip processors to mainframes.

- Radar/sonar: special purpose processors and array processors.

- Seismic/image processing: general purpose mainframes and array processors.

- Music/biomedical: special purpose hardware and micro- and minicomputers.

The general trend for implementing very high performance DSP has been through the use of special purpose hardware, especially where real-time performance requirements exist. For applications where processing flexibility is a major consideration, the trend has been away from large mainframe scientific processors toward the specialized attached processors that are optimized for high-speed arithmetic operations on arrays of data. This represents a first step toward multiple processor solutions to flexible real-time signal processing [Bowen82].

In general, there are two approaches to the design of a special-purpose computing system. One is to design a dedicated hard-wired system, which cannot be easily reprogrammed to perform other tasks besides the original task. Examples of such systems include FFT processors, and digital filters. On the other hand, a programmable system allows users to program the application tasks easily. Due to the greater flexibility offered by the programmable approach, we place more emphasis on the discussion of the programmable array processor systems. It should be noted that most of the discussions still apply to dedicated systems.

The design/implementation trade-offs for systolic/wavefront processors include the selection of fixed-point or floating-point arithmetic, bit-serial or bit-parallel computation, bit-serial or bit-parallel data communications, host interfacing, data I/O flow, interconnection networking, control methodologies, hardware support for partitioning, fault tolerance, software techniques, and many others. This chapter first addresses two levels of design and implementation considerations:

1. Processor level design: How can the systolic and wavefront arrays best be implemented in off-the-shelf or custom-designed hardware components?

2. Overall system design: How can these arrays be incorporated into an overall computing system?

7.2 Processor Level Implementation

Before we consider the design of an array processor system for DSP applications, we first consider the design of the processing elements, which will be used as the basic components. Unlike a general purpose microprocessor, the processing element of a DSP system has its own characteristics to meet the nature of computing tasks required for DSP applications. The key characteristics for DSP applications are adequate word length, fast multiply, accumulate, high-speed RAM, fast coefficient table addressing, and new sample-fetching mechanism.

The key to high performance of DSP hardware is to have the hardware processing elements organized in such a way that it matches the structure of the computation algorithm at hand. Hence, the challenge to the architect of a general DSP chip is to design one DSP architecture, which can be packed into a single IC chip and achieve a relatively good performance for a wide variety of DSP applications. The most important operations in DSP are addition, multiply and multiply-and-accumulate; therefore, it is important to design a fast arithmetic unit for these operations.

7.2.1   PE Architectural Considerations

7.2.1.1   Basic Building Blocks of a DSP Chip

The basic building blocks of a DSP chip are shown in Figure 7.1. The architecture design of a DSP chip involves the specification of these basic building blocks and the determination of the way in which they are interconnected [Jenq86]. The basic components are described as follows:

1. **Memory unit (MU)** The memory unit consists of *data memory* (DM) and *coefficient memory* (CM). DM stores data and some intermediate results, whereas CM stores the fixed coefficient constants, such as the filter coefficients and the sine and cosine tables of the FFT and the sinusoidal signal generation applications.

2. **Control Unit (CU)** The control unit consists of a *program counter* (PC), a *program memory* (PM), an *instruction decoder* (ID) and an *address computation unit* (ACU). The PC generates the address of the

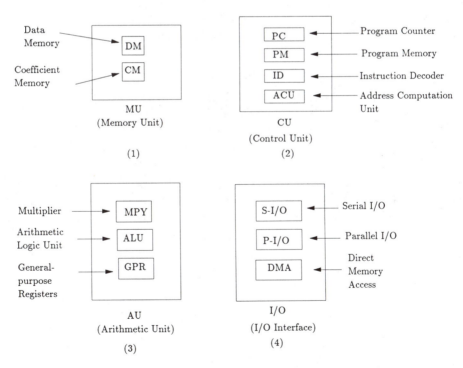

Figure 7.1: Basic functional components of a DSP chip.

instruction to be executed. The PM stores the program instructions and the ID decodes the instructions to generate the control signals to carry out the execution of the instruction. The ACU computes the addresses for the operands under the direction of the control signals generated by the ID.

3. **Arithmetic Unit (AU)**   The arithmetic unit consists of a *multiplier* (MPY), *arithmetic logic unit* (ALU) and *general purpose registers* (GPRs). The MPY is generally a parallel array multiplier in a DSP chip. The ALU performs addition, subtraction, shifting, and some logic functions. GPRs hold temporary data and are usually directly connected to the data bus and other basic building blocks. In the simplest case, there is just one register, which is called the accumulator.

4. **I/O Interface** The I/O interface includes the *serial I/O* (S-I/O) unit, the *parallel I/O* (P-I/O) unit, and the *direct memory access* (DMA) *controller*.

Let us now look more closely into separate components and list the requirements of some of the basic components of a DSP processor chip.

1. *Multiplier/ALU*   Since arithmetic operations dominate the execution of most DSP algorithms, a high-speed array multiplier is very desirable. Currently, the multiplication time is still the dominant factor in the determination of the instruction cycle time of a DSP chip. This implies that if the multiplication time can be further shortened beyond just the technology improvement, it will be a real challenge to the architecture design to keep the instruction cycle time as short as the multiplication time. An ALU with pre- and post-shifting capabilities would be useful for scaling operands to prevent the overflow in fixed point operations. Some nonlinear hardware operations, such as rounding saturation arithmetic operation, taking absolute value, and format conversion without extra instruction cycles, would be desirable for speed improvement.

2. *General purpose registers*   Having more than one general purpose register is convenient for handling temporary variables in the delay operation and the calculation of complex multiplications.

3. *Multiple buses*   In order to speed up execution, many parallel paths

exist, both inside and outside the processor, to enable parallel communications between memories and functional modules of the processor.

4. *Memory*   Since the signal processors are expected to execute all their programs during one sampling interval and because they require fast memories, which are expensive, they usually have a limited address space, the maximum size of which is determined by the ratio of the sampling interval and the processor cycle. Many DSP algorithms, such as FFTs and convolutions, require constant coefficients. If high throughput is required, it is a common practice to store the coefficients in a ROM. On the other hand, some arithmetic operations, such as multiplication, can also be implemented by looking up values in a table, which also requires the use of a ROM.

5. *Address computation unit*   To support a structured access, usually a separate address computation unit with various indexing capabilities is implemented in a DSP chip. The inclusion of a dedicated address computation unit has also contributed greatly to the shortening of the instruction cycle time. Some DSP chips go further, having a separate address computation unit dedicated to each memory unit, which gives even greater flexibility for structured data and coefficient accesses.

6. *I/O*   To support parallel processing, it is very important to include the I/O interface on the chip, since off-chip interfacing complicates the system design and slows down I/O bandwidth. Due to current pin limitations of packaging, a large number of parallel I/O channels do not seem to be feasible. As technology progresses, more parallel I/O support can be integrated with processing power into a single chip.

7.2.1.2   Performance Measures and Design Trade-off

Two major performance measures of a DSP-based system are precision and speed.

Precision   The achievable precision is dictated by the word size of the MUs and the AU. DSP algorithms are usually iterative, and the errors in the intermediate computations are accumulated. To avoid excessive errors, great accuracy must be maintained throughout execution. In real-time applications, it may be impossible to backtrack and correct errors once they occur. Hence accuracy becomes more crucial. Most of the commercial DSP processors of recent design have word lengths of at least 16 bits.

Speed The speed of a DSP chip depends on many factors, such as the instruction cycle time and the power of the instruction set supported by the architecture. If it takes N instructions to complete the execution of an algorithm, then the total time needed for this algorithm execution is $N \times t$, where t is the instruction cycle time. Notice that the ultimate speed performance measure is the product of N and t, so just a shorter instruction cycle time t alone is not enough. Therefore, a desirable architecture is the one that can support a powerful instruction set, such that most of the important DSP algorithms can be accomplished in a few instruction cycles, while having the shortest possible instruction cycle time achievable by the VLSI technology.

Currently, for most DSP chips, an instruction cycle time is dictated by the speed of the multiplier. Therefore, we should try to accomplish all other operations, such as instruction fetch, instruction decoding, operands (coefficients and data) fetch, and storing results back to DM, in a multiplication time. To achieve this, several architecture requirements are necessary.

- To support a highly pipelined operation, each building block operates in parallel; hence they should be carefully interconnected to avoid bus interference.

- A long instruction word is needed in order to decode the control signals quickly (i.e., in parallel) for each unit operating in parallel.

- It is even more desirable to have a separate data memory and coefficient memory units with the coefficient bus separated from the data bus and directly connected to one of the multiplier's register. In this way, two operands can be accessed in parallel.

- It is necessary to improve memory-AU bandwidth. The separation of PM and DM, the separation of the program bus from the data bus, and the dedicated address computation unit are all aimed to improve the memory-AU bandwidth.

Basic Design Trade-offs The factors influencing technology decisions and cost are power, area, speed and functionality. The following are the major trade-offs that are made during design of a DSP processor:

1. *Speed versus power* It is well known that, for a given technology, the faster a circuit, the more power it consumes, simply because the transistors conduct more frequently.

2. *Speed versus area*   Since the transistor loads in a circuit are capacitive, to achieve greater speed we must charge/discharge these capacitors using larger currents. This leads to the use of circuits of lower resistance, which is equivalent to using larger area per circuit.

3. *Functionality versus area*   The more functions implemented by a circuit then the more area is required because of the greater number of subcircuits and their interconnections.

4. *Memory trade-offs*   A signal processor must have a very fast memory to sustain high throughput, but a fast memory means high power consumption and large area as well as low decoding overhead. Hence it is desirable to use small memories with high speed.

5. *Number of connections*   The signal processor can be made faster if the information consumed (input) and produced (output) is transmitted in parallel. This means that a large number of connections with the outside world is required. For integrated circuits, this specifically means a large number of pins. With current technology, packaging can be more expensive than the actual chip, and the packaging cost depends heavily on the number of pins. A large number of pins requires a large number of I/O circuits on a chip; these circuits consume more power and slow down the operation.

7.2.1.3   Architectural Features

Here are some important architecture design considerations that are essential to achieve a desirable DSP architecture [Jenq86].

1. An AU capable of efficiently executing the multiply-and-accumulate operation is important, since most DSP algorithms are computationally intensive.

2. An architecture capable of supporting pipelining structure efficiently is important, since most DSP algorithms can be heavily pipelined.

3. The memory-AU bandwidth and the AU speed should be carefully balanced, such that the total system performance is not seriously degraded because of this bottleneck.

4. A good programming language is needed to support array processing.

5. Multiple I/O ports are needed to provide the mesh connections in the array.

6. Bit-serial versus bit-parallel designs must be considered. Bit-serial design is economical in terms of communication and is suitable for pipelining; on the other hand, bit-parallel design is faster, and requires more communication. [1]

Other commonly encountered design considerations in the PE architecture are briefly discussed next:

Harvard Architecture   Since the processors usually run fixed programs that operate on infinite streams of data, the program memory space is separated from the data memory space in order to facilitate instruction prefetching and data fetching in parallel.

Addressing Modes   Because of the localized nature of both data and instructions in DSP algorithms, few addressing modes are required, but the address calculation must be done with great speed. Pipelining the address calculation and/or parallel address determination are widely used technologies. Since high speed addressing units are required, the addressing modes are kept to an absolute minimum to simplify addressing unit design.

Multiple Field Assembly Programming   To ease the program development for a DSP chip based system, a multiple-field type of assembly language is more desirable than the traditional assembly language. In a DSP chip, there are several basic building blocks operating concurrently and somewhat independently, so a multiple-field type of assembly instruction can reveal the operation of the instruction more clearly and give greater flexibility in assembly programming. Several DSP chips (such as Bell DSP-1 and

---

[1] It is claimed [Denye83] that bit serial architectures are favored in terms of size, power, pin count and cost for the implementation of dedicated systems. Serial arithmetic offers a simple and efficient route to the implementation of a wide range of digital signal processors. The advent of primitive libraries and of supporting silicon compilers makes available a direct route into custom VLSI for a host of novel system developments. This certainly will make serial architectures more popular. On the other hand, bit-parallel systems are more favorable for a programmable system.

NEC μPD7720) use the multiple-field instruction format. Though the TMS 32010 does support parallel processing, this is not reflected in its assembly instructions. This fact makes it difficult to remember each micro operation performed in each processing unit in an instruction.

RISC versus CISC   Here, RISC stands for *reduced instruction set computer* and CISC stands for *complex instruction set computer (CISC)*. As discussed above, the speed performance can be achieved by [Henne84]:

1. Minimizing the clock cycle of the system. This implies both reducing the overhead on instructions as well as organizing the hardware to minimize the delays in each clock cycle.

2. Minimizing the number of cycles to perform each instruction. This minimization must be based on the expected dynamic frequency of instruction use.

These trade-offs and the observation that large architectures generate additional overhead have led to the RISC approach [Radin82], [Patte80], [Henne84]. This idea strongly contrasts to the traditional CISC approach, which is effectively an architecture with more powerful instructions. The pros and cons of these two approaches have been controversial.

A RISC is a machine with a simplified instruction set [Henne84]. The architectures that are generally considered to be RISC's are the Berkeley RISC I and II processors, the Stanford MIPS processor, and the IBM 801 processor (which is *not* a microprocessor). All three architectures have simpler instructions than most other machines and avoid features that require complex control structures. The most important features of RISC type architecture are: (1) Regularity and simplicity in the instruction set allow the use of the same, simple hardware units in a common fashion to execute almost all instructions. (2) Most instructions execute in one machine (or pipeline) cycle. These architectures are register-oriented: All operations on data objects are done in the registers. Only load and store instructions access memory. (3) Fixed-length instructions with a small variety of formats are used.

The advantages of RISC-style architecture come from a close interaction between architecture and implementation. The simplicity of the architecture leads to the simplicity of the implementation. The advantages of RISC include: fast instruction decoding, single cycle execution of most instructions,

and shorter clocking period. The potential disadvantages of the RISC architecture come from two areas: memory bandwidth and additional software
requirements.

However, one of the disadvantages of RISC architecture as used for
DSP applications is that in order to obtain fast multiply or multiply-and-
accumulate instructions, the instructions will probably need more than a
single cycle and a substantial amount of microcoding. Therefore, although
RISC is a good idea in simplifying the design of a VLSI processor, the design
of DSP processors may have to make a trade-off between RISC and CISC
approaches to meet the high throughput requirement.

7.2.2   Programmable Commercial DSP Chips

In the early 1980s, $3\mu m$ MOS technology made possible the fabrication of
single-chip programmable DSP devices that contained all of the basic ingre-
dients of DSP hardware on a single chip: the multiplier, data RAM, fast
ALU, program memory, and so on. Typically, those devices contained about
50,000 transistors which were connected together to execute instructions up
to a rate of 5 MIPS (million instructions per second). The microprocessor-
like architecture provided an additional major advantage of programmability.
It became possible to use the same device in multiple applications by simply
changing the program. The wide acceptance of the programmable approach
has resulted in the introduction of a second generation of devices based upon
$2\mu m$ MOS technology, integrating up to 150,000 transistors. These devices
are faster, contain more memory, and have even more microprocessor-like
features compared to the first generation of devices. The advantages in cost
and flexibility offered by those devices made possible the migration of DSP
technology into commercial systems. It produced the same kind of revolution in DSP system design as microprocessors did in the general-purpose
computing field.

Examples of commercially available VLSI chips worthy of consideration for array processor implementations are: TI TMS32010/32020, Fujitsu MB8764, INMOS transputer, NCR GAPP chip, NEC Data Flow chip
μPD7281, and Analog Devices ADSP-1XXX chip set.

7.2.2.1   Texas Instruments TMS32010/32020 Signal Processor

The functional block diagram of the TMS 32010 digital signal processor
[TI83] is shown in Figure 7.2. This chip was introduced in 1983, and was

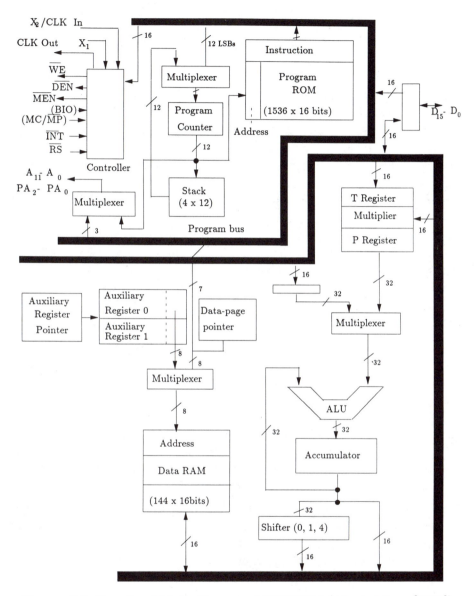

Figure 7.2: Functional block diagram of TMS32010 (adapted from [TI83]).

designed as a high speed digital controller with numerical capability. The TMS32010 uses a modified Harvard architecture. In a strict Harvard architecture, program and data buses are separated, which allows a full overlap of the instruction fetch and execution. In the TMS32010, paths are provided between the two buses to provide some flexibility. The main features of the TMS32010 are as follows:

Technology: The TMS32010 is fabricated in 2.7 μm NMOS and packaged in a 40-pin DIP. It has a 200 ns instruction cycle time at a 20 MHZ clock rate.

Control Unit: The TMS32010 has two versions; the first one is equipped with 1536 × 16 program ROM, whereas the second one has no internal program ROM. It has the capability of accessing up to 4K words of 16-bit-wide program memory at full speed if the external memory has an access time less than 100 ns. It has a program counter and a 4-level stack, which enables the user to perform subroutine calls, branches and interrupts. Two auxiliary registers, AR0 and AR1, selected by an auxiliary register (ARP) are used to perform the indirect addressing of the data RAM. The auxiliary registers can be made to auto increment/decrement after each instruction is executed. A 1-bit data page point is provided to combine with the 7-bit address in the instruction word to address directly the 144 × 16 data RAM. The two auxiliary registers serve many different functions. As discussed earlier, the lower 8 bits of the AR can be used for indirect addressing purposes. The lower 9 bits can also be used as a 9-bit loop counter. The upper 7 bits of the AR are not affected by any auto increment/decrement operation. Two auxiliary registers can also be used as general-purpose 16-bit registers for temporary storage use.

Arithmetic Unit: The TMS32010 has a 16 × 16 → 32-bit array multiplier, which can multiply two 16-bit 2's complement numbers to produce a 32-bit product in 200 ns. Three other arithmetic elements are the ALU, the accumulator, and two shifters. In a typical arithmetic operation, the data coming form the 16-bit data bus first passes through the barrel shifter, which can left-shift a word 0 to 15 bits, depending on the value specified in the instruction. The data then enter a 32-bit ALU, where they are loaded into or added to or subtracted from the 32-bit accumulator. Another shifter is provided to perform the scaling on the output of the accumulator as the data are being stored back to the data RAM. Although it seems, from the block diagram, that the architecture allows for the multiply-and-accumulate operation to be completed in one instruction, the instruction set does not support it. Hence, it takes two instruction cycles (i.e., 400 ns) to perform the multiply-and-accumulate operation. It is not surprising to see that the recently announced TMS32020 does have a one-instruction-cycle multiply-and-accumulate operation.

Memory Unit: All nonimmediate data operands are stored in the 144 × 16 data RAM. Two instructions, table read (TBLR) and table write (TBLW), are provided to transfer values from program memory to the on-chip data

RAM and from data RAM to program memory (presumably in the form of off-chip RAM), respectively.

I/O Unit: The two instructions IN and OUT allow users to transfer data between the peripheral and the data RAM. The three multiplexed least significant bits of the address bus are used as a port address by the IN and OUT instructions; hence up to eight I/O ports can be implemented with the TMS32010.

Development System: The availability and the good development support system may be two of the main reasons why the TMS32010 becomes the most successful DSP chip. For software development, it has a macro assembler/linker, a software library , and a simulator. For hardware support, it has the TMS32010 evaluation module, the TMS32010 emulator (XDS), and a TMS32010 analog interface board (AIB). It also has an active third-party support program. An extensive documentation support coupled with workshop offerings and the availability of public domain programs from universities make the TMS32010 the most popular DSP chip among industrial and academic users.

TI TMS32020   The TMS32020 signal processor was introduced in early 1985 [TI85], and is a direct upgrading from the TMS32010. The overall modified Harvard architecture remains unchanged. However, in almost every building block, performance is greatly improved. Here are some of the highlights.

Technology: The TMS32020 is fabricated in 2.4 μm NMOS and packaged in a 68-pin grid array. The instruction cycle time remains at 200 ns with 20 MHz clock rate.

Control Unit: The program counter is expanded from 12 bits to 16 bits; hence the addressing capability is expanded from 4K to 64K. A repeat counter is added to implement the repeat instruction. Also, a global memory allocation register is provided to facilitate the multiprocessor configuration. The address computation unit is greatly enhanced. The number of registers is increased from two to five, and a dedicated unsigned arithmetic unit for address calculation is added.

Arithmetic Unit: The major improvement of the AU is the implementation of the single-cycle instruction multiply-and-accumulate. Also, a shifter is added to the output of the product register.

Memory Unit: The memory unit is also greatly expanded. The TMS32020 provides three separate on-chip memory units: a 32 × 16 data RAM, a 256

× 16 data RAM, and a 256 × 16 data/program RAM. The data page pointer is also expanded from 1 bit to 9 bits.

I/O Unit: The TMS32020 provides up to 16 I/O ports, which doubles that of the TMS32010. A DMA capability is also provided. About 10 system I/O control registers are included in the new chip.

Instruction Set: Although the instruction word length remains at 16 bits, the instruction set is almost doubled (compared to TMS32010) due to the capability enhancement of all building blocks.

7.2.2.2 The Fujitsu MB8764 Signal Processor

The Fujitsu MB8764 is a very fast and very complex CMOS monolithic signal processor [Fujit84]. As Figure 7.3 shows, it consists of seven functional units,

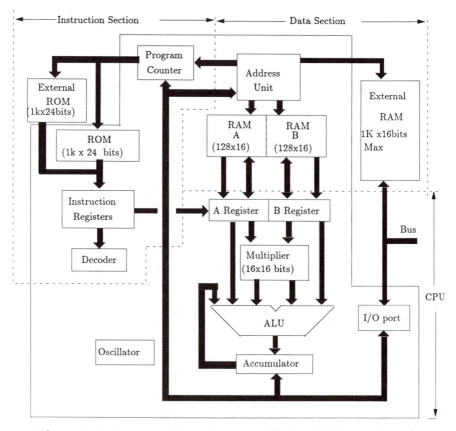

Figure 7.3: Functional block diagram of Fujitsu MB8764 (adapted from [Fujit84]).

the clock generator, the sequencer, the addressing section, the RAM, and the ALU section which contains the multiplier and the I/O section.

The sequencer contains the program counter, two loop counters, external program memory interface, and two instruction registers, which form a pipeline feeding the decoder section, which, in turn, controls the chip. Using the two instruction registers, instructions are fully decoded and operand addresses are calculated before the AU starts the actual computation. The addressing section contains two address ALUs so that both operand addresses can be calculated in parallel.

The RAM is divided into two 128-word (16-bit) RAM banks that are used to feed the inputs of the ALU simultaneously. The second bank can be extended using external RAM chips (up to 1K). The ALU consists of a 16×16 multiplier and an adder/accumulator, which is 26 bits wide, as well as a logic section. The cycle time of the ALU is 100 ns. The I/O section contains a DMA counter as well as address and data registers.

Because of the multiplicity of functional units, the Fujitsu chip can perform ALU operations in parallel with overhead operations such as address calculations and I/O, thus greatly speeding up the execution of a program.

The instruction word of the MB8764 is 24 bits wide and is split into two parts. The first part is the I/O and address-calculation instruction and the second part is the ALU instruction. The ALU instruction provides a variety of arithmetic and logical operations including multiply-and-add and right/left shifts. The address and I/O instruction provides simultaneous data transfer operations from both memories to the ALU and memory block transfers. Also a number of conditional and unconditional branch and flag operations are included.

7.2.2.3   INMOS's Transputer

INMOS's transputer chip (T414 or T424) is an Occam-language-based design, which provides hardware support for both concurrency and communication — the heart of array computing.

Figure 7.4 is taken from the INMOS advanced information document for the INMOS T424 transputer [Wils84a]. The chip is clearly a complete computer. Indeed, it has been designed so that its external behavior corresponds to the formal model of a process. It adopts the now-popular RISC architecture design. It has a 32 bit processor capable of 10 MIPS, 4 kbytes of 50 ns static RAM and, significantly, a variety of communications interfaces. These make it possible to construct various process networks. In particu-

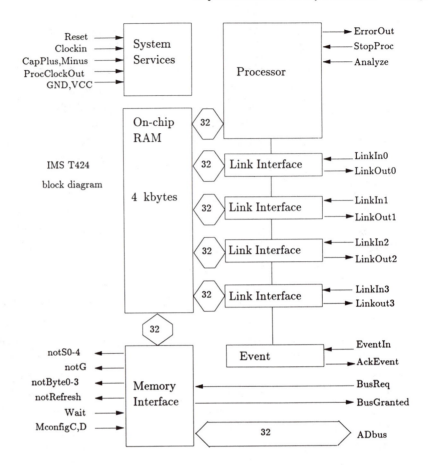

Figure 7.4: Functional block diagram of INMOS T424 transputer.

lar, the INMOS links are the hardware representation of the channels for process communication. They have a programmable data rate of up to 1.5 Mbytes/s. Data are transmitted as a sequence of bytes, each byte being acknowledged by the receiver before the next is transmitted. Since processes must be independent except while in communication, this protocol is essential for conformity with the computational model. It enables the processes to run asynchronously, only synchronizing when they need to communicate.

In addition to the INMOS links, there is a memory interface and an 8-bit bidirectional peripheral interface. The former can be used to extend the on-board memory to accommodate a large program or large amounts of data. It allows access of up to 6Gbytes of external direct address memory space, with a maximum data rate of 25M bytes/s. The latter is to provide communications with standard external devices and may be used to construct a broadcast facility.

The transputer should be seen then as a powerful building block from which new concurrent devices can be constructed. It does, of course, have significant limitations for many signal processing applications. For instance, it has no floating point processor and has only four neighbor connections. However, such objections may largely be overcome by the development of more specialized transputers. In any event, the ability to build, with standard components, a customized MIMD machine for prototype purposes will surely have a wide impact on the development of algorithms and architectures.

The Occam language [May84], [Wils84b] has been designed alongside the transputer and compiled Occam runs on the transputer as efficiently as its own native code. In keeping with the philosophy underlying the development of the transputer, Occam naturally describes a concurrent system as a set of independent processes that use locally defined variables and can communicate only via declared channels. Channel communication in Occam exactly mirrors communication via transputer links. In particular, there is a handshake protocol that allows data to pass only when the sender and receiver have both signified their readiness to each other. This imposes order, although there is no concept of global time in Occam, and all concurrent processes run asynchronously.

It is worth emphasizing the direct relationship between channel communications defined in Occam and implemented in transputer links and the communication protocol envisaged for the WAP. Transputer nets programmed in Occam may, quite naturally, be regarded as wavefront processors, given that they have a sufficiently simple data flow.

Tailoring The Transputer for DSP Applications   The transputer offers a 32-bit fixed-point word length, a precision adequate for most (but not for all) signal processing applications. As to the computing speed, arrays of transputers can be used to provide the concurrent computing performance very close to what is required in many signal processing applications. Take the FFT as an example: in order to cover the full audio spectrum up to 100 kHz, we need 6, 8, and 10 transputers for 64-, 256-, and 1024-point FFT computations, respectively.

A potential weak point of the transputer for DSP applications is that DSP often requires such features as fast and/or versatile arithmetic units (e.g., multiplier or CORDIC processor), and high-speed table addressing

[Wils84a]. Therefore, an upgraded and optimized DSP transputer chip will be of great interest to the DSP community. For DSP, a revised transputer chip should trade on-chip RAM for a fast multiplier, a fast shifter and, for specific applications, floating-point microcode. The current handshake links between PEs have to be retained and improved in speed — they will be essential for the construction of large-scale array processors. On the other hand, the use of internal parallelism (supported by the current transputer model) is somewhat unclear and depends greatly upon the class of applications intended.

To enhance the transputer's arithmetic speed, especially on multiplications, one could consider combining the transputer chip and a Weitek's floating-point arithmetic chip. The possibility of multilevel parallelism and pipelining (including both external and internal PE levels) are also worthy of further explorations.

7.2.2.4 NCR's GAPP Chip

For some low-precision digital and image processing applications, it is advisable to consider very simple processing primitives. A good example of a commercial VLSI chip is NCR's Geometric Arithmetic Parallel Processor, or GAPP (originally devised by W. Holsztynki), which is composed of a 6 × 12 arrangement of *single-bit* processor cells. Each of the 72 processor cells in the NCR45CG72 device contains an ALU, 128 bits of RAM and bidirectional communication lines connected to its four nearest neighbors: one to the north, east, south, and west [Gapp84] (see Figure 7.5). Each instruction is broadcast to all the PEs, making the array to perform like an SIMD machine.

Another architectural feature of GAPP is its cascadability. Many GAPP chips may be placed on a board to build up arbitrarily large arrays of processors in 6 × 12 increments. The packages are designed to interface directly.

The GAPP array can be programmed using low-level, macro assembly-type language or by using NCR's GAL language [Gapp85]. GAL (GAPP Algorithm Language) is a high level language designed to enhance the development of GAPP programs. The GAL compiler is a two phase compiler. The output of the first phase is an intermediate form with postfix notation, which is suitable for interpreting by a stack machine.

7.2.2.5 NEC Dataflow Chip μPD7281

NEC's dataflow chip [NEC85] is a dedicated, programmable image processor that features the dataflow architecture. It can help resolve a trade-off

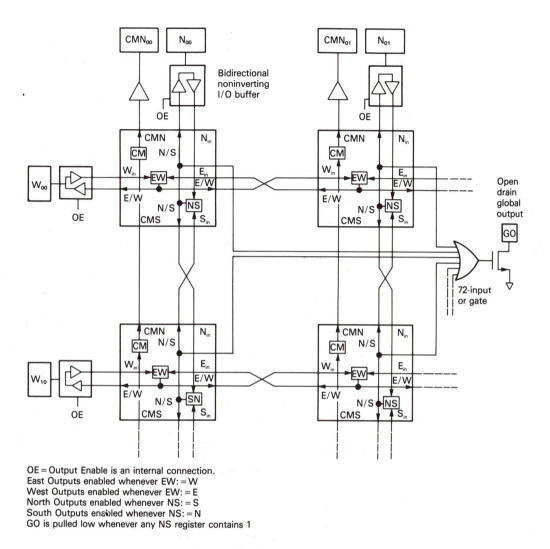

OE = Output Enable is an internal connection.
East Outputs enabled whenever EW: = W
West Outputs enabled whenever EW: = E
North Outputs enabled whenever NS: = S
South Outputs enabled whenever NS: = N
GO is pulled low whenever any NS register contains 1

Figure 7.5: NCR GAPP PE configuration and data bus identification (adapted from [Gapp84]).

between speed and flexibility. The NEC μPD7281 chip uses an internal circular pipeline and powerful instruction set to allow high end image processing. As a dataflow processor, the μPD7281 does not fetch instructions, it processes "tokens" that present operand data. A data flow architecture allows the processor to maximize efficiency in a variety of multiprocessing applications, because it does not need to load and store intermediate results to/from memory during computation.

As shown in Figure 7.6, this processing unit contains a 17 × 17 bit (including the sign bit) multiplier, an ALU, barrel-shifter, comparison, and bit-manipulation operations. Its internal circular pipeline allows the processing unit to operate nonstop at a 5-MHz rate. The following reasons relating to dataflow architectures, allow the image processing computer designer to take advantage of the high computational efficiency:

1. *It does not fetch instructions.* It has a program store that represents a form of dataflow graph. Each operand data is represented by a token. This token contains a data field and other control fields. During program execution, the dataflow processor learns exactly what type of operation will be performed on a token by referring to the dataflow graph image in its program store. When all operands or tokens are available, the dataflow processor sends the token and instructions to the processing unit. The processor also contains local memory to store tokens during the computations. Writes to, and reads from local memory are performed concurrently while the processing unit executes an instruction.

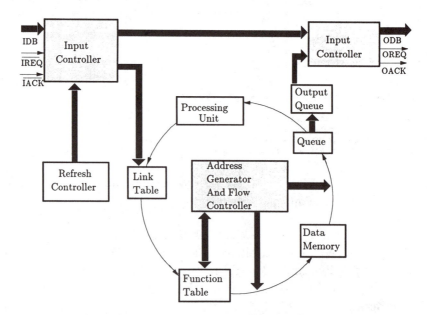

Figure 7.6: Functional block diagram of NEC μPD7281 chip (adapted from [NEC85]).

2. *It facilitates multiprocessing.* The processor communicates with the outside world through the I/O switch. When the host processor wants to communicate with the dataflow processor, the host sends a token. When a token enters the dataflow processor, the address field of the token is examined to determine whether or not the token is for that particular processor. If it is for that particular processor, the token is accepted. Otherwise, it is sent to the next processor. Clearly, tagging a data token with a specific address enables many dataflow processors to be connected in a cascade configuration for multiprocessing.

3. *It is data driven.* In each dataflow processor, program execution is driven only by the available data needed to execute the program. For example, to perform the $a = b+c$ operation, the processor must receive a token with the data b and a token with the data c before the addition operation can be performed. These two tokens may arrive in any order due to the processor's token matching capability. Token matching for tokens b and c can be done by examining each identifier field. Once the tokens are matched, they are sent to the processing unit for an addition operation. The resulting data are then tagged by the identifier a. If token a is to be matched with another token within the same processor, token a is stored in the temporary token storage memory until the partner token enters the processor. The processing unit can operate nonstop, assuming there are many operations other than $a = b + c$ to be performed.

Applications   The applications of the dataflow processing unit cover the fields of signal, sonar, and image processing. Such a unit implements basic functions such as digital filters, phase locked loops and FFTs. In addition, its programmability permits more specialized functions such as adaptive filters, synchronous video integrators, and signal search/recognition processors. The chip is well suited for image processing applications using algorithms involving two-dimensional convolution, enlarging, shrinking, and rotation, and pattern recognition for artificial intelligence (AI) applications. An example of implementing 2-D convolution using this chip is discussed in Section 8.3.

7.2.2.6   The Analog Devices Chip Set

The single chip DSP processors described previously have speed limitations because of the state of the art of integrated circuit fabrication. It is extremely difficult to integrate a very complex system on one chip and achieve great speed at the same time. To cover the applications where the ultimate in speed and/or precision is required, we are forced to use chip sets, each member of which implements a specific function in an optimal way. Such a chip family is the ADSP-1XXX from Analog Devices, Inc. It consists of six chips: a program sequencer, an address generator, an ALU-shift unit, an enhanced fixed-point multiplier/accumulator and two floating-point chips, a multiplier, and an ALU.

Using combinations of these chips, many levels of performance can be achieved, thus giving great flexibility to the designer and allowing implementation of very fast and complex signal processors. The detailed description of each chip is beyond the scope of this discussion.

7.2.2.7 Comparison of DSP Chips

Table 7.1 summarizes and compares the architectures presented.

7.2.3 Dedicated VLSI Chips

7.2.3.1 VHSIC Chips

Based on micro- or submicrometer technology, the VHSIC program in the United States is aimed at increasing the gate count to levels in excess of 30,000 gates per chip and simultaneously increasing the performance of the

Feature	32010/ 32020	MB8764	T424	GAPP	ADSP	uPD7281
Cycle (ns)	200	100	50	100	<100	200
Word (bits)	16	16	32	1	24	16
(bits-dynamic range)	32	26	32	1	40	32
Programmability	good	good	good	fair	fair	fair
Software base	large	med.	med.	med.	small	small
System price	low/ med.	med.	high	high	high	med.

Table 7.1: *Comparison of DSP chips.*

individual gates. This should allow development of complete systems such as FFT on single chips. The VHSIC effort has further significance in that it is focusing on ways to reduce the cost and development time for complex chips, which will result in increased opportunity for building advanced chip architectures as proposed here. Thus, we should be able to apply VHSIC components in a modular approach for real-time signal processing.

The VHSIC master schedule can be divided into three phases: Phase 1 chips from Honeywell, IBM, RCA, Texas Instruments, TRW, and other companies are now available and have been designed into many US DOD systems. Phase 1 of VHSIC was launched in 1981 to develop $1.25\mu m$ geometry chips with functional throughput rates 5×10^{11} gate-Hz/cm^2 and on-chip clock speeds of 25 MHz. This phase launched two new generations of high-speed military technology to meet the performance specifications of future defense systems. This work was intended to be a stepping stone to Phase 2 – military ICs with $0.5\mu m$ geometries having functional throughput rates of 10^{13} gate-Hz/cm^2 and on-chip clock speed of 100 MHz.

An example of applying VHSIC chip sets to the future antiradiation homing (ARH) missiles system is discussed here [Unive86]. Most processing currently done in missiles is analog. The limited amount of digital processing uses outdated medium scale integrated circuits (MSIC). It can be seen that VHSIC chips provide the computational power and design flexibility required for future ARH missiles. In addition, these chips operate at reasonably low power, considering the functions they perform. Improved package design now allows for the production of well over 100 chips per square inch. Another benefit of VHSIC is the high data throughput rates achieved when performing complex operations in a small number of clock cycles. Finally, VHSIC is designed to meet the reliability test and maintenance specifications required for military use. Table 7.2 summarizes the chips required for the future ARH missiles and some of the key specifications.

7.2.3.2   Commercial Chip: The Bit Slice Correlator

Bit level Systolic Arrays   As discussed in Section 4.3, the idea of using systolic arrays at the bit level was developed by McCanny and McWhirter [McCan82], [White86], who demonstrated that many of the components required in digital signal and information processing applications can be implemented as systolic arrays of bit level processing elements based on a gated full adder function. They argued that for computations involving sum of products operations, the demarcation between one systolic array element

VHSIC	Manu-facturer	Power(W)/chip	Description
Window addressable memory (WAM)	TRW	4.40	Matching radar signal type
Static random access memory (SRAM)	TI	0.45	Azimuth/elevation histogram storage
Matrix switch	TI	2.00	Switch between 2 SRAM banks
MAC	TRW	4.10	Extended Booth's multiplier
Register ALU	TRW	3.80	150 sets of 32 1-byte comparison
Microcontroller	TRW	3.00	Control the accessing of PROM
Address generator	TRW	4.10	Generating an address for accessing PROM
1750A data processor unit	TI	2.30	Filtering, tracking, classifying
General buffer unit	TI	2.00	Slow down the data

Table 7.2: *VHSIC chips for ARH missile system (adapted from [Unive86].*

and the next need not necessarily lie at the level of multiply-and-accumulate but at the level of a gated full adder, which is the bit level equivalent of an inner product step processor. Unless we are interested in specific intermediate results within these structures, there is no real need to draw artificial boundaries around specific groups of bit level cells and associate them with inner product processors at the word level provided that the individual carries can be taken into account. Problems could therefore be treated at the bit level from the outset and circuits could be constructed by tiling the silicon plane with an array of simple bit level PEs.

One example of a bit level systolic array is shown in Figure 7.7. This circuit can be used to compute matrix-vector multiplication of the form $\mathbf{y} = \mathbf{A} \, \mathbf{x}$, where each element of the matrix $\mathbf{A}$ takes the value 0 or 1 and can be represented by a single bit. The operation of the circuit is analogous to that of the word level array, the main difference being that the words x_i and y_i have been expanded in terms of their individual bits. Each inner product step processor is therefore replaced by a vertical column of gated

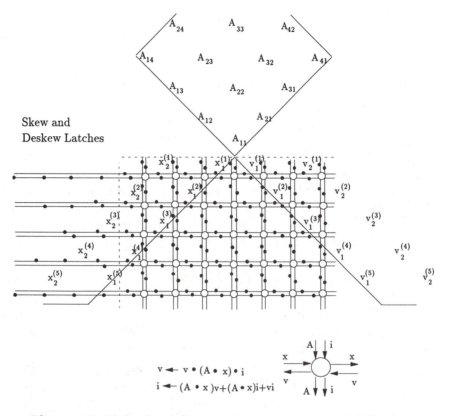

Figure 7.7: Bit level systolic array for matrix-vector multiplication (adapted from [White86]).

full adder cells whose logic function is also illustrated in Figure 7.7. Heavy dots have been used to represent latches and open circles denote the basic processing cells. Figure 7.7 depicts the case of a 4 × 4 matrix in which it is assumed that the incoming words are 3-bit two's complement numbers that have been sign extended to 5 bits (the range of the answer) before entering the array. The operation of this circuit is as follows: Data words, x_i, are input to the array from the left on every other cycle but with successive bits staggered. This is achieved by inputing parallel words through a wedge shaped arrangement of skew latches. The least significant bit, x_i^0, enters the array 1 clock cycle ahead of the next significant bit, and so on, and the bits move one cell to the right on every clock cycle. Bits representing elements in the matrix **A** move down through the array in the vertical direction, as shown. The output words, y_i, are initialized to zero on entering the array and move from right to left with their bits staggered in the manner indicated. This means that the kth bit of a word, y_i^k, meets all the terms required to form the sum

$$y_i^k = \sum_{j=1}^{n} A_{ij} x_j^k$$

The products $A_{ij} x_j^k$ are formed by passing these bits through an AND gate. Any carry bits that are generated in the course of the summation are latched vertically downward as shown. The carry save principle is therefore utilized which is the reason for the stagger on the bits of the words x_i and y_i. Having traversed the array, the output words have accumulated all the terms required to form a result, and one bit, y_i^k, emerges on every other cycle. The resulting word can then be deskewed by passing it through another wedge-shaped arrangement of latches.

One of the most useful applications of this circuit is as a digital corre-lator. In this case the matrix **A** is Toeplitz (i.e., all elements along all the diagonals are identical) and hence we can associate each column of cells with a fixed coefficient. A high-performance bit-sliced correlator based directly on this architecture has been designed by [Corry83] and further developed by Marconi Electronic Devices. The resulting chip is illustrated in Figure 7.8. It constitutes a 64-stage device in which the reference coefficients are either 0 or 1. It has been designed to handle 4 bit input data, although the input words are sign-extended to 10 bits so that the full range of any result can be accommodated within the device. The circuit therefore requires a 64 × 10 array of cells, but in practice this has been laid out as two separate sec-tions of dimensions 32 × 10. The chip has been designed and fabricated in CMOS/SOS technology. It comprises approximately 43,000 transistors and can handle data at sample rates up to 35 MHz. It consumes less than 250 mW at 20 MHz, 5 V operation and occupies an area of 7 mm × 7 mm. It has been also designed in such a way that it can be cascaded to increase the number of correlation stages, the reference word length, and the data word length.

7.2.3.3   VLSI Chip Sets

From the perspective of computer components, a typical set of primitive operations for array processors should include primitives for arithmetic and logic operations, data accessing and storage, control, I/O, and communica-

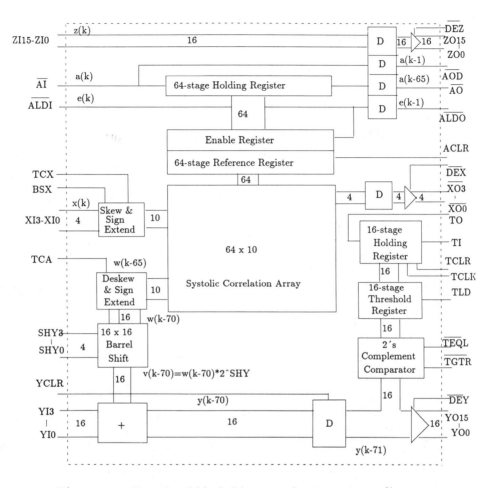

Figure 7.8: Functional block Diagram of a 64 × 10 systolic correlation array (adapted from [White86]).

tions. On the other hand, from DSP algorithm classifications, a basic set of modular operations (such as MAC), can be implemented in a silicon chip . These should be exploited in order to simplify the hardware module.

Based on a functional classification (independent of the device technological progress), we have the following categories:

1. *Arithmetic units*   e.g., adders, parallel multipliers, MACs, CORDIC (COordinate Rotation DIgital Computer) processors, residue number arithmetic units.

2. *Memory or storage*   e.g., RAM, ROM, table look-up, Stacks (LIFO, FIFO), buffer memory, cache.

3. *Communication/interconnection*   e.g., A/D, D/A converters, hand-shaking I/O ports, Direct Memory Access (DMA), packet-switching (e.g., 2 × 2 router), cross-bar, perfect shuffle.

4. *DSP functions (in chips only)* e.g., inner product, sorter, comparator, FIR, IIR filters, median filters, correlator, convolution, interpolation FFT, DFT, Hadamard transform.

7.3   Design of Arithmetic Units

This section concentrates on the design of the AU, which is the core of computations in a DSP processor chip. As mentioned in the earlier considerations, the AU is a very important component of any processor designed for signal processing.

The design of the AU depends upon the operations to be performed. Besides the multiply-and-accumulate operation, we note that in modern signal processing applications, such as filtering, and spectrum analysis, matrix rotation has become a popular computational tool. Examples are adaptive lattice filtering, pipelined Toeplitz solver, Givens' rotation for QR and Givens' rotations for eigenvalue decomposition. The operations that are needed for these computations are addition, subtraction, multiplication, division, square root, and hyperbolic functions. Once we have identified the operations to be performed by an AU we can design an AU that is fast, accurate, and requires a minimum of hardware.

In the following, we describe three types of AUs. The first one is the conventional multiplier-and-accumulator (MAC) based AU. The second is based on the residue number system, and the third is based on CORDIC.

7.3.1   Conventional MAC Design

In this subsection, conventional fixed-point and floating-point AUs are discussed. The main emphasis is on the design of the array multipliers, which is a crucial component for many signal processing applications.

7.3.1.1   Fixed-Point Array Multipliers

Due to today's VLSI technology, the array (or parallel) multiplier has become increasingly economic and popular. The following discussion on array multipliers are largely based on [YTHwa85]. The architecture of a straight-forward array multiplier is much the same as that of the usual paper-and-pencil algorithm. The array multiplier consists of two major sections, the generation of inner products, called the partial products, and the reduction of these intermediate results into a final product. In order to achieve a high speed multiplication, Wallace [Walla64] has indicated three areas where the processing performance can be improved.

1. Accelerate the formation of partial products.

2. Reduce the number of partial products.

3. Accelerate the addition of partial products. It can be further decomposed into two parts: (a) the acceleration of the reduction process of the partial product matrix and, and (b) the acceleration of the final addition stage.

Among these three areas, the first one − formation of partial products − is quite minimal relative to the other delay factors. Due to its concurrency, only one level of delay is introduced independent of the multiplication word length. Considerable attention must be put on the latter two terms which are the dominant factors of multiplication speed. In the following, we discuss methods to improve these two terms.

On the other hand, VLSI implementation requires that a good architecture should have the following two properties:

1. It should be implemented by only a few different types of simple cells (modularity).

2. It should have simple and regular data/control paths so that the cells can be connected by a network with local and regular interconnections (regularity).

The above considerations will be the deciding factors for selecting a design for array multipliers.

The Recoded Multiplication   A recoded algorithm recodes one of the two operands (one is called a multiplier and the other is called a multiplicand) in such a way that the number of partial products to be added together decreases; therefore, the computation is faster. We discuss the basic Booth algorithm [Booth51] and its modified version here. The original Booth algorithm did not deal with parallel multiplication but was aimed to improve the speed of the add-and-shift algorithm. Its principle comes from the idea of skipping arbitrarily long strings of 1s or 0s in the multiplier. This leads to a variable execution time of the multiplication. The "modified" Booth algorithm strictly considers a constant shift of groups of bits in the multiplier. Thus, a constant execution time is obtained, and the speedup factor depends on the number of bits shifted per step in the algorithm. The most conventional scheme is to consider a bit-pair in each step, i.e., bit-pair recoding. The multiplier bits are divided into 2-bit pairs, and 3 bits (a triplet) are scanned at a time, two bits form the present pair and the third bit (the overlap bit) from the high-order bit of the adjacent lower-order pair. After examining each bit-pair, the algorithm converts them into a set of 5 signed digits 0, +1, +2, −1 and −2. According to the Boolean truth table shown in Table 7.3, each recoded digit performs only a simplified processing on the multiplicand, such as add, subtract, or shift.

Multiplier bit triplet			The recorded operand	Remark
2^1	2^0	2^{-1}		
$i+1$	i	$i-1$		
0	0	0	0	no string
0	0	1	1A	end of string
0	1	0	1A	isolated 1
0	1	1	2A	end of string
1	0	0	-2A	begin of string
1	0	1	-A	end/begin of string
1	1	0	-A	begin of string
1	1	1	0	center of string

Table 7.3: *Truth table for the modified Booth algorithm with bit-pair recoding.*

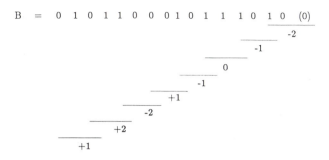

Figure 7.9: An example showing the bit-pair recoding scheme.

The multiplier bit-pair recoding scheme using the modified Booth algorithm can be easily comprehended from the example in Figure 7.9.

After recoding, only 8 signed digits are obtained. Note that the lowest-order bit pair is treated as if a zero was appended to the right of the LSB. In Figure 7.10, we demonstrate the multiplication of two signed numbers using the modified Booth algorithm.

The modified Booth algorithm can be applied to constant shifts of 3 or more bits, thus, even fewer partial products are obtained as compared to bit-pair recoding. However, in this case, we must have an extra precalculated operand available for new recoded terms, say, $3A$ or $-3A$. That means more hardware must be introduced. Therefore, in recoded multiplication, a recoding scheme of 3 or more bits does not make much sense for small multipliers, because the increased complexity is not fairly balanced by significant gains both in area and speed. An important advantage of the modified Booth algorithm is that it can be applied to 2's complement multiplication.

```
                 1  0  1  1  0  0  1  0              -78
              x) 1  0  0  1  1  1  0  1              -99
   ─────────────────────────────────────────────────────
 1  1  1  1  1  1  1  1  1  0  1  1  0  0  1  0        (+1)
 0  0  0  0  0  0  0  1  0  0  1  1  1  0             (-1)
 1  1  1  1  0  1  1  0  0  1  0  0                   (+2)
 0  0  1  0  0  1  1  1  0  0                         (-2)
 ─────────────────────────────────────────────────────
 0  0  0  1  1  1  1  0  0  0  1  0  0  0  1  0  1  0  1  0    7722
```

Figuer 7.10: Multiplication of two signed numbers using the modified Booth algorithm.

7.3.1.2   Cellular Array Multiplier

A cellular combinatorial array multiplier is simply the hardware equivalent of
the standard software multiplication. Similar to the software add-and-shift
algorithm, this particular multiplication algorithm generates the product
through a series of adds and shifts. For the hardware case, all shift operations
are hard-wired in. Thus the implementation of a multiplier reduces to a
matrix partial product generator and an array of adders, which perform the
summands addition to obtain the final result.

There are several widely used schemes to implement an array multiplier,
such as the carry save scheme, the Wallace tree, Dadda's scheme and the
matrix generation-reduction scheme. We discuss the carry save scheme, the
Wallace tree, and Dadda's scheme.

Carry Save Adder (CSA)   The term "carry save" comes from the way
the carries propagate in the partial products addition process. Since carry
outputs from adders are not fed into adders of the same level, they are
preserved and propagated diagonally to the next level. The schematic circuit
diagram of a 5 × 5 CSA array multiplier is shown in Figure 7.11.

In this scheme, the first stage for the reduction of n^2 summands uses
$n - 1$ half adders to generate the second least significant bit of the product.
The least significant bit is available without any transmission. Each one of

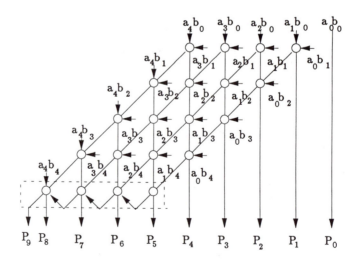

Figure 7.11: The carry save scheme array multiplier.

the following stages uses $n - 1$ full adders. In each stage, one bit of the product is generated and the number of partial products is reduced by one. At the last stage, we need $n - 2$ full adders and one half adder to accomplish the result. To speed up the operation, an $n - 1$ bit carry look ahead adder [Hwang79] may be used in the last stage.

Wallace Tree   Additional speed can be achieved by the use of a Wallace tree [Walla64], which is an interconnection of carry save adders that reduces n partial products to two operands. The principle of the Wallace tree is to use parallel processing of the n partial products. The Wallace scheme achieves the minimum number of addition stages, which is proportional to the logarithm of the number of bits in the multiplier, and hence achieves a minimum delay time.

Dadda's Counting Scheme   Dadda's scheme [Dadda65] reduces n partial products into two operands in a similar way as the Wallace tree does. In this scheme, parallel counters (p inputs, q outputs) are used to obtain the sum of 1s in each column of the summand matrix. Although a large parallel counter reduces the number of addition stages, more delay will be introduced due to the complexity of counters that necessarily have a large number of terminals. To minimize the delay, Dadda also introduced a procedure using carry save adders, i.e., $(3, 2)$ counters only. This procedure is optimal in the sense that it uses a minimum time delay to complete partial product addition stages.

Fast Addition Stage   After the process in a carry save adder array, the produced result is in a carry save form, and one more stage of summation is needed to obtain the final product. The conventional carry propagation adder would be inadequate here due to its limited speed of addition. Therefore, carry look-ahead techniques [Hwang79] are often used at this stage to speed up the summing process. The carry look ahead comes from the phenomenon that the carries can be generated or predicted before the corresponding sum bits for the particular stage have been obtained. These carries entering all the bit positions of a parallel adder are generated simultaneously by additional logic circuitry. This results in a constant addition time independent of the length of the adder. However, the large fan-in and fan-out restrictions prevent a direct extension of the carry look ahead adder for a large word length. To implement a large-size adder, a multilevel carry lookahead circuitry may be used. There are also other fast adders, such as carry selection adders.

7.3.1.3   Two's Complement Multiplication

In most DSP applications, 2's complement multiplication is often required. In contrast to the unsigned multiplication, the sign bit of each partial product must be extended to the full width of the final product. Additional summands are formed, and the shape of the multiplier becomes trapezoidal rather than rectangular. More carry save adders are required to combine these extra summands. In order to reduce this hardware overhead, a direct 2's complement multiplication approach was proposed. The basic idea is to evaluate the values of 2's complement numbers as positional numbers with a negatively weighted sign and positively weighted coefficient. Consider a 2's complement number $N = (a_{n-1}a_{n-2}\cdots a_1 a_0)_2$, where a_{n-1} is the sign bit. The value N_v of the number N can be represented as followed:

$$N_v = -a_{n-1} \times 2^{n-1} + \sum_{i=0}^{n-2} a_i \times 2^i \qquad (7.1)$$

Pezaris [Pezar71], and Baugh and Wooley [Baugh73] have each proposed famous algorithms for direct 2's complement multiplication. They speed up the multiplication process by eliminating the slow 2's complementing operations, although some drawbacks are also introduced. In the Pezaris array multiplier, four types of full adders with positively or negatively weighted inputs/outputs are needed. This increases the circuit complexity due to the irregularity in the basic cells. In Baugh-Wooley's multiplier shown in Figure 7.12, some improvements over the design of Pezaris have been achieved. The principle of this algorithm is that the signs of all summands are positive, thus allowing the array to be constructed entirely with conventional full adders. However, the negation of the summands are used in this case, which results in complexity in the circuit. Another disadvantage of the Baugh-Wooley array is that the recoded multiplication technique can not be applied in a straightforward way, such as a carry save adder array

7.3.1.4   VLSI fixed-point MAC Chip Examples

Despite the progress being made in boosting the performance of general-purpose DSP microprocessors, many DSP applications require even greater performance. For these applications, the use of discrete components offers a tenfold increase in performance over DSP microprocessors. The premier performer in a discrete chip set is the AU . The hardware multiplier and

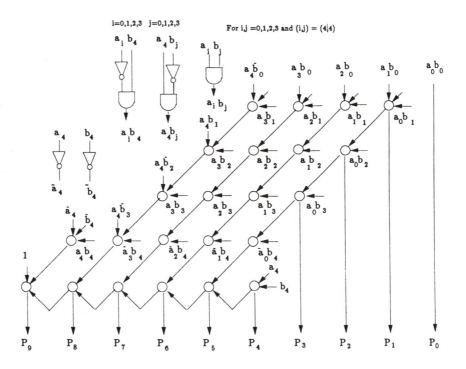

Figure 7.12: The Baugh-Wooley two's complement array multiplier.

ALU in this AU combine to execute the sum-of-products algorithm, found in most DSP and matrix calculations. In place of a separate multiplier and ALU, a single multiplier-accumulator (MAC) is often used. The accumulator performs a subset of the ALU functions, adding the result of the current multiplication to the result of the previous multiplication. Table 7.4 shows the CMOS fixed-point multipliers and MACs [Marri85].

7.3.2 Floating-Point Arithmetic

Fixed-point arithmetic handles small integers in an exact form for business or commercial applications. For scientific computing, we must round the numbers constantly in order to reduce the number of digits to a manageable amount. Fixed-point arithmetic presents some difficulties, such as limited range and rigid precision in handling scientific and engineering computations. The problems are caused primarily by the range, precision, and significance of digital numbers represented in a machine. In DSP applications, round-off

Comparison of CMOS Fixed-Point Multipliers and Multipliers/Accumulators					
Company/Device	12 x 12 Mul	12 x 12 MAC	16 x 16 Mul	16 x 16 MAC	24 x 24 Mul
Analog Devices	ADSP 1012 110 ns	ADSP 1009 130 ns	ADSP 1016 75 ns	ADSP 1010 85 ns	ADSP 1024 200 ns
TRW	None N/A	TMC 2009 135 ns	TMC 2161 45 ns	TMC 2210 50 ns	None N/A
AMD	None N/A	None N/A	AM 29C516 90 ns	None N/A	None N/A
Logic Device	LMU 12/13 80 ns	None N/A	LMU 16/17 90 ns	LMA 1010 75 ns	None N/A
Weitek	None N/A	None N/A	WTL 2516 55 ns	WTL 2010 75 ns	None N/A
IDT	IDT 7212/7213 45 ns	IDT 7209 55 ns	IDT 7216/7217 55 ns	IDT 7210/7243 65 ns	None N/A

Table 7.4: *Comparison of CMOS fixed-point multipliers and MACs.*

errors accumulate system noise, imprecisely placed filter poles and zeros alter the transfer function and possibly cause instability, and dynamic range constraints lead to saturation-induced signal distortion. Floating-point operations, however, can generally avoid these problems.

Floating-point arithmetic was proposed in the early 1940s to overcome the problems of fixed-point arithmetic. It has the advantages of large dynamic range and high precision. Even though it requires more hardware support, floating-point arithmetic has been universally accepted for high-speed scientific computations. There are two types of floating-point arithmetic: *unnormalized and normalized.* A normalized floating-point processor operates only with normalized floating-point numbers and enforces postnormalization steps on all intermediate and final results. In the following, we concentrate only on the normalized floating-point numbers.

A floating-point number consists of a pair of fixed-point numbers (M, E), in which M stands for mantissa and E stands for exponent. The value of the number (M, E) is $M \times B^E$, where B is a predetermined base. (Here we assume that $B = 2$.) Let (X_M, X_E) and (Y_M, Y_E) be two floating-point numbers. The basic formulas used to perform floating-point addition, subtraction, multiplication and division are as follows:

$$Addition: \quad X + Y = (X_M 2^{X_E - Y_E} + Y_M) \times 2^{Y_E} \quad where \ X_E \leq Y_E$$
$$Subtraction: \quad X - Y = (X_M 2^{X_E - Y_E} - Y_M) \times 2^{Y_E} \quad where \ X_E \leq Y_E$$
$$Multiplication: \quad X \times Y = (X_M \times Y_M) \times 2^{X_E + Y_E}$$
$$Division: \quad X \div Y = (X_M \div Y_M) \times 2^{X_E - Y_E}$$

Multiplication and division are relatively simple, since the mantissas and exponents can be processed independently. Floating-point multiplication requires a fixed-point multiplication of the mantissas and a fixed-point addition of the exponents. Floating-point division requires a fixed-point division involving mantissas and a fixed-point subtraction of exponents. Thus floating-point multiplication and division are not significantly more difficult to implement than the corresponding fixed-point operations. Floating-point addition and subtraction are complicated by the fact that the exponents of the two input numbers must be made equal before the mantissas can be added or subtracted. This involves a right-shifting the mantissa of the input number with the smaller exponent in order to make both exponents equal. Therefore, an extra step is needed for each of the four floating-point arithmetic operations in order to normalize the result.

7.3.2.1   IEEE Binary Floating-point Arithmetic Standard

There are two driving forces that are changing the floating-point hardware. First is VLSI technology. Second is the appearance of standards; with the IEEE binary floating point arithmetic standard (Task P754) being the dominant one. These standards should make the software more portable [Fandr85].

The IEEE floating-point standard is to standardize binary floating-point arithmetic for mini- and microcomputers [Coone80]. The purpose of the standard is to assure a uniform floating-point software environment for programmers. It may be implemented in software, hardware, or a combination of both. The standard precisely describes its data formats and the results of arithmetic operations. It must do so to be of use to the producers of microprocessor hardware and software, who cannot afford to provide the support software and personnel to perform conversions between systems conforming to a less rigid standard. Programs that now run in high-level languages like FORTRAN should be portable to a system with the new standard arithmetic at the cost of a modest amount of editing and a recompilation. The program should then execute with results almost certainly no worse than before.

7.3.2.2   VLSI floating-point MAC Chip Examples

In many DSP applications and scientific computations such as medical imaging, seismic energy exploration, electronic simulation, a large dynamic range and high precision are needed. Because of the speed requirement of these users, floating-point arithmetic hardware is essential.

In terms of precision, many DSP requirements can be satisfied by single precision (32-bit) floating-point whereas scientific applications need invariably double-precision (64-bit) or extended precision (80-bit) in IEEE P754 formats.

For 32-bit VLSI floating-point chips, there are currently three suppliers: Weitek, AMD and TRW. Weitek was the first one (1983) to introduce a chip-set (a floating-point multiplier WTL1032 and a floating-point ALU WTL1033), which has become a de facto standard for many applications such as graphics and FFTs. The performance is 10 MFLOPS per chip. Both AMD and Weitek have introduced 32-bit single-chip solutions with both multiplier and ALU on the same chip. AMD's 29325 offers single-cycle latency at 100ns. However, it cannot perform both multiply and add simultaneously and has a power dissipation close to 10W. The WTL3132 integrates register file, multiplier and adder on the same chip and offers 20 MFLOPS performance.

Both Weitek (WTL2264/2265) and Analog Devices (ADI3210/3220) offer 64-bit floating-point chip-sets that can perform 32-bit operations as well. The peak performance of WTL2264/65 is 10 MFLOPS for double-precision and 20 MFLOPS for single-precision. Currently in development by several companies are the integrated 64-bit chips that probably will appear in the market over the next couple of years.

7.3.3   Residue Number Arithmetic

According to [Taylo84], some of the well-known attributes of a fixed-radix weighted-number system are *algebraic comparison, dynamic range extension (i.e., add more digits), multiplication (division) by simple arithmetic shifts, and simplified overflow and sign detection.* The disadvantage of the fixed-radix, weighted-number system is that carry information must be passed from digits of lesser significance to those of greater significance. As a result, there is a slowdown of arithmetic related to the carry-management system used (e.g., ripple carry, look-ahead carry). Carry management can be accelerated but only at the expense of additional hardware. The time required to compute an n-ary function with gates restricted to p inputs is at least $\log_p n$.

On the other hand, the residue numbering system (RNS), which is based on the ancient Chinese remainder theorem, has attracted a great deal of attention recently. Digital systems constructed with residue AUs may play an important role in ultra-speed, dedicated, real-time systems that support pure parallel processing of integer-valued data. It is a "carry-free" system that performs addition, subtraction, and multiplication as concurrent (parallel) operations, side-stepping one of the principal arithmetic delays – managing carry information.

There have been attempts to use RNS in the 1960s [Szabo67]; however, the technology was insufficient to support the unique demands of the RNS. Since the mid 1970s, the microelectronic revolution has brought low-cost, high-performance RAM and ROM to replace the expensive and slow core that RNS once used. They provide the ideal technology for the residue arithmetic as a table-lookup operation. The RNS is now finding its way into many application areas involving digital filters and transforms. We first describe the basic theory of RNS.

7.3.3.1 Residue Number System

The RNS is defined in terms of a set of relatively prime moduli. If P denotes the moduli set, then

$$P = \{p_1, p_2, \ldots, p_L\}, \quad \mathrm{GCD}(p_i, p_j) = 1, \quad \text{for } i \neq j$$

Any integer in the residue class Z_M, where

$$M = p_1 \times p_2 \times \cdots \times p_L$$

has a unique L-tuple representation given by

$$X \rightarrow (X_1, X_2, \ldots, X_L)$$

where $X_i = X \bmod p_i$, and X_i is called the ith residue of X.

For a signed number system, any integer in $(-M/2, \ M/2)$, has an RNS L-tuple representation, where $X_i = X \bmod p_i$ if $X > 0$ and $(M - |X|) \bmod p_i$ otherwise. The signed RNS system is often referred as the symmetric system.

While RNS adds, subtracts, and multiplies efficiently, division is not a closed operation. If X, Y and Z have RNS representations given by

$$X \to (X_1, \cdots, X_L); \quad Y \to (Y_1, \cdots, Y_L); \quad Z \to (Z_1, \cdots Z_L)$$

and let "$\star$" denotes add, subtract, or multiply, the RNS version of the equation $Z = X \star Y$ satisfies:

$$Z \to (Z_1, \cdots, Z_L) = ((X_1 \star Y_1) \bmod p_1, \cdots, (X_L \star Y_L) \bmod p_L)$$

if Z belongs to Z_M. This equation indicates that the ith RNS digit, namely Z_i, is defined in terms of $(X_i \star Y_i) \bmod p_i$ only. That is, no carry information needs to be communicated between residue digits. The result is very high speed concurrent operations. This high speed makes the RNS very attractive for real-time applications. The following example illustrates this basic property.

Example 1: RNS Arithmetic

Let $P = (3, 4, 5)$ and $M = 60$; then note that:

$$4 \to (1, 0, 4); \quad 6 \to (0, 2, 1); \quad 10 \to (1, 2, 0); \quad 24 \to (0, 0, 4).$$

The addition, subtraction, and multiplication of these numbers in RNS are shown in Figure 7.13. From the example, it can be clearly seen that the RNS is carry-free for the above three operations.

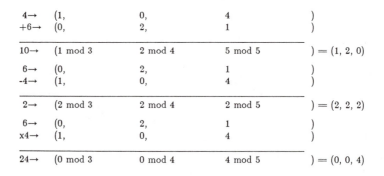

Figure 7.13: An example of the RNS arithmetic.

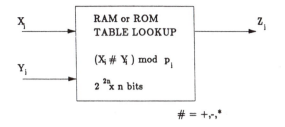

Figure 7.14: RNS arithmetic unit (adapted from [Taylo84]).

In order to exploit RNS parallelism, arithmetic units must efficiently and rapidly implement the modular statement $(X_i \star Y_i)$ mod p_i. Simple table-lookup operations can be used to replace the usual arithmetic algorithms. For example, the high-speed memory in Figure 7.14 can be programmed to output the value of $(X_i \star Y_i)$ mod p_i upon receipt of the address $[X_i : Y_i]$, which denotes the concatenation of X_i and Y_i. If p_i is bounded by 2^n for all i, then the concatenated address $[X_i : Y_i]$ is $2n$ bits wide. Therefore, RNS arithmetic can be performed as a table-lookup mapping, if the process does not exceed the address space of available memory.

7.3.3.2   Applications

In practice, the RNS is only useful when high data rates are required. Problems solved successfully with conventional arithmetic do not justify an RNS implementation. To be competitive, RNS systems must have arithmetic speeds on the order of hundreds of megahertz, and data acquisition and the accompanying decimal-to-residue conversion must be equally fast.

One of the most promising uses of the RNS is to implement digital filters or convolvers. In many filter applications, speed is of the essence and word lengths of the order of 16 to 32 bits are common. The following example [Nudd85] illustrates the use of RNS.

The residue number arithmetic is used for image analysis at Hughes Research Lab. This technique requires encoding the incoming data by division with prime numbers, as shown in Figure 7.15, and then working with only the remainders. In this case, for example, they use four prime numbers (31, 29, 23, and 19), each below 5 bits, and hence the residues or remainders are also below 5 bits. The arithmetic in each channel then proceeds directly, but with the provision that if any internal number exceeds the prime or base, it is reconverted to the remainder (by successive subtraction of the prime).

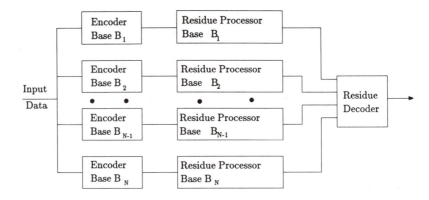

Figure 7.15: Hughes image analysis system (adapted from [Nudd85]).

The parallel channels can then be uniquely decoded to binary output data, providing an overall accuracy equivalent to the product of the bases (i.e., 392,863, or approximately 18 bits). The significant advantage of this approach is that all the arithmetic, including the initial encoding into residue notation, the internal calculations, and the final decoding, can be performed with small 5 × 32 bit lookup tables. Further, the machine can be made programmable by using high-speed RAMs, which are loaded with specific data to perform the various feature extraction functions. The RADIUS [Fouse79] machine that was built using these techniques is now operating at Hughes laboratory at essentially real-time rates. It consists of a number of special-purpose custom-built NMOS chips, which include the lookup tables and a number of modular arithmetic cells. The full machine performs operations of the form

$$y = \sum f_i(I_{ij})$$

where I_{ij} are the intensity values over a 5 × 5 element kernel and f_i is the polynomial functions of a single variable.

The effective throughput of the machine in this feature extraction mode is equivalent to 200×10^6 multiplications/s.

7.3.3.3   Table-lookup-based Approach

Due to the regularity of memory devices, it is very attractive to use more memories in a system design using VLSI technology. In an RNS system, when a large number is decomposed to several small fields, it is very convenient to implement the arithmetic operations of each field by table lookup. For example, we address such an approach for multiplication next.

A simple and regular solution to the multiplication problem can be solved by the table-lookup method. In principle, we can store all possible results of the product of input X and Y into ROM and use input data X and Y as the address to read the ROM and get the product. The problem is that the required memory space is too large for this simple but naive method. A solution to this problem is to observe the following equation:

$$X \times Y = \frac{(X+Y)^2 - (X-Y)^2}{4}$$

Therefore, we can store only the square of a number in a ROM and compute the product by reading twice the square of $(X+Y)$ and $(X-Y)$, then left-shift by two bits (divide by 4) the difference of the two readouts. Since the amount of storage required by the square is much less than the product and adder/subtracter and shifter are relatively inexpensive, it is reasonable to use this method for implementation.

7.3.4   CORDIC

Many popular algorithms such as DFT, FFT, complex filters, lattice filters and Givens' rotation based methods (e.g., QR decomposition) are readily described with rotations [Ahmed85], as mentioned in Sections 2.2 and 2.3. For this class of applications, the CORDIC processing units provides a convenient method for realizing the rotations [Volde59], [Walth71].

7.3.4.1   Basic CORDIC Scheme

The CORDIC scheme is an iterative scheme for computing generalized vector rotations rather than simple multiplication. The rotations can be described by the following equations:

$$\begin{bmatrix} x(n) \\ y(n) \end{bmatrix}$$

$$= \prod_{i=1}^{n} \cos(m^{\frac{1}{2}}\alpha_i) \begin{bmatrix} 1 & -m^{\frac{1}{2}}\sigma_i \tan(m^{\frac{1}{2}}\alpha_i) \\ m^{\frac{1}{2}}\sigma_i \tan(m^{\frac{1}{2}}\alpha_i) & 1 \end{bmatrix} \begin{bmatrix} x(0) \\ y(0) \end{bmatrix} \quad (7.2)$$

$$z(n) = z(0) + \sum_{i=1}^{n} \sigma_i \alpha_i \quad (7.3)$$

The rotation are called circular, hyperbolic, or linear according to whether $m = 1$, -1 or 0. The σ's are equal to 1 or -1 and are chosen to force either $z(n)$ equal to zero or to force $y(n)$ equal to zero. In the latter case, we call the operation a vectoring operation. Table 7.5 summarizes all the different possibilities for $m = 1$ (circular rotation), $m = -1$ (hyperbolic rotation), and $m = 0$ (linear rotation).

The angles α_i's should satisfy the following condition (so that either $y(n) \to 0$ or $z(n) \to 0$,) as the case may be):

$$\frac{1}{2}\alpha_{i-1} < \alpha_i < \alpha_{i-1} \tag{7.4}$$

These angles can be chosen such that we have aesthetic values like:

$$\tan(m^{\frac{1}{2}}\alpha_i) = 2^{-F_m(i)} - \delta_i 2^{-G_m(i)} \tag{7.5}$$

where $\delta_i = 0$ or 1 and $F_m(i)$ and $G_m(i)$ are nonnegative integers.

With this choice, Eq. 7.2 can be implemented using only adders and shifters (except for the scaling by $\prod_{i=1}^n \cos(m^{\frac{1}{2}}\alpha_i)$, which can be done by using the CORDIC with $m = 0$.

Note that, whereas there usually exists a more efficient scheme for a particular function (e.g., square root), the utility of the CORDIC lies in its generality and flexibility. A digital signal processor intended for a variety of applications can benefit significantly from such flexibility.

Their flexibility can be best utilized in array processing environments where different operations may be required along the same pipeline string. Note that the pipeline rate is often dictated by the worst (slowest) operation

m	$x(n)$	$y(n)$	$z(n)$
0	$x(0)$	$z(0)x(0) + y(0)$	0
0	$x(0)$	0	$z(0) - y(0)/x(0)$
1	$x(0)\cos(z(0)) - y(0)\sin(z(0))$	$x(0)\sin(z(0)) + y(0)\cos(z(0))$	0
1	$(x(0)^2 + y(0)^2)^{\frac{1}{2}}$	0	$\arctan(\frac{y(0)}{x(0)})$
-1	$x(0)\cosh(z(0)) + y(0)\sinh(z(0))$	$x(0)\sinh(z(0)) + y(0)\cosh(z(0))$	0
-1	$(x(0)^2 - y(0)^2)^{\frac{1}{2}}$	0	$\text{arctanh}(\frac{y(0)}{x(0)})$

Table 7.5: *Different cases of rotations corresponding to m = 0, 1 and −1.*

in the pipe, so it is very desirable to have certain uniformity of the operation times for the operations involved.

Once we have chosen a set of angles satisfying Eq. 7.4, the scaling factor

$$K_{m,n} = \prod_{i=1}^{n} \cos(m^{\frac{1}{2}}\alpha_i) \qquad (7.6)$$

is fixed and constant for all rotations. By proper choice of the $F_m(i)$ and $G_m(i)$, we can force $K_{m,n}$ to be a simple shift. A rotation is then a sequence of shift-adds and a final shift to implement the scaling (global scaling).

We can also use the first-order approximation

$$K_{m,n} = \prod_{i=1}^{n}(1 - m\epsilon_i(2^{-F_m(i)} - \delta_i 2^{-G_m(i)})), \qquad (7.7)$$

with $\epsilon_i \in \{0,1\}$ and have local scaling for each micro rotation. Because of Eq. 7.5, the scalings are again simple shift-add operations. Local scaling can be used to prevent overflow at any stage, but its implementation gives rise to a large silicon area. This global scaling, in combination with overflow control, may be a preferred alternative.

Since there is some freedom in choosing the set of angles $\{\alpha_i\}$, we can always find a set $\{\alpha_i\}$ within the constraints of Eqs. 7.4 and 7.5, with these additional constraints:

(1) $\{F_m(i), G_m(i)\} = \{F(i), G(i)\}$, i.e., the sequence of shift factors is the same for the 3 types of rotations.

(2) $K_{m,n}$ is a power of 2.

However, a set $\{\alpha_i\}$ for which $K_{m,n}$ is a power of 2, may not be a regular sequence. This will lead to a complicated controller (large silicon area) for the sequential form of implementation of Eq. 7.2 [Ahmed85].

On the other hand, implementing Eq. 7.2 as a pipe, where each section corresponds to a microrotation having its own hard-wired shift (Eq. 7.5), results in a more favorable architecture in terms of throughput and control provided that the first additional constraint is satisfied to keep complexity low. Moreover, a pipelined CORDIC is more desirable in high-throughput WAPs, especially in some partitioned arrays.

7.3.4.2   The Pipelined CORDIC

The pipelined CORDIC discussed here is described in detail in [Depre84]. Here we highlight only some basic properties of the pipeline.

The basic architecture of one section of the pipe is shown in Figure 7.16 for $\delta_i = 1, 0$ in Eq. 7.5.

For $\delta_i = 0$, the upper half repeats itself. In fact, for $\delta_i = 0$, we execute two microrotations in one section of the pipe. The σ_i ($\sigma_i = 1, -1$) can be readily derived from the sign bit of the y-entry in the case of 2's complement representation, i.e., $\sigma_i = \overline{msb}(y\text{-entry})$, where $\overline{x}$ denotes the boolean complement of x.

The direct implementation shown in Figure 7.16, however, suffers from a major drawback: the latency of a section for $\delta_i = 0$ depends on its index i. This happens because we have to wait for the first of the $(n - F(i))$ most significant bits of the y-entry (adders in the upper half of a section) to appear before the second microrotation (in the lower half of a section) can be started up to allow the carries of both halves to propagate in parallel. The last section of the pipe will have the largest latency and will therefore determine the throughput of the pipe.

To avoid this, the following scheme has been proposed in [Depre84] at the expense of some irregularity in the hardware.

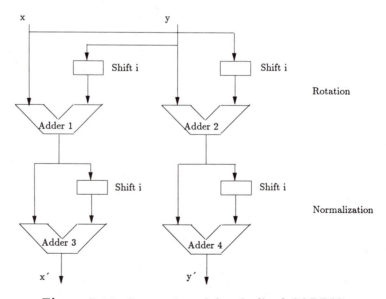

Figure 7.16: One section of the pipelined CORDIC.

7.3.4.3   The Pipelined CORDIC in Array Processors

The architecture shown in Figure 7.16 is not suitable for divisions (see Table 7.5), since the $z(i)$ part is not present. This is done to save silicon area. Since the remaining operations do not require $z(n)$ explicitly. Since the $\alpha_i's$ are fixed, the σ-sequence is used to represent $z(n)$. We need to store only 1 bit for each σ_i per pipe section or transmit it to a different pipe for starting up a rotation, thus allowing *bit level* pipelining (see Figure 7.17).

The pipelined CORDIC arithmetic unit is perhaps the best candidate for implementing arrays which perform circular and hyperbolic rotations as elementary operations. Examples are wavefront arrays for QR factorization, singular/eigenvalue decompositions, least squares solutions, solutions of systems of linear equations, and implementations of lattice filter algorithms. Because of its ability not only to perform different functions, but also to do function switching in a real pipelined fashion, it is also a favorable candidate for arrays which require their PEs to perform more than one function on the data after partitioning. Vectoring has the additional advantage that it derives its control (adding or subtracting) from the data. This implies that its

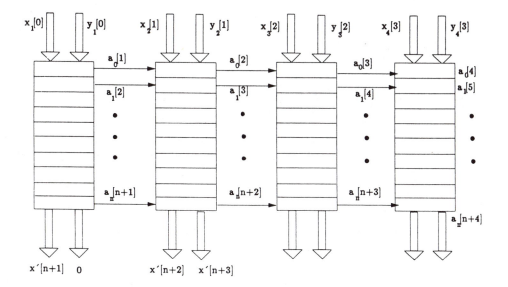

Figure 7.17: CORDIC wavefront processor with bit-level pipelining between modules. The timing of values relating to one instance of the algorithm is shown within the brackets [Dewil85].

controller can be kept to a minimum size. Also, function switching requires only one parameter, m, to be set. [2]

A unique situation for array processing is that the same angle is used in the PEs along the pipelining row. Therefore, it is desirable to pipe the angle information in a most effective way. Contrasting with the conventional CORDIC operation where the angle is given directly, here the angle information is created by the angle generator in terms of a special set of CORDIC control bits. For pipeline processing, it is much more effective to pipe those control bits along the row instead of the angle itself.

In terms of hardware complexity, speed and accuracy the pipelined CORDIC AU is less favorable. Compared to an array multiplier, it contains almost $\frac{1}{4}n^2$ more full adders (where n is either the number of microrotations or the word size). Its numerical accuracy depends on the chosen set, $\{\alpha_i\}$, of angles, the number of microrotations, and the accuracy of $K_{m,n}$ being a power of 2. Moreover, the architecture in Figure 7.16 does not allow either of the two operands in the multiplication to be larger than 1, since the data path is of fixed width.

For pure multiplication and division, the CORDIC AU cannot compete with the more conventional AUs. But for circular and hyperbolic rotation and vectoring, it is superior in speed to architectures consisting of table lookups and multiply/divide units. For a more thorough comparison in speed, area, and throughput of array multipliers and CORDIC AUs, we refer to [Ahmed85].

7.4   System Level Implementation

As mentioned in Section 6.2, the desirable features for an array processor system are: high-speed, flexibility, expandability, and inexpensiveness.

An overall system configuration is depicted in Figure 6.1, where the design considerations for the major components have been discussed. Therefore, we shall only provide a brief review here. Later in this section, we also elaborate on several types of interconnection networks, whose inclusion is becoming essential to many system designs.

---

[2]A very good example is the Givens' methods used in QR decomposition. We observe that although the rotation angle generations (used in the boundary PEs) and GRs (used in the interior PEs) are significantly different, they can both be executed by CORDIC units.

7.4.1   Overall System Architectural Considerations

An array processor system consists of four major components: host computer, interface system, PE arrays, and interconnection networks.

The *host computer* should provide batch data storage, management, and data formatting; determine the schedule program that controls the interface system and interconnection network, and generate and load object codes to the PEs. The system controller follows the schedule commands from the host and performs data rearrangement and direct data-transfer traffic.

The *interface unit*, connected to the host via the host bus, has the functions of down-loading and up-loading data. Based on the schedule program, the GCU monitors the interface system and interconnection network. The functions which are common in interfaces are DMA, buffering, (cache memory, if necessary), handling interrupt, and data and sequence control.

A *PE array* comprises a number of processor elements with local memory. The granularity of a PE is somewhat tied to the number of PEs in the array. We can roughly divide the arrays into three groups: (1) those with the number of processors on the order of ten, (2) thousands, and (3) millions. The processors in the first group are usually implemented on board(s). Processors in the second group are usually *single-chip*, and are compatible with many commercially available microprocessors. Processors in the third group are usually *many-per-chip* (and usually bit serial), and are compatible with custom VLSI designs. This classification is displayed in Figure 7.18 [Seitz84]. An example of smaller PE granularity is NCR's GAPP. The GAPP array lies in the intersection domain between the logic-enhanced-memory group and the computational array group as shown in the figure. Simple processor primitives are often preferred in many low-precision image processing applications. An example of a larger PE granularity is the INMOS' transputer. Many DSP applications requires fast multiply-and-accumulate, high-speed RAM, fast coefficient table addressing, and a possible choice appears to be an enhanced version of the transputer. Referring to Figure 7.18, a transputer array belongs largely to the micro-computer array domain. Because of the built-in asynchronous communication hardware, transputers are very suitable for the implementation of wavefront arrays.

Interconnection networks provide a set of mappings between processors and processors or between processors and memory modules to accommodate certain common global communication needs. Incorporating certain struc-

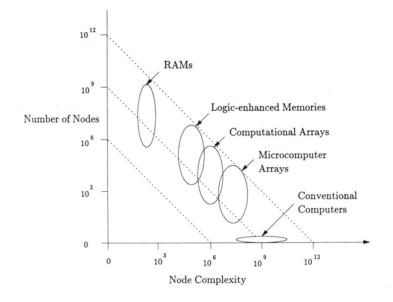

Figure 7.18: Different levels of granularity of PE in an array processor system, adapted from [Seitz84].

tured interconnections may significantly enhance the speed performance of the processor arrays. The topologies of interconnection networks are classified into three categories: *bus, static networks and dynamic networks*. Various kinds of interconnection networks will be discussed in more detail in the next subsection.

7.4.2   Interconnection Networks

Many recursive algorithms inherently require a global communication capability among the processing elements, which points to the need for interconnection networks. Interconnection networks provide a set of mappings between processors and memory modules to accommodate certain common global communication needs. The basic requirement of the interconnection network is to ensure that all messages can be successfully sent to their destinations as quickly as possible. The options and trade-offs in selecting appropriate interconnection networks to support array processors should be carefully studied. Some widely used examples are described as follows.

7.4.2.1   Bus-Oriented Networks

Single-bus or multiple-bus structures [Hwan84a] have been widely used for many years in conventional multiprocessor architectures. One of the examples is the Multimax machine by Encore Computer Corp. which has a 100 Mbyte/s bus and can be expanded from 2 to 20 microprocessors. Although bus-oriented networks can be made extremely powerful by using very high-speed buses, they are always limited to a certain number of processors by the bandwidth of the buses.

7.4.2.2 Static Networks

The static networks have interconnections which are hard-wired, the input and output are fixed without switching capability. Static networks can also be classified according to the dimensions required for layout. Typical examples are the one dimensional linear array; two dimensional ring, tree, star, mesh and hexagonal array, three dimensional completely connected chordal ring, 3 cube, and 3-cube-connected-cycle networks, and the D-dimensional hypercube array.

Tree and Fat-Tree A relatively inexpensive topology is the binary tree. The advantages of tree networks include short distance (order of $\log_2 N$) and efficient layout in two dimensions. The primary disadvantage is the communication bottleneck at the root of the tree.

The fat-tree [Leise85], which was proposed by Leiserson is a routing network based on a binary tree (see Figure 7.19). A set of n processors is located at the leaves of the fat-tree, and the internal nodes are switches.

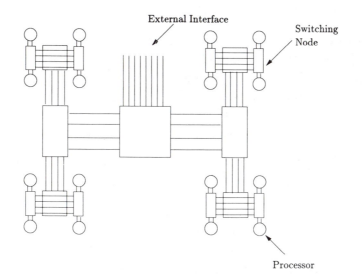

Figure 7.19: Fat-tree (adapted from [Leise85]).

Each edge of the underlying tree corresponds to two channels: one from parent to child, the other from child to parent. Going up the fat-tree, the number of edges connecting a node with its father increases, and hence the communication bandwidth increases. Leiserson also proved that for a given physical volume of hardware, no network is better than a fat-tree architecture for using resources.

Hypercube   One of the main commercial networks in the nonbus-oriented category is the hypercube. With this architecture, each processing unit, called a node, can communicate directly with its nearest neighbors in the n-dimensional space. For example, there are two nearest neighbors for each node in a 2-D hypercube, three for a 3-D system, and four for a 4-D machine. Intel's iPSC contains 32, 64, or 128 processors connected in a hypercube network by 10 Mbit/s links. Another hypercube machine is Ncube/ten by Ncube of Beaverton, Oregon., which can be expanded to 10-D, i.e., 1024 nodes. One drawback with this networking scheme is that if a processor needs to communicate with a node that is not one of its nearest neighbors, the data must be routed via intervening processors; this can slow down the overall processing rates if it occurs frequently.

7.4.2.3 Dynamic Networks

The dynamic networks are the networks with a capability to switch among different outputs at the same input. They can be classified into single-stage and multi-stage networks [Hockn83]. Cross-bar and shuffle networks are often adopted in array processor systems.

Crossbar switch   The crossbar switch provides complete connectivity for each node in the network because there is a separate bus associated with each node. The main advantage of using a crossbar switch is the ability to support simultaneous transfers for all participating network nodes. However, the hardware cost required to implement the switching matrix and to resolve conflicts of multiple requests is great, which limits the use of the crossbar switches.

Figure 7.20 shows a crossbar switch organization for multiprocessors [Enslo77].

Shuffle Network   The mathematical representation of a perfect shuffle network was discussed in Section 2.6. Perfect-shuffle networks have been the basis of many multistage interconnection networks. As mentioned earlier, FFT array processors can be implemented by the perfect-shuffle interconnection. Other applications of perfect-shuffle networks include bitonic sort-

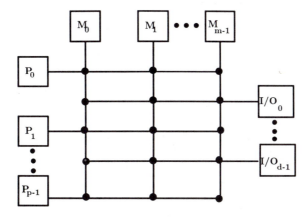

Figure 7.20: Crossbar switch for multiprocessors (adapted from [Enslo77]).

ing and polynomial evaluation [Stone71]. However, it can be shown that the perfect-shuffle network is not a very satisfactory network, as it leaves four disconnected subsets of processors. As a remedy, the perfect-shuffle exchange and perfect-shuffle near-neighbor networks have been proposed [Grosc79].

Single/Multiple Stages Interconnection Networks   Single-stage networks (such as the crossbar, hypercube and shuffle) consist of a fixed or single stage of switches. Thus in a P-processor array, the network consists of a single array of P multiway switches. Multiple-stage networks can be formed by a combination of some single-stage networks and can provide a cheaper alternative to the complete crossbar switch when a full connection network is required. The basic hardware configuration of a single-stage interconnection network is shown in Figure 7.21.

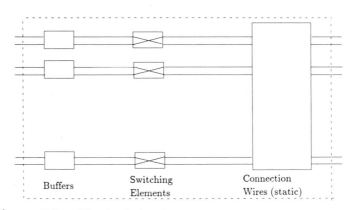

Figure 7.21: The basic hardware configuration of a single-stage network.

Single-stage networks are often limited in their connectivity because they can directly support only a limited number of permutations. More general permutations are established by stage-by-stage iterations. There are two primary approaches:

(1) *Recirculating single-stage* interconnection network (Figure 7.22(a)). As an example of recirculating single-stage interconnection network, if a network provides the single shift permutation α, then to perform k shifts, $\alpha^{(k)}$, requires k iterations through the network.

Another example is the FFT computation using the perfect-shuffle interconnection network shown in Figure 7.23 (a). In this case, a linear array of PEs computes one stage of the FFT, and intermediate results are passed to the next stage by the perfect-shuffle interconnection. It takes $\log_2 N$ stages

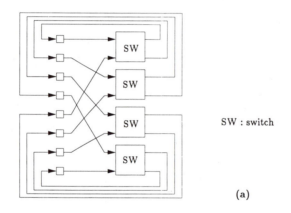

(a)

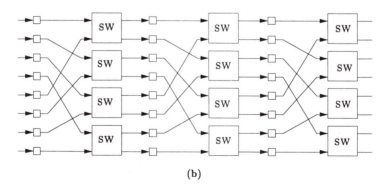

(b)

Figure 7.22: (a) Recirculating single-stage interconnection network. (b) Multistage interconnection network.

to compute the N-point FFT; therefore, $\log_2 N$ recirculation time are needed for this approach.

(2) *Multistage* interconnection network is realized by cascading many single-stage interconnection networks (Figure 7.22(b)).

Basically, a dynamic multistage network performs like a recirculating dynamic single-stage network, but has potentially better pipelinability.

Let us consider the above FFT example again. It is also possible to connect the $\log_2 N$ stages linear arrays through the perfect-shuffle network and build a multistage FFT array processor, as shown in Figure 7.23 (b). The advantage of this multistage array processor is that we can pipeline many FFT computations into this array and gain another speed-up of $\log_2 N$, although the latency of the FFT computation is still $\log_2 N$.

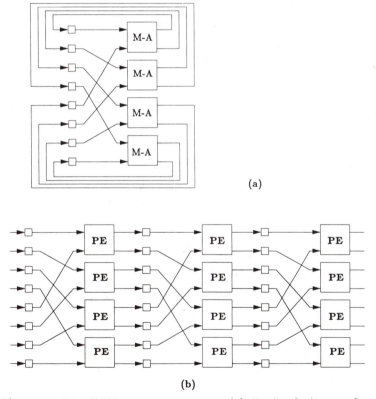

(a)

(b)

Figure 7.23: FFT array processor: (a) Recirculating perfect-shuffle connected FFT array processor. (b) Multistage perfect-shuffle connected FFT array processor.

7.4.2.4 Networks for Supporting Array-level Pipelining

A system may have one or more locally interconnected processor arrays. The concept of networking several processor arrays has now attracted a good deal of attention. For example, when a problem can be decomposed into several subproblems to be executed one after another, it is useful to have each subproblem executed in its own processor array, while utilizing the network to facilitate the data pipelining between the arrays. This suggests a pipelining scheme in the array level, which can increase the processing speedup by one more order of magnitude.

7.5 Examples of Array Processor Systems

In this section, we briefly describe several types of existing array processor systems, namely, SIMD arrays, systolic arrays, wavefront arrays, and hypercube computers. These systems are either under construction by research labs or commercial companies. They are considered to be playing a role in the future of array processor development. Other proposed array processor systems, such as multiple broadcast mesh array, reduced mesh array, and pyramid arrays, are also briefly discussed to show their unique architectures.

7.5.1 SIMD Array Processors

As mentioned in Section 1.1, SIMD computers are implemented as an array of arithmetic processors, with local connectivity between these and the local memory associated with each. Instructions are broadcast from a host, all processors executing the same instruction simultaneously. The conceptions of SIMD systems were developed in the late 1950s by Unger (1958) and later by Von Neumann (1966), but the first SIMD machine constructed for practical use was ILLIAC IV in the 1970s.

7.5.1.1 Binary Array Processors

A binary array processor (BAP) consists of a matrix of identical PEs, operating in SIMD mode. The main features of the BAP design are the bit-serial architecture of the PEs and the local interconnection scheme (cf. Figure 7.24). The bit serial architecture allows more flexible data formats (especially for fixed-point operations) and offers efficient memory and processing resource utilization. Several large scale binary array processors have been

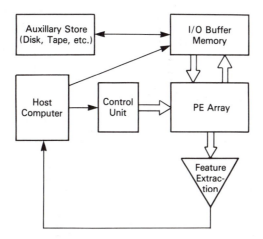

Figure 7.24: System organization of BAP.

developed, including the Massively Parallel Processor (MPP), and the Distributed Array Processor (DAP). A new addition to the BAP family is the NCR VLSI chip - Geometric-Arithmetic Parallel Processor (GAPP). [3]

Massively Parallel Processor (MPP)   The MPP is a 128 × 128 bit serial array processor constructed by Goodyear Aerospace in 1979 (see Figure 7.25). The PEs of the MPP have a number of interesting facets, including a variable-length shift register, multiplexed neighbor inputs and a data structure allowing array operations and data I/O to be overlapped. Eight such processors are integrated on a custom circuit interfaced to high speed, high-density RAMs. Because of the high speed of the custom circuits the memory access time dominates operations. The development of the MPP represents the state-of-the-art construction of large-scale SIMD computers, and are now being used by NASA for processing satellite images [Fouta85].

Distributed Array Processor (DAP)   The DAP was developed by International Computer Limited in the U.K.. The DAP can be constructed in groups of 16 PEs in various sizes, such as 32 × 32, 64 × 64, 128 × 128, and 256 × 256. Each PE of the DAP is a bit serial adder which sits on top of a block of RAM (4096 bits) and has direct connections to its four nearest neighbors in a square tessellation. An interesting feature consists of two

---

[3]In a sense, the Connection Machine (CM-1) [Hilli85] also belongs to this BAP family, but in our discussion it is considered to be in the class of hypercube computers.

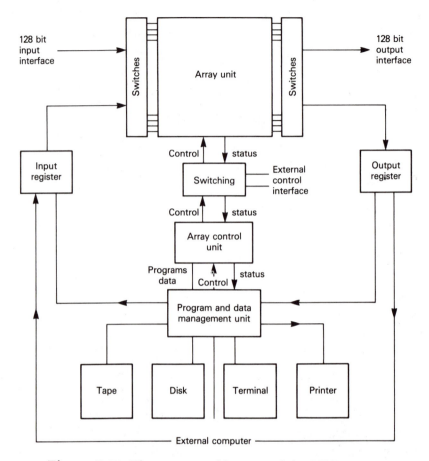

Figure 7.25: The system architecture of the MPP system.

orthogonal sets of communication lines which cross the array, making data shifting particularly easy. The host (ICL2900) provides the data I/O for the array (see Figure 7.26).

Comparing DAP and GAPP As we have discussed in Section 7.2.2.4 that the GAPP is a commercial VLSI chip, composed of a 6 by 12 arrangement of *single bit* processor cells (see Figure 7.5). In the GAPP, each instruction is broadcast to all the PEs, making the array perform in an SIMD mode. There are many similarities between the PE of the DAP and the much newer GAPP. The GAPP design more explicitly embodies the concepts of universality and of Boolean efficiency. The most important design difference is that much less memory (128 bits) is made available to each processor (in

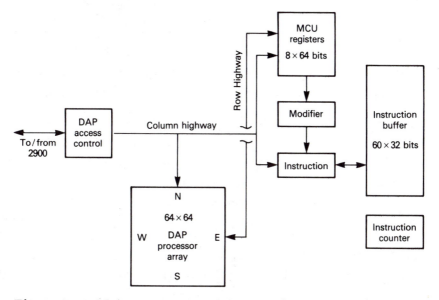

Figure 7.26: Major components of the DAP (adapted from [Hockn83]).

parallel) in the GAPP (compared to the 4096 bits available to each PE in the DAP). This results in an unprecedented density of 40,000 GAPP PEs in 60 cubic inches (with $3\mu m$ technology). On the other hand, the GAPP was originally designed only for low-precision digital and image processing applications. In contrast, the DAP can solve problems in, for example, numerical fluid dynamics, which would quickly exhaust the GAPP memory.

7.5.1.2   Cellular Array Processor

The cellular array processor (CAP) is a multi-bit SIMD processor, which has been designed with $1.25\mu m$ technology [Morto84]. Its chief innovations are that it is a multi-bit machine with a fault tolerant design that allows the use of 16 active processors on a chip, whereas the actual number of processors is 20. Thus, up to four of the processors can fail and the failure will be transparent to the users. The on-chip memory per PE is 4,000 bits and external memory can be multiplexed in.

The major functions of a PE in the CAP are controlled independently so that functions may run concurrently. These functions include (1) memory/common bus function, (2) arithmetic/logic function, (3) I/O function. Applications that seem well suited to the CAP are those that involve both parallelism and computational complexity or memory requirements to some degree. Examples include Kalman filtering and content addressable memories.

7.5.2　Systolic Array Processor: The Warp Machine

The Warp machine was designed and built by Carnegie-Mellon University with its industrial partners, as a powerful computing engine for many low-level signal and image processing tasks encountered in the vision area. To date, several copies of the Warp machine have been built. Passive navigation for mobile robots or autonomous vehicles were among the first demonstrations of the machine [Arnou85], [HTKun85], [Anna86b], [Anna86a].

The full-scale machine has a systolic array of 10 or more cells, each being a 10MFLOPS programmable processor. The array is integrated on a 68020-based multiprocessor host system that can run application code (under Unix BSD 4.2) and supply/receive data to/from the array and various I/O devices at a high rate.

The Warp is highly programmable; each cell in the array has a horizontal microengine, giving the user complete control over the various functional units. To overcome the complexity of this fine grain parallelism, CMU has developed a compiler to support high level language programming.

The Warp processor array supports two modes of programming. In the *systolic mode*, the array can efficiently implement a wide variety of systolic algorithms, including FFT, matrix multiplication, and 2-D convolution. In the *local mode*, the array can implement many nonsystolic algorithms, where each cell performs a complete computation on its own data independently from the rest.

The Warp machine is composed of three parts: the host, the Warp processor array, and the interface unit (IU), as shown in Figure 7.27.

7.5.2.1　The Warp Processor Array

As stated earlier, the Warp array is a programmable linear systolic array with identical cells called Warp cells, synchronized by a single global clock issued by the interface unit at a cycle time of 200 ns. Data flow through the array on two data paths (X and Y), whereas addresses and systolic control signals travel on the Adr path (as shown in Figure 7.28). Each Warp cell has its own program memory and sequence controller. The data path of a cell consists of two floating point processors: one multiplier and one ALU, a 64K-word

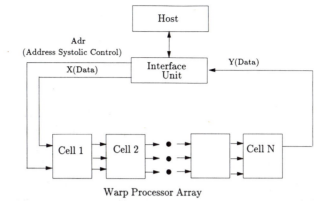

Figure 7.27: Configuration of the Warp machine (adapted from [Arnou85]).

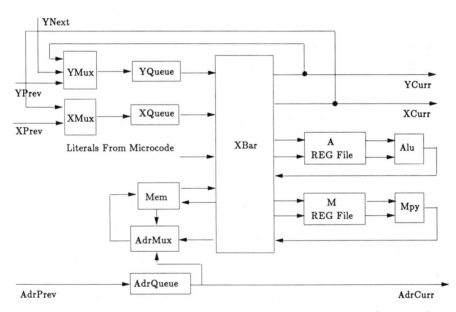

Figure 7.28: Configuration of the Warp cell (adapted from [Arnou85]).

memory for resident and temporary data, a queue for each communication channel, and a register file to buffer data for each arithmetic unit.

A Boundary Processor (BP) may be attached to one end of the Warp array. The BP has bidirectional links to the IU, to the first cell and to the last cell in the array. This ring connection makes it possible to recirculate data around the Warp array without involving the host. The BP contains a large memory unit to buffer data and a fast processing unit for the divide and square root operations. Its design is similar to the Warp cell, as depicted in Figure 7.28.

7.5.2.2 Host

The host is mainly responsible for all the on-line operations that are needed to provide the Warp array with necessary flow of data through the interface unit. The host can also perform computations that cannot be efficiently done on the Warp array. The host consists of a VME-based workstation (currently a Sun 2/160) that serves as the master controller of the Warp machine and an external host, so named because it is external to the workstation. The workstation provides a Unix environment for running application programs, and the external host provides a high data-transfer rate for communicating with the systolic array. The external host consists of three stand-alone 68020-based processors, which run without Unix support to avoid operating system overheads. Two of the stand-alone processors are *cluster processors*, responsible for transferring data to and from the Warp array, and the third one is a *support processor* that controls peripheral devices (e.g., the camera and the monitor) and handles interrupts originating in the clusters and the Warp array. A real-time operating system that allows scheduling events in sequence runs on the clusters and master.

7.5.2.3 Interface Unit

The interface unit (IU) handles all communication between the host and the Warp processor and provides all the control signals necessary to drive the Warp processor. As depicted in Figure 7.29, the IU has the following components:

1. Input/output FIFOs, each 32 bits wide and 512 bits deep, used to buffer data between the host and the Warp array.

2. Integer-to-floating-point and floating-point-to-integer converters.

3. Input/output crossbars, to allow the processor array to receive from or send to either of the two input/output ports of the IU.

4. Address generator to send 16-bit addresses to the Warp at the rate of one per 100 ns.

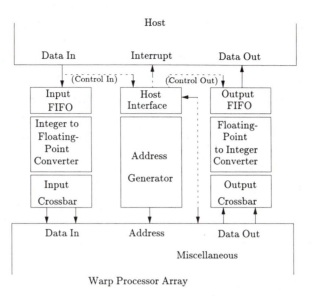

Figure 7.29: The Warp interface unit (adapted from [Arnou85]).

5. Host interface, containing three 32-bit registers: the status register, the control register, and the interrupt register. These registers can be read by the host at any time.

7.5.2.4   Performance

Several benchmark programs have been tested on the Warp machine. A 1024 point complex FFT can be processed by a 10-cell Warp at a rate of one every 0.6 ms. A 2-D discrete cosine transform on a 256×256 image can be processed in 13 ms. The Warp augmented by a boundary processor is effective for a common set of basic matrix operations. For example, the singular value decomposition of a 100×100 matrix can be processed in 1.4 seconds.

From the organization of the the Warp machine, we note that there is no interconnection network included in the design. The communication between processor array and the host is only through the boundary cells of the array. While this approach simplifies the communication problem between the array and the host, it is not flexible for other general communication needs. Consequently, the concurrency of the Warp is derived from its pipeline processing rather than parallel processing.

7.5.3   Wavefront Array Processor Systems

7.5.3.1   Memory Linked Wavefront Array Processor (MWAP)

The memory linked wavefront array processor (MWAP) [Dolec84] is a new WAP, developed at the Applied Physics Laboratory of the Johns Hopkins University, which employs VHSIC technology. Due to the use of the VHSIC chips, very high performance is achieved by the array. The entire processing system is connected as a ring network, with each node being a wavefront array. The MWAP uses a modular structure both for the processor system on a ring bus and for the PE. The basic system architecture (as shown in Figure 7.30) consists of a bus interface to a dual-port memory (which may be memory such as the TRW quad-port memory or TI SRAM memory, or a set of registers), multiple processing-element/dual-memory pairs, and a bus output interface.

Each MWAP PE, as shown in Figure 7.31, consists of a control unit, an instruction unit, an instruction cache, a block of memory-address registers,

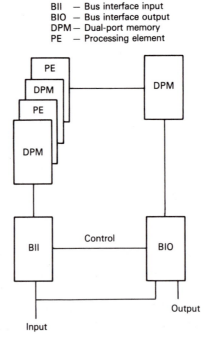

Figure 7.30: MWAP system interfaced with the ring bus (adapted from [Dolec84]).

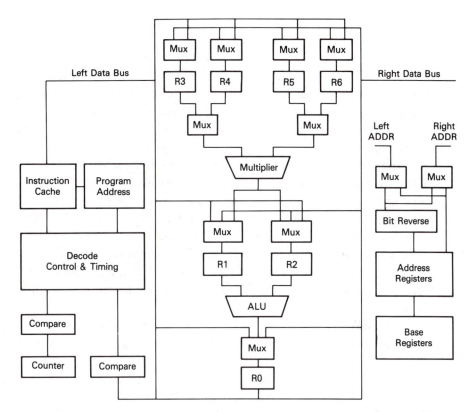

Figure 7.31: MWAP processing element (adapted from [Dolec84]).

a floating-point multiplier, and a floating-point ALU. Once the instruction cache is loaded, program and data memory are separate. The key to this architecture is the memory addressing structure. All memory addressing is done by reference to a memory address register, which can be read, read and then incremented, or read and then reset to a base address set up during program load. The PE can simultaneously read or write data memory in both directions. Synchronization of the PEs in the MWAP is done by the right and left control flags in the memory addressing unit.

A PE can always read the memory to either its right or left. However, it only can write into the memory when the control flag for that memory is reset. Each PE can set (S) or acknowledge (A), which resets the memory control flag in either memory. If the PE attempts a store to a memory that has its control flag set, program execution is suspended until the control flag is reset; then the store command is executed. In addition, the setting of a control flag in a memory causes an idle PE to begin program execution.

This memory control structure results in high-speed compact programs with complete MWAP dataflow control.

This combination of multiple address registers and memory not only provides global interconnections between PEs, it also permits efficient zero insertion for padding FFT inputs and interpolation/decimation filtering. The input memory to the MWAP is loaded with zeros once during initialization. Only data samples are written to the proper address during signal processing.

For signal processing, a serious consideration is the accuracy and speed of the arithmetic units. This leads to a trade-off between parallel versus serial and fixed-point versus floating-point arithmetic hardware. Parallel 24-bit floating-point arithmetic was chosen for the prototype MWAP design to allow a broad range of algorithms to be implemented and evaluated against the current Floating Point Systems AP120B array processor in the Sonar evaluation program analyzer (APAN). In addition, data scaling is simplified since the dynamic range of MWAP floating point numbers is 1.4×10^{-39} to 1.7×10^{38}.

The MWAP floating-point multiplier is capable of 10 million multiplications per second as opposed to 25 million per second for a fixed-point VHSIC multiplier. However, the Texas Instruments Static RAM (SRAM) VHSIC memory chip used in the current MWAP dual-port memory will support only about 10 million accesses per second per PE. Thus, for real-number filtering operations, memory speed and multiplication speed are matched. Complex FFTs require six left memory reads (and four right memory stores) with four multiplications, three adds, and three subtracts in each PE cycle, again matching memory speed and multiplication speed.

Note that the use of the TRW quad-port VHSIC memory chip in place of the SRAM would result in an MWAP design that could support 25 million multiplications per second. However, this would also result in 64 (1024 word × 4 bit) memory chips per PE compared to 4 (8194 word × 9 bit) chips using the SRAM.

7.5.3.2   STC-RSRE WAP System

Array computer design can be made easier if it is based on commercially available chips, at least when constructing experimental prototypes. These chips must provide, in addition to computing power, communication capabilities and interface mechanisms.

An Earlier Implementation   In the Royal Radar Signal Establishment (RSRE), U.K. [Broom85] a wavefront processor for the least-squares minimization algorithm was implemented on an array of microprocessors programmed in Occam. As the transputer was not yet available, transputer emulator boards were used to enable algorithms to be evaluated and performance assessments made. Each emulator consisted of a circuit board, containing an 8MHz Intel 8086 processor with transputer-like communication channels. Four ports operated asynchronously (i.e., handshaking) at 200K baud with system software ensuring that the channel facilities appeared the same as transputer silicon. A fifth channel was programmed to operate as an asynchronous RS232 link to enable programs to be loaded. The data storage with 8K bytes of EPROM was for the bootstrap loader and monitor. Additionally, each emulator had an Intel 8254 timer chip to allow Occam timing facilities to be implemented in hardware. The ratio of CPU execution time to I/O time was similar to that of the transputer with emulator operation at approximately one-fifth of the transputer speed.

A Reconfigurable Implementation   Recently, in order to demonstrate parallel signal processing, Standard Telecommunication Company (STC) and RSRE have jointly developed a wavefront array processor system for adaptive beamforming. The system is reconfigurable for many distributed array processing applications [Davie86].

PE Design   The PE design is based on the TMS32010, with additional hardware to provide multiple I/O ports, a floating-point capability, look-up table, localized control, and a bit serial diagnostic link. The PE also features both program memory ROM, containing fixed algorithmic code, and program RAM, giving the added flexibility of being able to down-load programs for other algorithms. The schematic diagram and its board implementation are shown in Figure 7.32 (a) and (b).

WAP System   When used for adaptive beamforming, the WAP system consists of 33 identical PEs, 21 of which are organized as a triangular wavefront array performing the adaptive beamforming function and the remaining 12 PEs performing data correction and other preprocessing functions. This is shown in Figure 7.33.

The final system is shown in Figure 7.34. The system is also supported with RF (radio frequency) /IF (intermediate frequency) front-end

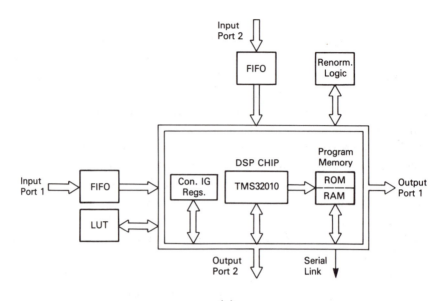

(a)

(b)

Figure 7.32: STC-RSRE WAP system, the PE Design: (a) Schematic diagram. (b) Processor board (adapted from [Davie86]).

520

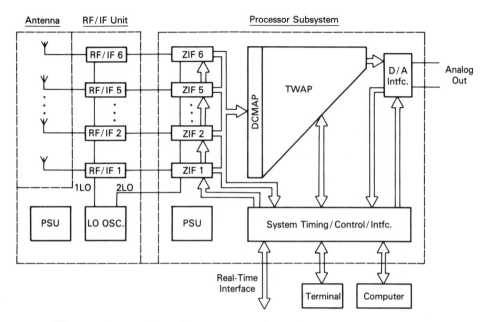

Figure 7.33: Overall system configuration of the STC-RSRE adaptive array system (adapted from [Davie86]).

Figure 7.34: The STC-RSRE WAP system (adapted from [Davie86]).

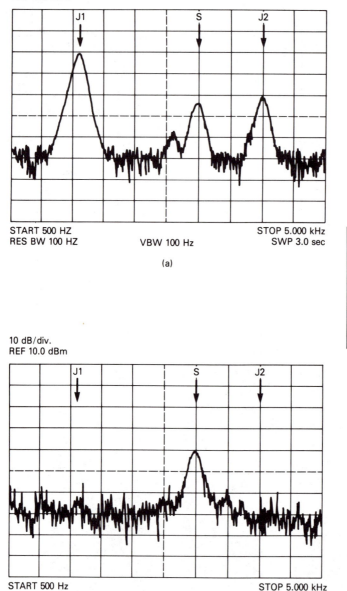

10 dB/div.
REF 10.0 dBm

START 500 HZ
RES BW 100 HZ
VBW 100 Hz
STOP 5.000 kHz
SWP 3.0 sec

(a)

6 ELEMENTS 3 SOURCES		
SOURCE	ANGLE	POWER (dB)
J1	−48°	−30
S	−15°	−50
J2	0°	−45
−60 dB NOISE FLOOR		

10 dB/div.
REF 10.0 dBm

START 500 Hz
RES BW 100 Hz
VBW 100 Hz
STOP 5.000 kHz
SWP 3.0 sec

(b)

Figure 7.35: (a) Input waveform before noise cancellation. (b) Output waveform after noise cancellation (adapted from [Davie86]).

subsystem, zero-IF receiver and A/D conversion. The processor system is housed in a racking system enclosure along with power supplies and fan cooling. Total power consumption is approximately 500W.

The system has been successfully applied to many real-time experiments. A jammer cancellation performance exceeding 50 dB for the case of a single jammer and exceeding 40 dB for the case of multiple jammers is expected. The input and output waveforms before and after noise cancellation are shown in Figure 7.35(a) and (b), respectively. In terms of throughput, a 10K samples/second rate is achieved and a much faster performance (up to 10M samples/second rate) is expected for the full-scale system utilizing VLSI node processors. This is discussed later.

7.5.3.3   WAP Based on A VLSI Node Processor

Any future realization of wavefront arrays for real-time signal processing requires the development of specialized VLSI processors to enable a more compact implementation of the system and to provide a throughput rate matched to modern radar, communications, and ECM (Electronic Counter Measure) systems. As a part of the UK VHPIC application demonstration program, recent work at STC has been directed towards the definition of a high-performance VLSI "Node Chip" intended as a programmable building block for the implementation of real-time WAP subsystems.

A number of key features have been included in the Node Chip specification:

- High-throughput, single-chip processor.

- On-board I/O control, program and data memory, arithmetic units, and parallel multiplier blocks.

- Full handshake control to allow simple interconnection with neighboring processors.

- Internal architecture optimized for floating point arithmetic.

- Programmability.

The Node Chip is a key component for fast front-end signal processing applications where the VLSI implementation provides processing throughput rates matched to the demanding requirements of modern radar and communication applications.

Two practical factors, i.e., the data flow and real-time functional control, are carefully considered in the implementation and design of this chip. Data flow within a wavefront array can be a complicated process where a high throughput rate is essential. Restrictions are placed on the amount of data that can be transferred between chips within a given algorithm. The Node Chip overlays computation and data flow by using FIFO buffering on both the input and output ports. Maximum data flow is therefore ensured by allowing computation and data flow to operate asynchronously.

An advanced level of dynamic functional control has also been necessary for the Node Chip concept. The application of sophisticated algorithms to real-time signal processing problems often requires that processing parameters be varied in accordance with the required system performance or environmental conditions (e.g., dynamic bandwidth control of a digital filter by rapid adjustment of the filter weighting coefficients). In addition to the usual data I/O ports on the Node Chip, there is also provision for the transfer of control information from an external supervisory processor. The control data is propagated in synchronism with data wavefronts through an array of processing nodes.

7.5.4   Hypercube Computers

A hypercube computer consists of a number of processors (or computers) connected by a hypercube network with bidirectional, asynchronous communication channels. More precisely, the hypercube node structure consists of an array of $N = 2^n$ processors, with each processor connected directly to its n neighbors placed in the n-dimensions of the cube (see Figure 7.36).

A convenient scheme for labeling the nodes at the $N = 2^n$ corners of a n-dimensional hypercube is to label each node by an n-bit binary number. The ith bit of the number represents the coordinate of that node in the ith dimension. The edges of the hypercube connect the nodes. In terms of the binary labeling scheme, nodes whose labels differ by only 1 bit are connected.

Figure 7.36: A 4-D hypercube computer architecture.

Another unique feature of a hypercube computer is that it uses *message passing* instead of shared variables for communication between concurrent processes. Message-passing machines are simpler and more economical than shared-storage machines; the greater the number of processors, the greater this advantage.

7.5.4.1   Cosmic Cube

The Caltech 64-node hypercube computer (Cosmic Cube) can execute up to 3 million floating-point operations per second [Seitz85]. The Cosmic Cube has no special programming notation. Process code is written in ordinary sequential programming languages (e.g., Pascal or C) and extended with statements or external procedures to control the sending and receiving of messages. These programs are compiled on other computers and loaded into and relocated within a node as binary code, data, and stack segments.

The present applications of the Cosmic Cube are mostly limited to compute-intensive tasks. I/O-intensive tasks require a great deal of I/O bandwidth for high throughput image processing applications.

7.5.4.2   Connection Machine

The Connection Machine is a new type of computing engine proposed by W. D. Hillis from Thinking Machines Corporation [Hilli85]. It provides a large number of very small processor/memory cells connected by a programmable communications network. Each cell is sufficiently small such that it is incapable of performing any meaningful computation on its own. Instead, multiple cells are connected together into data-dependent patterns, called *active data structures*, that both represent and process the data. The activities of these active data structures are controlled directly from outside the connection machine by a conventional host computer. A connection machine is connected to a conventional computer much like a conventional memory. However, it differs from a conventional memory in three respects. First, associated with each cell of storage is a processing cell that can perform local computations based on the information stored in that cell. Second, there exists a general intercommunication network that can connect all the cells in an arbitrary pattern. Third, there is a high-bandwidth I/O channel that can transfer data between the connection machine and peripheral devices at a much higher rate than would be possible through the host.

A prototype of the connection machine, called CM-1, has been constructed. The CM-1 contains $65,536 = 2^{16}$ processor cells, each with 4 Kbits memory and a simple serial arithmetic logic unit. The basic building blocks are custom CMOS chips, each containing 16 processor cells and one router unit. The router is responsible for routing messages between chips. The chips are physically connected as a 12-dimensional hypercube. Within a chip, the processors are connected in a 4 × 4 grid. Each processor cell can communicate directly with its north, east, west and south neighbors, without involving the router. This 2-D grid pattern can be extended across multiple chips. All processors execute instructions from a single stream generated by a microcontroller under the direction of a conventional host. The machine has a peak instruction rate (32-bit addition) of about 1000 MIPS. The block of the connection machine with host, processor/memory cells, communications network, and I/O is shown in Figure 7.37. There have been several benchmark programs running on the connection machine. We list some of

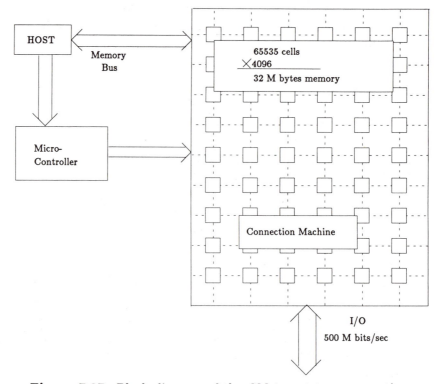

Figure 7.37: Block diagram of the CM-1 prototype connection machine (adapted from [Hilli85]).

the results [Frenk86]. The CM-1 can perform unstructured text retrieval in a 15,000-document database (about 40 Mbytes), achieving 80% recall rates compared to 20% for conventional systems. It processes visual information to produce contour maps at a rate of 7000 per hour, compared to 15 per hour on a conventional computer; it simulates VLSI chips composed of 8000 transistors, whereas conventional computers rarely simulate more than from 50 to 500; and it simulates fluid dynamics upto 8,000,000 particles.

7.5.4.3 FPS T Series

Floating Point Systems (FPS) has developed a homogeneous computer, the FPS T Series, based on the hypercube interconnection scheme [Gusta86]. The individual nodes are 64-bit floating-point computers that combine vector arithmetic, dual-port memory, and fast communication links between nodes. Each node is implemented on a single circuit board and can perform 64-bit floating-point arithmetic at a peak speed of 16 MFLOPS. Eight nodes are grouped together with a system node and disk support to form modules. These modules, housed in cabinet-sized packages, are capable of 128 MFLOPs peak performance and can be connected together to form larger systems. The largest system can have $2^{14} = 16,384$ nodes.

Node Processor Each node contains a control processor, floating- point arithmetic, dual-port memory and communication links to other nodes. The node configuration is shown in Figure 7.38. Each of the major elements of the node has been implemented with advanced, cost-effective VLSI technology. The control processor is the Inmos transputer T424, which is a 32-bit microprocessor and supports at least three times the speed of the Motorola 68020. The control processor executes system and user application code; it also serves to arrange vector operands to be sent to the vector arithmetic hardware. The main memory of each node consists of 1 Mbyte of dual-port dynamic RAM. The control processor and communication links read and write 32-bit words through a conventional random-access port, whereas the vector arithmetic unit makes use of a collection of *vector registers* closely coupled with main memory. The pipelined arithmetic hardware in the FPS T-series consists of a six-stage pipeline floating-point adder, five- or seven-stage pipeline floating-point multiplier, interconnection hardware, and some sequencing hardware. The adder and multiplier can each produce a 32- or 64-bit result every 125 ns, yielding a peak performance of 16 MFLOPS. Floating-point operations are performed using the proposed IEEE floating-point standard format. The communication channels are the four

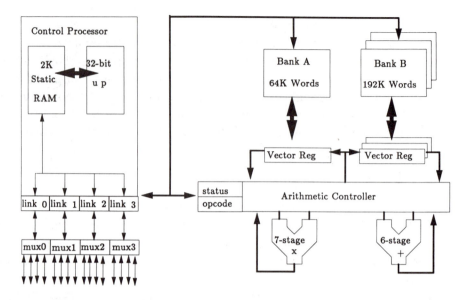

Figure 7.38: Node processor of the FPS T-Series system (adapted from [Gusta86]).

bidirectional serial links of the transputer, each with an effective rate of 0.5 Mbyte/s. Each link is multiplexed four ways to provide a total of 16 bidirectional sublinks per node. With software support, these sublinks divide the available bandwidth. Two sublinks are used for system communication, and two are utilized for mass storage and/or external I/O. Each node typically leaves 12 sublinks available for connection to other nodes.

System Description The nodes are connected into hypercubes. Eight nodes are combined with disk storage and a system board to form a *module*. The local internode communication bandwidth is 12 Mbit/s, whereas the system board can support 0.5 Mbit/s to an external connection. The system board provides I/O and management functions. It is connected to the nodes by a thread of communication links that traverses the eight processor nodes. The system boards are directly connected by communication links to form a *system ring* that is independent of the binary n-cube network. There are enough links per node to permit a 14-cube to be constructed as the largest T series configuration. Because the system is homogeneous, i.e., each module is identical and contains identical connections to other modules, programming is greatly simplified.

Programming Since FPS chose the transputer as the control processor, Occam is the natural choice of programming language. All features of the transputer can be directly accessed through Occam. Occam can describe the control and data flow for virtually any scientific computing algorithm and control the high-level operation of the vector arithmetic unit.

7.5.5   Other Proposed Array Processor Systems

Taking image processing applications as an example, we note that image processing problems actually cover two different domains: low-level image processing and high-level vision analysis. Low-level image processing involves algorithms for restoration, noise removal, geometrical correction, edge detection, or feature enhancement. For most of these operations, the critical performance criterion is the throughput rate. Systolic arrays appear to be very suitable for these operations. On the other hand, high-level vision analysis involves sophisticated feature extraction, feature classification, and scene analysis, which call upon many techniques developed in the field of pattern recognition or artificial intelligence. For such tasks, it appears that an MIMD computer structure is more suitable. Therefore, it is very promising to explore new MIMD architectures that incorporate VLSI structures into architectural design and are made more oriented to intelligent image processing and vision analysis applications.

7.5.5.1   Multiple Broadcast Mesh Array

Figure 7.39 illustrates an enhanced array processor with multiple broadcasting. An attractive capability of this architecture is that data in PEs of any row (column) can be transferred to PEs in other rows (columns) in one broadcasting step. Such transfers are often required in 2-D FFT or finding the maximum, median, or minimum points of an image. The data routing time can be reduced from $N^{\frac{1}{3}}$ steps when using a single global broadcasting bus to $N^{\frac{1}{6}}$ when using the multiple broadcasting arrays [Kung85].

7.5.5.2   Reduced Mesh Array

For some algorithms, the inherent parallelism is rather limited. This results in an inefficient utilization of the processors in a full-blown multiple broadcast mesh array. For such applications, it is cost-effective to reduce the number of processing elements without significantly hurting processing speed. This goal can be achieved using the concept of virtual processor arrays. More precisely, *only* the diagonal PEs (shown as square boxes in

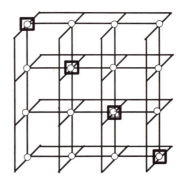

Figure 7.39: Multiple broadcasting and reduced mesh array. (a) An enhanced array processor with multiple broadcasting buses: a PE is assigned to every circle. (b) Reduced mesh array: if the PEs are assigned only to the square boxes (along the diagonal) and only storage units are assigned to the circles (adapted from [Kung85] and [Tseng85]).

Figure 7.39) will physically have a processing unit, whereas the remaining PEs in the array will be replaced by memory units. In other words, the diagonal PEs will be responsible for the processing loads originally belonging to all the PEs along the same column (or the same row).

This leads to the reduced-mesh multiprocessor proposed by [Tseng85]. This architecture matches with a variety of important parallel algorithms. The system consists of N processors, which share an orthogonally accessed memory mesh of N^2 modules in a conflict-free manner (see Figure 7.39). Obviously, global interconnection is required in order to support fast memory access or data transfer. Actually, this structure naturally permits operations in MIMD mode, thus it will be potentially more suitable for high level vision analysis tasks.

7.5.5.3 Pyramid Arrays

A pyramid can most simply be described as a stack of successively smaller arrays that are linked together by a tree [Uhr86]. Typically, a pyramidal computer system has several layers of PEs arranged in a square array of 2^n PEs; such as a 16 × 16 array for 256 PEs. The level above would contain 64 PEs in an 8 × 8 array, above which would be a 4 × 4 array, a 2 × 2 array, and a single PE at the apex of the pyramid. Other arrangements are possible, but this one represents a classical example. A PE can communicate with neighboring PEs on its own level, with a parent on the level above and, typically, with four children on the level below, except of course for the bottom layer as shown in Figure 7.40.

Each array in the pyramid can execute all array operations efficiently. In addition, passing information up and down between arrays needs greatly reduced logarithmic distances, rather than the linear distances needed within an array. This is the reason why a growing number of computer scientists believe that the hierarchical nature of processing in the pyramidal system is

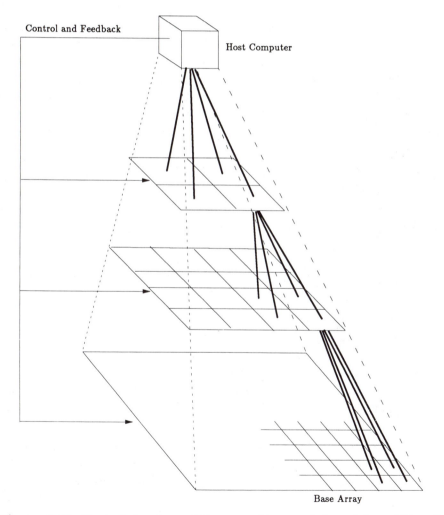

Figure 7.40: Typical structure of the pyramid array (adapted from [Uhr86]).

well adapted to image processing. The major purpose of a pyramid is not to communicate, but to compute. Rather than simply pass information, or average and compress information, moving up the pyramid, the processors can be programmed to process and transform that information in any way desired.

7.6   Concluding Remarks

This chapter presents the main issues concerning the processor and array implementations for SIMD, systolic, and wavefront arrays, with the focus placed on real-time signal/image processing applications. Throughput is the most critical factor in real time DSP systems. The features that sharply distinguish a generic DSP circuit from a general-purpose microprocessor are its multiply-accumulator (MAC) and multiple buses and memories (see Figure 7.41). With them the complete operation of a MAC may be executed in a single clock cycle (e.g., 60ns for 16 bit multiplication in AT&T's WEDSP16 DSP chip). Most DSP processors adopt the Harvard architecture, which allows the instruction and data to be fetched simultaneously. Some have a modified Havard architecture to allow storage of data in the program memory. In this way, static data like filter coefficients can be stored in the cheaper and slower program memory and then moved to the faster but smaller data memory when needed [Aliph87].

The changing CAD tools facilitate the implementation of dedicated systems. As the silicon feature size becomes smaller and integration scale grows larger, it will soon become practical to put the hardware for a PE in a single chip or a subsystem in a wafer. A large number of devices and functionalities can be integrated into a system. In this trend, the construction of dedicated array processors will eventually become very affordable and attractive. There also exist other device technologies, such GaAs devices, wafer scale integration, optical computing devices, which offer high-speed performance comparable to VLSI for special purpose supercomputing. These modern technologies along with VLSI will undoubtedly appear in many future supercomputing systems.

On the other hand, the advance of technology may also make the construction of programmable systems very attractive. They can provide fast processing speed and yet be applicable to a large class of applications. Thus, programmable systems will also be a major part in future supercomputer systems. Conceivably, a reconfigurable system can support and meet the needs

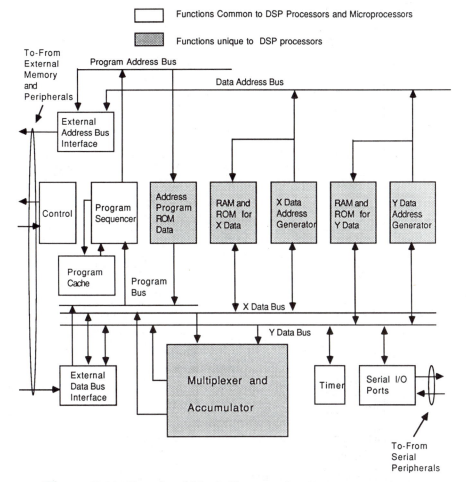

Figure 7.41: Functional block diagram of a digital signal processor. Note that the *blank* rectangles denote functions common to digital signal processors and microprocessors; the *shaded* rectangles denote functions unique to digital signal processors (adapted from [Aliph87]).

of many users. System capabilities may also be enhanced by expanding the existing systems with additional processing boards. The software requirements (such as compiler design) for such programmable, reconfigurable, or expandable array processor systems will be very demanding; hence the software tool developments for them will be vital for successful utilization of the systems.

7.7 Problems

1. *TMS32010 instruction set:* Some important instructions of TMS32010 are given below:

LAC Addr,\<shift\>	loads accumulator from loc. Addr shifted by \<shift\>
APAC	adds product to accumulator
SACH Addr,\<shift\>	stores the high 16 bits of the accumulator
SACL ADDR	stores low 16 bits of the accumulator
DMOV Addr	copies the contents of the memory location Addr to the memory location Addr+1.
MPY Addr	multiplies the contents of Addr with the T register
LARP r	auxiliary register r becomes the current pointer
LT Addr,\< r \>	load T register from Addr (optionally change AR)
LTA Addr	loads T register and accumulates previous product
LTD	performs a combined LTA and DMOV
ZAC	zero accumulator

(a) Using the instructions given in the table, implement a butterfly computation for the radix 2 FFT operation.

(b) Using the instructions given in the table, implement the equation $t = Aw + Bx + Cy + Dz$ in the fastest possible way, the coefficients A, B, C, D are fixed and known.

(c) Using the instructions given in the table, implement the equation $y(n) = Ax(n-1) + Bx(n-2) + Cx(n-3) + Dx(n-4)$.

2. *Characteristics of the TMS32010:*

(a) Discuss the usefulness of the following instructions for signal processing operations.

- REPEAT N : executing the next instruction N times (cf. TMS32020).

- MAC $A, B, n1, n2, n3$: multiplies A, B preshifted by $n1$ and $n2$ bits respectively and accumulates the result with post-shifting $n3$ bits.
- DMOVE *Addr*: copies the contents of the memory location *Addr* to the memory location *Addr* + 1.

(b) Why is it desirable to have post-increment and/or post-decrement addressing modes.

(c) Consider the FFT algorithm. Which instructions would be convenient to have in the instruction set of a general purpose signal processor in order to implement it.

3. *Two's complement arithmetic:* Prove that the negatively signed two's complement notation as defined in Eq. 7.1 is indeed a valid one such that $N_v + (-N_v) = 0$, where $-N_v$ is the value of $-N$, the two's complement of a number N (adapted from [Hwang79]).

4. *Residue number arithmetic:* Discuss the advantages and the disadvantages of using residue number arithmetic.

5. *Advantages and disadvantages of CORDIC:* Discuss the advantages and the disadvantages of using CORDIC arithmetic.

6. *Verifications of CORDIC:* Referring to Table 7.5, verify the followings:

(a) For $m = 0$, $z(n) \to 0$, show that $y(n) = z(0)x(0) + y(0)$.

(b) For $m = 1$, $z(n) \to 0$, show that $y(n) = x(0)\sin(z(0)) + y(0)\cos(z(0))$.

(c) For $m = -1$, $z(n) \to 0$, show that $y(n) = x(0)\sinh(z(0)) + y(0)\cosh(z(0))$.

7. *More verifications of CORDIC* Referring to Table 7.5, verify the followings:

(a) For $m = 0$, $y(n) \to 0$, show that $z(n) = z(0) - y(0)/x(0)$.

(b) For $m = 1$, $y(n) \to 0$, show that $z(n) = \arctan(\frac{y(0)}{x(0)})$.

(c) For $m = -1$, $y(n) \to 0$, show that $z(n) = \operatorname{arctanh}(\frac{y(0)}{x(0)})$.

8. *RISC architecture:* Recently, RISC architecture has attracted much attention. Discuss the advantages and disadvantages of a RISC architecture for real-time signal processing applications.

9. *Reduced mesh architecture:* Describe how to compute a 2-D FFT on a reduced mesh array processor?

10. *Bus Interconnection:*

    (a) Discuss the advantages and disadvantages of using (single or multiple) bus(es) for multiprocessor interconnection.

    (b) The bus contention problem happens when more than one processors try to access the bus at the same time. Describe at least three methods to resolve the bus contention problem.

11. *Bit-level systolic multipliers:* Due to their regularity and modularity, there are increasing interests in building bit-level systolic array multipliers. Start from the DG of multiplication of two unsigned integer numbers, design three systolic multipliers by different projections.

12. *Scaling in CORDIC technique:* In the CORDIC, a scaling factor $K_{m,n}$ (see Eq. 7.7) is required. Try to find a set of appropriate α_i's that can be implemented by simple shifts (see Eq. 7.6) and makes $K_{m,n} \approx 2$ for $m = 1$ and $n = 16$.

13. *CORDIC processors for QR triarray:* The triangular systolic array (*triarray*) for QR decomposition as discussed in Section 3.3 can be applied to implement very high speed signal processors for beamforming and Kalman filtering (cf. Section 8.3). In this triarray, both the the GG and GR operations are required, so the CORDIC technique are very attractive.

    (a) With reference to Table 7.5, when operating in the so-called vectoring mode, A CORDIC processor performs the operation for GG, i.e. calculates the rotation angles $\{\theta\}$ and the proper rotation. When it is in a rotation mode, it performs the GR operations. Determine the respective mode numbers m for both modes.

    (b) To further improve the processing speed, it is desirable to use bit-level pipelining technique. Try to design a bit-level CORDIC triarray.

Chapter 8

APPLICATIONS TO SIGNAL AND IMAGE PROCESSING

8.1 Introduction

Most of the basic algorithms and some advanced algorithms for signal/image processing are presented in Chapter 2. High-performance and real-time signal processing calls for several orders of magnitude of increase of current computational capability. High-speed VLSI arrays offer a very promising solution to this computational need and presage a major technological breakthrough in signal/image processing applications. In Chapters 3 to 7, it is demonstrated that the algorithm analysis in terms of parallelism and pipelinability dictates the mapping and design of special purpose array processors. *The theme of this chapter is to further demonstrate the top-down integrated design methodology by a variety of signal/image processing application examples.* The success of such a design process hinges upon an understanding of the signal/image formation process, the algorithm classes involved, and the specifications of the intended applicational systems.

Array Processors for Signal Processing The applications of digital signal processing cover speech, sonar, radar, seismic, weather, astronomical, biomedical, nuclear physics, structure analysis signal processing applications, and so on. For real-time signal processing, the data rates are usually very high and the computational requirements are extremely demanding. For example, the signal processing unit within the adaptive array processing system, SPAN-1 [Ashcr84] is prevailed by beamforming computations. Each beamformer can accept data from 52 sensors and form 64 beams. Sensor weights and delays are eight bit integers, and each beamformer should perform 200 million operations per second. The signal processing operations for the spectral analysis (e.g., digital filtering, FFT transformation, and frequency domain adaptive filtering) alone require a computation rate of 46 million multiplications per second. Only a very intimate interplay between VLSI array processors and modern signal processing techniques makes feasible the real time processing for the above and many other application domains.

Array Processors for Image Processing   Digital image processing constitutes a sophisticated and rapidly expanding field that has recently incorporated many modern image analysis and computer vision techniques. Thus the research activities dealing with images are now divided into two disciplines: image processing and image analysis [Chell85]. *Image processing* consists of enhancement, restoration, reconstruction, and coding, etc. *Image analysis*, on the other hand, deals with extraction of lines, curves and regions in images, classification of objects, texture analysis, analysis of moving objects, and scene analysis. Most image processing tasks are very time consuming. For example, low level operations, such as filtering or enhancement, typically require the order of some tens of machine instructions per pixel. A typical image obtained from a LANDSAT earth resources satellite is about 1000 × 1000 pixels/image. This implies a *computation requirement* of some tens of millions of instructions per image, not including the computation for any substantive higher level processing. If such simple low level operations are to be performed at a video rate, say 25 or 30 frames per second, this means a *throughput requirement* of about a billion instructions per second. In general, most real-time image processing throughput rates outstrip current parallel architectures. Thus image applications processing have long been (and will continue to be) a major driving force in the development of faster and more powerful parallel architectures [Offen85].

In the following, we first briefly preview basic application examples for some fundamental DSP operations, e.g., convolution/correlation, FFT, and matrix operations.

Example 1: Applications of Convolution and Correlation

As mentioned in Chapter 2, the convolution of two sequences $u(n)$ and $w(n)$ is given by $y(n) = u(n) * w(n)$. Depending on the type of filter $\{w(n)\}$ adopted, the convolution processing may represent image enhancement (e.g., in edge detection filtering) or signal smoothing (e.g., noise average filtering), or some other kinds of filtering effects (e.g., low-pass, high-pass, band-pass, band-reject filtering).

Convolution is also a useful tool for *linear interpolation* (see Section 8.4.3), which consists of taking a weighted sum of neighboring data to obtain a value for the data in between sampling instants. This technique enables the recovery of a band-limited continuous signal from samples of the signal, such as radar or tomographical image applications. Convolution is also applicable to *matched filtering*, a popular technique used in optimal receivers in telecommunication applications. In a digital matched filter, the received signal is convolved with a filter whose impulse response is the reversed waveform of the transmitted signal. Array architectures for 1-D convolution are discussed in Chapters 3 and 4. An array architecture for performing 2-D convolution will be treated later in Section 8.5.1.

Similar to the convolution operation, correlation is another useful operation that often arises in connection with time series analysis. It has a formula similar to convolution (see Chapter 2):

$$y(n) = \sum_{k=-\infty}^{\infty} u(k)w(n+k) = u(n) * w(-n)$$

and therefore, the array architecture for correlation is very similar to that for convolution.

The correlation operation has been widely used in *pattern recognition* and signal classification applications, where a signal $u(n)$ (assume normalized) with unknown time delay is received, and is compared with a pre-assumed pattern $w(n)$, which will produce a similarity measurement $y(n)$ useful for the determination of the compatibility of two signals. A special case of the correlation is the autocorrelation operation of a sequence $x(n)$, which is given by $y(n) = x(n) * x(-n)$. The autocorrelation function is useful in many applications, e.g., spectral estimation, adaptive filtering, data compression, etc.

Example 2: Applications of the Fast Fourier Transform

The FFT has been one of the most popular techniques in signal/image processing applications: e.g., Doppler frequency measurement in radar processing, audio/image restoration via deconvolution techniques, image reconstruction via Fourier slice theorem, speech parameters extraction via homomorphic processing, data compression via transform methods, etc. The computational efficiency of the FFT is also exploited in the implementation of a technique known as fast convolution or fast correlation. By using this transform method, the order of computation needed for convolution/correlation is reduced from $O(N^2)$ to $O(N \cdot \log_2 N)$.

For FFT computation, several array processor systems based on the perfect shuffle network are discussed in Chapter 3. Recent progress by TRW in the implementation of high-speed FFT and inverse FFT processors with state-of-the-art commercial and semi-custom VLSI circuits has been reported [Swart85]. The TRW processor can operate at a data rate of up to 40 MHz, and computes transforms of up to 16,384 points in length by means of a McClellan and Purdy radix 4 pipeline FFT algorithm [Rabin75]. The arithmetic is performed by commercial single-chip 22-bit floating-point adders and multipliers, whereas the interstage permutation is performed by delay commutators implemented with semicustom VLSI.

Example 3: Applications of Matrix Operations

It is observed that the computational requirement for many real-time signal/image processing tasks can be reduced to a common set of basic matrix operations. These include matrix-vector multiplication, matrix-matrix multiplication and addition, matrix inversion, the solution of systems of linear equations, least squares approximate solution of systems of linear equations, SVD of matrices, solution of eigensystems and so on [White85]. The theory and the design for array processors for these matrix operations are covered in Chapters 2, 3, 4, and 5.

A systolic linear algebraic processor, based primarily on the Faddeev algorithm [Nash83], implemented by Hughes research laboratory [Nash86], uses a 16×16 array of high performance VLSI custom designed PEs (MOP2) to achieve an overall peak system throughput of approximately 450 MOPs. Fixed point 32-bit multiplication and division time for each PE are estimated to be 750 nsec and 1 μsec. This architecture provides fast, numerically stable solutions to a basic set of linear system problems.

Key Techniques for Signal/Image Processing Applications Typical DSP techniques often find useful applications in many different signal/image processing areas. Table 8.1 lists some sample processing techniques and their corresponding signal/image applications. (An "X" indicates the specified algorithm is among the main ones used in the corresponding applications.) *Array processor design examples for several critical application domains constitute the main body of this chapter.* Specifically, *spectral estimation, speech processing, image processing,* and *image analysis* are treated respectively in Sections 8.2, 8.3, 8.4, and 8.5.

Applications Techniques	Adaptive Beamforming	Speech Processing	Image Processing	Image Analysis
FFT DFT	X	X	X	
Spectral Estimation	X	X	X	
Kalman Filtering	X			
Local Mask Operation			X	X
Data Compression		X	X	
Dynamic Programming		X		X
Relaxation Techniques			X	X

Table 8.1: Typical signal/image processing techniques and their applications.

8.2   Spectral Estimation, Beamforming, and Kalman Filtering

Spectral analysis is the most important technique for extracting information from the relevant data and, more specifically, for distinguishing and tracking signals of interest. The range of spectral estimation applications covers the fields of radar, speech, sonar, radio astronomy, microwave antenna arrays, geomagnetism, well logging, and so on [Marpl85]. For example, in radar ranging application, a "chirp" (linear frequency modulation) is emitted and the spectral estimation techniques can be used for differentiating the range (beat) frequencies based on the echo signal [Skoln62]. *In Section 8.2.1, key conventional and model-based spectrum estimation techniques are surveyed, and a systolic design for solving special linear systems incurred in least squares estimations is presented.*

One of the most important applications of spectral analysis is beamforming or phased array processing. Phased array processing deals with the processing of signals carried by propagating wave phenomena in radar, sonar, seismic, and communication systems. The received signal is obtained by means of sensors located at different locations in the field. An adaptive beamforming system can detect the external noise or interference sources (without prior knowledge of the signal/interference environment) and adaptively control the array beam pattern to suppress these noise sources and thereby extract useful characteristics of the source signal [Monzi80]. *Section 8.2.2 presents several least-squares beamforming techniques, including finding optimal weight values via the QR method, recursive QR method, and constrained least-squares array beamforming. For higher resolution (or super resolution) beamforming, several existing SVD based techniques are introduced.*

Kalman filtering has been successfully applied to many signal processing applications, including target prediction, target tracking, radar signal processing, on-board calibration of inertial systems, and in-flight estimation of aircraft stability and control derivatives. For most practical applications, the computations involved in both measurement and time updates are very demanding, and parallel architectures are critically required [Andre81], [Jover86], [MJChe86], [MJChe87]. By a proper reformulation of the original algorithm, the Kalman filter estimation can be solved by a least squares approach, which offers advantages in both numerical accuracy and computational efficiency. *Section 8.2.3 presents a triangular systolic array which can perform the recursive Kalman filtering operations including both the measurement and time updates.*

8.2.1   Array Processors for Spectral Estimation

8.2.1.1   Conventional Spectral Estimation

The conventional spectral estimation problem is to estimate the power spectrum shape of a discrete-time stochastic process, given a finite number of noisy measurements of the process or its first few covariance lags. Two spectral estimation techniques based on Fourier transform operations have evolved. One is based on an indirect approach via an autocorrelation estimate as popularized by Blackman and Tukey. The other is based on the direct approach via an FFT operation on the data, which is often referred to as the *periodogram* [Schus98].

The indirect approach due to Blackman and Tukey first estimates the autocorrelation over a finite segment and then takes the Fourier transform of the autocorrelation to obtain the power spectrum. It inherently assumes that the covariance lags outside that time sequence are zero.

$$P_{BT}(w) = \sum_{n=-M+1}^{M-1} r_{xx}(n)e^{-jwn} \tag{8.1}$$

Very often the autocorrelation function is weighted by a window function $W(n)$ (to smooth the transition) before the Fourier transform is taken.

The direct periodogram approach to the power spectral estimation is to use the magnitude squared value of the Fourier transform of the available data measurement.

$$P_{PER}(w) = \frac{1}{N} \parallel \sum_{n=0}^{N-1} x(n)e^{-jwn} \parallel^2 \tag{8.2}$$

Both of the above methods can be efficiently computed using the FFT algorithm. For high speed processing, array processors for computing the FFT are obviously required.

8.2.1.2   Model Based Spectral Estimation

In many modern applications, the spectral estimation has to be based on short data records (in radar, for example, only a few samples are available in each radar pulse). Then the above conventional spectral estimation methods become somewhat limited in that they do not provide sufficient resolution, due to the fundamental uncertainty limit bounded by the reciprocal of the covariance length used in the Fourier transform [Kung85a], [Kay81].

Since low-bias, low-variance, high resolution estimates are desired, additional constraints (or prior information) have to be incorporated to improve the resolution capability, when only a few data samples are available. In model-based methods, such prior information is exploited to improve the frequency resolution. In many applications, the underlying physical environment generating the signal can be efficiently modeled by a linear rational system of low order. Model-based methods extend the covariance sequence outside the given segment via certain recurrence relations determined by the model parameters. In other words, it completely specifies the infinite covariance extension and the corresponding power spectrum.

In general, a linear rational system can be characterized by a linear difference equation:

$$x(n) = \sum_{i=1}^{p} a_i x(n-i) + \sum_{i=1}^{q} b_i v(n-i) + v(n) \tag{8.3}$$

where the input is $\{v(n)\}$ and the output is $\{x(n)\}$. This model is known as an autoregressive moving average model, $\text{ARMA}(p, q)$. The transfer function $H(z)$ of this system is

$$H(z) = \frac{B(z)}{A(z)} \tag{8.4}$$

where

$$B(z) = 1 + \sum_{i=1}^{q} b_i z^{-i}$$

$$A(z) = 1 - \sum_{i=1}^{p} a_i z^{-i}$$

When the input $\{v(n)\}$ is white noise of variance σ^2, the power spectrum of $x(n)$ is

$$P(w) = S(z)|_{z=e^{jw}} \tag{8.5}$$

where

$$S(z) = \sigma^2 H(z) H(z^{-1}) = \sigma^2 \frac{B(z)B(z^{-1})}{A(z)A(z^{-1})} \tag{8.6}$$

The relationship of the ARMA parameters to the autocorrelation function can be established as follows. Multiplying both sides of Eq. 8.3 by $x^*(n-l)$ and taking expectation yields[1]

---

[1] where $x^*(i)$ stands for the complex conjugate element of $x(i)$.

$$r_{xx}(l) = \sum_{i=1}^{p} a_i r_{xx}(l-i) + \sum_{i=0}^{q} b_i r_{vx}(l-i) \tag{8.7}$$

where $l = 0, 1, \ldots, q$, and

$$r_{xx}(k) = \mathrm{E}\{x(n)x^*(k-n)\}$$

$$r_{vx}(k) = \mathrm{E}\{v(n)x^*(k-n)\}$$

Using the fact that the input $\{v(k)\}$ is white and that the system is causal, it can be shown that $r_{vx}(k) = 0$ for $k > 0$, therefore,

$$r_{xx}(l) = \sum_{i=1}^{p} a_i r_{xx}(l-i) \tag{8.8}$$

where $l = q+1, q+2, \ldots$,

The $\{a_i\}$ coefficients can be obtained by solving Eq. 8.8, which is basically a Toeplitz system of equations. The $\{b_i\}$ coefficients can then be obtained from Eq. 8.7.

Autoregressive Models   A more popular model used for spectral estimation is the special case of the ARMA model where $B(z)$ is constrained to be equal to 1. Then

$$x(n) = \sum_{i=1}^{p} a_i x(n-i) + v(n) \tag{8.9}$$

here output $x(n)$ is generated as a linear regression of its past values, and hence such a model is known as an AR model.

The recurrence relations satisfied by the covariance of this model are given by:

$$r_{xx}(l) = \sum_{i=1}^{p} a_i r_{xx}(l-i), \quad l > 0 \tag{8.10}$$

$$r_{xx}(0) = \sum_{i=1}^{p} a_i r_{xx}(-i) + \sigma^2 \tag{8.11}$$

Given M exact covariance lags $\{r(l)\}$, where $M > p$, the parameters $\{a_i\}$ of the AR(p) model can be estimated from the recurrence equation,

Eq. 8.10 for the first p values of l. They are known as the Yule-Walker equations. Thus the AR parameter estimation involves the solutions of a symmetric positive-definite Toeplitz system, which can be computed very efficiently using either the Levinson algorithm or the Schur algorithm. For the design of a pipelined architecture, the Schur algorithm proves to be better. A parallel architecture based on the algorithm is developed in the next section.

AR spectral estimation is the most popular spectral estimation in many signal/image processsing applications. For example, it can be used to model continuous wave radar clutter in order to classify the different forms of clutter as encountered in an air traffic control radar environment. (Clutter is essentially unwanted echoes on a radar display due to ground reflection of the transmitted wave, weather disturbances, and so on). The spectral density of a continuous wave radar clutter process has been shown to have a Gaussian shape, which can be closely represented by an AR process of relatively low order (e.g., third order). The use of AR spectral estimation for speech processing is one of the earliest uses of modern spectral estimation. The smoother appearance of the AR spectrum (as opposed to the periodogram) makes it feasible to locate and track the correct formant frequencies of speech signals. Using a short-term AR line tracker for passive sonar, some frequency variations not seen by the periodogram tracker can be picked up, which possibly indicate some useful information such as target movement.

Example 1: Systolic Array for Yule-Walker System Solver

As discussed previously, many spectral estimation methods involve solving a Toeplitz Yule-Walker system. To derive an appropriate systolic architecture for the Schur algorithm (cf., Section 2.3) a dependence graph analysis and the canonical mapping methodology can be adopted. The single assignment code for the Toeplitz system solver can be expressed as (for the notations, see Section 2.3):

$$v_k^{(1)} \;=\; u_k^{(1)} = t_k \; (0 \leq k \leq N)$$

For i from   1   to N

$$K^{(i+1)} \;=\; -v_i^{(i)}[u_1^{(i)}]^0$$

For k from 0 to N

$$v_k^{(i+1)} \;=\; v_k^{(i)} + K^{(i+1)} u_{k+1}^{(i)}$$

$$u_k^{(i+1)} \;=\; u_{k+1}^{(i)} + K^{(i+1)} v_k^{(i)} \tag{8.12}$$

The preceding single assignment algorithm can be localized via the localization procedure given in Section 3.3. The resulting localized DG for the parallel Schur algorithm is shown in Figure 8.1(a).

A pipelined computing structure for concurrent processing of symmetric Toeplitz system solutions is discussed next. Given the DG in Figure 8.1(a), an SFG can be derived by a simple projection (with default schedule). This leads to the bilinear SFG shown in Figure 8.1(b). The systolic design can be obtained by applying the systolization procedure proposed in Section 4.3.

The major computation in the algorithm lies in the linear combinations of two vectors in each recursion (cf., Eq. 8.12). During each recursion, the reflection coefficient is first computed in the divider cell and then propagated to the "lattice" cells (which perform Eq. 8.12) through the local interconnections. Upon the completion of the linear combination, the result in each upper PE is left-shifted to its immediate left neighbor, preparing for the next recursion, while the contents of the lower PEs, which correspond to a row of the **U** matrix, can be output and the next recursion is thus completed.

Based on the preceding formulation, it is clear that the computing time separating two consecutive recursions takes two time units (one for reflection-coefficient computation, the other for row operation). For N recursions,

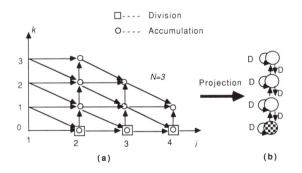

Figure 8.1: (a) DG for solving symmetric Toeplitz system by Schur algorithm. (b) SFG for parallel Schur algorithm.

this amounts to a total parallel processing time of $O(N)$ time units. This compares favorably with $O(N\log_2 N)$ as required in a pipelined architecture for the Levinson algorithm [Kung83b].

A complete Toeplitz system solver is composed of a pipelined lattice processor and a pipelined back-substitution processor. For a more detailed description and the VLSI design block diagram, the readers are referred to [Kung83b].

8.2.2 Array Processors for Beamforming

Beamforming is one of the main functions of a passive phased-array processing system. It involves forming multiple beams through applying appropriate delay and weighting elements to the signals received by the sensors. The purpose is to suppress unwanted jamming interferences and to determine the number of sources present in the medium and their characteristic parameters, such as direction and intensity. Among the important examples of beamforming applications are sonar for military surveillance and seismology exploration. The number of sensors in an array may vary from a small number up to thousands. Typical examples are channel equalization arrays (with about 50 sensors) and large planar sonar arrays (with up to 1000 sensors).

8.2.2.1 Linking Spectral Estimation and Beamforming

The foregoing methods used in modern spectrum estimation can be applied to the beamforming in phased-array processing applications, especially for linear and uniformly spaced sensor arrays (cf., Figure 8.2). The similarity be-

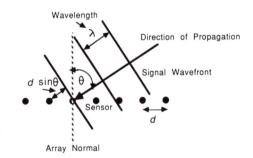

Figure 8.2: Incoming signal on a linear array of sensors.

tween spectral estimation and beamforming hinges upon *the duality between the problem of spectral-line estimation from a time series and that of angle estimation from data received by array sensors.* Here the angle θ, pointing to the direction of the source in beamforming is the dual of the frequency in spectral estimation. Correspondingly, the spatial samples provided by the sensors take the role of the time-samples.

For simplicity, it is assumed that a uniformly spaced linear array of sensors are used to measure the data. The role of a temporal correlation matrix $\mathbf{R}$ is now replaced by a spatial correlation matrix $\mathbf{S}$, where the elements of $\mathbf{S}$ represent correlation between data received at different sensors. The counterpart of the *power spectrum* is the spatial *directivity spectrum*, or *beamforming spectrum*, denoted by $P(\theta)$. Usually, the location of peaks in $P(\theta)$ provides an estimate of the source directions. In order to compute $P(\theta)$, a phase vector $\mathbf{p}(\theta)$ associated with the sensor array is introduced:

$$\mathbf{p}(\theta) = [1, \ e^{j2\pi f}, \ e^{2j2\pi f}, \ \ldots, \ e^{j2(N-1)\pi f}]^T$$

where $f = (d/\lambda)\sin\theta$, (see Figure 8.2). (Note that this form is valid only under a stationary environment.)

The conventional beamforming spectrum is given as

$$P_{BF}(\theta) = \mathbf{p}^*(\theta) \ \mathbf{S} \ \mathbf{p}(\theta) \tag{8.13}$$

Under the stationarity assumption, the spatial correlation matrix $\mathbf{S}$ is a Toeplitz matrix with entities $S(i,j) = S(i-j)$. Eq. 8.13 is reduced to the following:

$$P_{BF}(\theta) = \sum_{n=-(N-1)}^{N-1} W(n)S(n)e^{-j2\pi fn} \tag{8.14}$$

This exactly models the Fourier transform of the correlation function $S(n)$ with a Bartlett (triangular) window (see Problem 3) as mentioned in the Blackman-Tukey method (see Eq. 8.1).

Let us introduce two more popular methods for source direction finding. One is Capon's "maximum likelihood" method given as [Capon69]:

$$P_{ML}(\theta) = \frac{1}{\mathbf{p}^*(\theta)\mathbf{S}^{-1}\mathbf{p}(\theta)} \tag{8.15}$$

The other one is based on the "maximum entropy" method [Schmi79], and is given as follows:

$$P_{ME}(\theta) = \frac{1}{\mathbf{p}^*(\theta)\mathbf{c}\mathbf{c}^*\mathbf{p}(\theta)} \qquad \cdot \qquad (8.16)$$

where $\mathbf{c}$ is the first column of $\mathbf{S}^{-1}$ (to within a scaling factor) which may be obtained by solving the Yule-Walker equations.

It is important to note that *the above formulations can be naturally extended to the environment of irregularly spaced sensors array.* In this case, the phase vector $\mathbf{p}(\theta)$ will be dependent on the geometry of the sensors array. The spatial correlation matrix will not in general have a Toeplitz structure, i.e., the correlation between i-th and j-th sensors will in general differ from that between $(i+k)$-th and $(j+k)$-th sensors. However, the formulation in Eqs. 8.13, 8.15 and 8.16 remain almost the same.

8.2.2.2 Least Squares Based Beamforming

In general, a beamforming system has N sensor elements and a beam-pattern forming network comprising $N - 1$ weights that have to be determined in order to maximize the array response to the desired signals. The N-th (reference) sensor element is constrained to a constant value, e.g., 1. The function diagram of a typical beamforming system is shown in Figure 8.3.

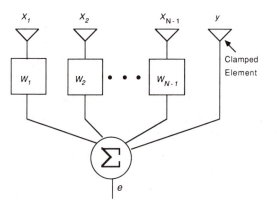

Figure 8.3: The functional diagram of an adaptive array system.

The error sequence of the array output in Figure 8.3 from the array at time instance t can be represented by [McWhi83]:

$$e(t) = \mathbf{x}(t)\mathbf{w} + y(t) \tag{8.17}$$

where the *row vector* $\mathbf{x}(t)$ has components describing the envelope of signals[2] received across the $N - 1$ sensor elements at time t, and $y(t)$ describes the signal received at the N-th reference element. The vector $\mathbf{w}$ specifies the values of weights which would be used in the filtering array.

The objective of an optimal beamforming system is to minimize the total error power via manipulation of the weight values subject to the clamped weight constraint. This can be considered as a generalization of the least-square error criterion used in the linear regression AR model in spectral estimation.

Under the assumption that the array size is large and the input signal is stationary, various "optimal" solutions for computing the weight vector have been proposed. However, in many practical systems, this assumption may not be valid. One way to overcome this difficulty is to use recursive adaptive updating, e.g., the gradient-based adaptive algorithms, such as the LMS algorithm. Unfortunately, they are not suitable for some critical high-performance applications because of the poor convergence for digital signal with broad dynamic range. The least squares based approach proposed by [Ward84], [McWhi86] may efficiently circumvent this difficulty.

8.2.2.3   Finding Optimal Weight via QR Method

The optimal array design problem can be expressed by the following least squares formulation (see Section 2.3). Let us define the $n \times (N - 1)$ matrix $\mathbf{X}_n$, and the $n \times 1$ vector $\mathbf{y}_n$ as following:

$$\mathbf{X}_n \equiv \begin{bmatrix} \mathbf{x}(1) \\ \mathbf{x}(2) \\ \cdot \\ \cdot \\ \cdot \\ \mathbf{x}(n) \end{bmatrix}, \qquad \mathbf{y}_n \equiv \begin{bmatrix} y(1) \\ y(2) \\ \cdot \\ \cdot \\ \cdot \\ y(n) \end{bmatrix}$$

---

[2]For simplicity, we are only dealing with k real data signal cases, However, our derivation can be easily modified for complex signals.

This leads to a matrix formulation for the residual vector $\mathbf{e}_n$:

$$\mathbf{e}_n = \mathbf{X}_n\mathbf{w}_n + \mathbf{y}_n \qquad (8.18)$$

The problem of optimal beamforming is to find the weight solution $\mathbf{w}_n$, such that the norm of vector $\mathbf{e}_n$,

$$\|\mathbf{e}_n\|_2 = \|\mathbf{X}_n\mathbf{w}_n + \mathbf{y}_n\|_2,$$

is minimized. The least-square solution for $\mathbf{w}_n$ may be obtained via a direct block inverse,

$$\mathbf{w}_n = (\mathbf{X}_n^T\mathbf{X}_n)^{-1}\mathbf{X}_n^T\mathbf{y}_n.$$

However, this approach incurs some numerical instability problem due to the squared condition number [Wilki65]. As discussed in Chapter 2, a popular numerical method to compute the least-square solution $\mathbf{w}_n$ is by QR decomposition, which consists of a sequence of unitary transformations applied to the measured data matrix $\mathbf{X}_n$ to transform it to a triangular matrix. More elaborately, applying the *orthogonal transformation*, $\mathbf{Q}_n$, to the righthand side of Eq. 8.18 we have:

$$\mathbf{Q}_n\mathbf{e}_n = \mathbf{Q}_n\mathbf{X}_n\mathbf{w}_n + \mathbf{Q}_n\mathbf{y}_n \qquad (8.19)$$

The orthogonal matrix $\mathbf{Q}_n$ is chosen to give

$$\mathbf{Q}_n\mathbf{X}_n = \begin{bmatrix} \mathbf{R}_n \\ \mathbf{0} \end{bmatrix}$$

where $\mathbf{R}_n$ is an $(N-1) \times (N-1)$ upper triangular matrix and $\mathbf{0}$ is a null matrix. Also,

$$\mathbf{Q}_n\mathbf{y}_n = \begin{bmatrix} \mathbf{b}_n \\ \mathbf{r}_n \end{bmatrix}$$

Therefore the system equation becomes:

$$\mathbf{R}_n\mathbf{w}_n + \mathbf{b}_n = \mathbf{0} \qquad (8.20)$$

which can be easily solved for $\mathbf{w}_{\text{opt}}$ (the minimum norm solution), using the method of back-substitution.

The systolic array for solving this least squares problem by using QR decomposition has been discussed in Chapter 3 (see Section 3.3), where a triangular systolic array with $\alpha = 1$ is shown to be preferrable.

8.2.2.4   Recursive QR Method

For the beamforming application, the triangular array offers yet another advantage that *it can cope with continuously arriving new data in a recursive updating manner.* The basic idea is to apply the algorithm in a recursive form whereby the triangular matrix $\mathbf{R}_n$ (cf. Eq. 8.20) can be updated as each new row of data enters the computation. By applying a sequence of orthogonal transformations to *nullify* the new data vector (e.g., $\mathbf{x}_{n+1}$ in Eq. 8.21) using the existing triangular matrix (e.g., $\mathbf{R}_n$ in Eq. 8.21), a new triangular matrix (e.g., $\mathbf{R}_{n+1}$ in Eq. 8.21) may be formed.

More precisely, if we append the new data vector obtained at the $(n + 1)$st time sample to the existing triangular system, we obtain:

$$
\begin{bmatrix} e'_1 \\ \cdot \\ \cdot \\ \cdot \\ e'_n \\ \hline e_{n+1} \end{bmatrix} = \begin{bmatrix} \mathbf{R}_n \\ \mathbf{0} \\ \hline \mathbf{x}_{n+1} \end{bmatrix} \mathbf{w}_{n+1} + \begin{bmatrix} \mathbf{b}_n \\ \mathbf{r}_n \\ \hline y_{n+1} \end{bmatrix} \tag{8.21}
$$

where $\mathbf{w}_{n+1}$ is the least squares solution using all $(n + 1)$ data samples and e_{n+1} is the desired residual from the array. The upper triangular system of Eq. 8.21 simply becomes the rows of $\mathbf{R}_n$ with the new data vector appended to the matrix. Representing the sequence of rotations by the orthogonal update transformation $\mathbf{Q}'_{n+1}$ then

$$
\begin{bmatrix} e''_1 \\ \cdot \\ \cdot \\ \cdot \\ e''_n \\ \hline e'_{n+1} \end{bmatrix} = \mathbf{Q}'_{n+1} \begin{bmatrix} e'_1 \\ \cdot \\ \cdot \\ \cdot \\ e'_n \\ \hline e_{n+1} \end{bmatrix} = \begin{bmatrix} \mathbf{R}_{n+1} \\ \mathbf{0} \\ \hline \mathbf{0} \end{bmatrix} \mathbf{w}_{n+1} + \begin{bmatrix} \mathbf{b}_{n+1} \\ \mathbf{r}_n \\ \hline b'_{n+1} \end{bmatrix} \tag{8.22}
$$

where $\mathbf{R}_{n+1}$ and $\mathbf{b}_{n+1}$ are the $(n+1)$th sample update of $\mathbf{R}_n$ and $\mathbf{b}_n$ respectively. The optimal weight vector $\mathbf{w}_{n+1}$ may be obtained by solving

$$\mathbf{R}_{n+1}\mathbf{w}_{n+1} + \mathbf{b}_{n+1} = \mathbf{0}. \tag{8.23}$$

In other words, in the optimal design Eq. 8.23 should be satisfied.

Example 2: Using Triarray Module for QR Operations.

A triangular array, abbreviated as *triarray*, as shown in Figure 8.4 provides a basic array module for performing the Givens rotations for the QR decomposition [HTKun81], [McWhi83]. Detailed derivations are already treated in Sections 2.2 and 3.3. For example, applying a cut-set systolization procedure to the SFG in Figure 3.33 will lead to the present systolic triarray design. In the triarray, the Givens generation (GG) operations are performed in the diagonal PEs, and the Givens rotation (GR) operations in all the other PEs.

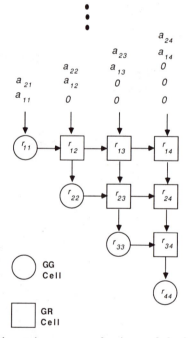

Figure 8.4: Using triarray as a basic module for performing the QR operations.

A very general application of the triarray is one involving the QR decomposition of the matrix $[\mathbf{R}^T \mathbf{A}^T]^T$ as shown below

$$\mathbf{Q} \begin{bmatrix} \mathbf{R} \\ \mathbf{A} \end{bmatrix} = \begin{bmatrix} \mathbf{R}' \\ \mathbf{0} \end{bmatrix} \tag{8.24}$$

where $\mathbf{R}$ is an upper triangular matrix and $\mathbf{A}$ is a full matrix. To use the triarray, we let the matrix $\mathbf{R}$ initially reside in the array and input the matrix $\mathbf{A}$ from the top row of the array. Then the matrix $\mathbf{A}$ can be nullified by proper Givens rotations with $\mathbf{R}$, as described below.

With reference to Figure 8.4, the first row of $\mathbf{A}$, $\mathbf{a}_1$, arrives at the top row of the triarray. PE(1,1) computes the proper Givens rotation parameters (so that the first element of $\mathbf{a}_1$ is nullified) and passes the parameters to the remaining PEs in the first row, where GRs are performed. The rotation results in two new rows. The first will reside in the original PEs. The other row (except the first *zero* element) will be propagated downward to the second row of PEs in the triarray, where a similar nullification procedure can be carried out. In this manner, when the (modified) $\mathbf{a}_1$ finally travels across the triarray, all of its elements are nullified. So are the other rows of $\mathbf{A}$ when they go through the triarray. The final data residing in the triarray will yield the elements of $\mathbf{R}'$.

If there is only one matrix $\mathbf{A}$ to be triangularized by the QR decomposition, then the same triarray can still be adopted by introducing a *zero* triangular matrix $\mathbf{R}$, i.e., $\mathbf{R} = 0$. In this case, the initial data residing the triarray are set to be zero, and the same computation process can be followed to yield the final triangular matrix $\mathbf{R}'$.

8.2.2.5   Direct Residual Extraction

In some beamforming applications, *the main objective is to compute the least squares residual* and the corresponding weight vector is not of direct interest. A slightly modified version of the above recursive QR algorithm allows the least squares residual to be produced at each stage of the recursive process [McWhi83]. The method is described below:

Based on Eq. 8.23, Eq. 8.22 can be reduced to

$$
\begin{bmatrix}
e'_1 \\
\cdot \\
\cdot \\
\cdot \\
e'_n \\
- - - \\
e_{n+1}
\end{bmatrix}
= \mathbf{Q}'^T_{n+1}
\begin{bmatrix}
0 \\
\cdot \\
\cdot \\
\cdot \\
\mathbf{r}_n \\
- - - \\
b'_{n+1}
\end{bmatrix}
\tag{8.25}
$$

Note that $\mathbf{Q}'^T_{n+1} = \mathbf{Q}'^{-1}_{n+1}$, due to the property of the orthogonal matrix. Based on Eq. 8.25, the least squares residual e_{n+1} can be derived as (see Problem 4):

$$
e_{n+1} = \{ \prod_{i=1}^{N-1} \cos\theta_i \} b'_{n+1}
\tag{8.26}
$$

where θ_i is the angles of successive Givens rotations which comprise the transformation matrix $\mathbf{Q}'_{n+1}$.

This modified version of the QR Triarray is shown Figure 8.5 [Ward84]. The same array configuration has been implemented as a wavefront array system (using the TMS32010 for the PE design), by STC and RSRE in the United Kingdom (see Section 7.5). When used for the beamforming operation, this WAP system consists of 33 identical PEs, 21 of which are organized as a triangular wavefront array performing the beamforming function and the remaining 12 PEs performing data correction and other preprocessing functions. This array can be used to extract the current least squares residual using the recursive QR updating method (cf. Eq. 8.26). This array can also be used in combination with additional back-substitution circuitry to compute the weight solution explicitly [Ward84].

8.2.2.6 Constrained Least Squares Array Beamforming

The constrained least squares filtering problem arises in sonar antenna arrays in which a desired signal with a known direction of arrival is contaminated by jammers or noise sources that arrive from different directions. *In order to take advantage of the prior information of signal and interference, the optimal spatial filter weights are chosen to minimize the total squared error*

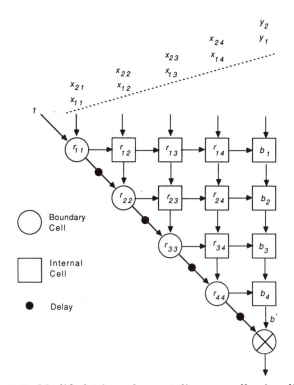

Figure 8.5: Modified triangular systolic array allowing direct extraction of least squares residuals.

output, while simultaneously preserving certain characteristics in the directions of arrival of the desired signal [Kalso85]. This problem is similar to the least squares problem with constraints on the weight vector [McWhi86], [Ward84].

The constrained least squares problem can be formulated as the following: Given an $n \times N$ ($n \geq N$) observation data matrix $\mathbf{X}$, find a weight vector $\mathbf{w}$ to minimize

$$\|\mathbf{X}\mathbf{w}\|_2$$

subject to the constraint(s)

$$\mathbf{C}\mathbf{w} = \mathbf{f}$$

where $\mathbf{w}$ is a column vector of length N (number of sensors), $\mathbf{C}$ is a constraint matrix of dimension $m \times N$ ($m < N$) and $\mathbf{f}$ is a column vector of length m. Note that the number of constraints is m.

This problem can be transformed to an unconstrained least squares estimation problem and thus can be solved by the method mentioned before. This transformation is well known as a generalized sidelobe canceller in adaptive antenna arrays [Griff82] and is derived next.

Any weight vector $\mathbf{w}$ that satisfies the constraint equation $\mathbf{Cw} = \mathbf{f}$, consists of two parts, a homogeneous solution $\mathbf{w}_h$ and a special solution $\mathbf{w}_q$ of the constraint equation. This means[3]

$$\mathbf{w} = \mathbf{w}_h + \mathbf{w}_q \qquad (8.27)$$

where $\mathbf{C} \, \mathbf{w}_h = 0$ and $\mathbf{C} \, \mathbf{w}_q = \mathbf{f}$.

Note that the special solution is independent of observation data and can be predetermined. The homogeneous solution $\mathbf{w}_h$ is in the null space of $\mathbf{C}$; therefore, it can be represented as a linear combination of the basis of that null space, which is of dimension $(N - m)$. Let the matrix $\mathbf{W}_s$ represent the basis of that null space, i.e., columns of $\mathbf{W}_s$ span the null space, then the homogeneous vector can be represented as

$$\mathbf{w}_h = \mathbf{W}_s \mathbf{w}_a$$

Consequently, any weight vector $\mathbf{w}$ that satisfies the constraint equation will be in the following form:

$$\mathbf{w} = \mathbf{W}_s \mathbf{w}_a + \mathbf{w}_q \qquad (8.28)$$

and

$$\mathbf{Xw} = \mathbf{XW}_s \mathbf{w}_a + \mathbf{Xw}_q \qquad (8.29)$$

Let us adopt the following notations:

$$\mathbf{e}' = \mathbf{Xw},$$

$$\mathbf{X}' = \mathbf{XW}_s,$$

and

$$\mathbf{y}' = \mathbf{Xw}_q,$$

---

[3]In general, a special solution to the constraint equation is not unique. In some cases, a minimal norm solution, which can be obtained by premultiplying the pseudo-inverse of $\mathbf{C}$ to $\mathbf{f}$, is preferred.

then Eq. 8.29 becomes

$$\mathbf{e}' = \mathbf{X}'\mathbf{w}_a + \mathbf{y}'. \tag{8.30}$$

Therefore, minimizing $\|\mathbf{X}\mathbf{w}\|_2$ is equivalent to minimizing

$$\|\mathbf{e}'\| = \|\mathbf{X}'\mathbf{w}_a + \mathbf{y}'\|_2,$$

which is now formulated as an unconstrained least squares problem. This means that the QR triarray described in Examples 2 can now be used. In the case that the residual vector $\mathbf{e}' = \mathbf{X}\mathbf{w}$ (instead of $\mathbf{w}$) is of interest, then the design for direct residual extraction is still applicable. The additional computation for the transformation involves only matrix-vector and matrix-matrix multiplication. The block diagram of the transformation in Figure 8.6 demonstrates the (rather simple) *preprocessing* operations required for converting a constrained least squares problem into an unconstrained one.

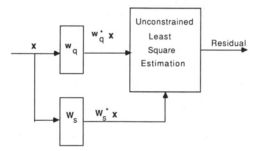

Figure 8.6: Transformation from a constrained least squares to an unconstrained least squares problem. Both the vector $\mathbf{w}_q$ and the matrix $\mathbf{W}_s$, which depend only on the constraints, can be pre-calculated.

8.2.2.7   Eigenvalue/SVD Decomposition Based Model

In high resolution spectral line estimation applications, the stochastic process is assumed to consist of p sinusoids corrupted by additive white noise. Such a process can be modeled as a special ARMA process. This model is equally suitable for high resolution direction finding and beamforming when linear and uniformly spaced sensor arrays are used. This model leads to an eigenvalue decomposition based spectral estimation method known as Pisarenko's method [Kay81]. (The mathematical derivation is somewhat involved and is omitted here.) The (minimum) eigenvalue of the data covariance matrix **R** can be computed efficiently by a modified Rayleigh quotient iteration scheme that exploits the Toeplitz structure of the matrix [YHHu85]. For numerical robustness, it is advisable to adopt SVD based approach – the Toeplitz approximation method (TAM) – originally proposed by Kung [Kung81], [Kung83d]. In simulation as well as real data, the TAM has demonstrated a superior numerical performance. For irregularly positioned sensor arrays, one has to resort to a more general SVD method, such as MUSIC (multiple signal classification method), in which all the eigenvectors corresponding to the smallest eigenvalues are used [Schmi79]. For real time applications, array processors for the SVD computations become a necessity. Some array processor designs for the SVD computation are discussed in the following example.

Wavefront Array for QR Iteration Eigenvalue Algorithm   According-ing to Parlett [Parle80], the QL and QR algorithms are the most effective ways of finding all the eigenvalues of a small symmetric matrix in a sequen-tial computing manner. A full matrix is first reduced to tridiagonal[4] form by a sequence of reflections or rotations and then the QR algorithm is used to swiftly reduce the off-diagonal elements until they are neligible (see Section 2.2.4). The algorithm repeatedly applies a complicated similarity transfor-mation to the result of the previous transformation, thereby producing a sequence of matrices that converges to a diagonal form. As the tridiago-nalization process requires $O(n^3)$ time on a sequential computing machine, a linear wavefront array with n PEs and $O(n^2)$ processing time has been proposed [Kung83a]. The iteration of the tridiagonal matrix is especially attractive for implementation by means of a bi-linear configuration of the wavefront array. Both the above operations can be performed using local communications only.

Systolic Array for Parallel Jacobi SVD Algorithm   A symmetric matrix $\mathbf{A} \in R^{n \times n}$ can be diagonalized by "zeroing" off-diagonal entries using the classical method of Jacobi which consists of using Givens rotations to perform the diagonalizing operation [Golub83], [Parle80]. The cyclic Jacobi method converges quadratically, although the performance is in general not as good as the QR algorithm; the Jacobi method requires as many operations as does the Schur algorithm using the symmetric QR algorithm. In certain

---

[4]A matrix $\mathbf{A} = \{a_{ij}\}$ is tridiagonal when $a_{ij} = 0$, for $|i - j| > 1$.

cases however it is advantageous to use the Jacobi method. For example, if $\mathbf{A}$ is close to the diagonal form, as is the case when a good approximate eigensystem is known, then the Jacobi method has an advantage over the QR algorithm.

Since the updates of Jacobi's rotations are mutually independent, this property can be exploited to maximize the concurrency in systolic arrays. A 2-D systolic array $(n/2 \times n/2)$ which exploits this feature has been proposed [Bren85a], [Bren85b]. This architecture ensures that the $n/2$ rotations of a given rotation set can be carried out concurrently. (This partition strategy is often used in chess and bridge tournaments, where it is used to ensure all players play with each other in a minimum number of blocks of games, i.e., all players are engaged in a game with no players meeting more than once.) A special array in which a PE communicates with its 8 nearest neighbours is adopted. The computation time is about $3sn$, where s is the number of iterations (usually about 6-10) required for the Jacobi method to converge.

8.2.3   Kalman Filtering for Least Squares Estimation

Kalman filtering has found frequent applications to many time varying (or time invariant) signal processing problems, such as those encountered in communications and control, radar and sonar processing and so on. A discrete time-varying recursive dynamic system can be represented as:

$$
\begin{aligned}
\mathbf{x}(k+1) &= \mathbf{F}(k)\mathbf{x}(k) + \mathbf{w}(k+1) \\
\mathbf{y}(k) &= \mathbf{C}(k)\mathbf{x}(k) + \mathbf{v}(k)
\end{aligned}
\tag{8.31}
$$

where $\mathbf{F}(k)$ and $\mathbf{C}(k)$ are coefficient matrices with dimension $n \times n$ and $m \times n$; $\mathbf{x}(k)$ and $\mathbf{w}(k+1)$ are the n-dimensional state vector and system noise vector respectively; $\mathbf{y}(k)$ and $\mathbf{v}(k)$ are the m-dimensional measurement vector and measurement noise vector respectively. The noise vectors have zero mean and known covariance matrices $\mathbf{R_w}(k+1)$ and $\mathbf{R_v}(k)$ respectively. The noise $\mathbf{w}$ is assumed to be uncorrelated with $\mathbf{v}$ (i.e. $E[\mathbf{w}(i)\mathbf{v}(j)] = 0$, for all i, j).

8.2.3.1   Least Squares Formulation of Kalman Filtering

The Kalman filtering is to compute the optimal prediction of $\mathbf{x}(k+1)$, denoted as $\hat{\mathbf{x}}(k+1)$, based on known measurements $\{\mathbf{y}(1), \mathbf{y}(2), \cdots, \mathbf{y}(k)\}$

[Kalma60]. The best prediction for the state vector in Eq. 8.31, based on a linear minimum variance criterion, can be solved recursively with computational complexity of $O(n^3)$. In the following, we reformulate the Kalman filtering algorithm into a classic least squares problem. For more details, see [Paige77], [MJChe86],[MJChe87].

For colored noise cases, where the covariance matrix, $\mathbf{R_w}(k+1)$ and $\mathbf{R_v}(k)$, are not identity matrices, a whitening procedure has to be first applied. The covariance matrices of the two noise vectors can be expressed as:

$$\begin{aligned} \mathbf{R_w^{-1}}(k+1) &= \mathbf{W}(k+1)^T\mathbf{W}(k+1) \\ \mathbf{R_v^{-1}}(k) &= \mathbf{V}(k)^T\mathbf{V}(k) \end{aligned}$$

where $\mathbf{W}(k+1)$ and $\mathbf{V}(k)$ are *upper triangular matrices* obtained from Cholesky decomposition [Golub83]. Applying premultiplication of the whitening operators $\mathbf{W}(k+1)$ and $\mathbf{V}(k)$ to Eq. 8.31 yields (see Problem 22):

$$\begin{aligned} \mathbf{W}(k+1)\mathbf{x}(k+1) &= \mathbf{W}(k+1)\mathbf{F}(k)\mathbf{x}(k) + \mathbf{W}(k+1)\mathbf{w}(k+1) \\ \mathbf{V}(k)\mathbf{y}(k) &= \mathbf{V}(k)\mathbf{C}(k)\mathbf{x}(k) + \mathbf{V}(k)\mathbf{v}(k) \end{aligned} \qquad (8.32)$$

By grouping together the state vectors up to stage k and forming a large vector $\mathbf{X}(k)$, Eq. 8.32 can be represented by the following matrix-vector form [Paige77]

$$\tilde{\mathbf{U}}(k) = \tilde{\mathbf{A}}(k)\mathbf{X}(k) + \tilde{\mathbf{Y}}(k) \qquad (8.33)$$

where

$$\begin{aligned} \mathbf{X}(k) &= [\mathbf{x}^T(1)\ \mathbf{x}^T(2)\ \dots\ \mathbf{x}^T(k)]^T \\ \tilde{\mathbf{U}}(k) &= [\tilde{\mathbf{w}}^T(1)\ \tilde{\mathbf{v}}^T(1)\ \tilde{\mathbf{w}}^T(2)\ \tilde{\mathbf{v}}^T(2)\ \dots \tilde{\mathbf{v}}^T(k-1)\ \tilde{\mathbf{w}}^T(k)]^T \\ \tilde{\mathbf{Y}}(k) &= [\mathbf{0}\ \tilde{\mathbf{y}}^T(1)\ \mathbf{0}\ \tilde{\mathbf{y}}^T(2)\ \dots\ \tilde{\mathbf{y}}^T(k-1)\ \mathbf{0}]^T \end{aligned}$$

and

$$\tilde{\mathbf{A}}(k) \;=\; \begin{bmatrix} \mathbf{W}(1) & & & & & \\ \tilde{\mathbf{C}}(1) & & & & \mathbf{0} & \\ \tilde{\mathbf{F}}(1) & \mathbf{W}(2) & & & & \\ & \tilde{\mathbf{C}}(2) & & & & \\ & \tilde{\mathbf{F}}(2) & \mathbf{W}(3) & & & \\ & & & \ddots & & \\ & & & & \mathbf{W}(k-1) & \\ & \mathbf{0} & & & \tilde{\mathbf{C}}(k-1) & \\ & & & & \tilde{\mathbf{F}}(k-1) & \mathbf{W}(k) \end{bmatrix}$$

where $\tilde{\mathbf{F}}(k) = -\mathbf{W}(k+1)\mathbf{F}(k)$, $\tilde{\mathbf{C}}(k) = \mathbf{V}(k)\mathbf{C}(k)$, $\tilde{\mathbf{w}}(k+1) = \mathbf{W}(k+1)\mathbf{w}(k+1)$, $\tilde{\mathbf{y}}(k) = -\mathbf{V}(k)\mathbf{y}(k)$, and $\tilde{\mathbf{v}}(k) = -\mathbf{V}(k)\mathbf{v}(k)$. In the above, it is assumed that $\mathbf{x}(0) = \mathbf{0}$ and therefore $\tilde{\mathbf{w}}(1) = \mathbf{W}(1)\mathbf{x}(1)$.

The noise vector $\tilde{\mathbf{U}}(k)$ in Eq. 8.33 is white with zero mean and an identity covariance matrix. Therefore the best estimate, $\hat{\mathbf{x}}(k)$, given $\{\mathbf{y}(1)\cdots\mathbf{y}(k-1)\}$, can now be solved as a least squares estimation problem using the QR decomposition method. Applying an orthogonal tranformation matrix $\mathbf{Q}$, with a dimension of $[(k-1)m+kn] \times [(k-1)m+kn]$ at stage k, for both sides of Eq. 8.33, gives:

$$\mathbf{Q}\tilde{\mathbf{U}}(k) = \mathbf{Q}\tilde{\mathbf{A}}(k)\mathbf{X}(k) + \mathbf{Q}\tilde{\mathbf{Y}}(k) \qquad (8.34)$$

where

$$[\mathbf{Q}\tilde{\mathbf{A}}(k)|\mathbf{Q}\tilde{\mathbf{Y}}(k)] = \left[\begin{array}{ccccc|c} \mathbf{R}_{11} & \mathbf{R}_{12} & & & & \mathbf{b}_1' \\ & \mathbf{R}_{22} & \mathbf{R}_{23} & \mathbf{0} & & \mathbf{b}_2' \\ & & \ddots & & & \vdots \\ & \mathbf{0} & & \mathbf{R}_{k-1,k-1} & \mathbf{R}_{k-1,k} & \mathbf{b}_{k-1}' \\ & & & & \mathbf{R}(k) & \mathbf{b}_k \\ \mathbf{0} & \mathbf{0} & \mathbf{0} & \cdots & \mathbf{0} & \mathbf{r}_1 \\ \vdots & \vdots & \vdots & \ddots & \vdots & \vdots \\ \mathbf{0} & \mathbf{0} & \mathbf{0} & \cdots & \mathbf{0} & \mathbf{r}_{k-1} \end{array} \right]$$

In the above expression the submatrices $\mathbf{R}_{ii}$ and $\mathbf{R}(k)$ are upper triangular; $\{\mathbf{b}_1', \ldots, \mathbf{b}_{k-1}', \mathbf{b}_k\}$ are n dimensional vectors; and $\{\mathbf{r}_1, \mathbf{r}_2, \ldots, \mathbf{r}_{k-1}\}$

are m dimensional residual vectors. Note also that under the orthogonal transformation, $\mathbf{Q}\tilde{\mathbf{U}}(k)$ remains white and therefore the best estimated state vector $\hat{\mathbf{x}}(k)$ depends only on the vector $\mathbf{b}_k$:

$$\hat{\mathbf{x}}(k) = -\mathbf{R}(k)^{-1}\mathbf{b}_k \qquad (8.35)$$

Since $\mathbf{R}(k)$ is an upper triangular matrix, Eq. 8.35 can be solved by back substitution.

8.2.3.2 Recursive Least Squares Estimation of Kalman Filter

At the next recursion, with the new measurement $\mathbf{y}(k)$, the updated system equation for estimating $\hat{\mathbf{x}}(k+1)$, is given by a modified matrix-vector form (see Problem 6):

$$\hat{\mathbf{U}}(k+1) = \hat{\mathbf{A}}(k+1)\mathbf{X}(k+1) + \hat{\mathbf{Y}}(k+1) \qquad (8.36)$$

where

$$\hat{\mathbf{U}}(k+1) \;\; = \;\; \begin{bmatrix} \mathbf{Q}_1\tilde{\mathbf{U}}(k) \\ --- \\ \tilde{\mathbf{v}}(k) \\ \tilde{\mathbf{w}}(k+1) \end{bmatrix}$$

and

$$[\hat{\mathbf{A}}(k+1)|\hat{\mathbf{Y}}(k+1)] =$$

$$\begin{bmatrix} \mathbf{R}_{11} & \mathbf{R}_{12} & & & & & | & \mathbf{b}'_1 \\ & \mathbf{R}_{22} & \mathbf{R}_{23} & & \mathbf{0} & & | & \mathbf{b}'_2 \\ & & \ddots & & & & | & \vdots \\ & \mathbf{0} & & \mathbf{R}_{k-1,k-1} & \mathbf{R}_{k-1,k} & & | & \mathbf{b}'_{k-1} \\ & & & & \mathbf{R}(k) & & | & \mathbf{b}_k \\ \mathbf{0} & \mathbf{0} & \cdots & \mathbf{0} & \mathbf{0} & \mathbf{0} & | & \mathbf{r}_1 \\ \vdots & \vdots & \ddots & \vdots & \vdots & \vdots & | & \vdots \\ \mathbf{0} & \mathbf{0} & \cdots & \mathbf{0} & \mathbf{0} & \mathbf{0} & | & \mathbf{r}_{k-1} \\ - & - & - & - & - & - & | & - \\ \mathbf{0} & \mathbf{0} & \cdots & \mathbf{0} & \tilde{\mathbf{C}}(k) & \mathbf{0} & | & \tilde{\mathbf{y}}(k) \\ \mathbf{0} & \mathbf{0} & \cdots & \mathbf{0} & \tilde{\mathbf{F}}(k) & \mathbf{W}(k+1) & | & \mathbf{0} \end{bmatrix}$$

To compute $\hat{\mathbf{x}}(k+1)$ by QR decomposition on $\tilde{\mathbf{A}}(k+1)$ we need only be concerned with the following $(2n+m) \times (2n+1)$ matrix,

$$\left[\begin{array}{ccc|c} \mathbf{R}(k) & \mathbf{0} & & \mathbf{b}_k \\ \tilde{\mathbf{C}}(k) & \mathbf{0} & & \tilde{\mathbf{y}}(k) \\ \tilde{\mathbf{F}}(k) & \mathbf{W}(k+1) & & \mathbf{0} \end{array}\right] \tag{8.37}$$

as opposed to the large $\tilde{\mathbf{A}}(k+1)$ matrix. Performing the QR decomposition on Eq. 8.37 using the orthogonal transformation matrix leads to the following:

$$\mathbf{Q}_1 \left[\begin{array}{ccc|c} \mathbf{R}(k) & \mathbf{0} & & \mathbf{b} \\ \tilde{\mathbf{C}}(k) & \mathbf{0} & & \tilde{\mathbf{y}}(k) \\ \tilde{\mathbf{F}}(k) & \mathbf{W}(k+1) & & \mathbf{0} \end{array}\right] = \left[\begin{array}{ccc|c} \mathbf{R}_{k,k} & \mathbf{R}_{k,k+1} & & \mathbf{b}'_k \\ \mathbf{0} & \mathbf{R}(k+1) & & \mathbf{b}_{k+1} \\ \mathbf{0} & \mathbf{0} & & \mathbf{r}_k \end{array}\right]$$
$$\tag{8.38}$$

where $\mathbf{R}_{k,k}$ and $\mathbf{R}(k+1)$ are $n \times n$ upper triangular matrices, and $\mathbf{r}_k$ is a residual vector.

Example 3: Systolic Array Design of Kalman Filter

Array processor designs for the Kalman filter have been proposed in [Andre81], [Jover86], [Sung86], [MJChe87]. The results in [Andre81] and [Jover86] handle only the measurement updating, while [Sung86] and [MJChe87] propose array designs for both the measurement and time updatings. However, all the previous methods fail to exploit the inherent triangular structure of the matrix $\mathbf{W}(k+1)$ in Eq. 8.37. Now we take advantage of this special structure to derive a more efficient systolic design.

Algorithm Description   The overall triangualrization procedure consists of two parts

1. Nullification of $[\tilde{\mathbf{C}}^T(k)\tilde{\mathbf{F}}^T(k)]^T$ by rotation with $\mathbf{R}(k)$ in a QR triarray (see Example 2).

2. Update $\mathbf{W}(k+1)$ while retaining its triangular structure.

On the surface, the first step appears to be straightforward. However, as shown in Figure 8.7, any rotation operations applied to each pair of rows

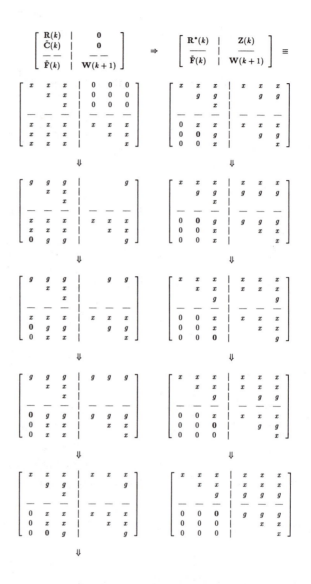

Figure 8.7: The progress of the nullification of the $\mathbf{F}(k)$ matrix, while retaining the triangular structure of $\mathbf{W}(k+1)$, where the *blank* elements are also regarded as 0. The g elements indicate the active row elements involved in that substep of the Givens rotation, and a **0** indicates where nullification is taking place.

566

in the matrices $\tilde{\mathbf{F}}(k)$ and $\mathbf{R}(k)$ are also applied to the corresponding pair of rows in the matrices $\mathbf{W}(k+1)$ and the zero matrix above it, denoted by $\mathbf{Z}(k)$. In order to guarantee a successful preservation of the upper triangular structure of $\mathbf{W}(k+1)$ in the second step, some processing ordering must be adopted when performing operations on $\tilde{\mathbf{F}}(k)$ in the first step. This can be illustrated by showing how the substeps progress and displaying their snapshots in Figure 8.7. We note that the $\tilde{\mathbf{F}}(k)$ matrix is nullified in a *bottom up* order, as shown in Figure 8.7, hence allowing $\mathbf{W}(k+1)$ to remain upper triangular throughout entire process.

Systolic Architecture Design   The above two-step procedure can be performed by a single QR triarray. The operation is divided into two phases: the processing on the $\tilde{\mathbf{C}}(k)$ and $\tilde{\mathbf{F}}(k)$ matrices is performed in the first phase (see Figure 8.8(a)); while the processing on $\mathbf{W}(k+1)$ is performed in the second phase (see Figure 8.8(b)).[5] The processing time for each recursion (i.e., both phases) is $4n + m$ [Kung87c]. The following describes how the triarray works:

1. **Operation on $\tilde{\mathbf{C}}(k)$:**   The new arriving matrix $\tilde{\mathbf{C}}(k)$ can be nullified by rotating with the resident triangular matrix $\mathbf{R}(k)$ (see Figure 8.8(a)).

   Operation on $\tilde{\mathbf{F}}(k)$:   As shown in Figure 8.8(a), the nullification on $\tilde{\mathbf{F}}(k)$ will continue right after (m time units) the operation on $\tilde{\mathbf{C}}(k)$ in a similar manner in the triarray. As discussed in Example 2, a diagonal PE performs GG (Givens generation) and then sends the GR (Givens rotation) parameters right-ward to the remaining PEs in the same row, where the GRs are performed. Most importantly, after $n+m$ time units, the parameters of the rotation angles $\{\theta_{ij}\}$, (i.e., $\{\cos\theta_{ij}, \sin\theta_{ij}\}$), start to emerge from the right side of the triarray and are stored in a data buffer to be used for the purpose of the second phase processing.

2. **Operation on $\mathbf{W}(k+1)$:**   We will prove in a moment that, right after the nullification of $\tilde{\mathbf{F}}(k)$ is completed at $t = 3n+m$, the matrix $\mathbf{W}(k+1)$ is already loaded into the triarray and the rotation parameters of $\{\theta_{ij}\}$

---

[5]To cope with the different kinds of processing in the two phases, a CORDIC based arithmetic unit design in the PE appears to be suitable.

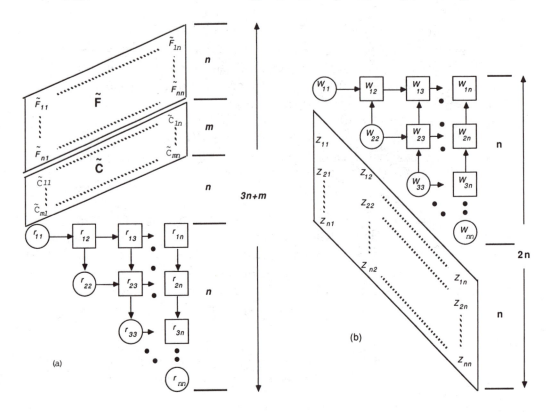

Figure 8.8: The two steps of one recursion of Kalman filter updating. (a) Using $\mathbf{R}(k)$ to nullify the $\tilde{\mathbf{C}}(k)$ and $\tilde{\mathbf{F}}(k)$ matrices. (b) Rotating $\mathbf{W}(k+1)$ and $\mathbf{Z}(k)$.

also become available for use. Given what we have just claimed, it is straightforward to perform the rotation operation on $\mathbf{W}(k+1)$ in a triarray. Note that the triarray should now provide upward and rightward communication channels, as opposed to the downward and rightward ones provided in the first phase processing.

Rotation Parameters: As shown in Figure 8.9(a), the parameters of the rotation angles, $\{\theta_{ij}\}$, emerge from the right side of the triarray in a skewed data pattern. The data along any anti-diagonal line will be sent to the same diagonal PE in the triarray. A natural design is to use a *data buffer* to store these parameters, as shown in Figure 8.9(b), with the top row of the data buffer directly linked to to the diagonal

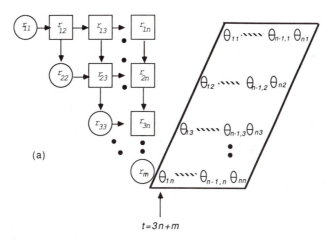

Figure 8.9: (a) The parameters of the rotation angles, $\{\theta_{ij}\}$, emerge from the right side of the triarray in a skewed data pattern. (b) *Data buffer* to store these parameters, with the top row directly linked to to the diagonal PEs of the triarray.

PEs of the triarray. At $t = 3n + m$, with all the parameters of $\{\theta_{ij}\}$ in place, they are sent row by row to the triarray to perform the Givens rotations on $\mathbf{W}(k + 1)$.

The overall data arrangements for the triarray are displayed in Figure 8.10. The time lag for the $\{\theta_{ij}\}$ parameters reflects the time for the triarray and buffer processing.

Loading $W(k+1)$:   As shown in Figure 8.10, $W(k+1)$ is loaded (from the top) into the triarray immediately after $\tilde{F}(k)$. This allows perfect synchrony in a timely engagement with the $\{\theta_{ij}\}$ parameters (from the left) and $Z(k)$ data (from the bottom), (see Figure 8.10).

Solution of $\hat{x}(k+1)$ The best prediction of the state vector $\hat{x}(k+1)$ is obtained by solving Eq. 8.35. This is easily done using the back substitution method on one additional linear array of length n (see Section 3.4).

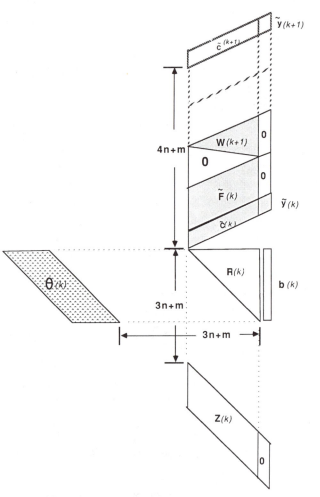

Figure 8.10: The overall data arrangements for one recursive updating in the Kalman filter algorithm.

Processing Time for One Recursion   There are two phases in the operation. The first phase takes $3n + m$ time units and the second phase takes $2n$ time units to complete their respective operations. However, immediately after n time units of the second phase processing, PE(1,1) becomes free and can be used to start processing the new recursion (for the $(k + 1)$st stage). Thus the processing time for each recursion is only $4n + m$ (instead of $5n + m$).

More elaborately, as shown in Figure 8.10, the k-th stage processing starts at $t = 0$ when $C_{m1}(k)$ enters the array, while the $(k + 1)$st stage processing (for $C_{m1}(k + 1)$) can start right after Z_{41} finishes its rotation at PE(1,1) at $t = 4n + m$. (Note that at this time PE(1,2) will still be performing the stage k computation but it will become available for the stage $k + 1$ at the next time unit when C_{m2} enters the array.) This amounts to $4n + m$ time units for processing one recursion [Kung87c].

8.3   Speech Processing

The main areas in speech processing are *speech analysis/synthesis, speech coding,* and *speech recognition.*

1. *Speech analysis* covers the techniques for speech waveform parameterization, speech enhancement, and spectral estimation for extracting formant frequencies. *Speech synthesis,* on the other hand, deals with the automatic generation of speech waveforms based on the parameters derived by speech analysis. *In Section 8.3.1, linear prediction techniques for speech analysis/synthesis are briefly discussed.*

2. *Speech coding* is used for efficient transmission, and storage of speech signals. *In Section 8.3.2, an example on array design for vector quantization speech coding is presented.*

3. *Speech recognition,* concerning the recognition of human speech, has found an increasing number of applications, such as automatic speech recognition systems and individual speaker recognition and identification [Bowen82], [Oppen78]. *In Section 8.3.3, dynamic time warping for speech recognition is treated in greater detail.*

8.3.1   Linear Prediction for Speech Analysis/Synthesis

Figure 8.11 illustrates a speech production mechanism. The system is excited by an impulse train for voiced speech or a random noise sequence for unvoiced speech. The time varying digital filter (vocal tract filter) with a "quasi-stationary" property can be treated as a time invariant filter in a short time interval (20-30 ms).

One of the most popular tools for *speech analysis* is the linear prediction method, which is based on the assumption that the vocal tract filter bears an underlying autoregressive (AR) model. Just like the AR spectral estimation, the linear prediction analysis provides estimates of the gain parameter, the pitch period, and the linear prediction parameters $\{a_i\}$. In estimating the filter coefficients it involves the computation of 8-12 lags of autocorrelation coefficients using short-term (20-30 ms) data samples and the solution of the corresponding Yule-Walker equations (see Section 8.2.1.2). Once the linear prediction parameters are obtained, the short-term power spectrum of the vocal tract filter can be obtained using Eq. 8.5 and Eq. 8.6 (with $B(z) = 1$). By tracking those peak responses (about 4 or 5) of the AR spectrum, the resonant frequencies (formants) of our vocal tract can be located and tracked. This approach has gradually become the predominant technique for analyzing the basic speech parameters, e.g., pitch, formants, spectra, vocal tract area functions, and for representing speech for low bit rate transmission or storage.

Speech synthesis is useful for many applications such as voice-response computer systems or speaking/reading aids for the handicapped. Speech

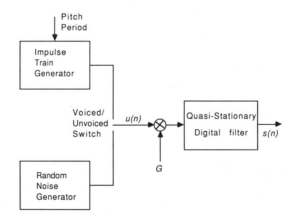

Figure 8.11: Block diagram of simplified model for speech production.

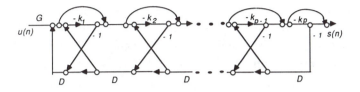

Figure 8.12: LPC speech synthesizer using lattice filter form.

can be synthesized from the linear prediction parameters in several different ways. A straightforward realization is to use the direct form filter to implement the AR model with the given linear prediction parameters. However, the direct form realization in general suffers from a high numerical sensitivity to the perturbation of coefficients. The numerically more appealing design is a lattice form structure as shown in Figure 8.12 [Rabin78], which is based on the so-called reflection coefficients, $\{K_i\}$. They may be derived from the linear prediction analysis, e.g., using the Schur algorithm (see Section 2.3).

8.3.2   Vector Quantization for Speech Coding

8.3.2.1   Coding Techniques for Data Compression

Data compression involves conversion of a (digital or analog) signal into a digital data stream of lower rate for communication over a digital communication channel or for storage in a digital memory. The main idea is to find an efficient representation of the signal by removing redundant information. Data compression plays a major role in achieving effective and yet reliable data transmission in digital communication systems. As an example, speech digitization by a PCM (pulse coded modulation) system causes a significant bandwidth expansion, (and thus no longer fits on ordinary telephone channels), so some form of data compression is needed for an efficient use of the communication channel.

Speech and image data compression are also known as *speech* and *image coding*. Both types of data have structural characteristics which permit very sophisticated coding for a high compression ratio.   In the following, three approaches to speech and image coding are reviewed: *predictive method*, *transform method*, and *vector quantization method*.

Data Compression via Predictive Method   Two of the most popular predictive coding schemes are waveform and parametric predictive coding. Waveform coding encodes the speech (image) signal directly on the time (space) domain or frequency domain. Some typical examples are differential PCM (DPCM), adaptive differential PCM (ADPCM). On the other hand, parametric coding is based on an assumed model, which dictates how the signals are correlated. One prominent example is linear prediction coding (LPC) based on the AR model, which is popular for speech coding due to its accurate estimates of the parameters and relatively low computation requirements. The LPC technique can also be applied for image coding by adopting an appropriate autoregressive model for 2-D images.

Data Compression via Transform Method   Data compression can be achieved by using a unitary transformation or orthogonal transform such that most information can be packed into a small number of samples, which are then quantized and coded. Two typical examples of transform methods for speech/image data compression are the *Fourier transform* and the *Hadamard transform*. It can be shown that as the length of transform increases, the Fourier transform is (asymptotically) equivalent to the optimal Karhunen-Loeve (KL) transform. From a computational standpoint, the Hadamard transform (see Chapters 2 and 6) offers a more efficient approach. The Hadamard matrix is composed of elements $+1$s and -1s only, thus no multiplications are needed. For an $N \times N$ picture, the Hadamard transform coefficients can be calculated with $O(N^2 log_2 N)$ additions and subtractions.

Data Compression via Vector Quantization   The idea of *vector quantization* (VQ) is adapted from a classical *pattern matching* technique. Its main attribute is the *potential to achieve very low data rates* to fit in with low bandwidth channels. The schematic diagram of VQ techniques is shown in Figure 8.13. The major burden associated with a VQ technique is the computationally intensive pattern-matching operations. In a VQ system, the input pattern (vector) is compared with elements in a set of pre-classified templates (codevectors), which form the database (codebook) for the reference patterns. The codevector that best approximates the input vector is chosen and the index of that codevector is used as the encoded output. The original signal may be (approximately) reconstructed later using the pattern in the codebook. This accomplishes the purpose of data compression.

For a typical LPC speech vocoding system, the input voice is segmented into frames and each frame is processed to extract out the LPC coefficients. The frame rate is about at 44 frames/sec. In each frame, 10 LPC parameters representing the vocal tract filter coefficients are extracted out, and 41 bits

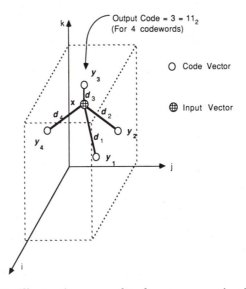

Figure 8.13: Illustrative example of vector quantization, dimension = 3.

are used to quantize these 10 parameters. This amounts to a 2400 bits/sec (bps) vocoding system. By the VQ technique, an 800 bps LPC vocoder can be developed [Wong82]. The key to attaining the 800 bps rate is the reduction from 41 to 10 bits (via VQ) in the coding of each set of the 10 LPC coefficients. This 10 bits coding is performed as follows: Firstly, 1024 codevectors ($2^{10} = 1024$) are collected in the codebook through a clustering strategy. This codebook is searched for the best match for each frame. Once this match is found, the corresponding (10-bit) index is the encoded output for that frame.[6]

8.3.2.2   Array Processors for Vector Quantization Coding

The pattern matching process in a VQ processor is often very computationally intensive. Even for simple distortion measures such as squared Euclidean distance (mean-squared error), the pattern-matching process is very demanding. Therefore a codebook of modest size will necessarily require a high-speed VQ-based coding architecture in order to perform pattern-matching efficiently [Cappe86].

---

[6] By splitting the whole codebook into *voiced* and *unvoiced* codebooks for different type of phonemes, better performance may be achieved.

Matrix-Vector Multiplication Formulation for VQ   Let $\mathbf{x}$ be a k-dimensional input vector (pattern), and $\{\mathbf{y}(n), \ n = 1, \ \ldots, \ N\}$ denote the N k-dimensional codevectors in the codebook. A simple and popular distortion measure is the mean-squared error[7]

$$D(\mathbf{x}, \ \mathbf{y}(n)) = \|\mathbf{x} - \mathbf{y}(n)\|^2 = \sum_{i=0}^{k-1} x_i^2 - 2 \sum_{i=0}^{k-1} y_i(n) x_i + \sum_{i=0}^{k-1} y_i^2(n) \quad (8.39)$$

where x_i and $y_i(n)$ are the i-th elements of the vectors $\mathbf{x}$ and $\mathbf{y}(n)$, respectively.

The first summation term in the Eq. 8.39 is constant for a given input vector $\mathbf{x}$ and its deletion does not affect the result of the codebook search. Therefore, VQ encoding based on searching for the minimum of $D(\mathbf{x}, \ \mathbf{y}(n))$ may be equivalently accomplished by the following two steps:

1. Compute

$$d_n = \sum_{i=0}^{k-1} z_{ni} x_i + c_n$$

   where $z_{ni} = -2 \ y_i(n)$, and $c_n = \sum_{i=0}^{k-1} y_i^2(n)$

2. Determine the index n, such that d_n is the minimum among $\{d_i\}$, and use it as the encoded output.

Note that $\{z_{ni}\}$ and $\{c_n\}$ may be precomputed and prestored for the VQ computation. The above can be equivalently given in a matrix-vector multiplication formulation. The corresponding two steps are

1. Compute

$$\mathbf{d} = \mathbf{Z}\mathbf{x} + \mathbf{c}$$

2. Determine

$$\text{Min } \{d_n\}$$

---

[7]Another common metric in vector quantization of LPC parameters is the well known Itakura-Saito distortion measure, based on a spectral-error matching function [Itaku75].

where $\mathbf{d}$ represents the *distortion vector*, $(\mathbf{d} = [d_1, \ d_2, \ \ldots, \ d_N]^T)$, and $\mathbf{c}$ represents the *initial vector*, $(\mathbf{c} = [c_1, \ c_2, \ \ldots, \ c_N]^T)$, and $\mathbf{Z} = \{z_{ni}\}$ is the $N \times N$ *codebook matrix* with each row containing one codevector of the codebook.

Example 1: Systolic Design for Vector Quantization

The DG for the above matrix-vector multiplication is shown in Figure 8.14. The the input vector $\mathbf{x}$ is propagated along the i-direction, and the elements $\{z_{ni}\}$ of the codebook matrix are preloaded to all the DG nodes. The initial values $\{c_i\}$ are input from the top row. The partial sums are accumulated along the vertical direction, and the final result $\{d_n\}$ are obtained at the k-th row, i.e., the second bottom-most row of the DG. To determine the minimum of $\{d_n\}$, all the values $\{d_n\}$ have to be compared across the $(k+1)$-th row of the DG. The index associated with the minimum is output from the $(k+1, \ N)$ node in the DG.

If the SFG projection of the DG is taken along the i-direction, then it leads to a linear array of size $k+1$ as shown in Figure 8.15. Functional descriptions of the two different types of PEs are presented in Figure 8.16. In this design, the speedup factor is somewhat limited because the number of PEs (equal to the dimension of the input vectors) is usually small (e.g., $k \approx 10$). If the projection along the j-direction is taken, it leads to an array of N PEs, then the speedup factor increases tremendously. The price is that

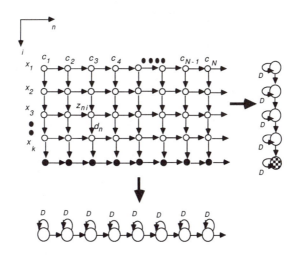

Figure 8.14: DG of matrix-vector multiplication with minimum element finding.

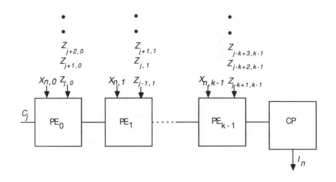

Figure 8.15: PE/CP array for VQ. The PEs operate on their three input words to produce the output $d = a * b + c$. The CP compares its d input with the contents of the minimum distortion register (MDR) and writes the most negative of these two words and the associated index back to the MDR on the subsequent cycle.

all the N PEs require both comparators and multiplier/adders. However, the partitioning scheme (see Section 6.3) may be applied whenever so desired.

Both systolic designs mentioned above may be easily extended to handle multiple input vectors situations. One way to achieve this is to use multiple systolic arrays to perform the VQ on the multiple input vectors in parallel. This of course speeds up the processing rate at the expense of hardware

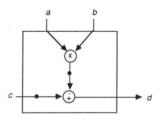

● = Unit Delay Element

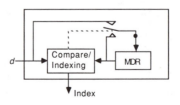

Figure 8.16: Function description of (a) PE structure, and (b) CP structure.

cost. The other approach is to repeatedly use the (single) systolic array by recycling the codebook matrix and block-pipelining the input vectors one after another (see Section 4.4).

8.3.3   Dynamic Time Warping for Speech Recognition

Speech recognition has applications including automobile manufacturing, banking, and voice query information systems. Some commercial systems are currently available [Rabin81]. Speech recognizers generally consist of the four functional blocks shown in Figure 8.17: a feature extractor, a pattern matcher, a reference vocabulary memory, and a decision maker [Bowen82]. Among them, the most crucial component is the pattern matcher, which involves the computation of a dissimilarity measure between a test pattern and a reference pattern. For example, in a machine-aided automobile inspection system, the reference patterns are derived from a number of keywords describing the parts of an automobile such as battery, tire, and brake. A set of characteristic parameters are extracted from the digitized waveforms of each spoken keyword. The complete sets are the reference patterns. The test pattern consists of a single set of parameters extracted from a word, such as the word "brake" spoken by the automobile inspector. This parameter set is matched with all the reference patterns to determine the most likely keyword. The pattern-matching task here is considerably more complicated than a pure template-matching procedure. The added difficulty lies in the requirement of compensating the variation of speaking rates, i.e., the time misalignment between the original training reference pattern and the new input test pattern. Dynamic time warping (DTW) may be used to efficiently compensate for such nonlinear temporal distortion [Sakoe78].

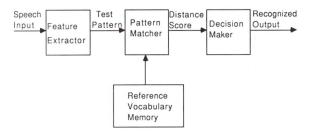

Figure 8.17: A complete speech-recognition system.

8.3.3.1 DTW Algorithm via Minimum Cost Path Formulation

In speech recognition, the input speech is in the from of m frames of a test pattern $\{t_i, i = 1, 2, \ldots, m\}$. These frames effectively form a set of characteristic parameters for the waveform. These parameters are extracted from intervals of speech data that are typically 15-30 ms, (it means that an isolated word of one second length speech has typically about 30-60 frames). In addition, we have Q sets of reference speech, each one consisting of n_q frames of reference patterns $\{r_{qj}, q = 1, 2, \ldots, Q, j = 1, 2, \ldots, n_q\}$. The task of the speech recognizer is to match the input test speech string with all the Q sets of reference speech to find the most similar pattern.

Suppose that we are to compare the test pattern with the q-th reference pattern, then a 2-D Cartesian grid of size $m \times n_q$ may be adopted (cf. Figure 8.18(a)). A prespecified cost d_{ij} is assigned to each node $V(i, j)$ of the

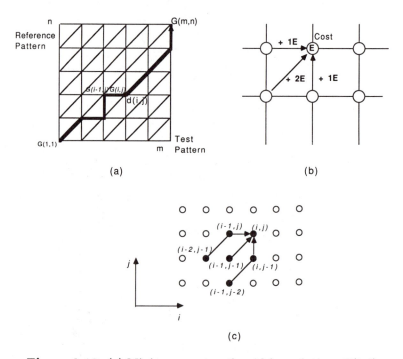

(a)

(b)

(c)

Figure 8.18: (a) Minimum-cost path grid formulation, $G(i, j)$ represents the minimum accumulated path cost from node $V(1, 1)$ to node $V(i, j)$. (b) Suppose $d(i, j) = E$, then the accumulated cost is either $1 \times E$ or $2 \times E$, depending on which arc the route travels through. (c) The valid paths to the point (i, j).

grid. In a pattern-matching system, as shown in Figure 8.18, the node cost $d(i, j)$ stands for the distance between reference frame $\mathbf{r}_{qj}$ and test frame $\mathbf{t}_i$. The nodes are connected by three types of arcs, *up*, *right*, and *diagonal*, each arc is assigned a prespecified weight, $\{w_1, w_2, w_3\}$, respectively. Based on the grid model, the DTW algorithm can be formulated by a *minimum cost path* problem in traveling through the 2-D grid from the lower left $V(1,1)$ to the upper right $V(m, n)$.

Unconstrained Formulation   The cost function of a path is determined not only by the cost in the visited nodes $d(i(k), j(k))$ but also the multiplicative weight associated with the entering arc (cf. Fig. 8.18(b)).

$$\textbf{Min} \ \sum_k d(i(k), \ j(k))w(k) \qquad (8.40)$$

where $V[i(k), j(k)]$ denotes the node visited at the k-th step, $d(i(k), j(k))$ denotes the node cost, and $w(k) = w_1, w_2, or w_3$, depending on the arc used to enter the node $V[i(k), j(k)]$.

In the DTW problem, Eq. 8.40 is used as the similarity measure for the matching with each of the reference patterns. If the frame parameters are LPC coefficients then the distance is usually the Itakura distance mentioned in the previous section. The arc weighting most popularly adopted is $w_1 = w_2 = 1$, and $w_3 = 2$ [Itaku75], [Sakoe78].

Local Continuity Constraint   To evaluate this function for all possible cost paths would in general be very computationally expensive. However, it is possible to impose constraints on the allowed path routes by taking into account the appropriate physical circumstance. problem. For example, a local continuity constraint when properly imposed might ensure that no more than two consecutive horizontal (or vertical) nodes can be visited. The physical interpretation is that no more than two replications of the same frame in a test (or reference) speech is allowed in the model. In the practical design, we also restrict the second derivative of the warping function so that the path does not orthogonally change its direction [Sakoe78]. Following the above constraints, there are only three valid paths to node $V(i, j)$ as shown in Figure 8.18(c);

1. The path directly coming from node $V(i-1, j-1)$.

2. The path from node $V(i - 2, \, j - 1)$ going through the intermediate node $V(i - 1, \, j)$.

3. The path from node $V(i - 1, \, j - 2)$ going through the intermediate node $V(i, \, j - 1)$.

Global Path Constraint   A global path constraint may be imposed by limiting the permissible path route in the grid plane of the reference vs. the test pattern to a diagonal stripe emanating from around the origin of the plane. This constrains the warping path to the diagonal region. The mathematical formulation for this global path constraint can be expressed as $|i(k) - j(k)| < B$, where $2B + 1$ is the effective matching area, i.e., the banded region. Hence if two patterns are deviating too much, then the minimum cost path will go beyond the banded diagonal region, and the patterns are judged to be no match. Thus no more computations are continued.

Example 2: Systolic Design for the DTW Algorithm

Let $G(i, j)$ denotes the accumulated cost of the minimal cost path up to any node $V(i, j)$. Note that in the selection of the path, no route reversing is permitted. Thus $G(i, j)$ can be calculated recursively as in the dynamic programming problem. This recursive procedure is discussed in this example. The overall cost $G_q(m, n)$ represents the measure for the dissimilarity between the two patterns.

Unconstrained Formulation   Here we shall look at the DG approach to the recursive DTW design. A solution to the *unconstrained DTW* problem (see Eq. 8.40) can be expressed as the recurrent equation shown in Eq. 8.41 [Sakoe78]. The DG for this problem is almost the same as Figure 8.18, with node $V(i, j)$ being reached from only three possible nodes: $(i, j - 1)$, $(i - 1, j - 1)$, $(i - 1, j)$. By dynamic programming, the single assignment code can be written as

$$G(i, j) \;=\; 0 \quad i \le 0 \text{ or } \; j \le 0$$

$$\textbf{For} \quad i \quad \textbf{from 1 to } m$$
$$\textbf{For} \quad j \quad \textbf{from 1 to } n$$

$$\mathbf{t}(i,j) = \mathbf{t}(i,j-1)$$
$$\mathbf{r}(i,j) = \mathbf{r}(i-1,j)$$
$$d(i,j) = \text{dist}\{\mathbf{t}(i,j), \mathbf{r}(i,j)\}$$

$$G(i,j) = \text{Min} \begin{cases} G(i,j-1) & + w_1 d(i,j) \\ G(i-1,j-1) & + w_3 d(i,j) \\ G(i-1,j) & + w_2 d(i,j) \end{cases} \qquad (8.41)$$

where $\mathbf{t}(i,1) = \mathbf{t}_i$ $i = 1, 2, \ldots, m$, and $\mathbf{r}(1,j) = \mathbf{r}_j$ $j = 1, 2, \ldots, n$ are the input frames to be time warped, and $\text{dist}\{\mathbf{t}(i,j), \mathbf{r}(i,j)\}$ is the distance measure between $\mathbf{t}(i,j)$ and $\mathbf{r}(i,j)$.

Local Continuity Constraint If the local continuity constraint previously mentioned is imposed, then the DG will be as shown in Figure 8.19, where a node $V(i,j)$ may be reached from and only from nodes $V(i-2,j-1)$, $V(i-1,j-1)$, and $V(i-1,j-2)$. The corresponding single-assignment formulation has to be modified as shown below. Note that, because of the local continuity constraint, we have two more weight values, (w_4 and w_5).

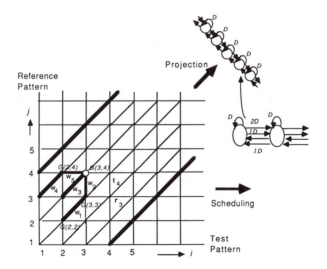

Figure 8.19: The DG of the DTW algorithm with local constraints and global constraint window.

$$G(i,j) \;=\; 0 \qquad i \le 0 \text{ or } j \le 0$$

For i **from 1 to** m

For j **from 1 to** n

$$\mathbf{t}(i,j) \;=\; \mathbf{t}(i,j-1)$$
$$\mathbf{r}(i,j) \;=\; \mathbf{r}(i-1,j)$$
$$d(i,j) \;=\; \text{dist}\{\mathbf{t}(i,j),\mathbf{r}(i,j)\}$$

$$G(i,j) = \text{Min} \begin{cases} G(i-2,j-1) \;+ w_1\, d(i-1,j) \;+ w_2\, d(i,j) \\ G(i-1,j-1) \;+ w_3\, d(i,j) \\ G(i-1,j-2) \;+ w_4\, d(i,j-1) \;+ w_5\, d(i,j) \end{cases} \tag{8.42}$$

Global Path Constraint   The global path constraint implies that the computational nodes are confined to within the strip $|i - j| < B$. To obtain high efficiency, it is more preferrable to project the DG onto an SFG in the $(1,1)$ direction. This yields a linear array configuration as shown in Figure 8.20. (Note that the function of each PE is based on Eq. 8.42.) After systolization of the SFG, a systolic design with local communication and pipelining period $\alpha = 2$ can be derived.

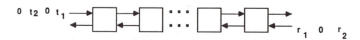

Figure 8.20: Linear array configuration of DTW algorithm.

It is also possible to further localize the DG by introducing some extra intermediate variables in the mathematical implementation of the single assignment code [Jutan84] (see Problem 9). There is still one remaining problem: the PEs work only half the time because of $\alpha = 2$. One possible solution to this is to process two reference patterns $\mathbf{r}$ and $\mathbf{r}'$ at the same time by interleaving them, which will make the overall performance twice as fast.

8.3.3.2   VLSI Implementation Examples

A VLSI systolic architecture based on the modified design (see Problem 9) has been implemented in France for various DTW algorithms [Jutan84]. Each PE computes a local distance and updates a global measure of dissimilarity. Each PE is made up of 1900 transistors using 2.5-μm NMOS technology. Twenty-five PEs and their local interconnections fit within a 35-mm^2 place of silicon that can be packaged in standard 40-pin packaging. A single chip can handle 300 words in real time. An array of twenty-two chips will recognize a syllable size pattern from a set of 6000 within 200 ms.

Systolic array designs for connected speech recognition based on the dynamic time warping and a probabilistic matching algorithm have also been proposed [Charo86]. A prototype programmable chip API89 is adopted to be a basic PE for the probabilistic matching algorithm. A ring array architecture has been designed by NTT in Japan, with the goal of realizing the DTW algorithm in a real-time speech-recognition hardware system. The design has the advantages of reduced number of PEs in the array, and sufficient throughput rates [Takah85]. A systolic array wafer-scale architecture for DTW operation is also developed at the Lincoln Laboratory for use in both isolated and connected word recognition. The array performs either LPC or spectrally based distance metrics. A restructurable VLSI (RVLSI) technology is used to implement the array with 65 PEs, using bit serial arithmetic, on 3-μm silicon technology. Speech recognition based on these RVLSI DTW wafers are projected to have real-time vocabularies as large as 12,000 words [Feldm84].

8.4   Image Processing

Image processing deals with deterministic and stochastic representation of images (image transforms and models), compression of the large amount of data in the images (image data compression), and improving the quality of the image by filtering and by removing any degradations present (image enhancement and restoration). The main areas in digital image processing are *enhancement, restoration, reconstruction* and *coding.*

1. *Image enhancement* encompasses methods which accentuate certain features of an image so that it becomes more suitable for subsequent image analysis. Examples are contrast and edge enhancement, pseudo-

coloring, filtering, etc. There are three fundamental types of enhancement methods: spatial, spectral, and temporal. *Spatial methods* take advantage of the geometric features of the image to improve its apparent quality. *Spectral methods* enhance the image by operating in the spectral domain. *Temporal methods* seek to utilize the difference of information between consecutive frames [Venet85]. *In Section 8.4.1, an example on median/rank-order filtering for image enhancement is discussed in greater detail.*

2. *Image restoration* seeks to restore an image from a degraded one, so that it is as close to the original as possible. It commonly employs tools derived from filtering and estimation theory. Some of the degradations to be dealt with are random noise, interference, geometrical distortion, field nonuniformity, contrast loss, and blurring. Restoration methods employ linear filtering techniques, such as inverse filtering, Wiener filtering, and Kalman filtering, as well as nonlinear filtering techniques, such as relaxation techniques, homomorphic filtering, maximum entropy methods, and Bayesian estimation methods. *In Section 8.4.2, an extensive discussion on a relaxation technique for image restoration is presented.*

3. *Image reconstruction* is a technique for constructing high-resolution images by processing data obtained from views of a target object from many different perspectives. For example, the problem of reconstructing 3-D objects from their 2-D projections has important applications to CAT (computer aided tomography), NMR (nuclear magnetic resonance), radioastronomical, and seismic applications. Digital convolution and Fourier transform techniques are extensively used in this area. *In Section 8.4.3, such techniques for image reconstruction are briefly discussed.*

4. *Image coding* deals with source encoding so as to compress the image by removing redundancies. It shares very similar techniques as speech coding. Therefore, the transform and predictive methods previously mentioned may also be applied to image coding. *For a more detailed design example, the reader is referred to Section 8.3.2.*

8.4.1   Median/Rank-Order Filtering for Image Enhancement

8.4.1.1   Median Filter for Video Signal Processing

Median filters are a special class of nonlinear filters useful in the removal of impulse noise from signals [Pratt78]. The idea of the median filter is very simple. Take a sampled signal of length L and slide it across a window of length $2N + 1$, the filter output at each window position is given the same position as the center sample of the window and is set equal to the median value of the $2N + 1$ samples in the window. Start-up and end effects are taken care of by appending samples of constant value to the beginning and the end of the data sequence. As an example consider the sequence

$$3 \quad 4 \quad 7 \quad 1 \quad 2$$

The median is found by first sorting this sequence and then choosing the center value of the sorted sequence. Hence the original center pixel value of the processing window (here of length five), "7", is replaced by the median of the sequence, "3". An attractive property of median filters for image processing is that it removes sparkle noise, while preserving edges. This means that while the traditional linear filtering is unable to preserve the scene changes, the median filtering has no such problems. Consider the example shown in Figure 8.21(a), where the scene change may be perceived as an edge, a median filter of window length 3 leads to a result shown in 8.21(b). This property may be exploited in frame to frame basis video signal processing, where scene changes need to be preserved.

8.4.1.2   Noise Removal via Median/Rank Order Filtering

One approach to the median filtering is to use systolic sorting arrays and to sort out the median value of a data window [Fishe82], [Oflaz83]. The window then slides one pixel further to let the sorter generate the next median value and so on. The bubble-sorter (see Chapter 3) can be adapted to realize such moving window characteristics for the median filtering. For example, the VLSI systolic chip designed in [Nicol85] adopts the bubble-sort algorithm to compute the median value at video rates. This kind of approach can be classified as being *window-oriented*, in which the pixels in every window are separately sorted. Most window-oriented designs actually use 2-D sorting

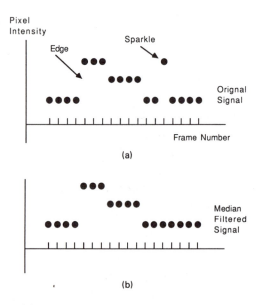

Pixel
Intensity

Figure 8.21: (a) Video signal with three level scene change on a
frame to frame basis. (b) Median filtered video signal (with filter
window length 3).

arrays. It is wasteful and is not appealing to large windows, because of the
large array size and low utilization rate.

Rank Order Filtering   A more efficient *pixel-oriented* approach is rank
order filtering, where the ranks of the intensities of the pixels are determined
in descending order. The pixel that has the required rank is extracted as
output, while the ranks of all but one of the pixels in the window are useful
for the rank-ordering of the next window.[8]

Consider a sequence of pixels and a 1-D window W of length k. As
the window slides into the next one, all the pixels except the departing
one appear again in the next window. Therefore, if the ranking in the
present window is evaluated and stored, the ranking in the next window
can be determined by a simple modification of the present ranking rather
than a reevaluation from scratch. With reference to Figure 8.22, suppose
that the window length is $k = 5$ and the present window considered is
$W_7 = [x_3 \ x_4 \ x_5 \ x_6 \ x_7]$. The next window $W_8 = [x_4 \ x_5 \ x_6 \ x_7 \ x_8]$ is

---

[8]This approach is applicable for extracting not only the median value but also data of
any specified rank.

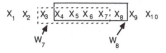

Figure 8.22: A window of length 5 for rank ordering and its next window.

obtained from W_7 by deleting the pixel x_3 and inserting the pixel x_8. Suppose that the ranks of pixels $\{x_4,\ x_5,\ x_6,\ x_7\}$ are previously evaluated in W_7, say, as $\{r_{i7},\ i = 4,\ 5,\ 6,\ 7\}$. Therefore, their new ranks $\{r_{i8}\}$ in the window W_8 can be determined by comparing the pixels $\{x_i,\ i = 4,\ 5,\ 6,\ 7\}$ with the departing pixel x_3 and the new arrival x_8, and accordingly modifying $\{r_{i7}\}$ into $\{r_{i8}\}$. This is called the PDI (pixel deletion/insertion) process. In the same process, the rank of pixel x_8, r_{88}, can also be obtained. Each window requires $O(k)$ comparisons for the pixel-oriented approach, compared with $O(k^2)$ comparisons for the window-oriented approach.

8.4.1.3   Systolic Array Design of a Rank Order Filter

Our discussion will be based on one example showing that the ranks in one window, e.g., $\{r_{i7}\}$, lead to the ranks in the next window $\{r_{i8}\}$. Then by induction, the overall procedure should follow. Suppose that after the 7-th recursion, the ranks $\{r_{37},\ r_{47},\ \ldots,\ r_{77}\}$ are available[9] and the data are arranged in a reversed order

$$x_7 \quad x_6 \quad x_5 \quad x_4 \quad x_3$$

Now the new input data x_8, x_9, $\ldots$ are available from the right side. Then the DG for the PDI procedure is shown in Figure 8.23(a), and the corresponding SFG obtained by projecting the DG in the vertical direction is shown in Figure 8.23(b). As shown in the DG, the PDI procedure consists of two processing phases (phases A and B). For the following illustration of the processing in the SFG, we note that, for the 8-th recursion, x_3 is the *deletion pixel* and x_8 is the *insertion pixel*.

Pixel Deletion Rank Updating   In phase A of the 8-th recursion, the data x_3 is propagated to the left, meeting all the other data $\{x_4\ x_5\ x_6\ x_7\}$

---

[9]Start-up can be taken care of by appending constant samples (e.g., 0's) to the beginning of the sequence.

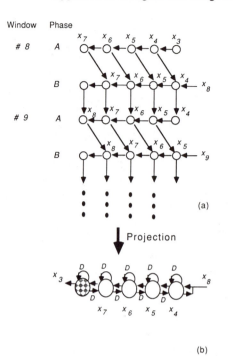

Figure 8.23: (a) DG for the pixel deletion/insertion procedure. (b) The corresponding SFG obtained by projecting DG in the vertical direction.

in their respective PEs. The PEs perform the following function:

$$\textbf{For} \quad i \quad = \quad 4, \ 5, \ 6, \ 7$$

$$\textbf{update} \quad r_{i7}$$

$$r'_{i7} \quad = \quad r_{i7} - 1 \qquad \textbf{if } x_i \geq x_3$$
$$r'_{i7} \quad = \quad r_{i7} \qquad \qquad \textbf{otherwise}$$

We note that $\{x_i, \ r'_{i7}\}$ should be moved to the right neighbor PE after each update to be ready for the phase B processing.

Pixel Insertion Rank Updating   In the phase B of the 8-th recursion, the data x_8 is propagated to left meeting all the other data $\{x_4 \ x_5 \ x_6 \ x_7\}$ in the array in their respective PEs. The PEs perform the following function:

$$\textbf{For} \quad i \quad = \quad 4, \ 5, \ 6, \ 7$$

$$\textbf{update} \quad r_{i8}$$

$$
\begin{aligned}
r_{i8} &= r'_{i7} + 1 && \textbf{if } x_i < x_8 \\
r_{i7} &= r'_{i7} && \textbf{otherwise}
\end{aligned}
$$

Rank Evaluation of New Pixel   Note that x_8 is the new arriving pixel, and its rank must be evaluated. This evaluation process can be accomplished while the pixel insertion process is taking place. The procedure is described as follows:

$$
\begin{aligned}
&\textbf{Set} && t_{83} && = \ 1 && \textit{initially} \\
&\textbf{For} && i && = \ 4, \ 5, \ 6, \ 7
\end{aligned}
$$

$$\textbf{update} \quad t_{8i}$$

$$
\begin{aligned}
t_{8i} &= t_{8,i-1} && \textbf{if } x_i < x_8 \\
t_{8i} &= t_{8,i-1} + 1 && \textbf{otherwise}
\end{aligned}
$$

$$
\begin{aligned}
&\textbf{propagate} && t_{88} && \textbf{to left} \\
&\textbf{Output} && r_{88} && = t_{87}
\end{aligned}
$$

Here, an auxiliary variable t_{8i} is introduced for the purpose of counting the number of pixels greater than or equal to x_8. After all the comparisons are performed, the rank for x_8 can be determined, $r_{88} = t_{87}$. We note that it is easy to combine the rank evaluation process with the pixel insertion process.

Rank Extraction of the Current Window   The rank extraction problem here is to determine the integer i^*, such that r_{i^*8} equals the desired rank and extract x_{i^*} from the array. For convenience of pipeline array design, it is proposed that the data x_{i^*}, once identified, should be propagated leftward and eventually output from the leftmost boundary PE. *In our design, the PDI, rank evaluation, and rank extraction processes are all embedded into one linear array* as shown in Figure 8.24.

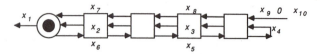

Figure 8.24: The overall systolic array for rank order filtering operations.

8.4.1.4 2-D Median Filtering

Image enhancement for video signal processing applications can make good use of 1-D median filtering performed on a frame to frame basis. However, for most other image processing applications, we still require filtering to take place in two dimensions. The key properties of the 2-D median filter are mostly preserved by using two separated 1-D median filters, one filter processes pixels along the rows and then the resulting image is processed by the other filter along the columns. Thus 2-D median filtering can virtually be performed by using two 1-D systolic arrays [Parke85] [Offen85].

8.4.2 Relaxation Technique for Image Restoration

8.4.2.1 Problem Formulation for MAP Estimator

Image restoration is to restore images from various degradation mechanisms, e.g., blurring, nonlinear deformation, multiplicative or additive noise. Mathematically, a degraded image **g** can be denoted by

$$\mathbf{g} \; = \; H(\mathbf{f}) \bigodot \mathbf{n} \tag{8.43}$$

where

- **f** is the image to be restored.

- H denotes the (linear or nonlinear) blurring/deformation transformation over a small window.

- $\odot$ denotes addition or multiplication operation.

- **n** is statistically independent noise.

If only a linear blurring mechanism and additive noise are considered, then the degraded image vector **g** in Eq. 8.43 can be rewritten as

$$\mathbf{g} \; = \; \mathbf{Hf} + \mathbf{n} \tag{8.44}$$

Image restoration is to recover the original non-blurred image $\mathbf{f}$, given the observed degraded image $\mathbf{g}$, the transformation function of H, and some statistical properties of the noise $\mathbf{n}$. Given the linear degradation formulation as shown in Eq. 8.44, the most straightforward method is to solve the least squares problem, i.e., to find an estimate $\hat{\mathbf{f}}$ which minimizes the total estimation error,

$$\min \ (\mathbf{g} - \mathbf{H}\hat{\mathbf{f}})^T(\mathbf{g} - \mathbf{H}\hat{\mathbf{f}}) \qquad (8.45)$$

In many cases, as we have discussed in Section 4.6, a priori information about the image properties (e.g., smoothness, intensity distribution) is known, then the following modified least squares formulation can be adopted:

$$\min \ (\mathbf{g} - \mathbf{H}\hat{\mathbf{f}})^T(\mathbf{g} - \mathbf{H}\hat{\mathbf{f}}) + \gamma(\mathbf{W}\hat{\mathbf{f}})^T(\mathbf{W}\hat{\mathbf{f}}) \qquad (8.46)$$

where the $\mathbf{W}$ matrix represents the intensity weighting for the overall smoothness measure of the image, and γ is a proper regularization parameter.

The above regularization formulation of image restoration can be directly solved using a neural network model (see Section 4.6). For dealing with a more general class of problems (e.g., nonlinear pixel deformation, multiplicative noise corruption and so on), the deterministic formulation can be generalized. One of the most popular statistical estimation methods for image restoration, called the maximum a posterior (MAP) approach based on Bayes' theorem, can be derived. Let $\mathbf{F}$ denote the set of all possible solutions of $\mathbf{f}$, and assume that $P(\mathbf{f}) > 0$ for all $\mathbf{f} \in \mathbf{F}$. The MAP solution $\hat{\mathbf{f}} \in \mathbf{F}$ is the one which maximizes the conditional probability:

$$P(\hat{\mathbf{f}}|\mathbf{g}) \propto P(\mathbf{g}|\hat{\mathbf{f}})P(\hat{\mathbf{f}}). \qquad (8.47)$$

Assume that there are N pixels in the image, and each pixel can have L intensity/color levels, then the size of the solution space is equal to L^N for such a discrete state case. Since there is no closed form solution to this problem, the computational load is clearly excessive. Therefore, iterative approaches which exploit the characteristics of locally dependent properties of the image model appears to be more attractive [Besag86].

8.4.2.2   Markov Random Field and Neighborhood System

Let the pixels of an image be represented by a set $S = \{s_1, s_2, ..., s_N\}$, and a possible solution be represented by $\mathbf{f} = \{x_s, \ s \in S\}$. It is popular and practical to assume that the image can be modeled after a Markov random field (MRF). An MRF is characterized by the property that [Besag74]

$$P(x_s|x_{S-\{s\}}) = P_s(x_s|x_{\partial s}) \tag{8.48}$$

where the ∂s is called the neighborhood of pixel site s.

First order and second order MRFs are the most common examples of MRF representations defined in a finite grid model. In the first order MRF, the neighborhood of a pixel s is defined as the set comprising its four nearest (north, south, east, and west) neighbors. The exceptions are the boundary pixels which have three neighbors, and the corners which have only two. In the second order MRF, the neighborhood of a pixel s is defined as the set comprising its eight nearest neighbors (north, south, east, west, north-east, north-west, south-east, and south-west).

8.4.2.3   Iterated Conditional Modes (ICM)

Iterated conditional modes (ICM), proposed in [Besag86], is a direct method to find a feasible Bayesian estimator of Eq. 8.47. The method updates the current solution $\hat{x}_s$ at pixel s by taking into account all the available information. For convenience of discussion, the observed image $\mathbf{g}$ is now represented by $\mathbf{g} = \{y_s, \ s \in S\}$. The best estimate of x_s, given y_s, and all the current estimates $\hat{x}_{S-\{s\}}$, can be obtained as the one which maximizes

$$P(x_s|y_s, \hat{x}_{S-\{s\}}) \propto P(y_s|x_s)P_s(x_s|\hat{x}_{\partial s}) \tag{8.49}$$

for all $s \in S$.

By applying this updating iteratively, $P(\hat{\mathbf{f}}|\mathbf{g})$ should increase (non-strictly) at each iteration and eventually converge to a steady state solution. The convergence speed is also very fast according to the simulation. The main drawback, however, of the ICM is that there is a high risk of converging to a local optimum. If this happens, it can not get out of the "trap" due to its nonnegotiable rule of non-decreasing $P(\hat{\mathbf{f}}|\mathbf{g})$ at each updating. The simulated annealing approach, originally proposed by Kirkpatrick et al. [Kirkp83] as a method for global optimization in combinatorial problems, provides a means to circumvent such a difficulty.

8.4.2.4   Simulated Annealing via Stochastic Relaxation

The essence of simulated annealing for image restoration and reconstruction is a stochastic relaxation algorithm which generates a sequence of images that converges at an appropriate pace to the MAP estimate. The MRF assumption is again very crucial for deriving an effective simulated annealing algorithm. Basically, an iterative change is made on every pixel. Furthermore, such a change can be modeled as a random process which depends on the pixels in its immediate neighborhood and a local conditional probability distribution.

Geman and Geman propose a method based on the simulated annealing method [Geman84], which considers the conditional probability of any state **f**, given **g**, as

$$P(\hat{\mathbf{f}}|\mathbf{g}) \propto \{P(\mathbf{g}|\hat{\mathbf{f}})P(\hat{\mathbf{f}})\}^{\frac{1}{T}} \tag{8.50}$$

There are various iterative methods of constructing an updating process, the method adopted in [Geman84] is based on the *Gibbs distribution* assumption. As we have mentioned in Section 2.5, an energy function E is defined on a finite set of states $\{f \in \mathbf{F}\}$. More precisely, a Gibbs distribution is a probability measure $P(\mathbf{f})$ with the following representation:

$$P(\mathbf{f}) = \frac{1}{Z}e^{-E(\mathbf{f})/T} \tag{8.51}$$

where Z is the normalization constant.

If we define a set of 2-D grid points as $\{s = (l, m)|\ s \in S\}$, and also let $s_0 = (l + 1, m)$ and $s_1 = (l, m + 1)$. According to the Hammersley-Clifford expansion [Besag74], $E(\mathbf{f})$ for the first order MRF can be written as:

$$E(\mathbf{f}) = \sum_s \alpha_s x_s \ + \ \sum_s \beta_s x_s x_{s_0} \ + \ \sum_s \gamma_s x_s x_{s_1} \tag{8.52}$$

where $\{\alpha_s\}$, $\{\beta_s\}$ and $\{\gamma_s\}$ are arbitrary set of functions. If we further assume homogeneity for an image, then the above formulation can be rewritten as

$$E(\mathbf{f}) = \alpha \sum_s x_s \ + \ \beta \sum_s x_s x_{s_0} \ + \ \gamma \sum_s x_s x_{s_1} \tag{8.53}$$

where α, β and γ are now constants, and are spatially invariant.

To simplify the global dependency implied by Gibbs distribution, we can again take advantage of the MRF model. Based on this, Eq. 8.50 for pixel s can be rewritten as

$$P_T(x_s|y_s, x_{S-\{s\}}) \propto \{P(y_s|x_s)P_s(x_s|x_{\partial s})\}^{\frac{1}{T}} \tag{8.54}$$

The detailed derivation can be found in [Besag74].

Equation 8.50 and Eq. 8.54 together yield the following iteration scheme which converges to a state $\mathbf{s}^*$, that minimizes $E(\mathbf{s})$. In this scheme the sequence $\mathbf{s}_1, \mathbf{s}_2, \ldots$, is formed by the following equations:

$$\mathbf{s}_{k+1} = \begin{cases} \mathbf{s}'_k, & \text{with probability } P_k \\ \mathbf{s}_k, & \text{otherwise} \end{cases}$$

where

$$P_k = \exp\left\{ \frac{-\text{Max}[\triangle E, 0]}{T_k} \right\}.$$

with $\triangle E = E(\mathbf{s}'_k) - E(\mathbf{s}_k)$.

8.4.2.5   Array Processors Design for Relaxation Algorithms

Simulated annealing is straightforward to implement for a locally dependent a priori distribution $\{P(x_s | x_{\partial s})\}$, since only y_s and the current estimate of the neighbors are required in order to update the pixel s. To ensure the convergence of the simulated annealing method by Geman and Geman, the pixels have to be updated under a prespecified schedule. More specifically, in the case of parallel processing, no two neighboring pixels should be simultaneously updated.

We note that the ICM and simulated annealing methods share the same parallel processing and data transaction structure in the updating process. Therefore, the array architecture to be discussed below is suitable for both methods. The only distinction is on the function inside the node which is dependent on the chosen relaxation algorithm. To allow efficient mapping of the simulated annealing algorithm onto array architectures, two important factors are crucial: (1) the order of neighborhood system and (2) the size of the blurring window.

First Order MRF   The simplest example is the case when there is no blurring, (nonlinear deformation on a single pixel is allowed), we shall consider the first order MRF. For this case, nodes in the DG can be embedded in a 3-D index space. The node with index (i, j, k) represents the k-th updating of a pixel point (i, j). The initial guess of the relaxation process is simply the input image $(i, j, 0)$. The DG is shown in Figure 8.25. In this graph, we have taken the maximum parallelism based on the "coding sets" updating procedure proposed in [Besag74]. (No two pixels which are neighbors should be updated at the same k index.)

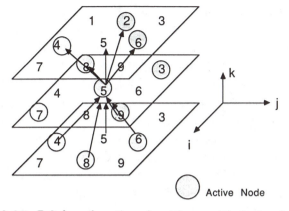

Figure 8.25: DG for relaxation algorithms, with first order MRF.

The single assignment code of this relaxation algorithm can be written as:

For k from 1 to ∞ step 2

$$x(i,j,k) \longleftarrow \begin{cases} \Phi\{x(i,j,k-1), x(i-1,j,k-1), \\ x(i,j-1,k-1), x(i+1,j,k-1), \\ x(i,j+1,k-1)\} & i+j = even \\[2mm] x(i,j,k-1) & i+j = odd \end{cases}$$

$$x(i,j,k+1) \longleftarrow \begin{cases} \Phi\{x(i,j,k), x(i-1,j,k), \\ x(i,j-1,k), x(i+1,j,k), \\ x(i,j+1,k)\} & i+j = odd \\[2mm] x(i,j,k) & i+j = even \end{cases}$$

The Φ function is dependent on the different formulation of the Gibbs distribution for different applications. The difference between the even and odd cases comes from the constraints that no two neighboring pixels should be simultaneously updated. Because of the non-deterministic nature of the size of the DG in the k direction (the iterative number axis), the only permissible projection direction is in the k direction with schedule planes lying on each ij-plane of different k indices. Note that each node in the SFG derived from this projection alternatively changes its function between an even and an odd index. For this, a special switch control is implemented inside the PEs. The resulting SFG and systolic array are shown in Figure 8.26. The systolic architecture is a mesh-connected array, with PEs being active for only 50% of the processing time due to the special data dependency. This inefficiency can be easily solved by PE sharing strategies.

The above type of architecture has been widely used in the parallel Jacobi method for solving 2-D elliptical PDE problems (also called the red/black partitioning approach [Kuo85]). In fact, by slightly modifying the DG, the Gauss-Siedel and Successive Over-Relaxation (SOR) algorithms for solving elliptical PDEs can be directly mapped onto 2-D systolic architectures (see Problem 13). Even more complicated *ad hoc* SOR algorithms, from which 1-D architecture gains more popularity, can be easily derived through the process of multiple projections from 3-D DGs to 1-D arrays (see Section 3.4).

Second Order MRF A more complicated case is the second order neighborhood system or where H is equal to a 3×3 convolving blurring matrix. By a similar idea, the maximum parallelism that can be implemented in an array is summarized by the coding pattern as shown in Figure 8.27. This implies that a mesh connected array with diagonal interconnections can be derived by projecting the DG in the k direction (see Problem 12). Note however, that each PE can be active only for 25% of the processing time. For higher order neighborhood systems or larger blurring window, it can be expected that the efficiency of 2-D array architectures will further degrade.

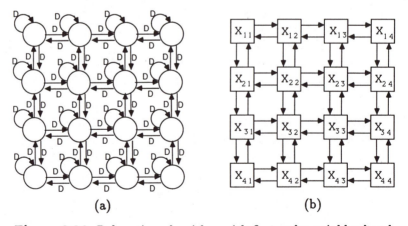

(a)   (b)

Figure 8.26: Relaxation algorithm with first order neighborhood system. (a) SFG array; (b) systolic array.

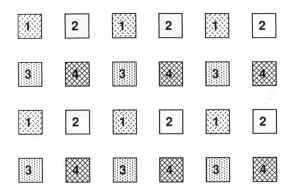

Figure 8.27: Maximum parallelism for updating as represented in "coding sets" [Besag74].

The inefficiency of PE utilization and the complicated interconnection patterns motivated us to explore a 1-D linear array [Kung86c]. For details, see Problem 12.

8.4.3   Interpolation Techniques for Image Reconstruction

Image reconstruction has many important applications such as computer aided tomography (CAT) and synthetic aperture radar (SAR). Tomography is used extensively for noninvasive medical examination of internal organs, e.g., the X-ray CAT scan technique enables the imaging of 2-D cross sections of solid objects and in nondestructive testing of manufactured items. Spotlight mode SAR imaging is a technique for producing high-resolution images from data collected via an airborne microwave radar, which illuminates the target from different perspectives [Fried83]. The CAT and SAR techniques share a very similar mathematical principle of image reconstruction, which is presented as follows.

8.4.3.1   Fourier Slice Reconstruction Algorithm

Let an object be represented by a two dimensional function $f(x,\ y)$, and a projection line L through the object be expressed as

$$L: \quad x\cos\theta + y\sin\theta = r.$$

where $0 \le \theta \le \pi$, and $-R \le r \le R$, see Figure 8.28(a). The integral of the function $f(x,\ y)$ along this line is

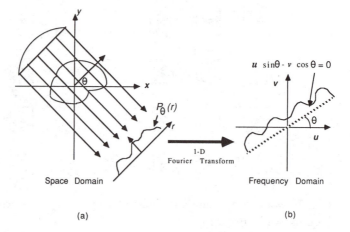

Figure 8.28: (a) A projection of a given object. (b) A slice of the Fourier transform of the object.

$$P_\theta(r) = \int_L f(x,\ y)ds = \int_{-\infty}^{\infty} \int_{-\infty}^{\infty} f(x,\ y)\delta(x\cos\theta + y\sin\theta - z)dxdy \quad (8.55)$$

where $\delta(w) = 1$, for $w = 0$, and zero elsewhere. The above integral is called the *Radon transform* and the function $P_\theta(r)$, $-R \leq r \leq R$, is called the *projection* of the object for a given θ.

Let $F(u,v)$ denote the 2-D Fourier transform of the object $f(x,y)$. The Fourier slice theorem is fundamental to a number of reconstruction techniques. It is based on an important relationship between the $F(u,v)$ and the Fourier transform of $P_\theta(r)$ [Rosen82]. More precisely, the Fourier transform of the projection $P_\theta(r)$, is equal to the values of $F(u,v)$ along the radial slice line $u \sin\theta - v \cos\theta = 0$, as shown in Figure 8.28(b). Theoretically speaking, if an infinite number of angles of projections were taken, then $F(u,v)$ would be obtainable at all points in the uv-plane. In the continuous domain, the image function $f(x,y)$ would then be recoverable by using the inverse Fourier transform of $F(u,\ v)$. In the digital domain, if the Fourier components $F(u,\ v)$ were known on a rectangular grid in the frequency plane, then $f(x,y)$ would be obtainable by performing the inverse Fourier transform:

$$f(x,y) = \frac{1}{A^2} \sum_{m=-N/2}^{N/2} \sum_{n=-N/2}^{N/2} F(\frac{m}{A},\ \frac{n}{A}) \exp[j2\pi(\frac{mx}{A} + \frac{ny}{A})] \quad (8.56)$$

for $-A/2 < x < A/2$ and $-A/2 < y < A/2$. Equation 8.56 can be effectively computed by the FFT algorithm, provided that the N^2 Fourier components on a rectangular grid $F(\frac{m}{A}, \frac{n}{A})$ are known.

8.4.3.2 Interpolation by means of Convolution

In practice, only a finite number of projections of an object can be taken. In such a case the function $F(u, v)$ is known only along a finite number of radial lines (i.e., finite number of projection angles $\{\theta_i\}$), as shown in Figure 8.28. In order to use Eq. 8.56, we must interpolate from these radial points to the points on a square grid. This may be accomplished by either 2-D or 1-D interpolation techniques. For simplicity, we discuss only 1-D interpolation technique here.

The basic idea of 1-D interpolation is to recover the band-limited continuous signal $g(y)$ from a finite set of samples of the signal, $[g(kY)]$ [Stark81]. Here, Y is the sampling interval. A popular scheme is the sinc function interpolation technique as described below:

$$g(y) = \sum_{k=-\infty}^{\infty} \frac{g(kY)\sin(\pi/Y)(y - kY)}{(\pi/Y)(y - kY)} \tag{8.57}$$

The representation of a continuous signal in the form of Eq. 8.57 is valid only for band-limited functions, and with Y chosen sufficiently small so that no aliasing occurs [Oppen75]. There are other possible interpolation functions, such as the Lagurre function and Legendre polynomials. An advantage of the sinc function is that, truncating Eq. 8.57 in fact represents a convolution operation. The design of systolic arrays for convolution should then follow easily.

8.4.4 Image Coding

Image coding uses similar techniques with speech coding (e.g., LPC, transform coding, or VQ coding). Consider a 256×256 image divided into 16×16 subpictures to be encoded using transform methods. After Fourier transformation, each subpicture has 256 Fourier coefficients, $|F(u, v)|$. The variance of each coefficient, $F(u, v)$, can be estimated by averaging $|F(u, v)|^2$ over all the subpictures. Based on a variance analysis the principal components can be extracted. In many cases only the highest ranking 64 Fourier components (based on the variance) are needed to represent the subpicture. This results in a significant data compression.

The LPC technique can also be applied for image coding by adopting an appropriate AR model for 2-D images. One application is synthetic texture generation with low bit rates [Gray84]. Another is the production of highly compressed image data for relatively crude background scenes in the remote guidance application. Predictive image coding techniques have generally low computational complexity which makes them attractive for hardware implementation. However, they are quite sensitive to channel noise (due to transmission or storage) as well as to changes in the statistical parameters of the image characteristics. In contrast, transform techniques have a high computational complexity but are quite robust with respect to channel noise and image characteristics [Jain85].

Another example of image coding is based on the VQ technique. Image VQ uses pixel intensity information directly to generate the codebook, unlike the VQ techniques for speech coding, where the LPC coefficients are used. The VQ operates on a small square sub-block of the image (typically of 9 to 16 pixels). Using a codebook of $2^6 = 64$ codewords, a bit rate of approximately 0.5 bit per pixel can be achieved. Better quality pictures could be obtained at the same bit rate by using larger block sizes. Different distance measures can also be applied for better performance. By splitting the codebook into *edge* and *texture* codebooks, the quality of reproduced image edge information may be substantially improved [Ramau83].

8.5   Image Analysis

Image analysis is concerned with extracting image information for high-level scene analysis methods and for classification of image segments or features. In image analysis, the input is an image, but the output could be simply a list of objects present or a list of features such as edges and curves, unlike image processing where both the inputs and outputs are images. The main techniques in image analysis include *feature extraction, shape analysis, pattern recognition*, and *scene analysis*.

1. *Feature extraction* involves converting images into simplified *maps* or *segments* and extracting primitive features such as edges, color, and texture contents from the segmented image. *In Section 8.5.1, array processors on edge detection for feature extraction are discussed in greater detail.*

2. *Shape analysis* involves interpretating the extracted features into known classified shape, and describing the image in terms of its parts and properties. *In Section 8.5.2, an example on Hough transform for line/curve detection is presented.*

3. *Pattern recognition* involves classifying an image into one of the pre-specified patterns based on its primitive parts and properties. *In Section 8.5.3, examples on template matching for pattern classification, and neural associative processing for pattern recognition are discussed.*

4. *Scene analysis* involves producing a concise description and meaningful interpretation of the classified image object. Techniques using artificial intelligence (AI) may be incorporated into scene analysis. *In Section 8.5.4, array design for region level operations is briefly surveyed.*

8.5.1   Edge Detection for Feature Extraction

Most of the useful information in an image is contained in those regions where a change of gray levels or colors occurs, i.e., at the edges. The information consists of the size of the transition (edge magnitude) and the direction in which the intensity changes most rapidly (edge direction). These quantities can be computed from the partial derivatives of image function $g(x, y)$. The information obtained from applying the derivative operators allows segmentation of an image, which is an important step toward object classification and identification. The extracted edge information can be used to determine the contour of objects, and under certain conditions, the topological features of 3-D objects can be determined. Edge information can also be used to evaluate threshold values to isolate specific image regions.

8.5.1.1   Edge Detection by means of 2-D Convolution

Two-dimensional convolution is the most common way for image edge detection. Assume that an image **g** is to be edge detected, the 2-D convolution involves convolving a small pattern (mask) **f** with the larger image **g**. This is done by moving the window pattern **f** over the image, as shown in Figure 8.29. At each point the convolution is computed:

$$c(m, n) = \sum_i \sum_j f(i, j) g(m - i, n - j) \tag{8.58}$$

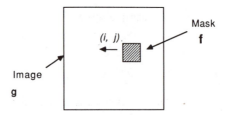

Figure 8.29: Moving windows for image-edge detection.

Various local operators have been widely used, such as Roberts operators, Laplacian operators, Prewitt operators, and the Sobel operators (see Figure 8.30). The Laplacian operators detect the edges as the local maxima of the image gradients. Therefore, zero-crossing of the second derivatives gives the edge points. Furthermore, the Laplacian operator can pick up edges from

$$W_{pz} = \begin{array}{ccc} -1 & 0 & 1 \\ -1 & 0 & 1 \\ -1 & 0 & 1 \end{array}$$

$$W_{py} = \begin{array}{ccc} 1 & 1 & 1 \\ 0 & 0 & 0 \\ -1 & -1 & -1 \end{array}$$

(a)

$$W_{sz} = \begin{array}{ccc} -1 & 0 & 1 \\ -2 & 0 & 2 \\ -1 & 0 & 1 \end{array}$$

$$W_{sy} = \begin{array}{ccc} 1 & 2 & 1 \\ 0 & 0 & 0 \\ -1 & -2 & -1 \end{array}$$

(b)

$$W_L = \begin{array}{ccc} 0 & 1 & 0 \\ 1 & -4 & 1 \\ 0 & 1 & 0 \end{array}$$

(c)

$$\begin{array}{ccccccccc}
0 & 0 & 0 & -1 & -1 & -1 & 0 & 0 & 0 \\
0 & 0 & 0 & -1 & -1 & -1 & 0 & 0 & 0 \\
0 & 0 & 0 & -1 & -1 & -1 & 0 & 0 & 0 \\
-1 & -1 & -1 & 4 & 4 & 4 & -1 & -1 & -1 \\
-1 & -1 & -1 & 4 & 4 & 4 & -1 & -1 & -1 \\
-1 & -1 & -1 & 4 & 4 & 4 & -1 & -1 & -1 \\
0 & 0 & 0 & -1 & -1 & -1 & 0 & 0 & 0 \\
0 & 0 & 0 & -1 & -1 & -1 & 0 & 0 & 0 \\
0 & 0 & 0 & -1 & -1 & -1 & 0 & 0 & 0 \\
\end{array}$$

(d)

Figure 8.30: Various windows used in edge detection: (a) Prewitt operators; (b) Sobel operators; (c) Laplacian operator; (d) expanded Laplacian operators.

all possible directions (isotropic) by a simple convolution with the image and potentially produce very thin edges.

Example 1: Array Design for 2-D Convolution

For 2-D convolution we define a $k_1 \times k_2$ matrix as a window. Since the size of an image may typically vary from 256×256 up to $8K \times 8K$ pixels, we immediately see the necessity of massive parallel processing. For real time applications, where the image is updated 30 to 60 times per second, the speed requirements are even more critical. In the following, a systolic design for the 2-D convolution is described.

DG Design Without loss of generality, assume that the window size is 3×3 and the image size is $N \times N$. Then Eq. 8.58 can be rewritten as:

$$c(m, n) = \sum_{i=0}^{2} \sum_{j=0}^{2} f(i, j) g(m - i, n - j) \tag{8.59}$$

Since there are four indices, i, j, m, n, in this equation, the DG for this algorithm is 4-D (four dimensional) and the SFGs obtained via projection are 3-D. Because of the difficulty in displaying the 4-D DG, we first decompose the double summation term in Eq. 8.59 into three terms

$$\begin{aligned} c(m, n) &= \sum_{j=0}^{2} f(0, j) g(m, n - j) + \sum_{j=0}^{2} f(1, j) g(m - 1, n - j) \\ &\quad + \sum_{j=0}^{2} f(2, j) g(m - 2, n - j) \end{aligned} \tag{8.60}$$

each summation term in Eq. 8.60 may be described by the 3-D DG as shown in Figure 8.31(a). The 3-D DG represents the image convolving with one row of the window pattern (1×3). Note that each 2-D ($3 \times N$) layer of the 3-D DG is basically a DG for 1-D convolution. The 4-D DG may be constructed by summing up the outputs of *three* such similar 3-D DGs.

SFG Design Let us first consider the design of a 3-D SFG with (approximate) size $3 \times 3 \times N$. This 3-D SFG may be obtained by first projecting each of the three 3-D DGs to one 2-D SFG (of size $3 \times N$) (see Figure 8.31(b))

and then combining these three 2-D SFGs by summing up their outputs
(see Figure 8.31(c)). If 2-D SFGs are preferred, then multiprojection can be
adopted to project the 3-D SFG to 2-D SFGs. Figure 8.31(d) shows a 2-D
SFG with size $3 \times N$. Different projection directions can also be applied if
desired. For example, it is apparent that a 3×3 SFG can also be obtained
by changing the direction of multi-projection.

Systolic/Wavefront Array Design   The SFGs in Figure 8.31(c) and (d)
can be easily systolized by applying the cut-set procedure. For example, in
the systolization of the 2-D SFG shown in Figure 8.31(d), two delays are

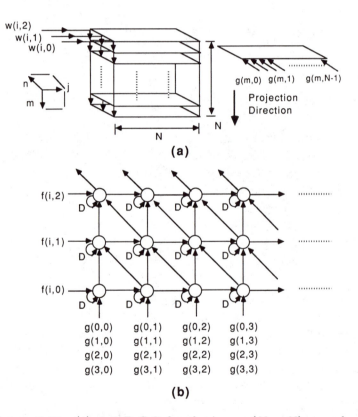

Figure 8.31: (a) A 3-D DG for the image $(N \times N)$ convolved
with one row of the window pattern (1×3). The 4-D DG may be
constructed by summing up the outputs of 3 such 3-D DGs; (b) A
2-D SFG obtained by projecting the 3-D DG along m-direction; (c)
The 3-D SFG obtained by summing up the outputs of 3 2-D SFGs;
and (d) The 2-D SFG obtained by adopting multiprojection.

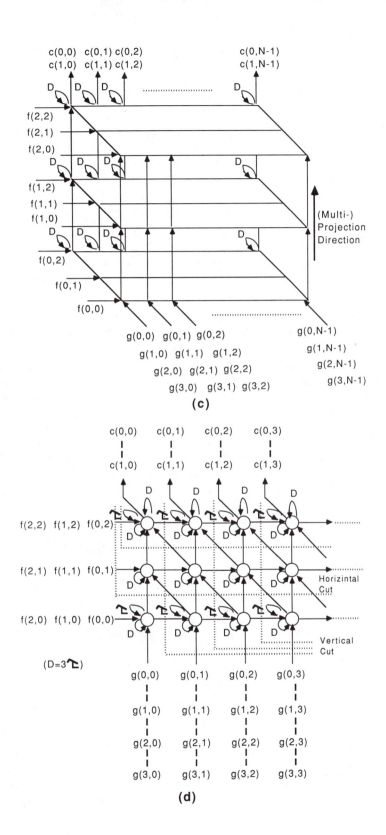

(c)

(d)

$(D=3$ $)$

607

added to each edge on the horizontal "cuts"; and then applying the vertical cuts, one delay can be transferred from each diagonal edge to the horizontal edges. A wavefront array for 2-D convolution based on the μPD7281 data flow processor has been proposed [Vlont87]. This design, with 86 ($\approx$ 256/3) processors, can calculate the pixel values of a 256-pixel column in 16.8 μsec, including I/O time. For the full 256 $\times$ 256 image, it takes 16.8 $\times$ 256 = 4.24 msec.

8.5.1.2   Use 1-D Arrays to Perform 2-D Edge Detection

In general the classic gradient or Laplacian operators perform poorly on noisy images. To improve the performance, one solution is to do averaging before the application of the operators. This results in an expanded Laplacian operator with 9 $\times$ 9 mask size (see Figure 8.30(d)), which can be used in noisy averaging and edge detecting. Marr and Hildreth also suggest a family of edge operators based on the zero-crossing of a generalized Laplacian [Marr80], which are also isotropic second derivative zero-crossing operators. Figure 8.32(a) shows an example of an 11 $\times$ 11 generalized Laplacian operator.   When a larger mask operator is adopted, the performance is in general improved but at the expense of higher hardware cost. To overcome this hardware problem, an alternative is proposed below.

Note that most of the large mask operators share a low rank property. For example, the expanded 9 $\times$ 9 Laplacian operator shown in Figure 8.30(d) has only rank *two*. The generalized 11 $\times$ 11 Laplacian operator shown in Figure 8.32(a) has numeric rank two, based on the distribution of their singular values as shown in Figure 8.32(b) (see Section 2.2.3). (Note that the first two singular values are significantly larger than the rest.)

Let p denote the numeric rank of a $k \times k$ mask operator W, e.g., $k = 9$ or 11 and $p = 2$ in the above examples. Then W can be approximated by a linear combination of p outer products, (see Eq. 2.17)

$$\mathbf{W} \approx \sum_{i=1}^{p} \sigma_i \mathbf{u}_i \mathbf{v}_i^T = \sum_{i=1}^{p} \hat{\mathbf{u}}_i \hat{\mathbf{v}}_i^T \qquad (8.61)$$

where $\mathbf{u}_i$ and $\mathbf{v}_i$ are the singular vectors of W, and $\hat{\mathbf{u}}_i$ and $\hat{\mathbf{v}}_i$ are the singular vectors scaled by the square root of the singular value σ_i. For example, the scaled singular vectors of the expanded 9 $\times$ 9 Laplacian operator are

```
-34.   -25.   -18.   -12.   -8.    -7.    -8.    -12.   -18.   -25.   -34.
-25.   -15.   -7.    -1.    3.     4.     3.     -1.    -7.    -15.   -25.
-18.   -7.    2.     8.     12.    14.    12.    8.     2.     -7.    -18.
-12.   -1.    8.     15.    20.    21.    20.    15.    8.     -1.    -12.
-8.    3.     12.    20.    24.    26.    24.    20.    12.    3.     -8.
-7.    4.     14.    21.    26.    27.    26.    21.    14.    4.     -7.
-8.    3.     12.    20.    24.    26.    24.    20.    12.    3.     -8.
-12.   -1.    8.     15.    20.    21.    20.    15.    8.     -1.    -12.
-18.   -7.    2.     8.     12.    14.    12.    8.     2.     -7.    -18.
-25.   -15.   -7.    -1.    3.     4.     3.     -1.    -7.    -15.   -25.
-34.   -25.   -18.   -12.   -8.    -7.    -8.    -12.   -18.   -25.   -34.
```

(a)

```
126.72 116.24   2.00   1.39   0.50   0.41   0.00   0.00   0.00   0.00   0.00
```

(b)

```
-34.18-24.89-17.70-12.05 -8.13 -7.10 -8.13-12.05-17.70-24.89-34.18
-24.89-15.05 -7.18 -1.04  3.04  4.24  3.04 -1.04 -7.18-15.05-24.89
-17.70 -7.18  1.44  8.12 12.42 13.80 12.42  8.12  1.44 -7.18-17.70
-12.05 -1.04  8.12 15.21 19.66 21.17 19.66 15.21  8.12 -1.04-12.05
 -8.13  3.04 12.42 19.66 24.15 25.72 24.15 19.66 12.42  3.04 -8.13
 -7.10  4.24 13.80 21.17 25.72 27.33 25.72 21.17 13.80  4.24 -7.10
 -8.13  3.04 12.42 19.66 24.15 25.72 24.15 19.66 12.42  3.04 -8.13
-12.05 -1.04  8.12 15.21 19.66 21.17 19.66 15.21  8.12 -1.04-12.05
-17.70 -7.18  1.44  8.12 12.42 13.80 12.42  8.12  1.44 -7.18-17.70
-24.89-15.05 -7.18 -1.04  3.04  4.24  3.04 -1.04 -7.18-15.05-24.89
-34.18-24.89-17.70-12.05 -8.13 -7.10 -8.13-12.05-17.70-24.89-34.18
```

(c)

Figure 8.32: The singular value decomposition of generalized Laplacian operator: (a) the original operator; (b) the singular values distribution; (c) the reconstructed operator using the first two ranks only.

$$\hat{\mathbf{u}}_1 = \begin{bmatrix} 1 & 1 & 1 & -4 & -4 & -4 & 1 & 1 & 1 \end{bmatrix}$$
$$\hat{\mathbf{v}}_1 = \begin{bmatrix} 0 & 0 & 0 & -1 & -1 & -1 & 0 & 0 & 0 \end{bmatrix}$$
$$\hat{\mathbf{u}}_2 = \begin{bmatrix} 0 & 0 & 0 & -1 & -1 & -1 & 0 & 0 & 0 \end{bmatrix}$$
$$\hat{\mathbf{v}}_2 = \begin{bmatrix} 1 & 1 & 1 & 0 & 0 & 0 & 1 & 1 & 1 \end{bmatrix}$$

Eq. 8.61 means that the 2-D edge detection can be implemented by a combination of cascaded 1-D column convolutions and 1-D row convolutions. More elaborately, the image is first processed by a row convolution then followed by a column convolution. Mathematically, let $\mathbf{W} * * \mathbf{g}$ denote the 2-D convolution of the mask operator $\mathbf{W}$ and the image $\mathbf{g}$, then based on Eq. 8.61

$$\mathbf{W} * * \mathbf{g} \approx \hat{\mathbf{u}}_1 *_c (\hat{\mathbf{v}}_1 *_r \mathbf{g}) + \hat{\mathbf{u}}_2 *_c (\hat{\mathbf{v}}_2 *_r \mathbf{g})$$

where $*_c$ and $*_r$ denote 1-D column convolution and 1-D row convolution respectively.

The systolic design for the cascaded row/column convolutions may be carried out by using the canonical mapping methodology. The detailed design process is left as an exercise. Since $p << k$ in general, this approach can yield a significant hardware saving compared with the 2-D convolution design.

8.5.2   Hough Transform for Line/Curve Detection

An edge-detected and thresholded image is often represented as a sequence of *coordinate* data pairs $\{x_k, y_k\}$, where x_k and y_k are nonnegative integers with the upper bound equal to the frame size. Using these coordinate pairs, the shape of the objects can thus be analyzed through the detection of object boundaries. The Hough transform method is a popular tool for determining all the possible straight lines or curves. It is based on the relationship between points on a line/curve and the parameters of that line/curve. So the processed data indicates the frequency of the image points being on a particular linear/curved boundary.

8.5.2.1   Hough Transform Formulations

Hough Transform for Line Detection   The general equation of a straight line can be represented by an (r, θ) pair:

$$x\cos\theta + y\sin\theta = r \tag{8.62}$$

where θ is the angle of the line and r is the distance from the origin to the line (see Figure 8.33). For the k-th selected image point (x_k, y_k) which satisfies Eq. 8.62, then it is considered to be on this line.

According to Eq. 8.62, a specific point in the image plane is mapped to a sinusoidal curve in the (r, θ) plane. From the dual perspective, all the

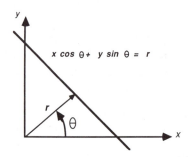

Figure 8.33: The (r, θ) representation of a straight line.

sinusoidal curves correspond to the selected image points on any straight line must *intersect in one common point* (r^*, θ^*). This in turn provides the key information about the actual parameters of the line through these image points [Nevat82].

The (r, θ) plane $(0 \leq r \leq R, \quad 0 \leq \theta \leq 2\pi)$ can be quantized into a $N \times M$ grid cells, i.e., r is quantized into N discrete values $(\{r_1, r_2, \ldots, r_N\})$, and θ into M discrete values $(\{\theta_1, \theta_2, \ldots, \theta_M\})$. These cells are represented by a 2-D array of accumulators $\{A(r_n, \theta_m)\}$. The procedure for implementing a Hough transform involves (1) performing the accumulator counts and (2) searching for the counts above threshold in the discrete (r_n, θ_m) plane. The two steps are described as follows:

1. First, all $\{A(r_n, \theta_m)\}$ are initially set to zero. For each point (x_k, y_k) (within the selected pixel points), determine those $\{(r_n, \theta_m)\}$ pairs which satisfy Eq. 8.62 within the limit of the quantization and increment the corresponding accumulators by

$$A(r_n, \theta_m) \leftarrow A(r_n, \theta_m) + 1$$

2. The final counts of the accumulators provide a measure of the number of points on the corresponding lines. Therefore, an accumulator which is higher than a certain threshold corresponds to collinear points in the image and indicates the presence of a line segment.

Hough Transform for Curve Detection   The line detection Hough transforms can be easily generalized to handle curve detection. It works for any curve $f(\mathbf{x}, \mathbf{a}) = 0$, where $\mathbf{a}$ is a parameter vector. For example, a circle can be parameterized by

$$(x - a)^2 + (y - b)^2 = r^2$$

The above two-step procedure for line detection can be modified to count a new accumulator array $\{A(a_n, b_m, r_l)\}$.

8.5.2.2   Systolic Array for Hough Transform

According to the procedure discussed above, for each of the given pixel points (x_k, y_k), there are many $\{(r_n, \theta_m)\}$ pairs satisfying Eq. 8.62. For each chosen quantized orientation angle θ_m, a corresponding quantized displacement r_q can be computed, then the corresponding accumulator $A(r_q, \theta_m)$ increments its counter. The algorithm leads directly to the linear array design shown in Figure 8.34.

In Figure 8.34 M values of θ_m are created by the the linear array from each edge pixel. The memory slot inside each PE corresponds to an accumulator cell $A(r_n, \theta_m)$. In these accumulator cells, a W register stores a quantized r and a C register stores the count of pixels contributing to the quantized r value.

Some variants of the Hough transform algorithms and the corresponding systolic designs are proposed [Chuan85] (see Problem 10). The new features added to the conventional design discussed above include: (1) the connectivity of the edge pixels can be checked and (2) the pixel counts may reveal the length of the line segments.

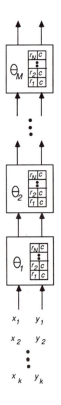

Figure 8.34: The systolic array for Hough transform.

8.5.3   Template Matching and Associative Pattern Recognition

8.5.3.1   Template Matching by means of 2-D Correlation

Template matching is the problem of matching a pattern (or a template) to a picture [Rosen82]. In the template matching problem, a digital image is represented as a 2-D array of nonnegative numbers. Each number represents the intensity of a single pixel of the picture. To determine the position at which a template **f** best matches a specific portion of a larger image **g**, the basic image correlation method may be adopted. The template **f** will traverse the image **g** (in a manner similar to the convolution operation as shown in Figure 8.29), and at each point compute the image correlation.

A simple distance measure for the dissimilarity between the template and a window of the image is the sum of the square of the differences. This measure requires the patterns to be matched to have the same intensity values. To guarantee this condition, a measure known as *normalized cross-correlation* $c(m, n)$ is introduced as:

$$
\begin{aligned}
c^2(m, n) &= \frac{(\sum_i \sum_j f(i, j) g(i + m, j + n))^2}{\sum_i \sum_j g^2(i + m, j + n)} \\[2mm]
&= \frac{b^2(m, n)}{a^2(m, n)}
\end{aligned}
\tag{8.63}
$$

at each pixel. Obviously, the point with the maximum $c^2(m, n)$ value determines where a template best matches the picture.

Example 2: Array Design for 2-D Normalized Cross Correlation

Here we first discuss the numerator part of the 2-D normalized cross correlation. The systolic array for image correlation utilizes the regularity of the algorithm, all operations will be pipelined in an attempt to keep all processors busy. Assume that the size of the image is $M \times N$, and the template is $m \times n$, then the array is configured as an $m \times n$ array of processors, as shown in Figure 8.35(a). Each PE holds a single coefficient of the template **f**. The array performs the correlation operation for pattern matching. For systolic correlation design, the filter coefficients are arranged in the reverse order as compared with the systolic convolution design in Figure 8.31(d). The first M rows of the image pass through the array from left to right with

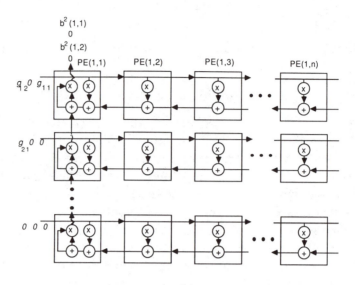

(a)

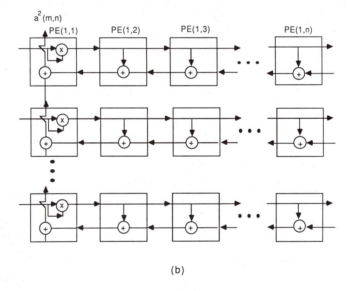

(b)

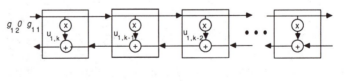

(c)

Figure 8.35: (a) Systolic array configuration for the numerator operations in 2-D correlation. (b) Denominator operations in 2-D correlations. (c) 1-D linear array for 2-D correlation operations.

614

an input data skew between each row because of the summing operation in the first column. The correlation measure emerges from the $(1, 1)$ processor. To implement Eq. 8.63, the correlation is multiplied with itself in $PE(1, 1)$ to yield the squared numerator value before it is pipelined out. The template to be matched is stored in the processor array as input to the multipliers, and the image data enters from the left and exits on the right in parallel motion. After all the N values of $b^2(1, n)$ have been computed, then the image is shifted up one row, and the same operations are executed again to calculate $b^2(2, n)$.

The computation of the denominator in Eq. 8.63 has a similar form to that of the numerator, except that the template elements $f(i, j)$ are all replaced by 1, and the image pixels $g(i, j)$ are substituted by the squared values for $1 \le i \le M$, and $1 \le j \le N$. Hence the array for computation of denominator has a similar structure to that of the numerator (see Figure 8.35(b)). $PE(1, 1)$ is configured to multiply the image pixel $g(i, j)$ with itself and then add the result to its partial sums. The squared value $g^2(i, j)$ is then pipelined to the rest of the PEs, where this value is added to the feedback partial sums. The final sum output from $PE(1, 1)$ will be $a^2(m, n)$. Comparison of the numerator and denominator array as shown in Figure 8.35(a) and Figure 8.35(b) indicates that they can both be performed using the same 2-D $m \times n$ array. Since the required inputs are the same, only one pass is required to load the input image pixels $g(i, j)$ for both the numerator and denominator calculation. Based on the observation, the systolic array design for the normalized cross correlation can thus be derived. The detailed derivation is left to the reader as an exercise (see Problem 17).

Example 3: Use 1-D Arrays to Perform Template Matching

As the template size increases, the array size of the above design will become too large. In most pattern recognition applications (e.g., texture element matching, or fixed font character recognition), the template has a large size but low numeric rank, which is conducive to the outer-product decomposition approach discussed next. This makes possible the use of 1-D correlation to accomplish the task of 2-D correlation (see Section 8.5.1).

First, we apply SVD rank reduction on $\mathbf{f}$ and derive an outer-product decomposition similar to that discussed in Section 8.5.1. The numerator correlation operation can then be computed by a 1-D correlation linear array as shown in Figure 8.35(c). After performing the row correlations of the image, we can then proceed to perform the column correlations. When p

linear arrays are available, where p equals the numeric rank of $\mathbf{f}$, then p 1-D correlations can be performed in parallel.

The denominator may also be computed as a 2-D correlation of the squared image intensity and a 2-D mask with all elements equal to 1. The 2-D mask can be represented as an outer-product

$$\mathbf{e} \, \mathbf{e}^T$$

where $\mathbf{e} = [1, 1, 1, ..., 1]^T$.

To implement the array, the input data is self-multiplied in the leftmost PE, then the squared value is input into the linear correlation array for the row correlation. The column convolution will be executed following the row convolution. The numerator and denominator correlation operations can be performed in parallel. The leftmost boundary PE is also equipped to carry out the division of the numerator by the denominator.

8.5.3.2   Neural Associative Processing for Pattern Recognition

An adaptive neural network recognizer is shown in Figure 8.36. Using the output values and labels to specify the correct class, internal parameters (or weights) in neural network recognizers are typically adapted or trained during use. This neural network recognizer can perform three different functions:

1. As a pattern classifier, identify the pattern that best represents an input pattern, which is corrupted by noise.

2. As a content-addressable or associative memory, which is useful for the recovery of the complete pattern based on the available partial input pattern.

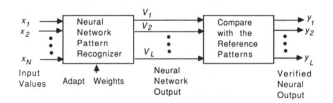

Figure 8.36: Block diagram of neural network pattern recognizer.

3. As a vector quantizer which compresses the amount of data without losing the critical information.

In speech/image pattern recognition, the neural network *recognizers* are non-parametric and can cope with less restrictive assumptions concerning the shape of underlying distributions than traditional statistical classifiers. They also exhibit a high degree of fault-tolerance and a great capacity for learning or adapting [Lippm87], [Atlas87], [Psalt85].

Example 4: Hopfield Model for Associative Pattern Recognizer

The discrete-state Hopfield model can be considered as an associative memory as discussed in Section 4.6; therefore, it may be adopted for speech/image pattern recognition applications [Atlas87], [Psalt85]. Briefly, the discrete-time transition of neuron i is formulated as (with $I_i = 0$):

$$U_i(k) = \sum_{j}^{N} T_{ij} V_j(k)$$

$$V_i(k+1) = step[U_i(k)] \tag{8.64}$$

where $step[x]$ is a unit step function, which is 1 for $x \geq 0$ and 0 for $x < 0$.

For the associative retrieval purpose, let the reference patterns be represented by $\mathbf{v}^m = [V_1^m, V_2^m, \ldots, V_N^m]$, where $m = 1, 2, \ldots, M$, then the synaptic strengths $\{T_{ij}\}$, where $1 \leq i, j \leq N$, are derived as follows:

$$T_{ij} = \left\{ \begin{array}{ll} \sum_{m=1}^{M}(2V_i^m - 1)(2V_j^m - 1) & i \neq j \\ 0 & i = j \end{array} \right. \tag{8.65}$$

where the states $\{V_i^m\}$ have binary values 0 or 1.

For pattern recognition, an unknown test pattern is initially input to an artificial neural network (ANN), such as the one proposed in Section 4.6. The network iterates following the transition rule given in Eq. 8.64. When the state of the neural network converges to a final state $\mathbf{v}^*$, then $\mathbf{v}^*$ is output from the network. In a pattern recognizer, this output $\mathbf{v}^*$ must be compared with the M reference patterns $\{\mathbf{v}^m, m = 1, 2, \ldots, N\}$ to determine if it exactly matches any of the reference patterns. If it does not match then a "no match" is declared. If it does, the output will be the matched reference pattern.

8.5.4   Region Level Operations for Scene Analysis

In scene analysis, regions are described in a concise way so that memory usage can be greatly reduced. Moreover, we can compute properties of regions and derive new regions from given ones by operating directly on the representations. Such operations can also be implemented in parallel, which requires a somewhat different design strategy since the representation is no longer array-like. The structure of parallel architectures becomes more ad hoc and more difficult to be systematically treated.

Some potential parallel array architectures for various types of region data representations have been proposed [Rosen85]. A ring array structure, for example, will be suitable for a region represented by a *border code*, which requires the coordinate of a starting point for each border, together with a sequence of codes defining the succession of moves from pixel to pixel around the border. A string of processor clusters (where each cluster contains the number of run lengths for a given row) will be suitable for a region represented by a *run length code*, which regards each row of the image that meets the regions as a succession of runs of 0's alternating with runs of 1's. A tree type of array processor can be used to perform operations on regions represented by a maximal-block structure, which is constructed by recursively subdividing into quadrants, subquadrants, and so on, until blocks of constant value are reached. For general-purpose region level operations, it may be suitable to consider some hiearchical multiprocessor systems, e.g., connection machines [Hilli85].

8.6   Concluding Remarks

The presentation of this chapter (and in fact the whole book) begins with *algorithm analysis* for the applications of interest and concludes with the VLSI array design. The signal and image processing applications generally call for algorithms which are deterministic in both time and space. This allows a unified theoretical framework for architectures and algorithm optimization for array processors. Special-purpose parallel processing systems are becoming a dominant and promising trend in future supercomputing technology. We will witness the construction of an increasing number of specialized array processors/computers tailored to execute important *kernel* algorithms from signal/image processing, numerical linear algebra, and scientific computing. Therefore, parallel algorithm analysis, design, and mapping methodologies onto VLSI array processors proposed in this book will play a useful role.

In the near future, we will face the challenge of modern and "intelligent" supercomputer systems, which are required to perform perception-level functions. These systems will be more adaptive to their environment and interact with users in more natural and efficient ways through the integration of logic circuitry and sensor stimulators. By launching novel computing concepts, these systems are also intended to solve otherwise "unsolvable" problems. From the perspective of such systems, the massively parallel array processors proposed in this book can only be regarded as basic "primitives", upon which novel computing (or thinking) machines are yet to be built.

In conclusion, future VLSI technologies promise an escalated computing capability. The overall performance is not only measured in terms of silicon area, circuit speed or computational power, but also characterized by the development of novel systems out of the interdisciplinary research, which is the main theme of this book. The design of VLSI array processors must be based on an integration of many technical disciplines including semiconductor physics, computer science, and electrical engineering. The domain of research subjects ranges from VLSI device technology, through algorithm design, architecture structure, and finally to practical applications. This book hopefully offers a useful first step towards this goal.

8.7 Problems

1. *Equivalence of two conventional spectral estimation methods:* In general, the two conventional spectral estimate $P_{BT}(w)$ and $P_{PER}(w)$ are not identical. Show that if the biased autocorrelation estimate $r_{xx}(n)$ given by Eq. 8.66 is used and as many lags as data samples ($M = N - 1$) are computed, then the Blackman-Tukey estimate and the periodogram estimate yield identical numerical results:

$$r_{xx}(m) = \frac{1}{N} \sum_{n=0}^{N-m-1} x(n)x(n+m) \qquad (8.66)$$

2. *Least squares solution:* If the $\mathbf{X}$ matrix in Eq. 8.18 is full rank, there are two popular approaches to solve the linear equations. (1) Use the QR block inverse method to solve for the $\mathbf{w}$ vector, as discussed in Section 8.2. (2) Apply the orthogonality principle [Stran80] and solve a normal equation for minimizing $\|\mathbf{e}\|_2$.

(a) What is the solution of $\mathbf{w}$ using the second approach?

(b) Show that this solution is equivalent to that derived by the QR method.

(c) What numerical drawback will the second approach lead to?

3. *Proof of Bartlett windowing*: As discussed in Section 8.2.2, the spatial directivity spectrum $P_{BF}(\theta)$ can be computed from

$$P_{BF}(\theta) = \sum_{n=-(N-1)}^{N-1} W(n)S(n)e^{-j2\pi fn}$$

Show that the weighting function, $W(n)$, is a triangular window.

4. *Proof of the recursive QR residual generation*: Show that the desired least squares residual e_{n+1} can be calculated by using the equation given in Eq. 8.26:

$$e_{n+1} = \{ \prod_{i=1}^{N-1} \cos\theta_i \} b'_{n+1}$$

Hint: The last row of matrix $\mathbf{Q}'^T_{n+1}$ given in Eq. 8.26 has the form $[\alpha|\mathbf{0}|\beta]$, where

$$\beta = \{ \prod_{i=1}^{N-1} \cos\theta_i \}$$

5. *Signal blocking matrix selection*: A generalized sidelobe-canceling structure that can be used to transform a constrained least squares problem to an unconstrained least squares problem is shown in Figure 8.6. One key element in this structure is the signal blocking matrix $\mathbf{W}_s$, as shown in Figure 8.6. Some important properties of the blocking matrix are:

$$\mathbf{W}_s = [\mathbf{c}_{m+1}, \mathbf{c}_{m+2}, ..., \mathbf{c}_n]$$

$$\mathbf{C} = [\mathbf{c}_1, \mathbf{c}_2, ..., \mathbf{c}_m]$$

where $\mathbf{c}_i^T \mathbf{c}_j = 0$, $1 \le i \le m$, $m+1 \le j \le n$

Based on the properties of the $\mathbf{W}_s$ matrix, various forms can be chosen, of which some are suitable for VLSI implementation because of their decomposibility. List two of the suitable candidates of $\mathbf{W}_s$, and explain how these matrices can be decomposed to make them implementable using VLSI arrays when n is very large.

6. *Proof of recursive formulation of Kalman filtering:* Prove that Eq. 8.36 is true.

7. *Bit-level systolic design:* In the design of VQ processors, we have outlined the systolic array to implement the matrix-vector multiplication and comparison operations. In order to achieve the maximum concurrency, a bit-level architecture can be used. Based on the given word-level architecture and canonical mapping methodology, design the bit level architecture for the matrix-vector multiplication and comparison operations [Cappe86]. **Hint:** Use the bit level DG projection design method as described in Chapter 4.

8. *Minimum-cost path finding:* A 4×4 grid with each intersection having a non-negative cost (d_{ij}) imposed is shown below.

$$
\begin{array}{cccc}
12 & 8 & 1 & 5 \\
5 & 2 & 10 & 7 \\
2 & 4 & 7 & 3 \\
9 & 16 & 3 & 1
\end{array}
$$

Supposed that you are asked to travel from the left lower corner to the right upper corner. The path allowed is arbitrary, i.e., without any local or global constraints, as described in Section 8.3.3. The multiplicative weights are $w_1 = w_2 = 1$, and $w_3 = 2$.

(a) What is the minimum cost path during the traveling?

(b) What is the minimum accumulated cost during the traveling?

   Hint: Determine stage by stage all the accumulating cost function G_{ij} for each grid point.

9. *Removal of the local constraints from DTW algorithm:* It is also possible to remove the local constraints and to localize the DG by introducing some extra intermediate variables $(G^*(i,j)$ and $G^\#(i,j))$ in the

mathematical implementation of the single assignment code [Jutan84]. The single assignment code can be rewritten as Eq. 8.67.

$$G(i,j) \;=\; 0 \qquad i \le 0 \; or \; j \le 0$$

For i **from 1 to** m

For j **from 1 to** n

$$
\begin{aligned}
\mathbf{t}(i,j) &= \mathbf{t}(i,j-1) \\
\mathbf{r}(i,j) &= \mathbf{r}(i-1,j) \\
d(i,j) &= \mathrm{dist}\{\mathbf{t}(i,j),\mathbf{r}(i,j)\} \\
G^{*}(i,j) &= G(i-1,j-1) + w_1\, d(i,j) \\
G^{\#}(i,j) &= G(i-1,j-1) + w_4\, d(i,j)
\end{aligned}
$$

$$
G(i,j) = \mathbf{Min}
\begin{cases}
G^{*}(i-1,j) & +\; w_2\, d(i,j) \\
G(i-1,j-1) & +\; w_3\, d(i,j) \\
G^{\#}(i,j-1) & +\; w_5\, d(i,j)
\end{cases}
\tag{8.67}
$$

Draw the DG of this modified DTW algorithm, and derive the corresponding systolic design. Compare this array design with the design given in Section 8.3.3.

10. *Array design for Hough transform:* In Section 8.5.2, we present a systolic design for the Hough transform, in which the r domain is uniformly quantized for different values of θ. Modify the linear systolic array so that the following two situations are allowable.

   (a) For different orientation angles θ_m, the quantization step size for r is different.

   (b) Dynamic allocation in the memory address of different quantization levels of r_{mn}, for the angle θ_m, based on the occurrence of that value.

11. *1-D and 2-D median filters:*

   (a) For the following 5×5 image, a 1-D median filter with window size 3 is used for removing the sparkle noise. Find the filtered image using this 1-D filter. (For a 1-D filter, row-wise and column-wise operations are applied separately to obtain the result.) For the purpose of processing, the gray levels of the pixels outside the image are assumed to be equal to the mean value of the entire image.

$$
\begin{array}{ccccc}
3 & 7 & 8 & 8 & 5 \\
5 & 2 & 4 & 6 & 7 \\
6 & 4 & 4 & 3 & 5 \\
9 & 1 & 3 & 6 & 6 \\
3 & 3 & 2 & 3 & 2
\end{array}
$$

   (b) Find the result if a 2-D 3×3 median filter is used.

12. *Second order MRF systolic design for SA:* As we have discussed in Section 8.4, the inefficiency of PE utilization and the complicated interconnection patterns for relaxation algorithms of second order MRF motivated us to resort to a 1-D linear array.

   (a) Design the DG for this second order MRF relaxation, using the maximum parallelism suggested by the "coding sets" updating idea (see Figure 8.27).

   (b) Find the resulting SFG by projecting in the k-direction. Verify that the efficiency of PE utilization is 25%.

   (c) How is the schedule of the SFG modified, so that it can be mapped onto a 1-D linear array with the highest parallelism?

   (d) What is the resulting 1-D systolic array derived from (c)?

13. *Partial differential equation (PDE) solver:* Partial differential equations arise in connection with various physical and geometrical problems, one very important equation is Poisson's equation,

$$
\nabla^2 U(x,y) = \frac{\partial^2 U(x,y)}{\partial x^2} + \frac{\partial^2 U(x,y)}{\partial y^2} = f(x,y)
$$

where $f(x, y)$ is a known input and $U(x, y)$ is the unknown solution. To derive solution $U(x, y)$ we need appropriate boundary conditions.

Consider the problem of solving Poisson's equation on a square $0 \leq x, y \leq 1$. Asssume that $f(x, y)$ is given for every $0 \leq x, y \leq 1$, and that $U(x, y)$ is known at the boundary of the square. The Poisson equation can be numerically approximated by using approximate expressions for the partial derivatives. If the square is discretized with step h in both directions, then it can be easily proved that for sufficiently small h

$$\frac{\partial^2 U(x, y)}{\partial x^2} \approx \frac{U(i+1, j) + U(i-1, j) - 2U(i, j)}{h^2}$$

and

$$\frac{\partial^2 U(x, y)}{\partial y^2} \approx \frac{U(i, j+1) + U(i, j-1) - 2U(i, j)}{h^2}$$

where $x = ih$ and $y = jh$.

(a) Prove the above difference equations and derive a *difference equation* that numerically approximates Poisson's equation.

(b) Assume that $h = 0.2$. The difference equation derived in part (a) can be written in a *matrix-vector* form as

$$\mathbf{Au} = \mathbf{b}$$

where

$$\mathbf{u} = [U(1,1), U(1,2), U(1,3), U(1,4), U(2,1), U(2,2), \ldots, U(4,4)]^T$$

What is the form of matrix $\mathbf{A}$ and vector $\mathbf{b}$.

(c) The *successive overrelaxation method (SOR)* mentioned in Chapter 2, can be adopted to solve this problem. Derive the iteration equation. Draw the 3-D DG for the derived iterative algorithm.

(d) Derive an SFG by projecting the DG derived in part (c). Which projection will you choose? Why? Specify the *schedule vector* and draw the resulting 2-D SFG.

(e) Derive a systolic array from the SFG developed in part (d) by using the cut-set systolization procedure with *optimum* time scaling. What is the best pipelining period α.

14. *Image reconstruction by ART*: The series expansion approach to X-ray image reconstruction from projection can be formulated in the following way: A Cartesian grid of pixels is introduced into the region of interest so that it covers the whole picture that has to be reconstructed. The pixels are numbered in some agreed manner, such as from 1 (top left corner) to N (bottom right corner pixel). The X-ray attenuation function is assumed to take a constant value f_j throughout the j-th pixel for $j = 1, 2, \ldots, N$. Assume that the intersection of the i-th ray with the j-th pixel, denoted by h_{ij}, for all $i = 1, 2, \ldots, M$ and $j = 1, 2, \ldots, N$, represents the weight of the contribution of the j-th pixel to the total attenuation along the i-th ray.

This problem can be formulated as a large set of linear equations $\mathbf{g} = \mathbf{Hf}$, where $\{g_i\}$ is the physical measurement of the attenuation of the i-th ray, which represents the line integral of the unknown attenuation function along the path of the ray. The matrix $\mathbf{H}$ is huge (typically of the order $10^5 \times 10^5$), sparse, and sometimes binary. Find a suitable iterative linear equation solver to solve this problem, and also propose an array architecture to implement the algorithm.

Hint: A very popular algorithm for solving this problem is called the algebraic reconstruction technique (ART); it is an iterative process that starts from an initial approximation $\mathbf{f}^0 \in \mathbf{R}^N$ to the image vector. In an iterative step, the current iterate $\mathbf{f}^k$ is refined (or say, corrected) to a new iterate $\mathbf{f}^{k+1}$ by taking into account only a single ray (i-th) and changing only the image values of the pixels that intersect this ray [Rosen82].

15. *Rank evaluation example*: Given the following input sequence, based on the rank evaluation formula (with window size 5) as described in Section 8.4,

$$3, 9, 7, 12, 6, 7, 15, 11, 4, 8, 9, 8, \ldots.$$

find the resulting rank evaluated sequence (r_5, r_6, r_7, ...).

16. *Edge-detection algorithm*: An image $g(m, n)$ is formed as the weighted difference of an original image, $f(m, n)$, and its Laplacian weighted image:

$$g(m, n) = cf(m, n) - (1 - c)\nabla^2 f(m, n)$$

where m, n = 0, 1, ..., $N-1$ and $0 \le c \le 1$, and ∇^2 is the Laplacian operator.

(a) Derive an expression for a discrete Laplacian operator as a 3 × 3 convolution mask.

(b) Show that $g(m,n)$ can be formed by a single convolution of $f(m,n)$ with a 3 × 3 convolution mask, and derive an expression for the mask.

(c) Write and simplify an expression for the frequency response of the filter in (b).

(d) Describe how the image $g(m,n)$ changes as function of parameter c.

17. *The numerator and denominator combined computation:* Based on the Eq. 8.63, we have discussed the array processor design for the numerator and denominator computation of normalized cross correlation as shown in Figure 8.35(a) and (b). How could the two arrays be combined into one array?

18. *Fourier slice theorem:* The Fourier slice theorem equates the 1-D Fourier transform of the projection $g(u,\theta)$ of an image $f(x,y)$ to a slice of the 2-D image Fourier transform $F(X,Y)$ along the line Y/X = $\tan \theta$.

(a) Consider the projection

$$g(x) = g(u,0) = \int_{-\infty}^{+\infty} f(x,y)\ dy$$

Prove that the 1-D Fourier transform of $g(x)$ equals $F(X,Y)$ along the line Y = 0.

(b) Generalize this result to a projection at any angle θ.

19. *Radon transform of uniform circular disc:* The 2-D Fourier transform, in polar coordinates, of a uniform circular disc has the form

$$F(\rho,\theta) = \frac{J_1(2\pi\rho)}{\rho}$$

where $J_1(x)$ denotes the first order Bessel function. Using the Fourier slice theorem, find an expression for the projections $g(u,\theta)$ of the disc.

20. *Euclidean distance formulation*: Calculating the Euclidean distances between many pairs of feature vectors is important in several pattern recognition applications. Develop a systolic/wavefront array for this computation using the canonical mapping methodology.

    Hint: Compare with the matrix-vector multiplication formulation of the speech vector quantization coding.

21. *Line thinning problem*: Thinning a detected line or curve requires the following operations: A line (or a curve) of width k is replaced by a line (curve) with width 1. The intensity of the new pixel value will be replaced by the maximum (or minimum) of the values in a certain specified neighborhood. Design a systolic/wavefront array for thinning an image contour. (Consider only the nearest four neighbors of a pixel [Nevat82].)

22. *Systolic Kalman filter for white additive noises $SKF - W$*: For most applications, it is common to adopt the assumption that both the system noise $\mathbf{w}(k + 1)$ and measurement noise $\mathbf{v}(k)$ are *white*, i.e.,

$$\mathbf{R_w}(k + 1) = diag[\sigma_{w1}, \sigma_{w2}, \ldots, \sigma_{wn}]$$
$$\mathbf{R_v}(k) = diag[\sigma_{v1}, \sigma_{v2}, \ldots, \sigma_{vm}].$$

    In this case, for a complete systolic Kalman filter design, the prewhitening operations can be reduced to a simple scaling operations [Kung87c]. Design the complete triarray system which performs the scaling and recursive updating operations.

23. *Systolic Kalman filter for color noises $SKF - C$*: In some applications, where the noise is *colored*, in addition to the recursive QR triangularization procedure as discussed in Section 8.2.3, a prewhitening procedure needs to be provided. The prewhitening procedure consists of the following two operations.

    (a) Reverse Cholesky decomposition (RCD) of $\mathbf{R_w}(k+1)$ and $\mathbf{R_v}(k)$.

    (b) Given the whitening operators, perform the prewhitening operations (PWO) on $\mathbf{F}(k)$, $\mathbf{C}(k)$, and $\mathbf{y}(k)$.

    A reverse Cholesky decomposition (RCD) of $\mathbf{A}$ is defined as

$$\mathbf{A}^{-1} = \mathbf{U}^T\mathbf{U}$$

where $\mathbf{U}$ is an upper triangular matrix.

(a) Design a systolic array for performing the RCD operation.

(b) Design a systolic array for performing the PWO operation.

Hint: The RCD operation can be rewritten as follows:

$$\mathbf{U}\mathbf{A} = \mathbf{U}^{-T}.$$

where

$$\mathbf{U} = \mathbf{U}'_1 \, \mathbf{U}'_2 \, \ldots \mathbf{U}'_n$$

with $\mathbf{U}'_k$ being given as

$$\mathbf{U}'_k = \begin{bmatrix} 1 & 0 & \cdots & u'_{1k} & 0 & \cdots & 0 \\ 0 & 1 & \cdots & u'_{2k} & 0 & \cdots & 0 \\ \vdots & \vdots & \ddots & \vdots & \vdots & \vdots & \vdots \\ 0 & 0 & \cdots & u'_{kk} & 0 & \cdots & 0 \\ 0 & 0 & \cdots & 0 & 1 & \cdots & 0 \\ \vdots & \vdots & \vdots & \vdots & \vdots & \ddots & \vdots \\ 0 & 0 & \cdots & 0 & 0 & \cdots & 1 \end{bmatrix}$$

and

$$\mathbf{U}'_n \mathbf{A}^{(n)} = \mathbf{A}^{(n-1)} = \begin{bmatrix} x & x & \cdots & x & 0 \\ x & x & \cdots & x & 0 \\ \vdots & \vdots & \ddots & \vdots & \vdots \\ x & x & \cdots & x & 0 \\ x & x & \cdots & x & g_n \end{bmatrix}$$

$$\mathbf{U}'_{n-1} \mathbf{A}^{(n-1)} = \mathbf{A}^{(n-2)} = \begin{bmatrix} x & x & \cdots & 0 & 0 \\ x & x & \cdots & 0 & 0 \\ \vdots & \vdots & \ddots & \vdots & \vdots \\ x & x & \cdots & g_{n-1} & 0 \\ x & x & \cdots & x & g_n \end{bmatrix}$$

$$\vdots \quad = \quad \vdots$$

$$\mathbf{U}_1'\mathbf{A}^{(1)} \;=\; \mathbf{U}^{-T} \;=\; \begin{bmatrix} g_1 & 0 & \cdots & 0 & 0 \\ x & g_2 & \cdots & 0 & 0 \\ \vdots & \vdots & \ddots & \vdots & \vdots \\ x & x & \cdots & g_{n-1} & 0 \\ x & x & \cdots & x & g_n \end{bmatrix}$$

where $\mathbf{A}^{(n)} = \mathbf{A}$, g_k is equal to the square root of $a_{kk}^{(k)}$ in matrix $\mathbf{A}^{(k)}$. By using above problem formulation, a triarray can be efficiently used for both RCD and PWO operations [Kung87c].

24. *Overall configuration for $SKF-C$:* Show that by using two RCD/PWO triarrays (with $O(n^2/2 + m^2/2)$ PEs), and one QR triarray (with $O(n^2/2)$ PEs), the overall array configuration for $SKF - C$ can be implemented as shown in Figure 8.37 [Kung87c].

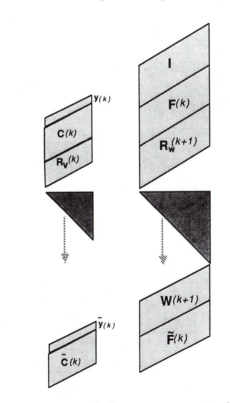

Figure 8.37: The overall data arrangements for one recursion of Kalman filter algorithm. (a) Two RCD/PWO triarrays (with $O(n^2/2 + m^2/2)$ PEs) for performing the prewhitening procedure.

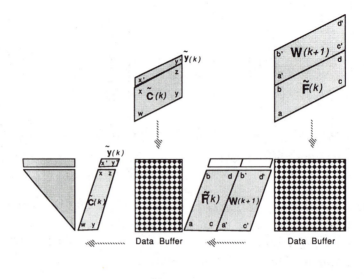

(b)

Figure 8.37:

(b) One QR triarray (with $O(n^2/2)$ PEs) for performing the recursive updating procedure. Note that data buffers are provided for corner turning the matrices.

Bibliography

[Ackle85]   D. H. Ackley, G. E. Hinton, and T. J. Sejnowski. A learning algorithm for Boltzmann machines. *Cognitive Science*, Vol. 9: pp. 147 – 169, 1985.

[Ahmed85]   H.M. Ahmed. Alternative arithmetic unit architectures for VLSI digital signal processors. In *VLSI and Modern Signal Processing*, Prentice Hall, Inc., Englewood Cliffs, NJ, 1985.

[Aho74]   A. V. Aho, J. E. Hopcraft, and J. D. Ullman. *The Design and Analysis of Computer Algorithms*. Addison-Wesley Publishing Co., Reading, Massachusetts, 1974.

[Aliph87]   A. Aliphas and J. A. Feldman. The versatility of digital signal processing chips. *IEEE Spectrum*, pp. 40–45, June, 1987.

[Allen85]   J. Allen. Computer architecture for digital signal processing. *Proceedings of the IEEE*, 1985.

[Ander77]   J. A. Anderson, J. W. Silverstein, S. A. Ritz, and R. S. Jones. Distinctive features, categorical perception, and probability learning: Some applications of a neural model. *Psych. Rev.*, 84, 1977.

[Andr76a]   H.C. Andrew and C.L. Patterson. Singular value decompositions (SVD) image coding. *IEEE Transactions on Communication*, pp. 425–437, April 1976.

[Andr76b]   H.C. Andrew and C.L. Patterson. Singular value decompositions and digital image processing. *IEEE Transactions on ASSP*, pp. 26–53, Feburary 1976.

[Andre81]   A. Andrew. Parallel processing of the Kalman filter. In *Proc. Int. Conf. Parallel Processing*, pp. 216–220, 1981.

[Anna86a]   M. Annaratone, E. Arnould, T. Gross, H. T. Kung, and M. Lam. Warp architecture and implementation. In *Proc. 13th Annual Inter. Symp. on Computer Architecture*, June 1986.

[Anna86b]   M. Annaratone, E. A. Arnould, H. T. Kung, and O. Menzilcioglu. Using warp as a supercomputer in signal processing. In *Proc. IEEE ICCASP'86*, pp. 2895–2898, Tokyo, Japan, April 1986.

[Arnou85]   E Arnould et al. A systolic array computer. In *Proc. IEEE ICASSP'85*, Tampa, FL, pp. 232–235, 1985.

[Arpin86]   D.A. Arpin and Y. Kim. Parsor: A parallel processor for sparse matrix solution by SOR iteration. In *Proc. of Int'l Conf. on Parallel Processing*, 1986.

[Ashcr84]   W. D. Ashcraft and H. M. South. Architecture and control of a distributed signal processor. In *Proc. IEEE ICASSP'84*, San Diego, Calif., pp 44.8.1 – 44.8.4, 1984.

[Atlas87]   L. E. Atlas, T. Homma, R. J. Marks II. A neural network model for vowel classification. Presented in *Proc. IEEE ICASSP'87*, Dallas, Texas, 27.15.1–27.15.4, 1987.

[Backu78]   J. Backus. Can programming be liberated from the Von Neumann style? A functional style and its algebra of programs. *Communications of the ACM*, 21: pp. 613–641, 1978.

[Balla82]   D. H. Ballard and C. M. Brown. *Computer Vision*. Prentice-Hall, Inc. N.J., 1982.

[Barnw83]   T. P. Barnwell and D. A. Schwartz. Optimal implementations of flow graphs on synchronous multiprocessors. In *IEEE Asilomar Conference*, Circuits and Systems, Pacific Grove, CA, November 1983.

[Batch80]   K. E. Batcher. Design of a Massively Parallel Processor. *IEEE Transactions on Computers*, C-28: pp. 836–840, 1980.

[Baugh73]   C. R. Baugh and B. A. Wooley. A two's complement parallel array multiplication algorithm. *IEEE Transaction on Computers*, C-22: pp. 1045–1047, December 1973.

[Bensc83]   N. F. Benschop. *MOS-LSI Digital Integrated Circuits and Systems*. Delft University, Delft, The Netherlands, 1983.

[Besag74]   J. Besag. Spatial interaction and statistical analysis of lattice system (with discussion). *J. Royal Statist. Soc., Series B*, 36: pp. 192–236, 1974.

[Besag86]   J. Besag. On the statistical analysis of dirty pictures. *J. Royal Statist. Soc., Series B*, 48(3): pp. 192–236, 1986.

[Booth51]   A. D. Booth. A signed binary multiplication technique. *Quart. Journ. Mech. and Appl. Math.,*, 4: pp. 236–240, 1951. Part 2.

[Bowen82]   B.A. Bowen and W.R. Brown. *VLSI Systems Design for Digital Signal Processing*. Volume 1 of *Design for Digital Signal Processing*, Prentice Hall, N.J., 1982.

[Bren85a]   R. P. Brent and F. T. Luk. The solution of singular-value and symmetric eigenvalue problems on multiprocessor arrays. *SIAM. J. SCI. STAT. Compu.*, Vol.6, No.1, Jan. pp 69-84, 1985.

[Bren85b]   R. P. Brent and F. T. Luk. Computatin of the singular value pecomposition using mesh-connected processors. *Journal of VLSI and Computer System*, Vol.1, No.3, pp 243-270, 1985.

[Breue77]   M. A. Breuer. Min-cut placement. *Journal of Design Automation and Fault Tolerant Computing*, October 1977.

[Broml84]   K. Bromley, J.J. Symanski, J.M. Speiser, and H.J. Whitehouse. Signal algorithms, architectures, and applications. In *SPIE short course notes*, 1984.

[Broom85]   D. S. Broomhead, J. G. Harp, J. G. McWhirter, K. J. Palmer, and J. G. B. Roberts. A practical comparison of the systolic and wavefront array processing architectures. In *Proc. IEEE ICASSP'85*, pp. 296 - 299, Tampa, FL, April 1985. Also in the Proceeding of IEEE Workshop on VLSI Signal Processing, Los Angeles., November 1984.

[Burst83]   M. Burstein and R. Pelavin. Hierarchical wire routing. *IEEE Transactions on Computer-Aided Design of Integrated Circuits and Systems*, October 1983.

[Cain84]   J. B. Cain and R. A. Kriete. A VLSI $r=1/2$, $k=7$ Viterbi decoder. *IEEE Proc. of the NAECON, Volume I*, pp. 20–28, May 1984.

[Capon69]   J. Capon. High resolution frequency wavenumber spectrum analysis. *Proc. IEEE*, 57: pp. 1408–1418, Aug., 1969.

[Cappe83]   P. R. Cappello and K. Steiglitz. Unifying VLSI array designs with geometric transformations. In *Int'l Conf. on Parallel Processing*, 1983.

[Cappe84]   P.R. Cappello etc. *VLSI Signal Processing*. IEEE PRESS, N.Y., 1984.

[Carai84]   C. Caraiscos and B. Liu. From digital filter flow-graphs to systolic arrays. Submitted to IEEE Transactions on ASSP, 1984.

[Cappe86]   P. Cappello, G. Davisdon, A. Gersho, C. Koc, and V. Somayazulu. A systolic vector quantization processor for real-time speech coding. In *In Proc. of IEEE ICASSP'86*, Tokyo, Japan, pp. 2143–2146, IEEE, 1986.

[Casas84]   D. Casasent. Acousto-optic linear algebra processors: architectures, algorithms, and applications. *Proceedings of the IEEE*, 72(7): pp. 831–849, July 1984.

[Chan86]   T. F. Chan and Y. Saad. Multigrid algorithms on the hypercube multiprocessor. *IEEE Transactions on Computers*, pp. 969–977, November 1986.

[Chang86]   C. Y. Chang and K. Yao. Viterbi decoding by systolic array. In *Workshop on VLSI Signal Processing*, IEEE Press, Los Angeles, CA, 1986.

[Chapm85]   R. Chapman, T. S. Durrani, and T. Willey. Design strategies for implementing systolic and wavefront arrays using Occam. In *Proc. IEEE ICASSP'85*, pp. 292–295, Tampa, FL, 1985.

[Charn86]   R. Charng and F. C. Lin. Linear Content-addressable Systolic Array for Sparse Matrix Multiplications. In *VLSI Signal Processing, II*, S. Y. Kung et al. (eds.), pp. 212-219, IEEE Press, November, 1986

[Charo86]   F. Charot, P. Frison, and P. Quinton. Systolic architectures for connected speech recognition. *IEEE Transactions on ASSP*, ASSP-34: pp. 765–779, August 1986.

[Chell85]   R. Chellappa, and A. A. Sawchuck. *Digital Image Processing and Analysis: Volume 1, 2. IEEE Computer Society*, 1985.

[Chen78]   T. C. Chen, V. Y. Lum, and C. Tung. The rebound sorter: an efficient sort engine for large files. In *Proc. 4th Int'l Conf. on Very Large Databases*, pp. 312–318, 1978.

[Chen83]   M. C. Chen. *Space-Time Algorithm: Semantics and Method-*

ology. PhD thesis, Computer Science Department, California Institute of Technology, 1983.

[Chen85a] M. C. Chen and C. Mead. Concurrent algorithms as space-time recursion equations. In *VLSI and Modern Signal Processing*, S.Y. Kung et al. (eds.), Prentice-Hall, Inc, 1985.

[Chen85b] M. C. Chen. *A Synthesis Method for Systolic Designs*. Technical Report 334, Yale University, January 1985.

[Chen85c] M. C. Chen. *Synthesing Systolic Designs*. Technical Report 374, Yale University, March 1985.

[Chen86] M. C. Chen. Synthesizing VLSI architectures: dynamic programming solver. In *International Conference on Parallel Processing*, pp. 776–784, Chicago, IL, August 1986.

[Cheng84] C. K. Cheng and E. S. Kuh. Module placement based on resistive network optimization. *IEEE Transactions on Computer-Aided Design of Integrated Circuits and Systems*, July 1984.

[Cheng86] K. H. Cheng and S. Sahni. A new VLSI system for adaptive recursive filtering. In *Proc. Int. Conf. on Parallel Processing*, pp. 387–389, 1986.

[Chuan85] Henry Y.H. Chunag and C. C. Li. A systolic array processor for straight line detection by modified Hough transform. *Proc. of CAPAIDM'85*, pp. 300–304, 1985.

[Commo71] F. Commoner and A. W. Holt. Marked directed graphs. *Journal of Computer and System Sciences*, pp. 511–523, 1971.

[Coone80] J. T. Coonen. An implementation guide to a proposed standard for floating-point arithmetic. *IEEE Computer Magazine*, pp. 68–79, January 1980.

[Corry83] A. Corry and K. Patel. Architecture of a CMOS correlator. In *Proc. IEEE Int. Symp. on Circuits and Systems*, pp 522–525, 1983.

[Creme76] A. B. Cremers and T. N. Hibbard. The semantic definition of programming languages in terms of their data spaces. *Informatik-Fachberichte*, 1: pp. 1–11, 1976. Springer-Verlag.

[Creme78] A. B. Cremers and T. N. Hibbard. Formal modeling of virtual machines. *IEEE Transactions on Software Eng.*, 4: pp. 426–436, 1978.

[Dadda65] L. Dadda. Some schemes for parallel multipliers. *Alta Frequenza*, 34: pp. 349–356, March 1965.

[Dahlq74] Germund Dahlqist. *Numerical Methods*. Prentice-Hall, 1974.

[Davie86] E. B. Davie, D. G. Higgins, and C. D. Cawthorn. An advanced adaptive antenna test-bed based on a wavefront array processor system. In *Proc. Inter. Workshop on Systolic Arrays, University of Oxford*, July 1986.

[Delos86]   J-M Delosme and I. C. F. Ispen. Efficient systolic arrays for the solution of Toeplitz systems: An illustration of a methodology for the construction of systolic architectures in VLSI. *International Workshop on Systolic Arrays, University of Oxford*, pp. F2, July 1986.

[Denni79]   J.B. Dennis. The varieties of data flow computers. In *Proc. 1st Int'l Conf. Distr. Comp. Sys.*, pp. 430–439, 1979.

[Denni80]   J.B. Dennis. Data flow supercomputers. *IEEE Computer Magazine*, pp. 48–56, November 1980.

[Denye83]   P. B. Denyer and D. Renshaw. Case studies in VLSI signal processing using a silicon compiler. In *Proc. IEEE ICASSP'83*, pp. 939–942, Boston, 1983.

[Denye85]   P. B. Denyer and D. Renshaw. *VLSI Signal Processing - A Bit-Serial Approach.* Addison-Wesley, 1985.

[Denye86]   P. B. Denyer. System compilers. In *Proc. IEEE VLSI Signal Processing Workshop*, 1986.

[Depre84]   E. Deprettere, P. Dewilde, and P. Udo. Pipelined CORDIC architecture for fast VLSI filtering and array processing. In *Proc. IEEE ICASSP'84*, pp. 41.A.6.1 – 41.A.6.4, San Diego 1984.

[Dew86]   P. M. Dew, and L. J. Manning. Comparison of systolic and SIMD architecture for computer vision computation. In *Proc. Inter. Workshop on Systolic Arrays, University of Oxford*, July 1986.

[Dewil85]   P. Dewilde, E. Deprettere, and R. Nouta. Parallel and pipelined implementation of signal processing algorithms. In *VLSI and Modern Signal Processing*, S. Y. Kung et al, editors, pp. 257, Prentice Hall, 1985.

[Dewil83]   P. Dewilde. Personal Communication, 1983.

[Dijks68]   E. W. Dijkstra. Cooperating sequential processes. In *Programming Languages*, F. Genuys, editor, pp. 43–112, Academic Press, 1968.

[Dolec84]   Q. E. Dolecek. Parallel processing systems for VHSIC. Technical Report APL 84-112 1984, The Johns Hopkins University, Also appears in VHSIC Applications Workshop, 1984.

[Dunca72]   D. B. Duncan and S. D. Horn. Linear dynamic recursive estimation from the viewpoint of regression analysis. *J. of ASA*, pp. 815–821, 1972.

[Dunlo85]   A.E. Dunlop and B.W. Kernighan. A procedure for placement of standard-cell VLSI circuits. *IEEE Transactions on Computer-Aided Design*, January 1985.

[Eichm85]   G. Eichmann and H. J. Gaulfield. Optical learning (inference) machines. *Applied Optics*, Vol. 24: 1985.

[Enslo77]   P. H. Enslow. Multiprocessor organization. *ACM Computing Surveys*, 9:103–129, March 1977.

[Fahlm87]   S. E. Fahlman and G. E. Hinton. Connectionist architectures for artificial intelligence. *IEEE Computer Magazine*, January 1987.

[Fandr85]   J. Fandrianto and B. Y. Woo. VLSI floating-point processors. In *Proc. Computer Arithmetic Conference*, pp. 93–100, 1985.

[Farha85]   N. H. Farhat, D. Psaltis, A. Prata and E. Paek. Optical implementation of the Hopfield model. *Applied Optics*, Vol. 24: pp. 1469– 1475, May 1985.

[Farle54]   B. G. Farley and W. A. Clark. Simulation of self-organization systems by digital computer. *IRE Transactions on Information Theory*, IT-4, 1954.

[Feng81]   T. Y. Feng A survey of interconnection networks. In *IEEE Computer*, pp. 12–27, December 1981.

[Feldm84]   J.A. Feldman, S.L. Garverick, F.M. Rhodes, and J.R. Mann. A wafer scale integration systolic processor for connected word recognition. In *Proc. IEEE ICASSP'84*, San Diego, pp. 25B.4.1–25B.4.4, 1984.

[Fettw76]   A. Fettweis. Realizability of digital filter networks. *AEU*, Band 30(Heft 2): pp. 90–96, 1976.

[Fishe82]   A. L. Fisher. Systolic algorithms for running order statistics in signal and image processing. *Journal of Digital Systems*, 4: pp. 251–264, 1982.

[Fishe83]   A.T. Fisher, et al. Design of the PSC: A programmable systolic chip. In R. Bryant, editor, *Third Caltech Conf. on VLSI*, Computer Science Press, 1983.

[Fish85a]   A. L. Fisher and H.T. Kung. Synchronizing large VLSI arrays. *IEEE Transactions on Computers*, pp. 734–740, August 1985.

[Fish85b]   A.L. Fisher and H.T. Kung. Special-purpose VLSI architectures: General discussions and a case study. In *VLSI and Modern Signal Processing*, pp. 153–169, Prentice-Hall, Inc., 1985.

[Floyd62]   R. W. Floyd. Algorithm 97: Shortest path. *Comm. ACM*, Vol. 5(6), June 1962.

[Flynn66]   M. J. Flynn. Very high speed computing systems. *Proceedings of the IEEE*, Vol. 54: pp. 1901–1909, 1966.

[Fort85a]   J. A. B. Fortes, K. S. Fu, and B. W. Wah. Systematic approaches to the design of algorithmic specified systolic arrays. In *Proc. IEEE ICASSP'85*, pp. 300–303, Tampa, Florida, March 1985.

[Fort85b]   J. A. B. Fortes and C. S. Raghavendra. Gracefully degradable processor arrays. *IEEE Transactions on Computers*, pp. 1033–1044, November 1985.

[Forte87]   J. Fortes. Algorithm Reconfiguration Techniques for Gracefully Degradable Processor Arrays. *Systolic Array*, eds. W. Moore, et al., Adam Hilger, 1987.

[Fouse79]   S. D. Fouse. et al. A VLSI architecture for pattern recognition using residue arithmetic. In *Proc. 6th Int. Conf. VLSI Architecture, Design and Fabrication*, pp. 65–90, Caltech, Pasadena, CA, 1979.

[Fouta85]   T. J. Fountain. A review of SIMD architectures. In J. Killter and M. J. B. Duff, editors, *Image Processing System Architectures*, John-Wiley & Sons Inc., 1985.

[Frank82]   M. Franklin and D. Wann. Asynchronous and clocked control structures for VLSI based interconnection networks. In *The 9-th Annual Symposium on Comput. Architecture*, April 1982. Austin, TX.

[Frenk86]   K. A. Frenkel. Evaluating two massively parallel machines. *Communications of the ACM*, August 1986.

[Fried83]   B. Friedlander, and J. Newkirk. A comparsion of two SAR processsing architectures for VLSI implementation. In *Proc. IEEE ICASSP'83*, Boston, pp: 919–922, 1983.

[Fujit84]   Fujitsu. DSP MB8764 hardware manual. 1984.

[Gajsk83]   D. Gajski, D. Kuck, D. Lawrie, A. Sameh. Cedar. *Technical Report, Department of Computer Science, University of Illinois, Urbana*, Feburary, 1983.

[Gajsk85]   D. D. Gajski. Silicon compilation. *VLSI System Design*, pp. 48–64, November 1985.

[Galla86]   N.C. Gallagher S.S.H. Naqvi and E.J. Coyle. An application of median filters to digital television. In *Proc. IEEE ICASSP'86*, Tokyo, Japan, pp. 2451–2454, 1986.

[Gapp84]   R.H. Davis and D. Thomas. Systolic array chip matches the pace of high-speed processing. *Electronic design*, pp. 207–218, October 1984.

[Gapp85]   NCR Company. NCR GAPP PC Development System. *NCR Documents No. NCR45GDS1*

[Geman84]   S. Geman and D. Geman. Stochastic relaxation, Gibbs distributions, and the Bayesian restoration of images. *IEEE Transactions on Pattern Analysis and Machine Intelligence*, Vol. 6: pp. 721–741, November 1984.

[Giord85]   Arthur A. Giordano and Frank M. Hsu. *Least Square Estimation with Applications to Digital Signal Processing*. Needham, Massachusetts, 1985.

[Gondr84]   M. Gondran, M. Minoux, and S. Vajda. *Graphs and Algorithms*. John Wiley & Sons, 1984.

[Golub83]   G.H. Golub. *Matrix Computation*. Johns Hopkins University Press, 1983.

[Goodm68]   J.W. Goodman. *Introduction to Fourier Optics*. McGraw Hill, N.Y., 1984.

[Goodm84]   J.W. Goodman, F.J. Leonberger, S.Y. Kung, and R.A. Athale. Optical interconnections for VLSI systems. *Proc. IEEE*, pp. 850–866, July 1984.

[Goto81]   S. Goto. An efficient algorithm for the two-dimensional placement problem in electrical circuit layout. *IEEE Transactions on Circuit and Systems*, January 1981.

[Gray84]   R.M. Gray. Vector quantization. *IEEE ASSP Magazine*, Vol. 2 (No. 2): pp. 4–29, April 1984.

[Griff82]   L.J. Griffiths and C.W. Jim. An alternative approach to linearly constrained adaptive beamforming. *IEEE Transactions on AP*, AP-30: pp. 27–34, January 1982.

[Grosc79]   G.E. Grosch. Performance analysis of Poisson solver on array computers. *Supercomputers, edited by C.R. Jesshope and R.W. Hockney*, Vol. 2: pp. 149–181, 1979.

[Guiba79]   L.J. Guibas, H.T. Kung, and C.D. Thompson. Direct VLSI implementation of combinatorial algorithms. In *Proc. Caltech Conference on VLSI*, L.A., 1979.

[Gusta86]   J. L. Gustafson, S. Hawkinson, and K. Scott. The architecture of a homogeneous vector supercomputer. *Journal of Parallel and Distributed Processing*, 1986.

[Hajek85]   B. Hajek. A tutorial survey of theory and applications of simulated annealing. In *Proc. of 24th Conference on Decision and Control*, pp. 755–760, 1985.

[Hayes78]   J.P. Hayes. *Computer Architecture and Organization*. McGraw Hill, N.Y., 1978.

[Hedlu82]   K.S. Hedlund and L. Snyder. Wafer scale integration of Configurable Highly Parallel (CHiP) processors. In *Conference on Parallel Processing*, pp. 262–264, 1982.

[Hedlu85]   K.S. Hedlund. WASP - a Wafer-Scale Systolic Processor. In *ICCD*, pp. 665-671, 1985.

[Helle84]   D. Heller. Partitioning big matrices for small systolic arrays. In *VLSI and Modern Signal Processing*, S. Y. Kung, et al, editors, Prentice Hall, 1984.

[Henne84]   J. F. Hennessy. VLSI processor architecture. *IEEE Transactions on Computers*, C-33(12): pp. 1221–1246, December 1984.

[Hilli85]   W. D. Hillis. *The Connection Machine*. Massachusetts Institute Technology Press, 1985.

[Hilli85]   W. D. Hillis. *The Connection Machine*. Massachusetts Institute Technology Press, 1985.

[Hoare78]   C. A. R. Hoare. Communicating sequential processes. *Communications of the ACM*, 21: pp. 666–677, 1978.

[Hockn83]   R.W. Hockney and C.R. Jesshope. *Parallel Computers*. Adam Hilger Ltd., Bristol, U.K., 1983.

[Hopfi82]   J. J. Hopfield. Neural network and physical systems with emergent collective computational abilities. In *Proc. Natl.. Acad. SCi. USA*, 1982.

[Hopfi84]   J. J. Hopfield. Neurons with graded response have collective computational properties like those of two-state neurons. In *Proc. Natl.. Acad. SCi. USA*, pp. 3088–3092, 1984.

[Hopfi85]   J. J. Hopfield and D. W. Tank. Neural computation of decision in optimization problems. *Biological Cybernetics*, 1985.

[Horii86]   S. Horii, M. Mubo, C. Ohara, K. Horiguchi, Y. Kuniyasu, E. Osaki, and Y. Ohshima. 32-bit image processor T9506 and its applications. In *Workshop on VLSI Signal Processing*, IEEE Press, Los Angeles, CA, 1986.

[Horik87]   S. Horiike, S. Nishida, and T. Sakaguchi. A design method of systolic arrays under the constraint of the number of the processors. In *Proc. of ICASSP 87*, pp. 764-767, 1987.

[Horni73]   J. J. Horning and B. Randell. Process structuring. *Comp. Surveys*, 5: pp. 5–30, 1973.

[Hsu83]   C.P. Hsu. General River Routing Algorithm. In *Proc. ACM/IEEE 20th Design Automation Conference*, 1983.

[HTKun78]   H.T. Kung and C.E. Leiserson. Systolic arrays (for VLSI). In *Sparse Matrix Symposium*, pp. 256–282, SIAM, 1978.

[HTKun81]   H. T. Kung and W. M. Gentleman. Matrix triangularization by systolic arrays. *Proc. SPIE, Real Time Signal Processing*, 1983.

[HTKun82]   H.T. Kung. Why systolic architectures? *IEEE, Computer*, 15(1), Jan 1982.

[HTKun83]   H.T. Kung and M. S. Lam. Fault-tolerant VLSI systolic arrays and two-level pipelining. In *Proc. SPIE, Real Time Signal Processing VI*, San Diego, CA, August 1983.

[HTKun84]   H. T. Kung and M. S. Lam. Wafer-scale integration and two-level pipelined implementations of systolic arrays. *J. Parallel and Distributed Computing*, 32–63, 1984.

[HTKun85]   H. T. Kung and J. A. Webb. Global operations on a systolic array machine. In *Proc. IEEE ICCD'85*, Oct 1985.

[HTKun86]   H. T. Kung. Memory requirements for balanced computer architectures. In *Proc. 13th Annual Int. Symp. on Computer Architecture*, pp. 49–54, June 1986.

[Hu82]   T. C. Hu. *Combinatorial Algorithms*. Addison-Wesley Publishing Company, 1982.

[Hu85]   T.C. Hu and E.S. Kuh. *Theory and Concepts of Circuit Layout*. IEEE Press, 1985.

[Huang84]   A. Huang. Architectural considerations involved in the design of an optical digital computer. *Proc. IEEE*, pp. 780–786, July 1984.

[Hud86]   P. Hudak. Parafunctional programming. *IEEE Computer Magazine*, August 1986.

[Hwang79]   K. Hwang. *Computer Arithmetic, principles, architecture, and design*. John Wiley and Sons, Inc., 1979.

[Hwang82]   K. Hwang and Y.H. Cheng. Partitioned matrix algorithms for

VLSI arithmetic systems. *IEEE Transactions on Computers*, C-31(12): pp. 1215–1224, December 1982.

[Hwan84a]   K. Hwang and F. Briggs. *Computer Architectures and Parallel Processing*. McGraw Hill, 1984.

[Hwang86]   K. Hwang and Joydeep Ghosh. *Supercomputer and Artificial Intelligence Machines*. Technical Report CRI-86-02, University of Southern California, February 1986.

[Hwang87]   K. Hwang and Joydeep Ghosh. *Supercomputer and Artificial Intelligence Machines*. Technical Report CRI-87-03, University of Southern California, January 1987.

[Itaku75]   F. Itakura. Minimum prediction residual principal applied to speech recognition. *IEEE Trans on ASSP*, pp. 67–72, February 1975.

[Jagad83]   J. V. Jagadish, T. Kailath, G. G. Mathews, and J. A. Newkirk. On pipelining systolic arrays. In *17th Asilomar conf. on Circuits, Systems, and Computers,*, November 1983. Pacific Grove, CA.; also On hardware description from block diagrams, In *Proc. IEEE ICASSP'84*, San Diego, CA., March 1984.

[Jain85]   R. Jain. *Image Processing*. In *Modern Signal Processing*, edited by T. Kailath, 1985.

[Jain86]   R. Jain et. al. Custom design of a VLSI PCM-FDM transmultiplexer from system specification to circuit layout using a computer-aided design system. *IEEE JSSC*, February 1986.

[Jenki87]   K. Jenkins and S. Sawchuck Personal communication, 1987.

[Jenq86]   Y. C. Jenq. 1986. Private Communication.

[Jessh85]   C. Jesshope and W. Moore. *Wafer Scale Integration*. Adam Hilger, Bristol and Boston, 1985.

[Jou86]   J. Y. Jou and J. A. Abraham. Fault-tolerant matrix arithmetic and signal processing on highly concurrent computing structures. *Proc. IEEE*, pp. 732–741, May 1986.

[Jover86]   J. M. Jover and T. Kailath. A parallel architecture for Kalman filter measurement update and parameter estimation. *Automatica*, 22: pp. 43–57, 1986.

[Jutan84]   F. Jutand, N. Demassieux, D. Vicard, and G. Chollet. VLSI architecture for DTW using systolic arrays. In *Proc. IEEE ICASSP'84*, pp. 34A.5.1– 34A.5.4, 1984.

[Kaila74]   T. Kailath. A view of three decades of linear filtering theory. *IEEE Transactions on Information Theory*, IT-20(2): pp. 145–181, Mar 1974.

[Kaila80]   T. Kailath. *Linear System*. Prentice-Hall, Inc., Englewood
              Cliffs, NJ, 1980.

[Kalma60]   R. E. Kalman. A new approach to linear filtering and predic-
              tion problems. *J. Basic Engineering*, 82: pp 35–45, 1960.

[Kalso85]   S. Kalson and K. Yao. A systolic array for linearly constrained
              least squares filtering. In *Proc. IEEE ICASSP'85*, pp. 977–
              980, 1985.

[Karp66]   Richard M. Karp and Raymond E. Miller. Properties of a
              model for parallel computations: determinancy, termination,
              queueing. *SIAM Jour. Applied Math.*, pp. 1390–1411, Novem-
              ber 1966.

[Karp67]   R. M. Karp, R. E. Miller, and S. Winograd. The organization
              of computations for uniform recurrence equations. *Journal of
              ACM*, 14(3): pp. 563–590, July 1967.

[Kay81]   S.M. Kay and S. L. Jr. Marple. Spectrum analysis - a modern
              perspective. *IEEE Proc.*, 69(11): pp. 1380–1498, Nov 1981.

[Keyes79]   R.W. Keyes. The evolution of digital electronics towards
              VLSI. *IEEE Transactions on Electron Devices*, ED-26(4):
              pp. 271–278, 1979.

[KHHua84]   Huang K. H. and Abraham J. A. Algorithm-based fault-
              tolerance for matrix operations. *IEEE Transactions on Com-
              puters*, pp. 518–528, June 1984.

[Kime86]   C. R. Kime. System diagnosis. In D. K. Pradhan, edi-
              tor, *Fault-Tolerant Computing, Theory and Techniques Vol.2*,
              Chapter 8, Prentice-Hall, 1986.

[Kirkp83]   S. Kirkpatrick, C.D. Gelatt Jr., and M.P. Vecci. Optimization
              by simulated annealing. *Science*, May 1983.

[Kisha86]   K. Jainandunsing. Optimal partitioning schemes for wave-
              front/systolic array processors. Technical Report April 28,
              1986, Delft University of Technology, Delft, The Netherlands,
              1986.

[Koch86]   C. Koch, J. Marroquin and A. Yuille. Analog "neuronal"
              networks in early vision. *Proc. of National Academy Science*,
              Vol. 83: pp. 4263–4267, 1986.

[Kohoe72]   T. Kohonen. Correlation matrix memories. *IEEE Transac-
              tions on Computer*, C-21, 1972.

[Koren81]   I. Koren. A reconfigurable and fault-tolerant multiprocessor
              array. In *8th Annual Symposium on Computer Architectures*,
              pp 425–442, 1981.

[Kuck80]   D. J. Kuck, R. H. Kuhn, B. Leasure, and M. Wolfe. The Structure of an Advanced Retargetable Vectorizer. In Proc. *COMPSAC'80*, 1980.

[Kung81]   S.Y. Kung. A Toeplitz approximation method and some applications. In Proc. *Int. Symp. Math. Theory Networks Syst.*, Santa Monica, Aug., 1981.

[Kung82a]   S.Y. Kung, K.S. Arun, R.J. Gal-Ezer, and D.V. Bhaskar Rao. Wavefront array processor: language, architecture, and applications. *IEEE Transactions on Computers, Special Issue on Parallel and Distributed Computers*, C-31(11): pp. 1054–1066, Nov 1982.

[Kung82b]   S. Y. Kung and R. J. Gal-Ezer. Synchronous vs. asynchronous computation in VLSI array processors. In *Proceedings, SPIE Conference*, 1982. Arlington, VA.

[Kung83a]   S. Y. Kung and R. J. Gal-Ezer. Eigenvalue, singular value and least square solvers via the wavefront array processor. In *Algorithmically Specialized Computer Organizations*, Academic Press, 1983.

[Kung83b]   S.Y. Kung and Y.H. Hu. A highly concurrent algorithm and pipelined architecture for solving Toeplitz systems. *IEEE Transactions on ASSP*, ASSP-31(No.1, pp.66-76), Feb. 1983.

[Kung83c]   S.Y. Kung. From transversal filter to VLSI wavefront array. In *Proc. Int. Conf. on VLSI 1983*, IFIP, Trondheim, Norway, 1983.

[Kung83d]   S. Y. Kung, K. S. Arun, and D. V. B. Rao. State space and SVD based approximation methods for the harmonic retrieval problem. *J. Optical Soc. of America*, 73: pp 1799–1811, Dec., 1983.

[Kung83d]   S.Y. Kung, D.V. Bhaskar Rao, and K.S. Arun. New state space and SVD based approximate modeling methods for harmonic retrieval. In *Spectral Estimation II*, pp. 266–271, Acoustic, Speech and Signal Processing, IEEE, Tampa, Fl, November 1983.

[Kung84a]   S.Y. Kung. On supercomputing with systolic/wavefront array processors. *Invited paper, Proceedings of the IEEE*, Vol. 72(7), July 1984.

[Kung84b]   S. Y. Kung, J. Annevelink, and P. Dewilde. Hierarchical iterative flow-graph design for VLSI array processors. In *IEEE Workshop on VLSI Signal Processing*, L.A., Nov 1984.

[Kung85]   S.Y. Kung, J. Annevelink, P. Dewilde, and S. C. Lo. Hierar-
             chical iterative flow-graph design for VLSI array processors.
             In *IEEE ICASSP 85*, Tampa, Florida, March 1985.

[Kung85a]   S. Y. Kung, D. V. Bhaskar, and K. S. Arun. Spectral Es-
             timation: From Conventional Methods to High Resolution
             Modelling Methods. In *VLSI and Modern Signal Processing*,
             S.Y. Kung et al. (eds.), Prentice-Hall, Inc, 1985.

[Kung86a]   S. Y. Kung, P. S. Lewis, and S. C. Lo. On optimally mapping
             algorithms to systolic arrays with application to the transitive
             closure problem. In *Proc. 1986 IEEE Int. Sym. on Circuits
             and Systems*, pp 1316–1322, 1986.

[Kung86b]   S. Y. Kung, S. C. Lo, and P. S. Lewis. Timing analysis and op-
             timization of VLSI data flow arrays. In *Proc.ICPP'86*, 1986.

[Kung86c]   S. Y. Kung, J. N. Hwang, and S. C. Lo. Mapping digital signal
             processing algorithms onto VLSI systolic/wavefront arrays.
             In *Proc. 12th Annual Asilomar Conf. on Signals, Systems
             and Computers*, pp. 6–12, November 1986.

[Kung86d]   S.Y. Kung, C.W. Chang, and C.W. Jen. Real-time reconfig-
             uration for fault-tolerant VLSI array processors. Proc. Real-
             Time Systems Symposium, pp. 46 – 54, 1986.

[Kung87a]   S. Y. Kung, P. S. Lewis, and S. N. Jean. Canonic and gen-
             eralized mapping from algorithms to arrays - a graph based
             methodology. In *Proc. of the Hawaii Inter. Conf. on System
             Sciences*, Vol. 1, pp. 124 – 133, January, 1987.

[Kung87b]   S. Y. Kung and J. N. Hwang. Systolic design of electronic
             neural networks. Sumitted to *26th IEEE Conf. on Decision
             and Control*, Los Angeles, December, 1987.

[Kung87c]   S. Y. Kung and J. N. Hwang. A new systolic array for Kalman
             Filtering. Sumitted to *IEEE Trans. on Acoustics, Speech, and
             Signal Processing*, 1987.

[Kuo85]   C. C. Kuo, B. C. Levy, B. R. Musicus. A local relaxation
             method for solving elliptic PDEs on mesh-connected arrays.
             SIAM, J. Sci. Stat. Comp., July, 1987.

[Kumar85]   V. K. P. Kumar and C. S Raghavendra. Array processor with
             multiple broadcasting. In *12th Symp. Computer Architecture*,
             pp. 2–10, 1985.

[Lawri75]   D. H. Lawrie. Access and alignment of data in an array com-
             puter. *IEEE Trans. on Computer*, C-24: pp. 1145–1155, 1975.

[Lehma77]   D. H. Lehmann. Algebraic structures for transitive closure. *Theoretical Computer Science*, (4): pp. 59–76, 1977.

[Leise81]   C.E. Leiserson. *Area-Efficient VLSI Computation*. PhD thesis, Carnegie-Mellon University, Pittsburgh, Penn., October 1981.

[Leis83a]   C.E. Leiserson, F.M. Rose, and J.B. Saxe. Optimizing synchronous circuitry by retiming. In *Proceedings, Caltech VLSI Conference*, Pasadena, CA, 1983.

[Leis83b]   C.E. Leiserson and J.B. Saxe. Optimizing synchronous systems. *J. of VLSI and Computer Systems*, 1(1): pp. 41–67, 1983.

[Leise85]   C. E. Leiserson. Fat-trees: universal networks for hardware efficient supercomputing. In *Proc. IEEE Int. Conf. on Parallel Processing*, pp 393–402, 1985.

[Lenfa78]   J. Lenfant. Parallel permutations of data: A Benes network control algorithm for frequently used permutations. *IEEE Trans. on Computer*, C-27: pp. 637-647, July, 1978.

[Levin47]   N. Levinson. The Wiener RMS (root-mean-square) error criterion in filter design and prediction. *J. Math. Phys.*, 25: pp. 261–278, Jan 1947.

[Lev-A83]   H Lev-Ari. *Modular computing networks: a new methodolgy for analysis and design of parallel algorithms/architectures*. Internal Report, Technical Memo, ISI-29, Integrated Systems Inc., Palo Alto, Ca., 1983.

[Lewis86]   P. S. Lewis and S. Y. Kung. Dependence graph based design of systolic arrays for the Algebraic Path Problem. In *Proc. 12th Annual Asilomar Conf. on Signals, Systems and Computers*, November 1986.

[Li85]   G. J. Li and B. W. Wah. The design of optimal systolic arrays. *IEEE Transactions on Computers*, Vol. C-34(1), January 1985.

[Lin85]   F.C. Lin and Wu. Systolic arrays for transitive closure algorithms. *International Symposium on VLSI Systems and Designs, Taipei*, May 1985.

[Ling86]   F. Ling, D. Manolakis, and J.G. Proakis. A flexible, numerically robust array processing algorithm and its relationship to the Givens transform. In *Proc. IEEE ICASSP'86*, Tokyo, Japan, pp. 2127–2130, 1986.

[Lippm87]   R. P. Lippmann. An introduction to computing with neural nets. *IEEE ASSP magazine*, 4: pp 4–22, April, 1987.

[Mada85]   H. Mada. Architecture for optical computing using holographic associative memories. *Applied Optics*, Vol. 24: 1985.

[Mahr84]   B. Mahr. Iteration and summability in semirings. *Annals of Discrete Mathematics 19*, Dec 1984.

[Marke76]   J. D. Markel and A. H. Gray. *Linear Prediction of Speech.* Spring-Verlag, 1976.

[Marpl85]   S. L. Jr. Marple. *Spectral Estimation and its Applications.* In *Modern Signal Processing*, edited by T. Kailath, 1985.

[Marr80]   D. Marr and E. Hildreth. Theory of edge detection. *Proceedings of Royal Society of London*, 207: pp. 187–217, 1980.

[Marri85]   K. E. Marrin. VLSI and software move DSP techniques into mainstream. *Computer Design*, pp. 69–87, September 1985.

[May84]   M. D. May and R. J. B. Taylor. Occam - An overview. *Microprocessor & Microsystems*, pp. 73–39, August 1984.

[McCan82]   J.V. McCanny and J. G. McWhirter. Implementation of signal processing functions using 1-bit systolic arrays. *Electron. Letts.*, (18): pp. 241–243, 1982.

[McCan84]   J. V. McCanny, K. W. Wood, J. G. McWhirter, and C. J. Oliver. The relationship between word and bit level systolic arrays as applied to matrix and matrix multipication. In *Proc. SPIE, Tech. Symp. Real Time Signal Processing VI*, 1983.

[McCan86]   J. V. McCanny, , and J. G. McWhirter. The derivation and utilization of bit level systolic array architectures. In *Proc. of the International Workshop on Systolic Arrays, University of Oxford*, pp. F1.1–F1.12, July 1986.

[McWhi83]   J.G. McWhirter. Recursive least-squares minimization using a systolic array. In *SPIE, In Proc. Real Time Signal Processing VI*, pp 105–110, SPIE, 1983.

[McWhi86]   P.J. Hargrave C.R. Ward and J. McWhirter. A novel algorithm and architecture for adaptive digital beamforming. *IEEE Transactions on Antennas and Propagation*, pp. 338–346, March 1986.

[Mead80]   C. Mead and L. Conway. *Introduction to VLSI Systems.* Addison-Wesley, 1980.

[Meind86]   J. Meindl. Informal discussion: microprocessors in the year 2001. In *Proc. Int. Solid-State Circuits Conference*, 1986.

[Melhe86]   R. G. Melhem. Irregular wavefronts in data-driven data-dependent computations. *Technical Report, University of Pittsburgh*, June 1986.

[Miran84]   W. L. Miranker. Space-time representations of computational structures. *Computing*, 1984.

[Mital76]   K. V. Mital. *Optimization Methods*. John Wiley and Sons, New York, 1976.

[MJChe86]   M. J. Chen and K. Yao. On realization of least-squares estimation and Kalman filtering by systolic arrays. In *Proc. of Intl. Workshop on Systolic Arrays*, Oxford, July 1986.

[MJChe87]   M. J. Chen and K. Yao. On realization and implementation of Kalman filtering by systolic array. In *Proc. of John Hopkins Workshop*, 1987.

[Moldo83]   D. I. Moldovan. On the design of algorithms for VLSI systolic arrays. *Proceedings of the IEEE*, Vol. 71(1), January 1983.

[Moldo86]   D. I. Moldovan and J. A. B. Fortes. Partitioning and mapping of algorithms into fixed size systolic arrays. *IEEE Transactions on Computers*, 35(1): pp. 1–12, January 1986.

[Monzi80]   R. A. Monzingo and T. W. Miller. *Introduction to Adaptive Arrays*. Wiley-Interscience, 1980.

[Moore84]   W.C. Moore and K. Steiglitz. Efficiency of parallel processing in the solution of Laplace equation. In *Fifth IMACS International Symposium on Computer Methods for Partial Differential Equations*, 1984.

[Moore86]   W. R. Moore. A review of fault-tolerant techniques for the enhancement of integrated circuit yield. *Proc. IEEE*, pp. 684–698, May 1986.

[Moore87]   W. Moore and A. McCabe and R. Urquhart. *Systolic Arrays*. Adam Hilger, 1987.

[Morto84]   S. Morton. A fault tolerant bit parallel cellular array processor. Reprint, ITT advanced technology center, 1984.

[Murat77]   T. Murata and K. Onaga. Deadlocks in capacitated computation networks. In *Proc. 12th Midwest Symp. on Circuits and Systems*, pp 99–103, August 1977.

[Murog82]   S. Muroga. *VLSI System Design*. John Wiley and Sons, 1982.

[Nakan72]   K. Nakano. Associatron - a model of associative memory. *IEEE Transactions on SMC*, Vol. SMC-2, 1972.

[Nash83]   J. G. Nash, and S. Hansen. Modified Faddeev algorithm for matrix manipulation. In *Proc. SPIE Conf.*, San Diego, Aug., 1983.

[Nash86]   J. G. Nash, K. W. Przytula, and S. Hansen. Systolic/celluar processor for linear algebraic operations. *IEEE Workshop on VLSI Signal Processing*, L.A., Nov., 1986.

[Nevat82]   R. Nevatia. *Machine Perception*. Prentice-Hall, Inc., 1982.

[NEC85]   NEC Electronics. uPD7281 User's Guide. 1985.

[Nicol85]   D. Nicolas, J. Francis, S. Marc, and D. Michel. VLSI architecture for a one chip video median filter. In *Proc. IEEE ICASSP'85*, Tampa, FL., pp. 26.7.1–26.7.4, 1985.

[Nudd85]   G. R. Nudd and J. G. Nash. Application of concurrent VLSI systems to two-dimensional signal processing. In *VLSI and Modern Signal Processing*, Chapter 17, pp 307–325, Prentice Hall, 1985.

[O'Don85]   M. J. O'Donnell. *Equational Logic as a Programming Language*. MIT Press, 1985.

[Oflaz83]   K. Oflazer. Design and implementation of a single chip 1-D median filter. *IEEE Transactions on ASSP*, October 1983.

[Offen85]   I. Offen and R.J. Raymond Jr. *VLSI Image Processing*. McGraw-Hill, 1985.

[O'Kee86]   M. T. O'Keefe and J. A. B. Fortes. A comparative study of two systematic design methodologies for systolic arrays. In *International Conference on Parallel Processing*, pp 672–675, Chicago, IL, August 1986.

[Onaga86]   K. Onaga and T. Takechi. On design of rotary array communication and wavefront-driven algorithms for solving large-scale band-limited matrix equations. Technical Report of Computer Science group 1986, Hiroshima University, Japan, 1986.

[Oppen75]   A. Oppenheim and R. Schafer. *Digital Signal Processing*. Prentice-Hall, Inc., Englewood Cliffs, NJ, 1975.

[Oppen78]   A. Oppenheim. *Applications of Digital Signal Processing*. Prentice Hall, N.J., 1978.

[Ore62]   O. Ore. *Theory of Graphs*. American Mathematical Society, Providence, Rhode Island, 1962.

[Paige77]   C. C. Paige and M. A. Saunders. Least squares estimation of discrete linear dynamic systems using orthogonal transformation. *SIAM J. Numer. Anal.*, 14: pp. 180–193, 1977.

[Papad83]   C. H. Papadimitriou and K. Steiglitz. *Combinatorial Optimization: Algorithms and Complexity.* Prentice-Hall, 1983.

[Parke85]   I.N. Parker. VLSI architecture. In *VLSI Image Processing*, R. J. Offen, editor, Chapter 3, pp. 99–127, 1985.

[Parle80]   B. N. Parlett. *The Symmetric Eigenvalue Problem.* Prentice-Hall, Inc., N.J., 1980.

[Parna72]   D. L. Parnas. A technique for software module specification with examples. *Communications of the ACM*, 15: pp. 330–336, 1972.

[Patte80]   D.A. Patterson and D.R. Ditzel. The Case for the Reduced Instruction Set Computer. *Computer Architecture News*, 8: pp. 25–33, October 1980.

[Patte81]   D.A. Patterson and C.H. Sequin. A VLSI RISC. *Computer, IEEE*, 14, September 1981.

[Peter81]   J. L. Peterson. *Petri Net Theory and The Modeling of Systems.* Prentice-Hall, Englewood Cliffs, N.J.07632, 1981.

[Pezar71]   S.D. Pezaris. A 40 ns 17-bit-by-bit array multiplier. *IEEE Transactions on Computers*, C-20(4): pp. 442–447, April 1971.

[Porte86]   W. A. Porter, and J. L. Aravena. Array architectures for estimation and control applications: An introduction and overview. Sumitted to *25th IEEE Conf. on Decision and Control*, Athens, 1986.

[Pount86]   D. Pountain. *A Tutorial Introduction to Occam Programming.* INMOS, 1986.

[Pratt78]   W. K. Pratt. *Digital Image Processing.* Wiley, 1978.

[Prepa84]   F. P. Preparata. VLSI algorithms and architectures. In M. P. Chytil and V. Koubek, editors, *Lecture Notes in Computer Science 176*, Springer-Verlag, 1984.

[Psalt84]   D. Psaltis. Two-dimensional optical processing using one-dimensional input devices. *Proc. IEEE*, pp. 962–974, July 1984.

[Psalt85]   D. Psaltis and N. Farhat. Optical information processing based on an associative-memory model of neural nets with thresholding and feedback. *Optics Letters*, 10: pp. 98–99, Jan., 1985.

[Quint84]   P. Quinton. Automatic synthesis of systolic arrays from Uni-
              form Recurrent Equations. In *Proceedings of 11th Annual
              Symposium on Computer Architecture*, pp. 208–214, 1984.

[Rabae86]   J.M. Rabaey and R. W. Broderson. Experiences with auto-
              matic generation of audio band digital signal processing cir-
              cuits. In *Proc. IEEE ICASSP'86*, 1986.

[Rabin75]   L. R. Rabiner and B. Gold. *Theory and Application of Digital
              Signal Processing*. Prentice-Hall, Inc., Englewood Cliffs, N.J.,
              1975.

[Rabin78]   L.R. Rabiner and R.W. Schafer. *Digital Processing of Speech
              Signals*. Prentice-Hall Inc., 1978.

[Rabin81]   L. R. Rabiner and S. E. Levinson. Isolated and connected
              word recognition – theory and selected application. *IEEE
              Transactions on Communication*, Vol. Com-29(No. 5): pp
              621–658, May 1981.

[Radin82]   G. Radin. The 801 minicomputer. In *Proc. Symp. Archi-
              tectural Support for Programming Languages and Operating
              Systems*, pp. 39–47, Ass. Comput. Mach., Palo Alto, CA.,
              March 1982.

[Raffe85]   J. I. Raffel, A. H. Anderson, et al. A wafer-scale digital in-
              tegrator using restructurable VLSI. *IEEE Journal of Solid-
              State Circuits*, Feburary 1985.

[Ragha83]   R. Raghavan and S. Sahni. Single row routing. *IEEE Trans-
              actions on Computers*, March 1983.

[Ramam80]   C. V. Ramamoorthy and S. Gary. Performance evaluation
              of asynchronous concurrent systems using Petri Nets. *IEEE
              Trans. Software Eng.*, pp. 440–449, September 1980.

[Ramau83]   B. Ramamurthi and A. Gersho. Image coding using seg-
              mented codebooks. *Proc. Intel's Picture Coding Symposium*,
              March 1983.

[Ramch73]   C. Ramchandani. *Analysis of Asynchronous Concurrent Sys-
              tems by Petri Nets*. PhD thesis, Dept. of Electrical Engi-
              neering, Massachusetts Institute of Technology, Cambridge,
              Massachusetts, July 1973.

[Rande82]   B. Randell and P. C. Treleaven. *VLSI Architecture*. Prentice
              Hall, 1982.

[Rao85]   S. K. Rao. *Regular Iterative Algorithms and Their Implema-
              tions on Processor Arrays*. PhD thesis, Stanford University,
              Stanford, California, 1985.

[Reite68]   R. Reiter. Scheduling parallel computations. *Journal of ACM*, pp. 590–599, October 1968.

[Rich83]   Elaine Rich. *Artificial Intelligence.* McGraw-Hill, 1983.

[Rober86]   Y. Robert and D. Trystram. An orthogonal systolic array for the Algebraic Path Problem. In *Inter. Workshop on Systolic Arrays*, 1986.

[Rosbe83]   A. L. Rosenberg. The Diogenes approach to testable fault-tolerant arrays of processors. *IEEE Transactions on Computers*, pp. 902–910, October 1983.

[Rosbl58]   F. Rosenblatt. The perception: A probablistic model for information storage and organization in the brain. *Psych. Rev.*, Vol. 65, 1958.

[Rosen82]   A. Rosenfeld and A.C. Kak. *Digital Picture Processing Vol. I, II.* Academic Press, 1982.

[Rosen85]   A. Rosenfeld. Parallel Algorithms for Image Analysis. In *VLSI and Modern Signal Processing*, S.Y. Kung et al. (eds.), Prentice-Hall, Inc, 1985.

[Rote85]   G. Rote. A systolic array algorithm for the Algebraic Path Problems (shortest paths, matrix inversion). *Computing*, (34): pp. 192–219, 1985.

[Rumel86]   D. E. Rumelhart, J. L. McClelland, and the PDP Research Group. *Parallel Distributed Processing: Exploration in the Microstructure of Cognition.* MIT Press, Cambridge, Massachusetts, 1986.

[Saal86]   Ing. R. Saal. *A linear systolic array for computation matrices.* Technical Report, Institute on Network and Circuit Theory, Technical University of Munich, Germany, 1986.

[Sakoe78]   H. Sakoe and S. Chiba. Dynamic programming optimization for spoken word recognition. *IEEE Transactions on ASSP*, pp. 43–49, August 1978.

[Sami83]   M. G. Sami and R. Stefanelli. Reconfigurable architecture for VLSI processor array. In *National Computer Conference*, pp. 565–577, 1983.

[Sami86]   M. Sami and R. Stefanelli. Reconfigurable architectures for VLSI processing arrays. *Proc. IEEE*, pp. 712–722, May 1986.

[Schmi79]   R. Schmidt. Multiple emitter location and signal parameter estimation. *Proc. RADC Spectral Estimation Workshop*, Rome, pp. 243–258, N.Y., 1979.

[Schus98] A. Schuster. On the investigation of hidden periodicities with application to a supposed 26 day period of meterological phenomena. *Terrest. Magn.*, pp. 13–41, March 1898.

[Seitz84] C. Seitz. Concurrent VLSI architectures. *Invited paper, IEEE Transactions on Computer*, C-33, December 1984.

[Seitz85] C.L. Seitz. The cosmic cube. *Communication of the ACM*, pp. 22–33, January 1985.

[Seitz80] C. Seitz. System timing. Chapter 7 in Introduction to VLSI systems by Mead and Conway, 1980.

[Seitz84] C. Seitz. Concurrent VLSI architectures. *Invited paper, IEEE Transactions on Computer*, C-33, December 1984.

[Shim81] Y.S. Shim and Z.H. Cho. SVD pseudo inversion image reconstruction. *IEEE Transactions on ASSP*, pp. 904–909, August 1981.

[Siege82] L.S. Siegel, H.J. Siegel, and A.E. Feather. Parallel processing approaches to image correlation. *IEEE Transactions on Computers,*, C-31(3): pp. 208–218, March 1982.

[Siewi82] D. P. Siewiorek and R. S. Swarz. *The Theory and Practice of Reliable System Design.* Digital Press, 1982.

[Singl67] R. C. Singleton. A method for computing the fast Fourier transform with auxiliary memory and limited high-speed storages. *IEEE Transactions on Audio Eletroaccoust.*, AU-15: 91–97, June 1967.

[Sked85a] S. Skedzielewski and J. Glauert. IF1, an intermediate form for applicative languages. Technical Manual, Lawerence Livermore Laboratory, 1985.

[Sked85b] S. Skedzielewski and J. Glauert. SISAL reference manual. Technical Manual, Lawerence Livermore Laboratory, 1985.

[Smith76] B. T. Smith et al. *Lecture Notes in Computer Science, Vol. 6, Ed. 2: Matrix Eigensystem Routines, EISPACK Guide.* Springer Verlag, N.Y., 1976.

[Skoln62] M. I. Skolnik. *Introduction to Radar Systems.* McGraw-Hill, N.Y., 1962.

[Snyde82] L. Snyder. Introduction to the configurable, highly parallel computer. *IEEE Computer Magazine*, pp. 47–56, January 1982.

[Soman85] A. K. Somani, V. K. Agarwal, and D. Avis. A generalized theory for system level diagnosis. In *Proc. ICCD'85*, pp. 707–711, 1985.

[Stark81]   A. Stark. An investigation of computerized tomography by direct Fourier inversion and optimum interpolation. *IEEE Transactions on Biomedical Engineering*, 28: pp. 496–505, 1981.

[Stein61]   K. Steinbuch. The learning matrix. *Kybernetik*, 1961.

[Stewa73]   G.W. Stewart. *Introduction to Matrix Computations*. Academic Press, 1973.

[Stone71]   H. S. Stone. Parallel processing with the perfect shuffle. *IEEE Transactions on Computers*, C-20(No.2 , pp.153-161), Feb. 1971.

[Stran80]   G. Strang. *Linear Algebra and Its Applications, Second Edition*. Academic Press, 1980.

[Sung86]   T. Y. Sung and Y. H. Hu. VLSI implementation of real-time Kalman filter. In *Proc. IEEE ICASSP'86*, Tokyo, Japan, pp. 2223–2226, 1986.

[Swart85]   E.E. Swartzlander. High speed FFT processor implementation. Appeared in VLSI Signal Processing edited by P. Cappello et al., IEEE Press, 1985.

[Swart86]   E.E. Swartzlander Jr. *VLSI Signal Processing Systems*. Kluwer Academic Publishers, 1986.

[Szabo67]   N. Szabo and R. Tanaka. *Residue Arithmetic and Its Applications to Computer Technology*. McGraw Hill, 1967.

[Taha82]   Hamdy A. Taha. *Operations Research: An Introduction, Third Edition*. New York: Macmillan, 1982.

[Takah85]   J. Takahashi. A ring processor architecture for highly parallel Dynamic Time Warping. *IEEE Transactions on ASSP*, 1985. Submitted to be published.

[Taked86]   M. Takeda and J.W. Goodman. Neural networks for computation: number representations and programming complexity. *Applied Optics*, Vol. 25: pp. 3033–3046, September 1986.

[Tani78]   Katsuji Tani and Tadao Murata. Scheduling parallel computations with storage constraints. In *Proc. 12th Asilomar Conf. Circuits, Systems, and Computers*, pp. 736–743, November 1978.

[Tank86]   D. W. Tank and J. J. Hopfield. Simple "neural" optimization networks: An A/D converter, signal decision circuit, and a linear programming circuit. *IEEE Trans. on Circuits and Systems* Vol. 33: pp. 533–541, 1986.

[Taylo84]   F. J. Taylor. Residue arithmetic: A tutorial with examples. *IEEE Computer Magazine*, pp. 50–62, May 1984.

[Thomp79]   C. D. Thompson. Area-time complexity for VLSI. *Proc. Eleventh Annual ACM Symposium on the Theory of Computing*, pp. 81–88, 1979.

[Thomp83]   C. D. Thompson. The VLSI complexity of sorting. *IEEE Transactions on Computers*, C-32: pp. 1171–1184, December 1983.

[Ting83]   B.S. Ting and B.N. Tien. Routing techniques for gate array. *IEEE Transactions on Computer-Aided Design of Integrated Circuits and Systems*, October 1983.

[TI83]   Texas Instruments. TMS32010 user's guide. 1983.

[TI85]   Texas Instruments. TMS32020 user's guide. 1985.

[Tseng85]   P. S. Tseng, K. Hwang, and V. K. Prasanna Kumar. A VLSI-based multiprocessor architecture for implementing parallel algorithms. In *International Conference on Parallel Processing*, Chicago, IL, August 1985.

[Uhr86]   L. Uhr. Parallel, hierarchical software/hardware pyramid architectures. Computer Sciences Technical Report #646, Computer Sciences Dept, University of Wisconsin, Madison, 1986.

[Ullma84]   J. D. Ullman. *Computational Aspects of VLSI*. Computer Science Press, 1984.

[Unive86]   Johns Hopkins University. Designing a VHSIC-based signal processor for future anti-radiation homing missiles. Technical Report of Applied Physics Laboratory, April, 1986.

[Urquhar]   R. B. Urquhart and D. Wood. Systolic matrix and vector multiplication methods for signal processing. *IEE Proc. Pt F*, pp. 623–631.

[Venet85]   A. N. Venetsanopoulos. Digital image processing and analysis. In T.S. Durrani and J.L. Lacoume, editors, *Signal Processing*, North Holland, 1985.

[Vlont87]   J. Vlontzos. A wavefront array processor using dataflow processing elements. In *Proc. Intl' Conf. on Supercomputing*, Athens, Greece, June, 1987.

[Volde59]   J. E. Volder. The CORDIC trigonometric computing technique. *IRE Transactions on Electron. Comput.,*, EC-8(3): pp. 330–334, September 1959.

[Walla64]   C. S. Wallace. A suggestion for a fast multiplier. *IEEE Transactions on Electronic Computers*, pp. 14–17, February 1964.

[Walth71]   J. S. Walther. A unified algorithm for elementary functions. In *Proc. AFIPS Conf.*, pp. 379–385, 1971.

[Ward84]   C. R. Ward, A. J. Robson, P. J. Hargrave, and J. G. McWhirter. Application of a systolic array to adaptive beamforming. *IEE Proc.*, 131: pp. 638–645, October 1984.

[Ware84]   F. Ware. Fast 64-bit chip set gangs up for double-precision floating-point work. *Electronics*, pp. 99–103, July 1984.

[Warsh62]   S. Warshall. A theorem on boolean matrices. *J. ACM*, Vol. 9(1), January 1962.

[White86]   J. C. White. et al. A high speed CMOS/SOS implementation of a bit level systolic correlator. In *IEEE ICASSP'86*, pp. 1161–1164, Tokyo, Japan, April 1986.

[White85]   H. J. Whitehouse. *Signal Processing Technology*. In *Modern Signal Processing*, edited by T. Kailath, 1985.

[Wilki65]   J. H. Wilkinson. *The Algebraic Eigenvalue Problem*. Oxford University Press, London, 1965.

[Wils83]   P. Wilson. Occam architecture eases system design. *Computer Design*, November, 1983.

[Wils84a]   P. Wilson. Thirty-two bit micro supports multiprocessing. *Computer Design*, pp. 143–150, June 1984.

[Wils84b]   P. Wilson. Digital signal processing with the IMS T424 Transputer. In *Proceedings, Record 4, Wescon/84*, L.A., October 1984.

[Wong82]   D.Y. Wong, B.H. Juang, and A.H. Gray. An 800 bit/s vector quantization LPC vocoder. *IEEE, Transaction on ASSP*, pp. 770–779, October 1982.

[Wong85]   Yiwan Wong and J-M Delosme. Optimal systolic implementations of N-dimensional recurrences. *ICCD*, pp. 618–621, 1985.

[YHHu85]   Y.H. Hu and S.Y. Kung. Toeplitz Eigensystem Solver. *IEEE Transactions on Acoustics, Speech, and Signal Processing*, Vol. 33: pp 1264–1271, 1982.

[Yoshi82]   T. Yoshimura and E.S. Kuh. Efficient algorithms for channel routing. *IEEE Transactions on Computer-Aided Design of Integrated Circuits and Systems*, January 1982.

[YTHwa85]   Y.T. Hwang. The design and implementation of a 16-bit CMOS LSI multiplier. Master Thesis, The Institute of Electronics, Chiao Tung University, Taiwan, 1985.

INDEX

A

E

F

G